Mathematical Statistics
and Data Analysis

THE WADSWORTH & BROOKS/COLE STATISTICS/PROBABILITY SERIES

Series Editors
Ole Barndorff-Nielsen, *Århus University*
Peter J. Bickel, *University of California, Berkeley*
William S. Cleveland, *AT&T Bell Laboratories*
Richard M. Dudley, *Massachusetts Institute of Technology*

R. Becker, J. Chambers, *Extending the S System*
R. Becker, J. Chambers, *S: An Interactive Environment for Data Analysis and Graphics*
P. Bickel, K. Doksum, J. Hodges, Jr., *Festschrift for Erich L. Lehmann*
G. Box, *The Collected Works of George E. P. Box, Volumes I and II*, G. Tiao, editor-in-chief
L. Breiman, J. Friedman, R. Olshen, C. Stone, *Classification and Regression Trees*
J. Chambers, W. Cleveland, B. Kleiner, P. Tukey, *Graphical Methods for Data Analysis*
F. Graybill, *Matrices with Applications in Statistics, Second Edition*
L. Le Cam, R. Olshen, *Proceedings of the Berkeley Conference in Honor of Jerzy Neyman and Jack Kiefer, Volumes I and II*
J. Rice, *Mathematical Statistics & Data Analysis*
J. Romano, A. Siegel, *Counterexamples in Probability and Statistics*
J. Tukey, *The Collected Works of J. W. Tukey*
    Volume I: Time Series 1949–1964, edited by D. Brillinger
    Volume II: Time Series 1965–1984, edited by D. Brillinger
    Volume III: Philosophy and Principles of Data Analysis: 1949–1964, edited by L. Jones
    Volume IV: Philosophy and Principles of Data Analysis: 1965–1986, edited by L. Jones
    Volume V: Graphics: 1965–1985, edited by W. Cleveland

Mathematical Statistics and Data Analysis

JOHN A. RICE
University of California, San Diego

Wadsworth & Brooks/Cole Advanced Books & Software
Pacific Grove, California

Wadsworth & Brooks/Cole Advanced Books & Software
A Division of Wadsworth, Inc.

Printed in the United States of America
10 9 8 7 6 5 4 3 2 1

Library of Congress Cataloging-in-Publication Data

Rice, John A., [date]–
    Mathematical statistics and data analysis.

    (Wadsworth and Brooks/Cole statistics/probability
series)
    Bibliography: p.
    Includes Index.
    1. Statistics.   I. Title.   II. Series: Wadsworth
and Brooks/Cole statistics/probability series.
QA276.12.R53   1987   519.5   87-8047
ISBN 0-534-08247-5

Sponsoring Editor: *John Kimmel*
Marketing Representative: *Sue Lasbury*
Editorial Assistants: *Corinne Gonyier, Maria Tarantino*
Production Editors: *Steve Bailey, Phyllis Larimore*
Production Associate: *Dorothy Bell*
Manuscript Editor: *Jane Hoover*
Interior Design: *Katherine Minerva*
Cover Design: *Lois Stanfield*
Art Coordinator: *Lisa Torri*
Interior Illustration: *Carl Brown*
Typesetting: *Asco Trade Typesetting Limited, Hong Kong*
Printing and Binding: *Arcata Graphics, Fairfield, Pennsylvania*

(Credits continue on page 595.)

We must be careful not to confuse data with the abstractions we use to analyze them.

WILLIAM JAMES (1842–1910)

PREFACE

Intended Audience

This text is intended for juniors, seniors, or beginning graduate students in statistics, mathematics, natural sciences, and engineering as well as for adequately prepared students in the social sciences and economics. At the University of California at San Diego, we teach a three-quarter course, spending the first quarter on probability and the latter two on statistics. A year of calculus, including Taylor Series and multivariable calculus, and an introductory course in linear algebra are prerequisites.

Why Another Book?

This book is a consequence of my having taught a course at this level several times and having come away each time feeling rather dissatisfied with what had been accomplished. It seems to me that the usual approach is a low-level introduction to a graduate course on optimality theory (which most students never take). This approach does not give students an overview of what statistics is really about. Rather, the overemphasis on optimality theory and on the abstract mathematical aspects of statistics presents a somewhat rigid and formal picture and gives students little feeling for the role actually played by statistical methods in scientific investigations.

I have tried to write a book that reflects my view of what a first, and for many students last, course in statistics should be. Such a course should include some traditional topics in mathematical statistics (such as methods based on likelihood), topics in descriptive statistics and data analysis with special attention to graphical displays, aspects of experimental design, and realistic applications of some complexity. It should also reflect the quickly growing use of computers in statistics. These themes, properly interwoven, can give students a view of the nature of modern statistics. The alternative of teaching two separate courses, one on theory and one on data analysis, seems to me artificial. Furthermore, many students take only one course in statistics and do not have time for two or more.

In order to draw these themes together, I have tried to write a book closely tied to the practice of statistics. It is in the analysis of real data that one sees the roles played by both formal theory and informal data analytic methods. I have organized this book around various kinds of problems that entail the use of statistical methods and have included many real examples to motivate and introduce the theory. Among the advantages of such an approach are that theoretical constructs are presented in meaningful contexts, that they are gradually supplemented and reinforced, and that they are integrated with more informal methods. This is, I think, a fitting approach to statistics, the historical development of which has been spurred on primarily by practical needs rather than abstract or aesthetic considerations. At the same time, I have not shied away from using the mathematics that the students are supposed to know.

Brief Outline

A complete outline can be found, of course, in the table of contents. Here I will just highlight some points and indicate various curricular options for the instructor.

The first six chapters contain an introduction to probability theory, particularly those aspects most relevant to statistics. Chapter 1 introduces the basic ingredients of probability theory and elementary combinatorial methods from a non-measure theoretic point of view. In this and the other probability chapters, I have tried to use real-world examples rather than balls and urns whenever possible.

The concept of a random variable is introduced in chapter 2. I chose to discuss discrete and continuous random variables together, instead of putting off the continuous case until later. Several common distributions are introduced. An advantage of this approach is that it provides something to work with and develop in later chapters.

Chapter 3 continues the treatment of random variables by going into joint distributions. The instructor may wish to skip lightly over Jacobians; this can be done with little loss of continuity, since they are utilized rarely in the rest of the book. The material in Section 3.7 on extrema and order statistics can be omitted if the instructor is willing to do a little backtracking later.

Expectation, variance, covariance, conditional expectation, and moment-generating functions are taken up in chapter 4. The instructor may wish to pass lightly over conditional expectation and prediction, especially if he or she does not plan to cover sufficiency later. The last section of this chapter introduces the δ method, or the method of propagation of error. This method is used several times in the statistics chapters.

The law of large numbers and the central limit theorem are proved in chapter 5 under fairly strong assumptions.

Chapter 6 is a compendium of the common distributions related to the normal and sampling distributions of statistics computed from the usual normal random sample. I don't spend a lot of time on this material here but do develop the necessary facts as they are needed in the statistics chapters. It is useful for students to have these distributions collected in one place.

Chapter 7 is on survey sampling, an unconventional, but in some ways natural, beginning to the study of statistics. Survey sampling is an area of statistics with which most students have some vague familiarity, and a set of fairly specific, concrete statistical problems can be naturally posed. It is a context in which, historically, many important statistical concepts have developed, and it can be used as a vehicle for introducing concepts and techniques that are developed further in later chapters, for example:

- The idea of an estimate as a random variable with an associated sampling distribution
- The concepts of bias, standard error, and mean squared error
- Confidence intervals and the application of the central limit theorem
- An exposure to notions of experimental design via the study of stratified estimates and the concept of relative efficiency
- Calculation of expectations, variances, and covariances

One of the unattractive aspects of survey sampling is that the calculations are rather grubby. However, there is a certain virtue in this grubbiness, and students are given practice in such calculations. The instructor has quite a lot of flexibility as to how deeply to cover this chapter. The sections on ratio estimation and stratification are optional and can be skipped entirely or returned to at a later time without loss of continuity.

Chapter 8 is concerned with parameter estimation, a subject that is motivated and illustrated by the problem of fitting probability laws to data. The method of moments and the traditional method of maximum likelihood are developed. The concept of efficiency is introduced, and the Cramer–Rao Inequality is proved. Section 8.7 introduces the concept of sufficiency and some of its ramifications. The material on the Cramer–Rao lower bound and on sufficiency can be skipped;

to my mind, the importance of sufficiency is usually overstated. Section 8.6.1 (the negative binomial distribution) can also be skipped.

Chapter 9 is an introduction to hypothesis testing with particular application to testing for goodness of fit, which ties in with chapter 8. (This subject is further developed in chapter 11.) Informal, graphical methods are presented here as well. Several of the last sections of this chapter can be skipped if the instructor is pressed for time. These include Section 9.7 (the Poisson dispersion test), Section 9.8 (hanging rootograms), and Section 9.10 (tests for normality).

A variety of descriptive methods are introduced in chapter 10. Many of these techniques are used in later chapters. The importance of graphical procedures is stressed, and notions of robustness are introduced. The placement of a chapter on descriptive methods this late in a book may seem strange. I have chosen to do so because descriptive procedures usually have a stochastic side and, having been through the three chapters preceding this one, students are by now better equipped to study the statistical behavior of various summary statistics (for example, a confidence interval for the median). If the instructor wishes, the material on survival and hazard functions can be skipped.

Classical and nonparametric methods for two-sample problems are introduced in chapter 11. The concepts of hypothesis testing, first introduced in chapter 9, are further developed. The chapter concludes with some discussion of experimental design and the interpretation of observational studies.

These first eleven chapters are the heart of an introductory course; the theoretical constructs of estimation and hypothesis testing have been developed, graphical and descriptive methods have been introduced, and aspects of experimental design have been discussed.

The instructor has much more freedom in selecting material from chapters 12 through 15. In particular, it is not necessary to proceed through these chapters in the order in which they are presented.

Chapter 12 treats the one-way and two-way layouts via analysis of variance and nonparametric techniques. The problem of multiple comparisons, first introduced at the end of chapter 11, is discussed.

Chapter 13 is a rather brief treatment of the analysis of categorical data. Likelihood ratio tests are developed for homogeneity and independence.

Chapter 14 concerns linear least squares. Simple linear regression is developed first and is followed by a more general treatment using linear algebra. I have chosen to employ matrix algebra but have kept the level of the discussion as simple and concrete as possible, not going beyond concepts typically taught in an introductory one-quarter course. In particular, I have not developed a geometric analysis of the general linear model or made any attempt to unify regression and analysis of variance. Throughout this chapter, theoretical results are balanced by more qualitative data analytic procedures based on analysis of residuals.

Chapter 15 is an introduction to decision theory and the Bayesian approach. I believe that students are most likely to appreciate this material after having been exposed to the classical material developed in earlier chapters.

Computer Use and Problem Solving

A VAX and a SUN were used in working the examples in the text and are used by my students on most problems involving real data. My students and I have used both S and Minitab; other packages could be used as well. I have not discussed any particular packages in the text and leave the choice of what, if any, package to use up to the instructor. However, a floppy disk containing the data sets in the text will be available to instructors who have adopted it. Contact the publisher for further information.

This book includes a fairly large number of problems, some of which will be quite difficult for students. I think that problem solving, especially of nonroutine problems, is very important. The course as I teach it includes three hours of lecturing per week and a one-hour section for problem solving and instruction on the use of the computer.

Acknowledgments

I am indebted to a large number of people who contributed directly and indirectly to this book. Earlier versions were used in courses taught by Richard Olshen, Yosi Rinnot, Donald Ylvisaker, Len Haff, and David Lane, who made many helpful comments. Students in their classes and in my own had many constructive comments. Teaching assistants, especially Joan Stansiwalis, Roger Johnson, Terri Bittner, and Peter Kim, worked through many of the problems and found numerous errors. Many reviewers provided useful suggestions: Rollin Brant, University of Toronto; George Casella, Cornell University; Howard B. Christensen, Brigham Young University; David Fairley, Ohio State University; Peter Guttorp, University of Washington; Hari Iyer, Colorado State University; Douglas G. Kelly, University of North Carolina; Thomas Leonard, University of Wisconsin; Albert S. Paulson, Rensselaer Polytechnic Institute; Charles Peters, University of Houston, University Park; Andrew Rukhin, University of Massachusetts, Amherst; Robert Schaefer, Miami University; and Ruth Williams, University of California, San Diego. Richard Royall and W. G. Cumberland kindly provided the data sets used in chapter 7 on survey sampling. Several other data sets were brought to my attention by statisticians at the National Bureau of Standards, where I was fortunate to spend a year while on sabbatical. I deeply appreciate the patience, persistence, and faith of my editor, John Kimmel, in bringing this project to fruition. Of course, I bear final responsibility for remaining errors, flaws, and blemishes.

John A. Rice

CONTENTS

1
Probability   1

2
Random Variables 31

3
Joint Distributions 65

4
Expected Values 104

5

Limit Theorems   155

6

Distributions Derived from the Normal Distribution   168

7

Survey Sampling   175

8
Estimation of Parameters and Fitting of Probability Distributions   222

9
Testing Hypotheses and Assessing Goodness of Fit   270

10
Summarizing Data   312

11
Comparing Two Samples   347

12
The Analysis of Variance   396

13
The Analysis of Categorical Data   434

14

Linear Least Squares　453

15

Decision Theory and Bayesian Inference　511

Mathematical Statistics
and Data Analysis

1

Probability

1.1 Introduction

The idea of probability, chance, or randomness is quite old, whereas its rigorous axiomatization in mathematical terms occurred relatively recently. Many of the ideas of probability theory originated in the study of games of chance. In this century, the mathematical theory of probability has been applied to a wide variety of phenomena; the following are some representative examples:

- Probability theory has been used in genetics as a model for mutations and ensuing natural variability.
- The kinetic theory of gases has an important probabilistic component.
- In designing and analyzing computer operating systems, the lengths of various queues in the system are modeled as random phenomena.
- There are highly developed theories that treat noise in electrical devices and communication systems as random processes.
- Many models of atmospheric turbulence use concepts of probability theory.
- In operations research, the demands on inventories of goods are often modeled as random.

- Actuarial science, which is used by insurance companies, relies heavily on the tools of probability theory.
- Probability theory is used to study complex systems and improve their reliability, such as in modern commercial or military aircraft.

The list could go on and on.

This book develops the basic ideas of probability and statistics. The first part explores the theory of probability as a mathematical model for chance phenomena. The second part of the book is about statistics, which is essentially concerned with procedures for analyzing data, especially data that in some vague sense have a random character. In order to comprehend the theory of statistics, you must have a sound background in probability.

1.2   Sample Spaces

Probability theory is used as a model for situations for which the outcomes occur randomly. Generically, such situations are called "experiments," and the set of all possible outcomes is the **sample space** corresponding to an experiment. The sample space is denoted by Ω, and a generic element of Ω is denoted by ω. The following are some examples.

EXAMPLE A.   Driving to work, a commuter passes through a sequence of three intersections with traffic lights. At each light, he either stops, s, or continues, c. The sample space is the set of all possible outcomes:

$$\Omega = \{ccc, ccs, css, csc, sss, ssc, scc, scs\}$$

where csc, for example, denotes the outcome that the commuter continues through the first light, stops at the second light, and continues through the third light. $\square$

EXAMPLE B.   The number of jobs in a print queue of a mainframe computer may be modeled as random. Here the sample space can be taken as

$$\Omega = \{0, 1, 2, 3, \ldots\}$$

that is, all the nonnegative integers. In practice, there is probably an upper limit, N, on how large the print queue can be, so instead the sample space might be defined as

$$\Omega = \{0, 1, 2, \ldots, N\}$$ $\square$

EXAMPLE C.   Earthquakes exhibit very erratic behavior, which is sometimes modeled as random. For example, the length of time between successive earthquakes in a particular region that are greater in magnitude than a given threshold may be regarded as an experiment. Here Ω is the set of all nonnegative real numbers:

$$\Omega = \{t | t \geq 0\} \qquad \square$$

We are often interested in particular subsets of Ω, which in probability language are called **events**. In Example A, the event that the commuter stops at the first light is the subset of Ω denoted by

$$A = \{sss, ssc, scc, scs\}$$

(Events, or subsets, are usually denoted by italic uppercase letters.) In Example B, the event that there are fewer than five jobs in the print queue can be denoted by

$$A = \{0, 1, 2, 3, 4\}$$

The algebra of set theory carries over directly into probability theory. The union of two events, A and B, is the event C that either A occurs or B occurs or both occur: $C = A \cup B$. For example, if A is the event that the commuter stops at the first light (listed above) and if B is the event that he stops at the third light,

$$B = \{sss, scs, ccs, css\}$$

then C is the event that he stops at the first light or stops at the third light and consists of the outcomes that are in A or in B or in both:

$$C = \{sss, ssc, scc, scs, ccs, css\}$$

The intersection of two events, $C = A \cap B$, is the event that both A and B occur. If A and B are as listed above, then C is the event that the commuter stops at the first light and stops at the third light and thus consists of those outcomes that are common to both A and B:

$$C = \{sss, scs\}$$

The complement of an event, A^c, is the event that A does not occur and thus consists of all those elements in the sample space that are not in A. The complement of the event that the commuter stops at the first light is the event that he continues at the first light:

$$A^c = \{ccc, ccs, css, csc\}$$

You may recall from previous exposure to set theory that rather mysterious set called the empty set, usually denoted by $\varnothing$. The empty set is the set with no elements; it is the event with no outcomes. For example, if A is the event that the commuter stops at the first light and C is the event that he continues through all three lights, $C = \{ccc\}$, then A and C have no outcomes in common, and we can write

$$A \cap C = \varnothing$$

In such cases, A and C are said to be **disjoint**.

Venn diagrams, such as those in Figure 1-1, are often a useful tool for visualizing set operations.

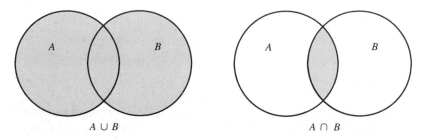

Figure 1-1. Venn diagrams of $A \cup B$ and $A \cap B$.

The following are some laws of set theory.

Commutative Laws:

$$A \cup B = B \cup A$$

$$A \cap B = B \cap A$$

Associative Laws:

$$(A \cup B) \cup C = A \cup (B \cup C)$$

$$(A \cap B) \cap C = A \cap (B \cap C)$$

Distributive Laws:

$$(A \cup B) \cap C = (A \cap C) \cup (B \cap C)$$

$$(A \cap B) \cup C = (A \cup C) \cap (B \cup C)$$

Of these, the distributive laws are the least intuitive, and you may find it instructive to illustrate them with Venn diagrams.

1.3 Probability Measures

A **probability measure** on Ω is a function P from subsets of Ω to the real numbers that satisfies the following axioms:

1. $P(\Omega) = 1$.
2. If $A \subset \Omega$, then $P(A) \geq 0$.
3. If A_1 and A_2 are disjoint, then

$$P(A_1 \cup A_2) = P(A_1) + P(A_2)$$

More generally, if $A_1, A_2, \ldots, A_n, \ldots$ are mutually disjoint, then

$$P\left(\bigcup_1^\infty A_i\right) = \sum_{i=1}^\infty P(A_i)$$

The first two axioms are rather obvious. Since Ω consists of all possible outcomes, $P(\Omega) = 1$. The second axiom simply states that a probability is nonnegative. The third axiom states that if A and B are disjoint—that is, have no outcomes in common—then $P(A \cup B) = P(A) + P(B)$, and also that this property extends to limits. For example, the probability that the print queue contains either one or three jobs is equal to the probability that it contains one plus the probability that it contains three.

The following properties of probability measures are consequences of the axioms.

PROPERTY A. $P(A^c) = 1 - P(A)$. This property follows since A and A^c are disjoint with $A \cup A^c = \Omega$ and thus, by the first and third axioms, $P(A) + P(A^c) = 1$. In words, this property says that the probability that an event does not occur equals one minus the probability that it does occur.

PROPERTY B. $P(\emptyset) = 0$. This property follows from Property A since $\emptyset = \Omega^c$. In words, this says that the probability that there is no outcome at all is zero.

PROPERTY C. If $A \subset B$, then $P(A) \leq P(B)$. This property follows since B can be expressed as the union of two disjoint sets:

$$B = A \cup (B \cap A^c)$$

Then, from the third axiom,

$$P(B) = P(A) + P(B \cap A^c)$$

and thus

$$P(A) = P(B) - P(B \cap A^c) \leq P(B)$$

This property states that if B occurs whenever A occurs, then $P(A) \leq P(B)$. For example, if whenever it rains (A) it is cloudy (B), then the probability that it rains is less than or equal to the probability that it is cloudy.

PROPERTY D. (Addition Law) $P(A \cup B) = P(A) + P(B) - P(A \cap B)$. To see this, we decompose $A \cup B$ into three disjoint subsets, as shown in Figure 1-2:

$$C = A \cap B^c$$

$$D = A \cap B$$

$$E = A^c \cap B$$

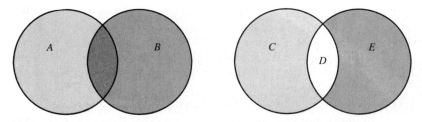

Figure 1-2. Venn diagram illustrating the addition law.

We then have, from the third axiom,

$$P(A \cup B) = P(C) + P(D) + P(E)$$

Also, $A = C \cup D$, and C and D are disjoint; so $P(A) = P(C) + P(D)$. Similarly, $P(B) = P(D) + P(E)$. Putting these results together, we see that

$$P(A) + P(B) = P(C) + P(E) + 2P(D)$$

$$= P(A \cup B) + P(D)$$

or

$$P(A \cup B) = P(A) + P(B) - P(D)$$

This property is easy to see from the Venn diagram in Figure 1-2. If $P(A)$ and $P(B)$ are added together, $P(A \cap B)$ is counted twice.

EXAMPLE A. Suppose that a fair coin is thrown twice. Let A denote the event of heads on the first toss and B the event of heads on the second toss. The sample space is

$$\Omega = \{hh, ht, th, tt\}$$

We assume that each elementary outcome in Ω is equally likely and has probability $\frac{1}{4}$. $C = A \cup B$ is the event that heads comes up on the first toss or on the second toss. Clearly, $P(C) \neq P(A) + P(B) = 1$. Rather, since $A \cap B$ is the event that heads comes up on the first toss and on the second toss,

$$P(C) = P(A) + P(B) - P(A \cap B) = .5 + .5 - .25 = .75 \qquad \square$$

1.4 Computing Probabilities: Counting Methods

Probabilities are especially easy to compute for finite sample spaces. Suppose that $\Omega = \{\omega_1, \omega_2, \ldots, \omega_N\}$ and that $P(\omega_i) = p_i$. To find the probability of an event A, we simply add the probabilities of the ω_i that constitute A.

EXAMPLE A.   Suppose that a fair coin is thrown twice and the sequence of heads and tails is recorded. The sample space is

$$\Omega = \{hh, ht, th, tt\}$$

As in the previous example, we assume that each outcome in Ω has probability .25. Let A denote the event that at least one head is thrown. Then $A = \{hh, ht, th\}$, and $P(A) = .75$. $\qquad \square$

This is a simple example of a fairly common situation. The elements of Ω all have equal probability; so if there are N elements in Ω, each of them has probability $1/N$. If A can occur in any of n mutually exclusive ways, then $P(A) = n/N$, or

$$P(A) = \frac{\text{number of ways } A \text{ can occur}}{\text{total number of outcomes}}$$

Note that this formula holds only if all the outcomes are equally likely. In Example A, if only the number of heads were recorded, then Ω would be $\{0, 1, 2\}$. These outcomes are not equally likely, and $P(A)$ is not $\frac{2}{3}$.

The preceding example is a very simple case. To compute probabilities for more complex situations, we must develop systematic ways of counting outcomes, which is the subject of the next two sections.

1.4.1   The Multiplication Principle

The following is a statement of the very useful multiplication principle.

MULTIPLICATION PRINCIPLE.   If one experiment has m outcomes and another experiment has n outcomes, then there are mn possible outcomes for the two experiments.

Proof. Denote the outcomes of the first experiment by $a_1, \ldots, a_m$ and the outcomes of the second experiment by $b_1, \ldots, b_n$. The outcomes for the two experiments are the ordered pairs (a_i, b_j). These pairs can be exhibited as the entries of an $m \times n$ rectangular array, in which the pair (a_i, b_j) is in the ith row and the jth column. There are mn entries in this array.   □

EXAMPLE A.   Playing cards have 13 face values and 4 suits. There are thus $4 \times 13 = 52$ face-value/suit combinations.   □

EXAMPLE B.   A class has 12 boys and 18 girls. The teacher selects one boy and one girl to act as representatives to the student government. She can do this in any of $12 \times 18 = 216$ different ways.   □

EXTENDED MULTIPLICATION PRINCIPLE.   If there are p experiments and the first has n_1 possible outcomes, the second $n_2, \ldots$, and the pth n_p possible outcomes, then there are a total of $n_1 \times n_2 \times \cdots \times n_p$ possible outcomes for the p experiments.

Proof.   This can be proved from the multiplication principle by induction. We have seen that it is true for $p = 2$. Assume that it is true for $p = q$—that is, that there are $n_1 \times n_2 \times \cdots \times n_q$ possible outcomes for the first q experiments. To complete the proof by induction, we must show that it follows that the property holds for $p = q + 1$. We apply the multiplication principle, regarding the first q experiments as a single experiment with $n_1 \times \cdots \times n_q$ outcomes, and conclude that there are $(n_1 \times \cdots \times n_q) \times n_{q+1}$ outcomes for the $q + 1$ experiments.   □

EXAMPLE C.   An eight-bit binary word is a sequence of eight digits, of which each may be either a zero or a one. How many different eight-bit words are there?
There are two choices for the first bit, two for the second, etc., and thus there are

$$2 \times 2 \times 2 \times 2 \times 2 \times 2 \times 2 \times 2 = 2^8 = 256$$

such words.   □

EXAMPLE D.   A DNA molecule is a sequence of four types of nucleotides, denoted by A, G, C, and T. The molecule can be millions of units long and can thus encode an enormous amount of information. For example, for a molecule 1 million units long, there are 4^{10^6} different possible sequences. This is a staggeringly large number having nearly a million digits. An amino acid is coded for by a sequence of three nucleotides; there are $4^3 = 64$ different codes, but there are only 20 amino acids since some of them can be coded for in several ways. A protein molecule is composed of as many as hundreds of amino acid units, and thus there are an incredibly large number of possible proteins. For example, there are 20^{100} different sequences of 100 amino acids.   □

1.4.2 Permutations and Combinations

A permutation is an ordered arrangement of objects. Suppose that from the set $C = \{c_1, c_2, \ldots, c_n\}$ we choose r elements and list them in order. How many ways can we do this? The answer depends on whether we are allowed to duplicate items in the list. If no duplication is allowed, we are **sampling without replacement**. If duplication is allowed, we are **sampling with replacement**. We can think of the problem as that of taking labeled balls from an urn. In the first type of sampling, we are not allowed to put a ball back before choosing the next one, but in the second, we are. In either case, when we are done choosing, we have a list of r balls ordered in the sequence in which they were drawn.

The extended multiplication principle can be used to count the number of different samples possible for a set of n elements. First, suppose that sampling is done with replacement. The first ball can be chosen in any of n ways, the second in any of n ways, etc., so that there are $n \times n \times \cdots \times n = n^r$ samples. Next, suppose that sampling is done without replacement. There are n choices for the first ball, $n - 1$ choices for the second ball, $n - 2$ for the third, ..., and $n - r + 1$ for the rth. We have just proved the following proposition.

PROPOSITION A. For a set of size n and a sample of size r, there are n^r different ordered samples with replacement and $n(n - 1)(n - 2) \cdots (n - r + 1)$ different ordered samples without replacement. ∎

COROLLARY A. The number of orderings of n elements is $n(n - 1)(n - 2) \cdots 1 = n!$. ∎

EXAMPLE A. How many ways can five children be lined up?

This corresponds to sampling without replacement. According to Corollary A, there are $5! = 5 \times 4 \times 3 \times 2 \times 1 = 120$ different lines.   □

EXAMPLE B. Suppose that from ten children, five are to be chosen and lined up. How many different lines are possible?

From Proposition A, there are $10 \times 9 \times 8 \times 7 \times 6 = 30{,}240$ different lines.   □

EXAMPLE C. In some states, license plates have six characters: three letters followed by three numbers. How many distinct such plates are possible?

This corresponds to sampling with replacement. There are $26^3 = 17{,}576$ different ways to choose the letters and $10^3 = 1000$ ways to choose the numbers. Using the multiplication principle again, we find there are $17{,}576 \times 1000 = 17{,}576{,}000$ different plates.   □

EXAMPLE D. If all sequences of six characters are equally likely, what is the probability that the license plate for a new car will contain no duplicate letters or numbers?

Call the desired event A; Ω consists of all $17{,}576{,}000$ possible sequences. Since these are all equally likely, the probability of A is the ratio of the number of ways that A can occur to the total number of possible outcomes. There are $26 \times 25 \times 24 = 15{,}600$ ways to choose the letters without duplication (doing so corresponds to sampling without replacement) and $10 \times 9 \times 8 = 720$ ways to choose the numbers without duplication. From the multiplication principle, there are $15{,}600 \times 720 = 11{,}232{,}000$ nonrepeating sequences. The probability of A is thus

$$P(A) = \frac{11{,}232{,}000}{17{,}576{,}000} = .64 \qquad \square$$

EXAMPLE E.   (Birthday Problem) Suppose that a room contains n people. What is the probability that at least two of them have a common birthday?

This is a famous problem with a counterintuitive answer. Assume that every day of the year is equally likely to be a birthday, disregard leap years, and denote by A the event that there are at least two people with a common birthday. As is sometimes the case, it is easier to find $P(A^c)$ than to find $P(A)$. This is because A can happen in many ways, whereas A^c is much simpler. There are 365^n possible outcomes, and A^c can happen in $365 \times 364 \cdots \times (365 - n + 1)$ ways. Thus,

$$P(A^c) = \frac{365 \times 364 \times \cdots \times (365 - n + 1)}{365^n}$$

and

$$P(A) = 1 - \frac{365 \times 364 \times \cdots \times (365 - n + 1)}{365^n}$$

The following table exhibits the latter probabilities for various values of n:

n	$P(A)$
4	.016
16	.284
23	.507
32	.753
40	.891
56	.988

From the table, we see that if there are only 23 people, the probability of at least one match exceeds .5. $\qquad \square$

EXAMPLE F.   How many people must you ask in order to have a 50 : 50 chance of finding someone who shares your birthday?

Suppose that you ask n people; let A denote the event that someone's birthday is the same as yours. Again, it is easier to work with A^c. The total number of outcomes is 365^n, and the total number of ways that A^c can happen is 364^n. Thus,

$$P(A^c) = \frac{364^n}{365^n}$$

and

$$P(A) = 1 - \frac{364^n}{365^n} = 1 - \left(1 - \frac{1}{365}\right)^n$$

In order for the latter probability to be .5, n should be 253, which may seem counterintuitive.   □

We now shift our attention from counting permutations to counting combinations. Here we are no longer interested in ordered samples, but in the constituents of the samples regardless of the order in which they were obtained. In particular, we ask the following question: If r objects are taken from a set of n objects without replacement and disregarding order, how many different samples are possible? From the multiplication principle, the number of ordered samples equals the number of unordered samples multiplied by the number of ways to order each sample. Since the number of ordered samples is $n(n - 1) \cdots (n - r + 1)$ and since each sample can be ordered in $r!$ ways, the number of unordered samples is

$$\frac{n(n - 1) \cdots (n - r + 1)}{r!} = \frac{n!}{(n - r)!r!}$$

This number is also denoted as $\binom{n}{r}$. We have proved the following proposition.

PROPOSITION B.   The number of unordered samples of r objects from n objects without replacement is $\binom{n}{r}$.

The numbers $\binom{n}{k}$, called the **binomial coefficients**, occur in the expansion

$$(a + b)^n = \sum_{k=0}^{n} \binom{n}{k} a^k b^{n-k}$$

In particular,

$$2^n = \sum_{k=0}^{n} \binom{n}{k}$$

This latter result can be interpreted as the number of subsets of a set of n objects. We just add the number of subsets of size 0 (with the usual convention that $0! = 1$) and the number of subsets of size 1 and the number of subsets of size 2, etc.

EXAMPLE G.   From ten boys, five are to be chosen for a team. In how many ways can this be done?

Here we are not interested in order. Proposition B tells us that the desired number is $\binom{10}{5} = 252$. ▢

EXAMPLE H. In the practice of quality control, only a fraction of the output of a manufacturing process is sampled and examined, since it may be too time-consuming and expensive to examine each item or because sometimes the testing is destructive. Suppose that there are n items in a lot and a sample of size r is taken. There are $\binom{n}{r}$ such samples. Now suppose that the lot contains k defective items. What is the probability that the sample contains exactly m defectives?

Clearly, this question is relevant to the efficacy of the sampling scheme, and the most desirable sample size can be determined by computing such probabilities for various values of r. Call the event in question A. The probability of A is the number of ways A can occur divided by the total number of outcomes. To find the number of ways A can occur, we use the multiplication principle. There are $\binom{k}{m}$ ways to choose the m defective items in the sample from the k defectives in the lot, and there are $\binom{n-k}{r-m}$ ways to choose the $r - m$ nondefective items in the sample from the $n - k$ nondefectives in the lot. Therefore, A can occur in $\binom{k}{m}\binom{n-k}{r-m}$ ways. Thus, $P(A)$ is the ratio of the number of ways A can occur to the total number of outcomes, or

$$P(A) = \frac{\binom{k}{m}\binom{n - k}{r - m}}{\binom{n}{r}}$$ ▢

EXAMPLE I. (Capture/Recapture Method) The so-called capture/recapture method is sometimes used to estimate the size of a wildlife population. Suppose that 10 animals are captured, tagged, and released. On a later occasion, 20 animals are captured, and it is found that 4 of them are tagged. How large is the population?

We assume that there are n animals in the population, of which 10 are tagged. If the 20 animals captured later are taken in such a way that all $\binom{n}{20}$ possible groups are equally likely (this is a big assumption), then the probability that 4 of them are tagged is

$$\frac{\binom{10}{4}\binom{n - 10}{16}}{\binom{n}{20}}$$

Clearly, n cannot be precisely determined from the information at hand, but it can be estimated. One method of estimation, called **maximum likelihood**, is to choose that value of n that makes the observed outcome most probable. (The method of maximum likelihood is one of the main subjects of a later chapter in this text.)

To find the maximum likelihood estimate, suppose that, in general, t animals are tagged. Then, of a second sample of size m, r tagged animals are recaptured. We estimate n by the maximizer of the likelihood

$$L_n = \frac{\binom{t}{r}\binom{n-t}{m-r}}{\binom{n}{m}}$$

To find the value of n that maximizes L_n, consider the ratio of successive terms, which after some algebra is found to be

$$\frac{L_n}{L_{n-1}} = \frac{(n-t)(n-m)}{n(n-t-m+r)}$$

This ratio is greater than 1; i.e., L_n is increasing if

$$(n-t)(n-m) > n(n-t-m+r)$$

$$n^2 - nm - nt + mt > n^2 - nt - nm - nr$$

$$mt > nr$$

$$\boxed{\frac{mt}{r} > n}$$

Thus, L_n increases for $n < mt/r$ and decreases for $n > mt/r$; so the value of n that maximizes L_n is the greatest integer not exceeding mt/r.

Applying this result to the data given above, we see that the maximum likelihood estimate of n is 50. This estimate has some intuitive appeal, as it equates the proportion of tagged animals in the second sample to the proportion in the population:

$$\frac{4}{20} = \frac{10}{n} \qquad \qquad \square$$

Proposition B has the following extension.

PROPOSITION C. The number of ways that n objects can be grouped into r classes with n_i in the ith class, $i = 1, \ldots, r$, and $\sum_{i=1}^n n_i = n$ is

$$\binom{n}{n_1 n_2 \cdots n_r} = \frac{n!}{n_1! n_2! \cdots n_r!}$$

Proof. This can be seen by using Proposition B and the multiplication principle. (Note that Proposition B is the special case for which $r = 2$.) There are $\binom{n}{n_1}$ ways to choose the objects for the first class. Having done that, there are $\binom{n-n_1}{n_2}$ ways of choosing the objects for the second class. Continuing in this manner, there are

$$\frac{n!}{n!(n-n_1)!} \frac{(n-n_1)!}{(n-n_1-n_2)!n_2!} \cdots \frac{(n-n_1-n_2-\cdots-n_{r-1})!}{0!n_r!}$$

choices in all. After cancellation, this yields the desired result.   ☐

EXAMPLE J.   In how many ways can $n = 2m$ people be paired and assigned to m courts for the first round of a tennis tournament?

In this problem, $n_i = 2$, $i = 1, \ldots, m$, and, according to Proposition C, there are

$$\frac{(2m)!}{2^m}$$

assignments.

One has to be careful with problems such as this one. Suppose it were asked how many ways $2m$ people could be arranged in pairs without assigning the pairs to courts. Since there are $m!$ ways to assign the m pairs to m courts, the preceding result should be divided by $m!$, giving

$$\frac{(2m)!}{m!2^m}$$

pairs in all.   ☐

1.5   Conditional Probability

The definition and use of conditional probability can best be introduced with an example. Digitalis therapy is often beneficial to patients who have suffered congestive heart failure, but there is the risk of digitalis intoxication, a serious side effect that is, moreover, difficult to diagnose. To improve the chances of a correct diagnosis, the concentration of digitalis in the blood can be measured. Beller et al. (1971) conducted a study of the relation of the concentration of digitalis in the blood to digitalis intoxication in 135 patients. Their results are simplified slightly in the table below, where the following notation is used:

$T+$ = high blood concentration (positive test)

$T-$ = low blood concentration (negative test)

$D+$ = toxicity (disease present)

$D-$ = no toxicity (disease absent)

	D +	D −	Total
T +	25	14	39
T −	18	78	96
Total	43	92	135

Thus, for example, 25 of the 135 patients had a high blood concentration of digitalis and suffered toxicity.

We can introduce probability at this stage in one of two ways. We can either consider selecting a patient at random from the 135 or assume that the relative frequencies in the study roughly hold in some larger population of patients. (Making inferences about the frequencies in a large population from those observed in a small sample is a statistical problem, which will be taken up in a later chapter of this book.) Converting the frequencies in the table above to proportions (relative to 135), which we will regard as probabilities, we obtain the following table:

	D +	D −	Total
T +	.185	.104	.289
T −	.133	.578	.711
Total	.318	.682	1.000

From the table, $P(T+) = .289$ and $P(D+) = .318$, for example. Now if a doctor knows that the test was positive (that there was a high blood concentration), what is the probability of disease (toxicity) given this knowledge? We can restrict our attention to the first row of the table, and we see that of the 39 patients who had positive tests, 25 suffered from toxicity. We denote the probability that a patient shows toxicity given that the test is positive by $P(D+|T+)$, which is called the **conditional probability** of $D+$ given $T+$.

$$P(D+|T+) = \frac{25}{39} = .640$$

Equivalently, we can calculate this probability as

$$P(D+|T+) = \frac{P(D+ \cap T+)}{P(T+)}$$

$$= \frac{.185}{.289} = .640$$

Thus, we see that the unconditional probability of $D+$ is .318, whereas the conditional probability $D+$ given $T+$ is .640. Therefore, knowing that the test is positive makes toxicity more than twice as likely. What if the test is negative?

$$P(D-|T-) = \frac{.578}{.711} = .848$$

For comparison, $P(D-) = .682$. Two other conditional probabilities from this example are of interest: The probability of a false positive is $P(D-|T+) = .360$, and the probability of a false negative is $P(D+|T-) = .187$.

In general, we have the following definition.

DEFINITION.   Let A and B be two events with $P(B) \neq 0$. The conditional probability of A given B is defined to be

$$P(A|B) = \frac{P(A \cap B)}{P(B)}$$

The idea behind this definition is that if we are given that the event B occurred, the relevant sample space becomes B rather than Ω and that conditional probability is a probability measure on B. In the digitalis example, to find $P(D+|T+)$, we restricted our attention to the 39 patients who had positive tests. In order for this new measure to be a probability measure, it must satisfy the axioms, and it can be shown that in fact it does.

In some situations, $P(A|B)$ and $P(B)$ can be found rather easily, and we can then find $P(A \cap B)$.

MULTIPLICATION LAW.   Let A and B be events and assume $P(B) \neq 0$. Then

$$P(A \cap B) = P(A|B)P(B)$$

The multiplication law is often useful in finding the probabilities of intersections, as the following examples illustrate.

EXAMPLE A.   An urn contains three red balls and one blue ball. Two balls are selected without replacement. What is the probability that they are both red?

Let R_1 and R_2 denote the events that a red ball is drawn on the first trial and on the second trial, respectively. From the multiplication law,

$$P(R_1 \cap R_2) = P(R_1)P(R_2|R_1)$$

$P(R_1)$ is clearly $\frac{3}{4}$, and if a red ball has been removed on the first trial, there are two red balls and one blue ball left. Therefore, $P(R_2|R_1) = \frac{2}{3}$. Thus, $P(R_1 \cap R_2) = \frac{1}{2}$.   □

EXAMPLE B.   Suppose that if it is cloudy (B), the probability that it is raining (A) is .3, and that the probability that it is cloudy is $P(B) = .2$ The probability that it is cloudy and raining is

$$P(A \cap B) = P(A|B)P(B) = .3 \times .2 = .06 \qquad \square$$

Another useful tool for computing probabilities is provided by the following law.

LAW OF TOTAL PROBABILITY.   Let $B_1, B_2, \ldots, B_n$ be such that $\bigcup_{i=1}^{n} B_i = \Omega$ and $B_i \cap B_j = \emptyset$ for $i \neq j$, with $P(B_i) > 0$ for all i. Then, for any event A,

$$P(A) = \sum_{i=1}^{n} P(A|B_i)P(B_i)$$

Proof.   Before going through a formal proof, it is helpful to state the result in words. The B_i are mutually disjoint events whose union is Ω. To find the probability of an event A, we sum the conditional probabilities of A given B_i, weighted by $P(B_i)$. Now, for the proof, we first observe that

$$P(A) = P(A \cap \Omega)$$

$$= P\left(A \cap \left(\bigcup_{i=1}^{n} B_i\right)\right)$$

$$= P\left(\bigcup_{i=1}^{n} (A \cap B_i)\right)$$

Since the events $A \cap B_i$ are disjoint,

$$P\left(\bigcup_{i=1}^{n} (A \cap B_i)\right) = \sum_{i=1}^{n} P(A \cap B_i)$$

$$= \sum_{i=1}^{n} P(A|B_i)P(B_i) \qquad \square$$

The law of total probability is useful in situations where it is not obvious how to calculate $P(A)$ directly but in which $P(A|B_i)$ and $P(B_i)$ are more straightforward, such as in the following example.

EXAMPLE C.   Referring to Example A, what is the probability that a red ball is selected on the second draw?

The answer may or may not be intuitively obvious—that depends on your intuition. On the one hand, you could argue that it is "clear from symmetry" that $P(R_2) = P(R_1) = \frac{3}{4}$. On the other hand, you could say that it is obvious that a red ball is likely to be selected on the first draw, leaving fewer red balls for the second draw, so that $P(R_2) < P(R_1)$. The answer can be derived easily by using the law of total probability:

$$P(R_2) = P(R_2|R_1)P(R_1) + P(R_2|B_1)P(B_1)$$

$$= \tfrac{2}{3} \times \tfrac{3}{4} + 1 \times \tfrac{1}{4} = \tfrac{3}{4}$$

where B_1 denotes the event that a blue ball is drawn on the first trial. □

As another example of the use of conditional probability, we consider a model that has been used for occupational mobility.

EXAMPLE D. Suppose that occupations are grouped into upper (U), middle (M), and lower (L) levels. U_1 will denote the event that a father's occupation is upper-level; U_2 will denote the event that a son's occupation is upper-level, etc. (the subscripts index generations). Glass and Hall (1954) compiled the following statistics on occupational mobility in England and Wales:

	U_2	M_2	L_2
U_1	.45	.48	.07
M_1	.05	.70	.25
L_1	.01	.50	.49

Such a table, which is called a matrix of transition probabilities, is to be read in the following way: If a father is in U, the probability that his son is in U is .45, the probability that his son is in M is .48, etc. The table thus gives conditional probabilities; for example, $P(U_2|U_1) = .45$. Examination of the table reveals that there is more upward mobility from L into M than from M into U. Suppose that of the fathers' generation, 10% are in U, 40% in M, and 50% in L. What is the probability that a son in the next generation is in U?

Applying the law of total probability, we have

$$P(U_2) = P(U_2|U_1)P(U_1) + P(U_2|M_1)P(M_1) + P(U_2|L_1)P(L_1)$$

$$= .45 \times .10 + .05 \times .40 + .01 \times .50 = .07$$

$P(M_2)$ and $P(L_2)$ can be worked out similarly. □

Continuing with Example D, suppose we ask a different question: If a son has occupational status U_2, what is the probability that his father had occupational status U_1? Compared to the question asked in Example D, this is an "inverse" problem; we are given an "effect" and are asked to find the probability of a particular "cause." In situations like this, Bayes' Rule, which we state below, is useful. Before stating the rule, we will see what it amounts to in this particular case.

We wish to find $P(U_1|U_2)$. By definition,

$$P(U_1|U_2) = \frac{P(U_1 \cap U_2)}{P(U_2)}$$

$$= \frac{P(U_2|U_1)P(U_1)}{P(U_2|U_1)P(U_1) + P(U_2|M_1)P(M_1) + P(U_2|L_1)P(L_1)}$$

Here we used the multiplication law to reexpress the numerator and the law of total probability to restate the denominator. The value of the numerator is $P(U_2|U_1)P(U_1) = .45 \times .10 = .045$, and we calculated the denominator in Example D to be .07, so we find that $P(U_1|U_2) = .64$. In other words, 64% of the sons who are in upper-level occupations have fathers who were in upper-level occupations.

We now state Bayes' Rule.

BAYES' RULE. Let A and $B_1, \ldots, B_n$ be events where the B_i are disjoint, $\bigcup_{i=1}^{n} B_i = \Omega$, and $P(B_i) > 0$ for all i. Then

$$P(B_j|A) = \frac{P(A|B_j)P(B_j)}{\sum_{i=1}^{n} P(A|B_i)P(B_i)}$$

The proof of Bayes' Rule follows exactly as in the discussion above.

EXAMPLE E. Diamond and Forrester (1979) applied Bayes' Rule to the diagnosis of coronary artery disease. A procedure called cardiac fluoroscopy is used to determine whether there is calcification of coronary arteries and thereby to diagnose coronary artery disease. From the test, it can be determined if 0, 1, 2, or 3 coronary arteries are calcified. Let T_0+, T_1+, T_2+, T_3+ denote these events. Let $D+$ or $D-$ denote the event that disease is present or absent, respectively. Diamond and Forrester presented the following table, based on medical studies:

i	$P(T_i+|D+)$	$P(T_i+|D-)$
0	.42	.96
1	.24	.02
2	.20	.02
3	.15	.00

According to Bayes' Rule,

$$P(D+|T_i+) = \frac{P(T_i+|D+)P(D+)}{P(T_i+|D+)P(D+) + P(T_i+|D-)P(D-)}$$

Thus, if the initial probabilities $P(D+)$ and $P(D-)$ are known, the probability that a patient has coronary artery disease can be calculated.

Let us consider two specific cases. For the first, suppose that a male between the ages of 30 and 39 suffers from nonanginal chest pain. For such a patient, it is known from medical statistics that $P(D+) \approx .05$. Suppose that the test shows that no arteries are calcified. From the equation above,

$$P(D+|T_0+) = \frac{.42 \times .05}{.42 \times .05 + .96 \times .95} = .02$$

It is unlikely that the patient has coronary artery disease. On the other hand, suppose that the test shows that one artery is calcified. Then

$$P(D+|T_1+) = \frac{.24 \times .05}{.24 \times .05 + .02 \times .95} = .39$$

Now it is more likely that this patient has coronary artery disease, but by no means certain.

As a second case, suppose that the patient is a male between ages 50 and 59 who suffers typical angina. For such a patient, $P(D+) = .92$. For him, we find that

$$P(D+|T_0) = \frac{.42 \times .92}{.42 \times .92 + .96 \times .08} = .83$$

$$P(D+|T_1) = \frac{.24 \times .92}{.24 \times .92 + .02 \times .08} = .99$$

Comparing the two patients, we see the strong influence of the prior probability, $P(D+)$. □

1.6   Independence

Intuitively, we would say that two events, A and B, are independent if knowing that one had occurred gave us no information about whether the other had or had not occurred; that is, $P(A|B) = P(A)$ and $P(B|A) = P(B)$. Now, if

$$P(A) = P(A|B) = \frac{P(A \cap B)}{P(B)}$$

then

$$P(A \cap B) = P(A)P(B)$$

We will use this last relation as the definition of independence. Note that it is symmetric in A, and B and does not require the existence of a conditional probability; that is, $P(B)$ can be 0.

DEFINITION. A and B are said to be independent events if $P(A \cap B) = P(A)P(B)$.

EXAMPLE A.   A card is selected randomly from a deck. Let A denote the event that it is an ace and D the event that it is a diamond. Knowing that the card is an ace gives no information about its suit. Checking formally that the events are independent, we have $P(A) = \frac{4}{52} = \frac{1}{13}$ and $P(D) = \frac{1}{4}$. Also, $A \cap D$ is the event that the card is the ace of diamonds and $P(A \cap D) = \frac{1}{52}$. Since $P(A)P(D) = (\frac{1}{4}) \times (\frac{1}{13}) = \frac{1}{52}$, the events are in fact independent.   □

EXAMPLE B.   A system is designed so that it fails only if a unit and a backup unit both fail. Assuming that these failures are independent and that each unit fails with probability p, the system fails with probability p^2. If, for example, the probability that any unit fails during a given year is .1, then the probability that the system fails is .01, which represents a considerable improvement in reliability.   □

Things become more complicated when we consider more than two events. For example, suppose we know that events A, B, and C are pairwise independent (any two are independent). We would like to be able to say that they are all independent based on the assumption that knowing something about two of the events does not tell us anything about the third, for example, $P(C|A \cap B) = P(C)$. But as the following example shows, pairwise independence does not guarantee mutual independence.

EXAMPLE C.   A fair coin is tossed twice. Let A denote the event of heads on the first toss, B the event of heads on the second toss, and C the event that exactly one head is thrown. A and B are clearly independent, and $P(A) = P(B) = P(C) = .5$. To see that A and C are independent, we observe that $P(C|A) = .5$. But

$$P(A \cap B \cap C) = 0 \neq P(A)P(B)P(C) \qquad\qquad □$$

To encompass situations such as that in Example C, we define a collection of events, $A_1, A_2, \ldots, A_n$, to be **mutually independent** if for any subcollection, $A_{i_1}, \ldots, A_{i_m}$,

$$P(A_{i_1} \cap \cdots \cap A_{i_m}) = P(A_{i_1}) \cdots P(A_{i_m})$$

EXAMPLE D.   Consider a circuit with three relays (Figure 1-3). Let A_i denote the event that the ith relay works, and assume that $P(A_i) = p$ and that the relays are mutually independent. If F denotes the event that current flows through the circuit, then $F = A_3 \cup (A_1 \cap A_2)$ and, from the addition law and the assumption of independence,

$$P(F) = P(A_3) + P(A_1 \cap A_2) - P(A_1 \cap A_2 \cap A_3) = p + p^2 - p^3 \qquad □$$

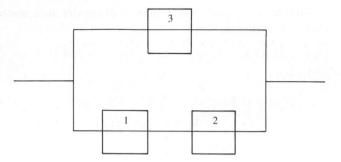

Figure 1-3.   Circuit with three relays.

EXAMPLE E.   Suppose that a system consists of components connected in series, so the system fails if any one component fails. If there are n mutually independent components and each fails with probability p, what is the probability that the system will fail?

It is easier to find the probability of the complement of this event; the system works if and only if all the components work, and this situation has probability $(1 - p)^n$. The probability that the system fails is then $1 - (1 - p)^n$. For example, if $n = 10$ and $p = .05$, the probability that the system works is only $.95^{10} = .60$, and the probability that the system fails is .40.

Suppose, instead, that the components are connected in parallel, so the system fails only when they all fail. In this case, the probability that the system fails is only $.05^{10} = 9.8 \times 10^{-14}$.   □

Calculations like those in Example E are made in reliability studies for systems consisting of quite complicated networks of components. The absolutely crucial assumption is that the components are independent of one another. Theoretical studies of the reliability of nuclear power plants have been criticized on the grounds that they have incorrectly assumed independence of the components.

1.7 Concluding Remarks

This chapter provides a simple axiomatic development of the mathematical theory of probability. Some subtle issues that arise in a careful analysis of infinite sample spaces have been neglected. Such issues are typically addressed in graduate-level courses in measure theory and probability theory. Certain philosophical questions have also been avoided. One might ask what is meant by the statement "The probability that this coin will land heads up is $\frac{1}{2}$." Two commonly advocated views are the **frequentist approach** and the **Bayesian approach**. According to the frequentist approach, the statement means that if the experiment were repeated many, many times, the long-run average number of heads would tend to $\frac{1}{2}$. According to the Bayesian approach, the statement is a quantification of the speaker's uncertainty about the outcome of the experiment and thus is a personal or subjective notion; the probability that the coin will land heads up may be different for different speakers, depending on their experience and knowledge of the situation. There has been vigorous and occasionally acrimonious debate among proponents of various versions of these points of view.

In this and ensuing chapters, there are many examples of the use of probability as a model for various phenomena. In any such modeling endeavor, an idealized mathematical theory is hoped to provide an adequate match to characteristics of the phenomenon under study. The standard of adequacy is relative to the field of study and the modeler's goals.

1.8   Problems

1. A coin is tossed three times, and the sequence of heads and tails is recorded.
    (a) List the sample space.
    (b) List the elements that make up the following events: (1) $A =$ at least two heads, (2) $B =$ the first two tosses are heads, (3) $C =$ the last toss is a tail.
    (c) List the elements of the following events: (1) A^c, (2) $A \cap B$, (3) $A \cup C$.
2. Two six-sided dice are thrown sequentially, and the face values that come up are recorded.
    (a) List the sample space.
    (b) List the elements that make up the following events: (1) $A =$ the sum of the two values is at least 5, (2) $B =$ the value of the first die is higher than the value of the second, (3) $C =$ the first value is 4.
    (c) List the elements of the following events: (1) $A \cap C$, (2) $B \cup C$, (3) $A \cap (B \cup C)$.
3. An urn contains three red balls, two green balls, and one white ball. Three balls are drawn without replacement from the urn, and the colors are noted in sequence. List the sample space. Define events A, B, and C as you wish and find their unions and intersections.
4. Draw Venn diagrams to illustrate DeMorgan's Laws:

$$(A \cup B)^c = A^c \cap B^c$$

$$(A \cap B)^c = A^c \cup B^c$$

5. Let A and B be arbitrary events. Let C be the event that either A occurs or B occurs, but not both. Express C in terms of A and B using any of the basic operations of union, intersection, and complement.
6. Verify the following extension of the addition rule (a) by an appropriate Venn diagram and (b) by a formal argument using the axioms of probability and the propositions in this chapter.

$$P(A \cup B \cup C) = P(A) + P(B) + P(C)$$
$$- P(A \cap B) - P(A \cap C) - P(B \cap C)$$
$$+ P(A \cap B \cap C)$$

7. Prove Bonferroni's Inequality:

$$P(A \cap B) \geq P(A) + P(B) - 1$$

8. Prove that

$$P\left(\bigcup_{i=1}^{n} A_i\right) \leq \sum_{i=1}^{n} P(A_i)$$

9. The first three digits of a university telephone exchange are 452. If all the sequences of the remaining four digits are equally likely, what is the probability that a randomly selected university phone number contains seven distinct digits?

10. In a game of poker, 5 players are each dealt 5 cards from a 52-card deck. How many ways are there to deal the cards?

11. In a game of poker, what is the probability that a five-card hand will contain (a) a straight (five cards in unbroken numerical sequence), (b) four of a kind, and (c) a full house (three cards of one value and two cards of another value)?

12. The four players in a bridge game are each dealt 13 cards. How many ways are there to do this?

13. How many different meals can be made from four kinds of meat, six vegetables, and three starches if a meal consists of one selection from each group?

14. How many different letter arrangements can be obtained from the letters of the word *statistically*, using all the letters?

15. In acceptance sampling, a purchaser samples 4 items from a lot of 100 and rejects the lot if one or more are defective. Graph the probability that the lot is accepted as a function of the percentage of defective items in the lot.

16. A lot of n items contains k defectives, and m are selected randomly and inspected. How should the value of m be chosen so that the probability that at least one defective item turns up is .90? Apply your answer to (a) $n = 1000$, $k = 10$, and (b) $n = 10,000$, $k = 100$.

17. A committee consists of five Chicanos, two Asians, three blacks, and two whites.
    (a) A subcommitte of four is chosen at random. What is the probability that all the ethnic groups are represented on the subcommittee?
    (b) Answer the question for part (a) if a subcommittee of five is chosen.

18. A deck of 52 cards is shuffled thoroughly. What is the probability that the four aces are all next to each other?

19. A fair coin is tossed five times. What is the probability of getting a sequence of three heads?

20. A deck of cards is shuffled thoroughly, and n cards are turned up. What is the probability that a face card turns up? For what value of n is this probability about .5?

21. How many ways are there to place n indistinguishable balls in n urns so that exactly one urn is empty?

22. If n balls are distributed randomly into k urns, what is the probability that the last urn contains j balls?

23. A woman getting dressed up for a night out is asked by her significant other to wear a red dress, high-heeled sneakers, and a wig. In how many orders can she put on these objects?

24. The game of Mastermind starts in the following way: One player selects four pegs, each peg having six possible colors, and places them in a line. The second player then tries to guess the sequence of colors. What is the probability of guessing correctly?

25. If a five-letter word is formed at random (meaning that all sequences of five letters are equally likely), what is the probability that no letter occurs more than once?

26. How many ways are there to encode the 26-letter English alphabet into 8-bit binary words (sequences of 8 zeros and ones)?

27. A poker player is dealt three spades and two hearts. He discards the two hearts and draws two more cards. What is the probability that he draws two more spades?

28. A group of 60 second graders is to be randomly assigned to two classes of 30 each. (The random assignment is ordered by the school district to ensure against any bias.) Five of the second graders, Marcelle, Sarah, Michelle, Katy, and Camarin, are close friends. What is the probability that they will all be in the same class? What is the probability that exactly four of them will be? What is the probability that Marcelle will be in one class and her friends in the other?

29. Six male and six female dancers perform the Virginia reel. This dance requires that they form a line consisting of six male/female pairs. How many such arrangements are there?

30. A wine taster claims that he can distinguish four vintages of a particular Cabernet. What is the probability that he can do this by merely guessing? (He is confronted with four unlabeled glasses.)

31. An elevator containing five people can stop at any of seven floors. What is the probability that no two people get off at the same floor? Assume that the occupants act independently and that all floors are equally likely for each occupant.

32. Prove the following identity:

$$\sum_{k=0}^{n} \binom{n}{k}\binom{m-n}{n-k} = \binom{m}{n}$$

(*Hint*: How can each of the summands be interpreted?)

33. Prove the following two identities both algebraically and by interpreting their meaning combinatorially.

(a) $\dbinom{n}{r} = \dbinom{n}{n-r}$

(b) $\dbinom{n}{r} = \dbinom{n-1}{r-1} + \dbinom{n-1}{r}$

34. A child has six blocks, three of which are red and three of which are green. How many patterns can she make by placing them all in a line? If she is given three white blocks, how many total patterns can she make?

35. A drawer of socks contains seven black socks, eight blue socks, and nine green socks. Two socks are chosen in the dark.
    (a) What is the probability that they match?
    (b) What is the probability that a black pair is chosen?

36. How many ways can 11 boys on a soccer team be grouped into 4 forwards, 3 midfielders, 3 defenders, and 1 goalie?

37. A software development company has three jobs to do. Two of the jobs require three programmers, and the other requires four. If the company employs ten programmers, how many different ways are there to assign them to the jobs?

38. In how many ways can 12 people be divided into three groups of 4 for an evening of bridge? In how many ways can this be done if the 12 consist of six pairs of partners?

39. Show that if the conditional probabilities exist, then

$$P(A_1 \cap A_2 \cap \cdots \cap A_n)$$
$$= P(A_1)P(A_2|A_1)P(A_3|A_1 \cap A_2) \cdots P(A_n|A_1 \cap A_2 \cap \cdots A_{n-1})$$

40. Urn A has three red balls and two white balls, and urn B has two red balls and five white balls. A fair coin is tossed; if it lands heads up, a ball is drawn from urn A, and otherwise a ball is drawn from urn B.
    (a) What is the probability that a red ball is drawn?
    (b) If a red ball is drawn, what is the probability that the coin landed heads up?

41. Urn A has four red, three blue, and two green balls. Urn B has two red, three blue, and four green balls. A ball is drawn from urn A and put into urn B, and then a ball is drawn from urn B.
    (a) What is the probability that a red ball is drawn from urn B?
    (b) If a red ball is drawn from urn B, what is the probability that a red ball was drawn from urn A?

42. An urn contains three red and two white balls. A ball is drawn, and then it and another ball of the same color are placed back in the urn. Finally, a second ball is drawn.

    (a) What is the probability that the second ball drawn is white?

    (b) If the second ball drawn is white, what is the probability that the first ball drawn was red?

43. A fair coin is tossed three times.

    (a) What is the probability of two or more heads given that there was at least one head?

    (b) What is this probability given that there was at least one tail?

44. Two dice are rolled, and the sum of the face values is six. What is the probability that at least one of the dice came up a three?

45. Answer Problem 44 again given that the sum is less than six.

46. Suppose that 5 cards are dealt from a 52-card deck and the first one is a king. What is the probability of at least one more king?

47. A fire insurance company has high-risk, medium-risk, and low-risk clients, who have, respectively, probabilities .02, .01, and .0025 of filing claims within a given year. The proportions of the numbers of clients in the three categories are .10, .20, and .70, respectively. What proportion of the claims filed each year come from high-risk clients?

48. This problem introduces a simple meteorological model, more complicated versions of which have been proposed in the literature. Consider a sequence of days, and let R_i denote the event that it rains on day i. Suppose that $P(R_i|R_{i-1}) = \alpha$ and $P(R_i^c|R_{i-1}^c) = \beta$. Suppose further that only today's weather is relevant to predicting tomorrow's; that is, expressions like $P(R_i|R_{i-1} \cap R_{i-2} \cap \cdots \cap R_0)$ are equivalent to $P(R_i|R_{i-1})$.

    (a) If the probability of rain today is p, what is the probability of rain tomorrow?

    (b) What is the probability of rain the day after tomorrow?

    (c) What is the probability of rain n days from now? What happens as n approaches infinity?

49. This problem continues Example D of Section 1.5 and concerns occupational mobility.

    (a) Find $P(M_1|M_2)$ and $P(L_1|L_2)$.

    (b) Find the proportions that will be in the three occupational levels in the third generation. In order to do this, assume that a son's occupational status depends on his father's status but that given his father's status, it does not depend on his grandfather's.

50. A couple has two children. What is the probability that both are girls given that the oldest is a girl? What is the probability that both are girls given that one of them is a girl?

51. There are three cabinets, A, B, and C, each of which has two drawers. Each drawer contains one coin; A has two gold coins, B has two silver coins, and C has one gold and one silver coin. A cabinet is chosen at random, one drawer is opened, and a silver coin is found. What is the probability that the other drawer in that cabinet contains a silver coin?

52. A teacher tells three boys, Drew, Chris, and Jason, that two of them will have to stay after school to help her clean erasers and that one of them will be

able to leave. She further says that she has made the decision as to who will leave and who will stay at random by rolling a special three-sided Dungeons and Dragons die. Drew wants to leave to play soccer and has a clever idea about how to increase his chances of doing so. He figures that one of Jason and Chris will certainly stay and asks the teacher to tell him the name of one of the two who will stay. Drew's idea is that if, for example, Jason is named, then he and Chris are left and they each have a probability .5 of leaving; similarly, if Chris is named, Drew's probability of leaving is still .5. Thus, by merely asking the teacher a question, Drew will increase his probability of leaving from $\frac{1}{3}$ to $\frac{1}{2}$. What do you think of this scheme?

53. A box has three coins. One has two heads, one has two tails, and the other is a fair coin with one head and one tail. A coin is chosen at random, is flipped, and comes up heads.
   (a) What is the probability that the coin chosen is the two-headed coin?
   (b) What is the probability that if it is thrown another time it will come up heads?
   (c) Answer part (a) again, supposing that the coin is thrown a second time and comes up heads again.

54. A factory runs three shifts. In a given day, 1% of the items produced by the first shift are defective, 2% of the second shift's items are defective, and 5% of the third shift's items are defective. If the shifts all have the same productivity, what percentage of the items produced in a day are defective? If an item is defective, what is the probability that it was produced by the third shift?

55. Suppose that chips for an integrated circuit are tested and that the probability that they are detected if they are defective is .95 and the probability that they are declared sound if in fact they are sound is .97. If .5% of the chips are faulty, what is the probability that a chip that is declared faulty is sound?

56. Show that if $P(A|E) \geq P(B|E)$ and $P(A|E^c) \geq P(B|E^c)$, then $P(A) \geq P(B)$.

57. Suppose that the probability of living to be older than 70 is .6 and the probability of living to be older than 80 is .2. If a person reaches her 70th birthday, what is the probability that she will celebrate her 80th?

58. If B is an event, with $P(B) > 0$, show that the set function $Q(A) = P(A|B)$ satisfies the axioms for a probability measure. Thus, for example,

$$P(A \cup C|B) = P(A|B) + P(C|B) - P(A \cap C|B)$$

59. Show that if A and B are independent, then A and B^c as well as A^c and B^c are independent.

60. Show that $\varnothing$ is independent of A for any A.

61. Show that if A and B are independent, then

$$P(A \cup B) = P(A) + P(B) - P(A)P(B)$$

62. If A is independent of B and B is independent of C, then A is independent of C. Prove this statement, or give a counterexample if it is false.

63. If A and B are disjoint, can they be independent?

64. If $A \subset B$, can A and B be independent?

65. Show that if A, B, and C are independent, then $A \cap B$ and C are independent and $A \cup B$ and C are independent.

66. Suppose that n components are connected in series. For each unit, there is a backup unit, and the system fails if and only if both a unit and its backup fail. Assuming that all the units are independent and fail with probability p, what is the probability that the system works? For $n = 10$ and $p = .05$, compare these results with those of Example E in Section 1.6.

67. A system has n independent units, each of which fails with probability p. The system fails only if k or more of the units fail. What is the probability that the system fails?

68. What is the probability that the following system works if each unit fails independently with probability p (see Figure 1-4)?

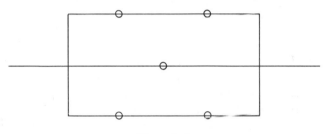

Figure 1-4

69. This problem deals with an elementary aspect of a simple branching process. A population starts with one member; at time $t = 1$, it either divides with probability p or dies with probability $1 - p$. If it divides, then both of its children behave independently with the same two alternatives at time $t = 2$. What is the probability that there are no members in the third generation? For what value of p is this probability equal to .5?

70. Here is a simple model of a queue. The queue runs in discrete time ($t = 0, 1, 2, \ldots$), and at each unit of time the first person in the queue is served with probability p and a new person arrives with probability q. At time $t = 0$, there is one person in the queue. Find the probabilities that there are 0, 1, 2, 3 people in line at time $t = 2$.

71. A player throws darts at a target. On each trial, independently of the other trials, he hits the bull's-eye with probability .05. How many times should he throw so that his probability of hitting the bull's-eye at least once is .5?

72. This problem and the next introduce some aspects of a simple genetic model. Assume that genes in an organism occur in pairs and that each member of the pair can be either of the types a or A. The possible genotypes of an organism are then AA, Aa, and aa (Aa and aA are equivalent). When two organisms mate, each independently contributes one of its two genes; either one of the pair is transmitted with probability .5.

(a) Suppose that the genotypes of the parents are AA and Aa. Find the possible genotypes of their offspring and the corresponding probabilities.

(b) Suppose that the probabilities of the genotypes AA, Aa, and aa are p, $2q$, and r, respectively, in the first generation. Find the probabilities in the second and third generations, and show that these are the same. This result is called the Hardy–Weinberg Law.

(c) Compute the probabilities for the second and third generations as in part (b) but under the additional assumption that the probabilities that an individual of type AA, Aa, or aa survives to mate are u, v, and w, respectively.

73. Many human diseases are genetically transmitted (for example, hemophilia or Tay–Sachs disease). Here is a simple model for such a disease. The genotype aa is diseased and dies before it mates. The genotype Aa is a carrier but is not diseased. The genotype AA is not a carrier and is not diseased.

(a) If two carriers mate, what are the probabilities that their offspring are of each of the three genotypes?

(b) If the male offspring of two carriers is not diseased, what is the probability that he is a carrier?

(c) Suppose that the nondiseased offspring of part (b) mates with a member of the population for whom no family history is available and who is thus assumed to have probability p of being a carrier (p is a very small number). What are the probabilities that their first offspring has the genotypes AA, Aa, and aa?

(d) Suppose that the first offspring of part (c) is not diseased. What is the probability that the father is a carrier in light of this evidence?

2

Random Variables

2.1 Discrete Random Variables

A random variable is essentially a random number. We will be interested in random numbers that are determined by experiments. As motivation for a definition, let us consider an example. A coin is thrown three times, and the sequence of heads and tails is observed; thus,

$$\Omega = \{hhh, hht, htt, hth, ttt, tth, thh, tht\}$$

Examples of random variables associated with Ω are (1) the total number of heads, (2) the total number of tails, and (3) the number of heads minus the number of tails. Each of these is a real-valued function defined on Ω; that is, each is a rule that assigns a real number to every point $\omega \in \Omega$. Since the outcome in Ω is random, the corresponding number is random as well.

In general, a random variable is a function from Ω to the real numbers. Since the outcome of the experiment for which Ω is the sample space is random, the number produced by the function is random as well. It is conventional to denote random variables by italic uppercase letters from the end of the alphabet. For example, we might define X to be the total number of heads in the experiment described above. A **discrete random variable** is a random variable that can take on only a finite or at most a countably infinite number of values. The random variable X just defined is a discrete random variable since it can take on only the values 0, 1, 2, and 3. For an example of a random variable that can take on a countably infinite number of values, consider an experiment that consists of tossing a coin until a head turns up and defining Y to be the total number of tosses. The possible values of Y are 0, 1, 2, 3, ... In general, a countably infinite set is one that can be put into one-to-one correspondence with the integers.

If the coin is fair, then each of the outcomes in Ω above has probability $\frac{1}{8}$, from which the probabilities that X takes on the values 0, 1, 2, and 3 can be easily computed:

$$P(X = 0) = \tfrac{1}{8}$$

$$P(X = 1) = \tfrac{3}{8}$$

$$P(X = 2) = \tfrac{3}{8}$$

$$P(X = 3) = \tfrac{1}{8}$$

Generally, the probability measure on the sample space determines the probabilities of the various values of X; if those values are denoted by $x_1, x_2, \ldots$, then there is a function p such that $p(x_i) = P(X = x_i)$ and $\sum_i p(x_i) = 1$. This function is called the **probability mass function**, or the **frequency function**, of the random variable X. Figure 2-1 shows a graph of $p(x)$ for the coin tossing experiment. The

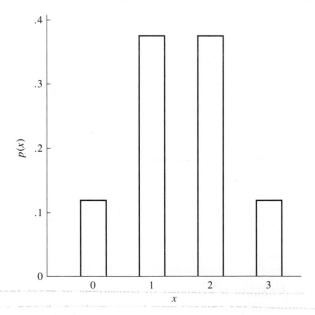

Figure 2-1.   A probability mass function.

frequency function describes completely the probability properties of the random variable.

In addition to the frequency function, it is sometimes useful to use the **cumulative distribution function (cdf)** of a random variable, which is defined to be

$$F(x) = P(X \le x), \qquad -\infty < x < \infty$$

Cumulative distribution functions are usually denoted by uppercase letters and frequency functions by lowercase letters. Figure 2-2 is a graph of the cumulative distribution function of the random variable X of the preceding paragraph. Note that the cdf jumps wherever $p(x) > 0$ and that the jump at x_i is $p(x_i)$.

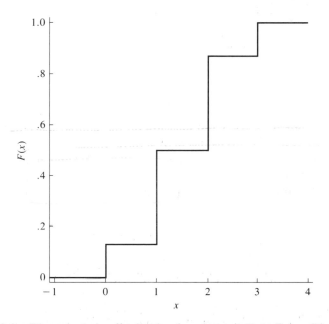

Figure 2-2.   The cumulative distribution function corresponding to Figure 2-1.

Chapter 3 will cover in detail the joint frequency functions of several random variables defined on the same sample space, but it is useful to define here the concept of independence of random variables. In the case of two discrete random variables X and Y, taking on possible values $x_1, x_2, \ldots$ and $y_1, y_2, \ldots$, X and Y are said to be independent if, for all i and j,

$$P(X = x_i \text{ and } Y = y_j) = P(X = x_i)P(Y = y_j)$$

The definition is extended to collections of more than two discrete random variables in the obvious way; for example, X, Y, and Z are said to be mutually independent if, for all i, j, and k,

$$P(X = x_i, Y = y_j, Z = z_k) = P(X = x_i)P(Y = y_j)P(Z = z_k)$$

We next discuss some common discrete distributions that arise in applications.

2.1.1   Bernoulli Random Variables

A Bernoulli random variable takes on only two values: 0 and 1, with probabilities $1 - p$ and p, respectively. Its frequency function is thus

$$p(1) = p$$

$$p(0) = 1 - p$$

$$p(x) = 0, \quad \text{if } x \neq 0 \text{ or } x \neq 1$$

An alternative and sometimes useful representation of this function is

$$p(x) = \begin{cases} p^x(1 - p)^{1-x}, & \text{if } x = 0 \text{ or } x = 1 \\ 0, & \text{otherwise} \end{cases}$$

If A is an event, then the **indicator random variable**, I_A, takes on the value 1 if A occurs and the value 0 if A does not occur:

$$I_A(\omega) = \begin{cases} 1, & \text{if } \omega \in A \\ 0, & \text{otherwise} \end{cases}$$

I_A is a Bernoulli random variable. In applications, Bernoulli random variables often occur as indicators. A Bernoulli random variable might take on the value 1 or 0 according to whether a guess was a success or a failure.

2.1.2   The Binomial Distribution

Suppose that n independent experiments, or trials, are performed, where n is a fixed number, and that each experiment results in a "success" with probability p and a "failure" with probability $1 - p$. The total number of successes, X, is a binomial random variable with parameters n and p. For example, a coin is tossed 10 times and the total number of heads is counted ("head" is identified with "success").

The probability that $X = k$, or $p(k)$, can be found in the following way: Any particular sequence of k successes occurs with probability $p^k(1 - p)^{n-k}$, from the multiplication principle. The total number of such sequences is $\binom{n}{k}$, since there are $\binom{n}{k}$ ways to assign k successes to n trials. Thus,

$$p(k) = \binom{n}{k} p^k(1 - p)^{n-k}$$

Two binomial frequency functions are shown in Figure 2-3. Note how the shape varies as a function of p.

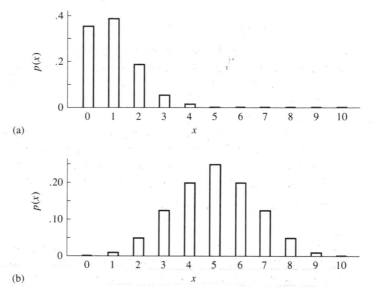

(a)

(b)

Figure 2-3. Binomial frequency functions, (a) $n = 10$ and $p = .1$ and (b) $n = 10$ and $p = .5$.

EXAMPLE A. Tay–Sachs disease is a rare but fatal disease of genetic origin occurring chiefly in infants and children, especially those of Jewish or eastern European extraction. If a couple are both carriers of Tay–Sachs disease, a child of theirs has probability .25 of being born with the disease. If such a couple has four children, what is the frequency function for the number of children that will have the disease?

We assume that the four outcomes are independent of each other, so, if X denotes the number of children with the disease,

$$p(k) = \binom{4}{k}.25^k.75^{4-k}, \quad k = 0, 1, 2, 3, 4$$

These probabilities are given in the following table:

k	$p(k)$
0	.316
1	.422
2	.211
3	.047
4	.004

□

EXAMPLE B.   If a single bit (0 or 1) is transmitted over a noisy communications channel, it has probability p of being incorrectly transmitted. To overcome this problem, the bit is transmitted n times, where n is odd. A decoder at the receiving end, called a majority decoder, decides that the correct message is that carried by a majority of the received bits. Under a simple noise model, each bit is independently subject to being corrupted with the same probability p. The number of bits that are in error, X, is thus a binomial random variable with n trials and probability p of success on each trial (in this case, and frequently elsewhere, the word *success* is used in a generic sense; here a success is an error). Suppose, for example, that $n = 5$ and $p = .1$. The probability that the message is correctly received is the probability of two or fewer errors, which is

$$\sum_{k=0}^{2} \binom{n}{k} p^k (1 - p)^{n-k} = p^0(1 - p)^5 + 5p(1 - p)^4 + 10p^2(1 - p)^3$$

$$= .9914$$

This indicates a considerable improvement in reliability.   □

A random variable with a binomial distribution can be expressed in terms of independent Bernoulli random variables, a fact that will be quite useful for analyzing some properties of random variables in later chapters of this book. Specifically, let $X_1, X_2, \ldots, X_n$ be independent Bernoulli random variables with $P(X_i = 1) = p$. Then $Y = X_1 + X_2 + \cdots + X_n$ is a binomial random variable.

2.1.3   The Geometric and Negative Binomial Distributions

The **geometric distribution** is also constructed from independent Bernoulli trials, but from an infinite sequence. On each trial, a success occurs with probability p, and X is the total number of trials up to and including the first success. In order that $X = k$, there must be $k - 1$ failures followed by a success. Therefore, from the independence of the trials, we have

$$p(k) = P(X = k) = (1 - p)^{k-1}p, \quad k = 1, 2, 3, \ldots$$

Note that these probabilities sum to 1:

$$\sum_{k=1}^{\infty} (1 - p)^{k-1}p = p \sum_{j=0}^{\infty} (1 - p)^j = 1$$

EXAMPLE A.   The probability of winning in a certain state lottery is said to be about $\frac{1}{9}$. If it is exactly $\frac{1}{9}$, the distribution of the number of tickets a person must purchase up to and including the first winning ticket is a geometric random variable with $p = \frac{1}{9}$. Figure 2-4 shows the frequency function.   □

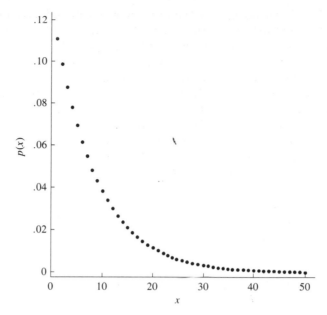

Figure 2-4.   The probability mass function of a geometric random variable with $p = \frac{1}{9}$.

The **negative binomial distribution** arises as a generalization of the geometric distribution. Suppose that a sequence of independent trials is performed until there are r successes in all; let X denote the total number of trials. To find $P(X = k)$, we can argue in the following way: Any particular such sequence has probability $p^r(1 - p)^{k-r}$, from the independence assumption. The last trial is a success, and the remaining $r - 1$ successes can be assigned to the remaining $k - 1$ trials in $\binom{k-1}{r-1}$ ways. Thus,

$$P(X = k) = \binom{k - 1}{r - 1} p^r (1 - p)^{k-r}$$

It is sometimes helpful in analyzing properties of the negative binomial distribution to note that a negative binomial random variable can be expressed as the sum of r independent geometric random variables: the number of trials up to and including the first success plus the number of trials after the first success up to and including the second success, ... plus the number of trials from the $(r - 1)$th success up to and including the rth success.

EXAMPLE B.   Continuing Example A, the distribution of the number of tickets purchased up to and including the second winning ticket is negative binomial:

$$p(k) = (k - 1)p^2(1 - p)^{k-2}$$

This frequency function is shown in Figure 2-5.   □

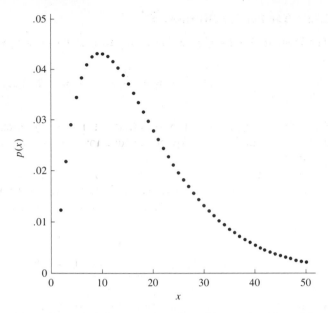

Figure 2-5. The probability mass function of a negative binomial random variable with $p = \frac{1}{9}$ and $r = 2$.

The definitions of the geometric and negative binomial distributions vary slightly from one textbook to another; for example, instead of X being the total number of trials in the definition of the geometric distribution, X is sometimes defined as the total number of failures.

2.1.4   The Hypergeometric Distribution

The **hypergeometric distribution** was introduced in chapter 1 but was not named there. Suppose that an urn contains n balls of which r are black and $n - r$ are white. Let X denote the number of black balls drawn when taking m balls without replacement.

$$P(X = k) = \frac{\binom{r}{k}\binom{n - r}{m - k}}{\binom{n}{m}}$$

X is a hypergeometric random variable with parameters r, n, and m. We have seen applications of the hypergeometric distribution to quality control and the capture/recapture method in Examples II and I in Section 1.4.2.

2.1.5 The Poisson Distribution

The **Poisson frequency function** with parameter λ $(\lambda > 0)$ is

$$P(X = k) = \frac{\lambda^k}{k!} e^{-\lambda}, \quad k = 0, 1, 2, \ldots$$

Since $e^\lambda = \sum_{k=0}^{\infty} (\lambda^k / k!)$, it follows that the frequency function sums to 1. Figure 2-6 shows four Poisson frequency functions. Note how the shape varies as a function of λ.

The Poisson distribution can be derived as the limit of a binomial distribution as the number of trials, n, approaches infinity and the probability of success on each trial, p, approaches zero in such a way that $np = \lambda$. The binomial frequency function is

$$p(k) = \frac{n!}{k!(n - k)!} p^k (1 - p)^{n-k}$$

Setting $np = \lambda$ allows this function to be written as

$$p(k) = \frac{n!}{k!(n - k)!} \left(\frac{\lambda}{n}\right)^k \left(1 - \frac{\lambda}{n}\right)^{n-k}$$

$$= \frac{\lambda^k}{k!} \frac{n!}{(n - k)!} \frac{1}{n^k} \left(1 - \frac{\lambda}{n}\right)^n \left(1 - \frac{\lambda}{n}\right)^{-k}$$

As $n \to \infty$,

$$\frac{\lambda}{n} \to 0,$$

$$\frac{n!}{(n - k)! n^k} \to 1,$$

$$\left(1 - \frac{\lambda}{n}\right)^n \to e^{-\lambda}$$

and

$$\left(1 - \frac{\lambda}{n}\right)^{-k} \to 1$$

We thus have that

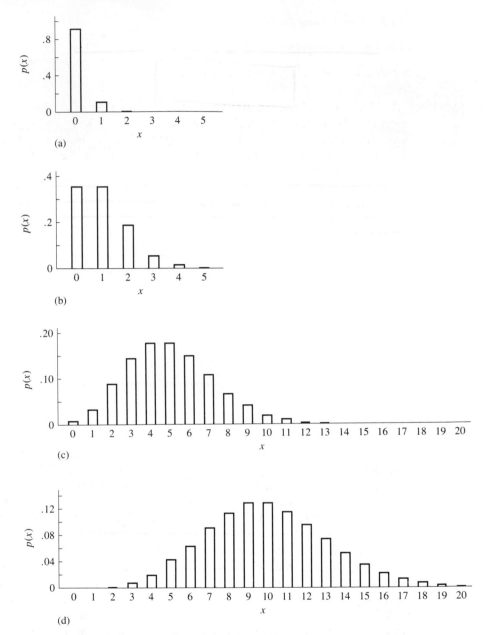

Figure 2-6. Poisson frequency functions, (a) $\lambda = .1$, (b) $\lambda = 1$, (c) $\lambda = 5$, (d) $\lambda = 10$.

$$p(k) \rightarrow \frac{\lambda^k e^{-\lambda}}{k!}$$

which is the Poisson frequency function.

EXAMPLE A. Two dice are rolled 100 times, and the number of double sixes, X, is counted. The distribution of X is binomial with $n = 100$ and $p = \frac{1}{36} = .0278$. Since n is large and p is small, we can approximate the binomial probabilities with $\lambda = np = 2.78$. The exact binomial probabilities and the Poisson approximations are shown in the following table:

k	Binomial probability	Poisson approximation
0	.0596	.0620
1	.1705	.1725
2	.2414	.2397
3	.2255	.2221
4	.1564	.1544
5	.0858	.0858
6	.0389	.0398
7	.0149	.0158
8	.0050	.0055
9	.0015	.0017
10	.0004	.0005
11	.0001	.0001

The approximation is quite good.   ☐

The Poisson frequency function can be used to approximate binomial probabilities for large n and small p. More important, it suggests how Poisson distributions can arise in practice. Suppose that X is a random variable that equals the number of times some event occurs in a given interval of time. Heuristically, let us think of dividing the interval into a very large number of small subintervals of equal length, and let us assume that the subintervals are so small that the probability of more than one event in a subinterval is negligible relative to the probability of one event, which is itself very small. Let us also assume that the probability of an event is the same in each subinterval and that whether an event occurs in one subinterval is independent of what happens in the other subintervals. The random variable X is thus nearly a binomial random variable, with the subintervals constituting the trials, and, from the limiting result above, X has nearly a Poisson distribution.

The preceding argument is not formal, of course, but merely suggestive. But, in fact, it can be made rigorous. The important assumptions underlying it are that what happens in one subinterval is independent of what happens in any other subinterval, that the probability of an event is the same in each subinterval, and that events do not happen simultaneously. The same kind of argument can be made if we are concerned with an area or a volume of space rather than with an interval on the real line.

The Poisson distribution is of fundamental theoretical and practical importance. It has been used in many areas, including the following:

- The Poisson distribution has been used in the analysis of telephone systems. The number of calls coming into an exchange during a unit of time might be modeled as a Poisson variable if the exchange services a large number of customers who act more or less independently.
- One of the earliest uses of the Poisson distribution was to model the number of alpha particles emitted from a radioactive source during a given period of time.
- The Poisson distribution has been used as a model by insurance companies. For example, the number of freak accidents, such as falls in the shower, for a large population of people in a given time period might be modeled as a Poisson distribution, since the accidents would presumably be rare and independent (provided there was only one person in the shower).
- The Poisson distribution has been used by traffic engineers as a model for light traffic. The number of vehicles that pass a marker on a roadway during a unit of time can be counted. If traffic is light, the individual vehicles act independently of each other. In heavy traffic, however, one vehicle's movement may influence another's, so the approximation might not be good.

EXAMPLE B.   This amusing classical example is from von Bortkiewicz (1898). The number of fatalities that resulted from being kicked by a horse was recorded for 10 corps of Prussian cavalry over a period of 20 years, giving 200 corps-years worth of data. These data and the predictions from a Poisson model with $\lambda = .61$ are displayed in the table below. (Methods for fitting a probability law to data and for testing goodness of fit will be covered later in this book. Thus, discussion of how the number $\lambda = .61$ was determined must be postponed.)

Number of deaths per year	Observed	Fitted
0	109	108.7
1	65	66.3
2	22	20.2
3	3	4.1
4	1	.6

☐

The Poisson distribution often arises from a model called a **Poisson process** for the distribution of random events in a set S, which is typically one-, two-, or three-dimensional, corresponding to time, a plane, or a volume of space. Basically, this model states that if $S_1, S_2, \ldots, S_n$ are disjoint subsets of S, then the numbers of events in these subsets, $N_1, N_2, \ldots, N_n$, are independent random variables that follow Poisson distributions with parameters $\lambda|S_1|, \lambda|S_2|, \ldots, \lambda|S_n|$, where $|S_i|$ denotes the measure of S_i (length, area, or volume, for example). The crucial assumptions here are that events in disjoint subsets are independent of each other and that the Poisson parameter for a subset is proportional to the subset's size. Later, we will see that this latter assumption implies that the average number of events in a subset is proportional to its size.

EXAMPLE C.   Suppose that an office receives telephone calls as a Poisson process with $\lambda = .5$ per minute. The number of calls in a 5-minute interval follows a Poisson distribution with parameter $\omega = 5\lambda = 2.5$. Thus, the probability of no calls in a 5-minute interval is $e^{-2.5} = .082$. The probability of exactly one call is $2.5e^{-2.5} = .205$.   ☐

EXAMPLE D.   Figure 2-7 shows four realizations of a Poisson process with $\lambda = 25$ in the unit square, $0 \le x \le 1$ and $0 \le y \le 1$. It is interesting that the eye tends to

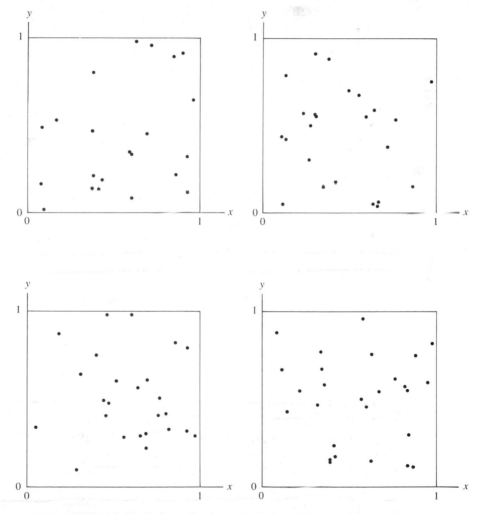

Figure 2-7.   Four realizations of a Poisson process with $\lambda = 25$.

perceive patterns, such as clusters of points and large blank spaces. But by the nature of a Poisson process, the locations of the points have no relationship to one another, and these patterns are entirely due to chance.   ☐

2.2   Continuous Random Variables

In applications, we are often interested in a random variable that can take on a continuum of values rather than a countably infinite number. For example, a model for the lifetime of an electronic component might be that it is random and can be any positive real number. For a continuous random variable, the role of the frequency function is taken by a **density function**, $f(x)$, which has the properties that $f(x) \geq 0$, f is piecewise continuous, and $\int_{-\infty}^{\infty} f(x)\,dx = 1$. If X is a random variable with a density function f, then for any $a < b$ the probability that X falls in the interval (a, b) is the area under the density function between a and b:

$$P(a < X < b) = \int_a^b f(x)\,dx$$

EXAMPLE A.   A **uniform random variable** on the interval $[0, 1]$ is a model for what we mean when we say "choose a number at random between 0 and 1." Any real number in the interval is a possible outcome, and the probability model should have the character that the probability that X is in any subinterval of length h is equal to h. The following density function does the job:

$$f(x) = \begin{cases} 1, & 0 \leq x \leq 1 \\ 0, & x < 0 \text{ or } x > 1 \end{cases}$$

This is called the **uniform density** on $[0, 1]$. The uniform density on a general interval $[a, b]$ is

$$f(x) = \begin{cases} 1/(b - a), & a \leq x \leq b \\ 0, & x < a \text{ or } x > b \end{cases}$$ ☐

One consequence of this construction is that the probability that a continuous random variable X takes on any particular value is 0:

$$P(X = a) = \int_a^a f(x)\,dx = 0$$

Although this may seem strange initially, it is really quite natural. If the uniform random variable of Example A had a positive probability of being any particular number, it should have the same probability for any number in $[0, 1]$, in which case the sum of the probabilities of any countably infinite subset of $[0, 1]$ would be infinite. If X is a continuous random variable, then

$$P(a < X < b) = P(a \le X < b) = P(a < X \le b)$$

Note that this is not true for a discrete random variable.

For small δ, if f is continuous at x,

$$P\left(x - \frac{\delta}{2} \le X \le x + \frac{\delta}{2}\right) = \int_{x-\delta/2}^{x+\delta/2} f(u)\, du \approx \delta f(x)$$

Therefore, the probability of a small interval around x is proportional to $f(x)$. It is sometimes useful to employ differential notation: $P(x \le X \le x + dx) = f(x)\, dx$.

The cumulative distribution function is often employed in finding probability properties of a continuous random variable and is defined as

$$F(x) = P(X \le x) = \int_{-\infty}^{x} f(u)\, du$$

The cdf is a nondecreasing function with $\lim_{x \to -\infty} F(x) = 0$ and $\lim_{x \to \infty} F(x) = 1$. The cdf can be used to evaluate the probability that X falls in an interval:

$$P(a \le X \le b) = F(b) - F(a)$$

From the fundamental theorem of calculus, if f is continuous at x, $f(x) = F'(x)$.

EXAMPLE B.   From Example A, we see that the cdf of a uniform random variable on $[0, 1]$ is

$$F(x) = \begin{cases} 0, & x \le 0 \\ x, & 0 \le x \le 1 \\ 1, & x \ge 1 \end{cases}$$ $\square$

Suppose that F is strictly increasing on some interval I and that $F = 0$ to the left of I and $F = 1$ to the right of I; I may be unbounded. Under this assumption, the inverse function F^{-1} is well-defined; $x = F^{-1}(y)$ if $y = F(x)$. The pth **quantile** of the distribution F is defined to be that value x_p such that $F(x_p) = p$, or $P(X \le x_p) = p$. Under the assumption stated above, x_p is uniquely defined as $x_p = F^{-1}(p)$; see Figure 2-8. Special cases are $p = \frac{1}{2}$, which corresponds to the **median** of F, and $p = \frac{1}{4}$ and $p = \frac{3}{4}$, which correspond to the lower and upper **quartiles** of F.

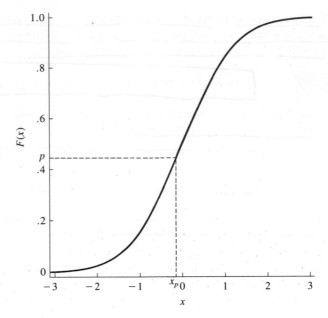

Figure 2-8.   A cdf F and F^{-1}.

EXAMPLE C.   Suppose that $F(x) = x^2$ for $0 \le x \le 1$. This statement is shorthand for the more explicit statement

$$F(x) = \begin{cases} 0, & x \le 0 \\ x^2, & 0 \le x \le 1 \\ 1, & x \ge 1 \end{cases}$$

To find F^{-1}, we solve $y = F(x) = x^2$ for x, obtaining $x = F^{-1}(y) = \sqrt{y}$. The median is $F^{-1}(.5) = .707$, the lower quartile is $F^{-1}(.25) = .50$, and the upper quartile is $F^{-1}(.75) = .866$.   □

We next discuss some density functions that commonly arise in practice.

2.2.1   The Exponential Density

The exponential density function is

$$f(x) = \begin{cases} \lambda e^{-\lambda x}, & x \ge 0 \\ 0 & x < 0 \end{cases}$$

Like the Poisson distribution, the exponential density depends on a single parameter, $\lambda > 0$, and it would therefore be more accurate to refer to it as the family of exponential densities. Several exponential densities are shown in Figure 2-9. Note that as λ becomes larger, the density drops off more rapidly.

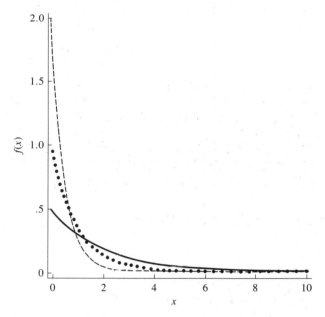

Figure 2-9. Exponential densities with $\lambda = .5$ (solid), $\lambda = 1$ (dotted), and $\lambda = 2$ (dashed).

The cumulative distribution function is easily found:

$$F(x) = \int_0^x \lambda e^{-\lambda u}\, du = \begin{cases} 1 - e^{-\lambda x}, & x \geq 0 \\ 0 & x < 0 \end{cases}$$

The median of an exponential distribution, η, is readily found from the cdf:

$$1 - e^{-\lambda \eta} = \tfrac{1}{2}$$

from which we have

$$\eta = \frac{\log 2}{\lambda}$$

The exponential distribution is often used to model lifetimes or waiting times, in which context it is conventional to replace x by t. Suppose that we consider modeling the lifetime of an electronic component as an exponential random variable, that the component has lasted a length of time s, and that we wish to calculate the probability that it will last at least t more time units; that is, we wish to find $P(T > t + s \mid T > s)$:

$$P(T > t + s | T > s) = \frac{P(T > t + s \text{ and } T > s)}{P(T > s)}$$

$$= \frac{P(T > t + s)}{P(T > s)}$$

$$= \frac{e^{-\lambda(t+s)}}{e^{-\lambda s}}$$

$$= e^{-\lambda t}$$

We see that the probability that the unit will last t more time units does not depend on s. The exponential distribution is consequently said to be **memoryless**; it is clearly not a good model for human lifetimes since the probability that a 16-year-old will live at least 10 more years is not the same as the probability that an 80-year-old will live at least 10 more years. It can be shown that the exponential distribution is characterized by this memoryless property—that is, the memorylessness implies that the distribution is exponential. It may be somewhat surprising that a qualitative characterization, the property of memorylessness, actually determines the form of this density function.

The memoryless character of the exponential distribution follows directly from its relation to the Poisson process. Suppose that events occur in time as a Poisson process with parameter λ and that a given event occurs at time t_0. Let T denote the length of time until the next event. The density of T can be found as follows:

$$P(T > t) = P(\text{no events in } (t_0, t_0 + t))$$

Since the number of events in the given interval follows a Poisson distribution with parameter λt, this probability is $e^{-\lambda t}$, and thus T follows an exponential distribution with parameter λ. We can continue in this fashion. Suppose that the next event occurs at time t_1; the distribution of time until the third event is again exponential by the same analysis and, from the independence property of the Poisson process, is independent of the length of time between the first two events. Generally, the times between events of a Poisson process are independent, identically distributed, exponential random variables.

EXAMPLE A.   Muscle and nerve cell membranes contain large numbers of channels through which selected ions can pass when the channels are open. Using sophisticated experimental techniques, neurophysiologists can measure the resulting current that flows through a single channel, and experimental records often indicate that a channel opens and closes at seemingly random times. In some cases, simple kinetic models predict that the duration of the open time should be exponentially distributed.

Figure 2-10, from Montal (1985), shows the agreement of theory and experiment for measurements taken on a sodium channel from a rat's brain. The figure

Nice Exposition Read again

shows the cumulative number of openings whose duration exceeded t as a function of t. For an exponential model, $P(T \geq t) = e^{-\lambda t}$. Therefore, of n openings, we would expect about $ne^{-\lambda t}$ of them to have durations greater than t. On the figure, this model is compared to the experimental data. The quantitative assessment of goodness of fit is a statistical problem that will be taken up later in this book. For now, it is sufficient to note that there appears to be some deviation from the model for small and large values of t. The deviation for small values of t can perhaps be partially accounted for by the limited time resolution of the measurement apparatus, which means that some openings of short duration are being missed. □

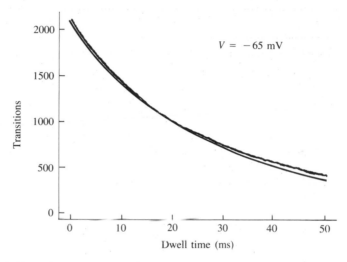

Figure 2-10. Fit to an exponential distribution of the durations of openings of a sodium channel.

2.2.2 The Gamma Density

The gamma density function depends on two parameters, α and λ:

$$g(t) = \frac{\lambda^{\alpha}}{\Gamma(\alpha)} t^{\alpha-1} e^{-\lambda t}, \quad t \geq 0$$

For $t < 0$, $g(t) = 0$. In order that the density be well-defined and integrate to 1, $\alpha > 0$ and $\lambda > 0$. The gamma function, $\Gamma(x)$, is defined as

$$\Gamma(x) = \int_{0}^{\infty} u^{x-1} e^{-u} \, du, \quad x > 0$$

Some properties of the gamma function are developed in the problems at the end of this chapter.

Note that if $\alpha = 1$, the gamma density coincides with the exponential density. The parameter α is called a **shape parameter** for the gamma function, and λ is called a **scale parameter**. Varying α changes the shape of the density, whereas varying λ corresponds to changing the units of measurement (say, from seconds to minutes) and does not affect the shape of the density.

Figure 2-11 shows several gamma densities. Gamma densities provide a fairly flexible class for modeling nonnegative random variables.

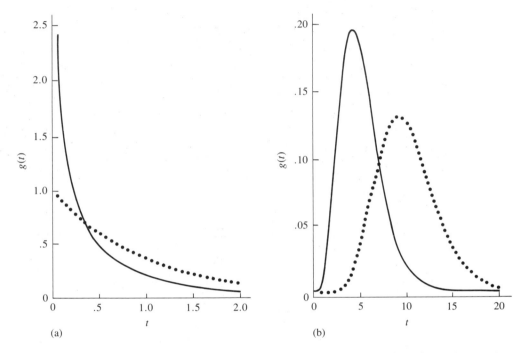

Figure 2-11. Gamma densities, (a) $\alpha = .5$ (solid) and $\alpha = 1$ (dotted) and (b) $\alpha = 5$ (solid) and $\alpha = 10$ (dotted); $\lambda = 1$ in all cases.

EXAMPLE A. The patterns of occurrence of earthquakes in terms of time, space, and magnitude are very erratic, but attempts are sometimes made to construct probabilistic models for these events. The models may be used in a purely descriptive manner or, more ambitiously, for purposes of predicting future occurrences and consequent damage.

Figure 2-12 shows the fit of a gamma density and an exponential density to the observed times separating a sequence of small earthquakes (Udias and Rice, 1975). The gamma density clearly gives a better fit ($\alpha = .509$ and $\lambda = .00115$). Note that an exponential model for interoccurrence times would be memoryless; that is, knowing that an earthquake had not occurred in the last t time units would tell us nothing about the probability of occurrence during the next s time units. The gamma model does not have this property. In fact, although we will not show this, the gamma model with these parameter values has the character that there is a large likelihood that the next earthquake will immediately follow any given one and this likelihood decreases monotonically with time. □

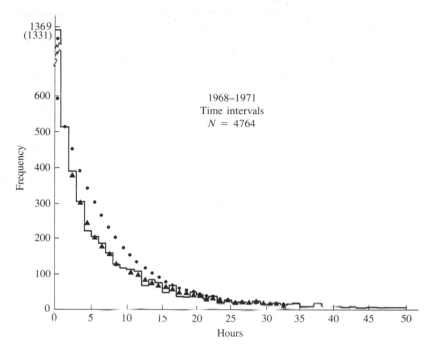

Figure 2-12. Fit of gamma distribution (triangles) and of exponential distribution (circles) to times between microearthquakes.

2.2.3   The Normal Distribution

The normal distribution plays a central role in probability and statistics, for reasons that will become apparent in later chapters of this book. This distribution is also called the Gaussian distribution after Carl Friedrich Gauss, who proposed it as a model for measurement errors. The central limit theorem, which will be discussed in chapter 6, justifies the use of the normal distribution in many applications. Roughly, the central limit theorem says that if a random variable is the sum of a large number of independent random variables, it is approximately normally distributed. The normal distribution has been used as a model for such diverse phenomena as a person's height, the distribution of IQ scores, and the velocity of a gas molecule. The density function of the normal distribution depends on two parameters, μ and σ ($-\infty < \mu < \infty, \sigma > 0$):

$$f(x) = \frac{1}{\sigma\sqrt{2\pi}}e^{-(x-\mu)^2/2\sigma^2}, \quad -\infty < x < \infty$$

The cdf cannot be evaluated in closed form from this density function (the integral that defines the cdf cannot be evaluated by an explicit formula but must be found numerically).

As shorthand for the statement "X follows a normal distribution with parameters μ and σ," it is convenient to use $X \sim N(\mu, \sigma^2)$. From the form of the density function above, we see that the density is symmetric about μ [$f(x - \mu) = f(-(x - \mu))$], where it has a maximum, and that the rate at which it falls off is determined by σ. Figure 2-13 show several normal densities. The special case for which $\mu = 0$ and $\sigma = 1$ is called the **standard normal density**. Its cdf is denoted by Φ and its density by ϕ (not to be confused with the empty set). The relationship between the general normal density and the standard normal density will be developed in the next section.

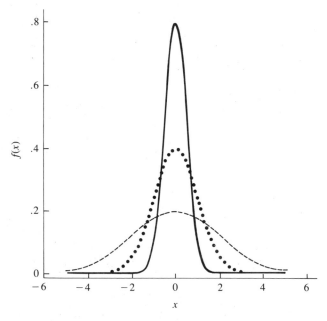

Figure 2-13. Normal densities, $\mu = 0$ and $\sigma = .5$ (solid), $\mu = 0$ and $\sigma = 1$ (dotted), and $\mu = 0$ and $\sigma = 2$ (dashed).

EXAMPLE A.   Acoustic recordings made in the ocean contain substantial background noise. In order to detect sonar signals of interest, it is useful to characterize this noise insofar as possible. In the Arctic, much of the background noise is produced by the cracking and straining of ice. Veitch and Wilks (1985) studied recordings of Arctic undersea noise and characterized the noise as a mixture of a Gaussian component and occasional large-amplitude bursts. Figure 2-14 is a trace of one recording that includes a burst. Figure 2-15 shows a Gaussian distribution fit to observations from a "quiet" (nonbursty) period of this noise.   □

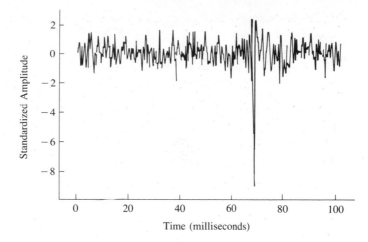

Figure 2-14. A record of undersea noise containing a large burst.

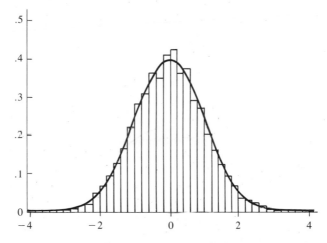

Figure 2-15. A histogram from a "quiet" period of undersea noise with a fitted normal distribution.

EXAMPLE B. Turbulent air flow is sometimes modeled as a random process. Since the velocity of the flow at any point is subject to the influence of a large number of random eddies in the neighborhood of that point, one might expect from the central limit theorem that the velocity would be normally distributed. Van Atta and Chen (1968) analyzed data gathered in a wind tunnel. Figure 2-16, taken from their paper, shows a normal distribution fit to 409,600 observations of one component of the velocity; the fit is remarkably good. □

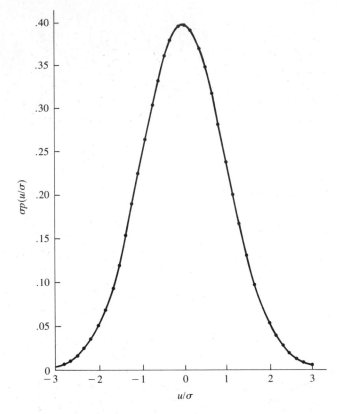

Figure 2-16. A normal distribution (solid line) fit to 409,600 measurements of one component of the velocity of a turbulent wind flow.

2.3   Functions of a Random Variable

When X is a continuous random variable with the density function $f(x)$, it is often necessary to find the density function of $Y = g(X)$, where g is a given function. Often, the density and cdf of X are denoted by f_X and F_X and those of Y by f_Y and F_Y. In order to illustrate techniques for solving such a problem, we first develop some useful facts about the normal distribution.

Suppose that $X \sim N(\mu, \sigma^2)$ and that $Y = aX + b$, where $a > 0$. The cumulative distribution function of Y is

$$F_Y(y) = P(Y \le y)$$

$$= P(aX + b \le y)$$

$$= P\left(X \le \frac{y-b}{a}\right)$$

$$= F_X\left(\frac{y-b}{a}\right)$$

Thus,

$$f_Y(y) = \frac{d}{dy}F_X\left(\frac{y-b}{a}\right)$$

$$= \frac{1}{a}f_X\left(\frac{y-b}{a}\right)$$

Up to this point, we have not used the assumption of normality at all, so this result holds for a general continuous random variable, provided that F_X is appropriately differentiable. If f_X is a normal density function with parameters μ and σ, we find, after substitution,

$$f_Y(y) = \frac{1}{a\sigma\sqrt{2\pi}}\exp\left[-\frac{1}{2}\left(\frac{y-b-a\mu}{a\sigma}\right)^2\right]$$

From this, we see that Y follows a normal distribution with parameters $a\mu + b$ and $a\sigma$.

The case for which $a < 0$ can be analyzed similarly (see Problem 47 in the end-of-chapter problems), yielding the following proposition.

PROPOSITION A. If $X \sim N(\mu, \sigma^2)$ and $Y = aX + b$, then $Y \sim N(a\mu + b, a^2\sigma^2)$.

This proposition is quite useful for finding probabilities from the normal distribution. Suppose that $X \sim N(\mu, \sigma^2)$ and we wish to find $P(a < X < b)$ for some numbers a and b. From the proposition, $Z = (X - \mu)/\sigma$ follows a standard normal distribution whose cdf is Φ. Therefore,

$$F_X(x) = P(X \le x)$$

$$= P\left(\frac{X-\mu}{\sigma} \le \frac{x-\mu}{\sigma}\right)$$

$$= P\left(Z \le \frac{x-\mu}{\sigma}\right)$$

$$= \Phi\left(\frac{x-\mu}{\sigma}\right)$$

and

$$P(a < X < b) = F_X(b) - F_X(a)$$

$$= \Phi\left(\frac{b - \mu}{\sigma}\right) - \Phi\left(\frac{a - \mu}{\sigma}\right)$$

Thus, probabilities for general normal random variables can be evaluated in terms of probabilities for standard normal random variables. This is quite useful, since tables need only be made up for the standard normal distribution rather than separately for every μ and σ.

EXAMPLE A.   Let $X \sim N(\mu, \sigma^2)$, and find the probability that X is a distance less than σ away from μ; that is, find $P(|X - \mu| < \sigma)$.

This probability is

$$P(-\sigma < X - \mu < \sigma) = P\left(-1 < \frac{X - \mu}{\sigma} < 1\right)$$

$$= P(-1 < Z < 1)$$

where Z follows a standard normal distribution. From tables of the standard normal distribution, this last probability is

$$\Phi(1) - \Phi(-1) = .68$$

Thus, a normal random variable is within 1 standard deviation of its mean with probability .68.   □

We now turn to another example involving the normal distribution.

EXAMPLE B.   Find the density of $X = Z^2$ where $Z \sim N(0, 1)$.

Here, we have

$$F_X(x) = P(X \le x)$$

$$= P(-\sqrt{x} \le Z \le \sqrt{x})$$

$$= \Phi(\sqrt{x}) - \Phi(-\sqrt{x})$$

We find the density of X by differentiating the cdf:

$$f_X(x) = \tfrac{1}{2} x^{-1/2} \phi(\sqrt{x}) + \tfrac{1}{2} x^{-1/2} \phi(-\sqrt{x})$$

$$= x^{-1/2} \phi(\sqrt{x})$$

In the last step, we used the symmetry of ϕ. Evaluating the last expression, we find

$$f_X(x) = \frac{x^{-1/2}}{\sqrt{2\pi}} e^{-x/2}, \quad x \geq 0$$

Comparing this to the gamma density function, we see that it is a gamma density with parameters $\alpha = \lambda = \frac{1}{2}$. This density is also called the **chi-square density** with 1 degree of freedom. As an interesting by-product of the calculation, we see that, since the density integrates to 1, $\Gamma(\frac{1}{2}) = \sqrt{\pi}$.   □

As one more example, let us consider the following.

EXAMPLE C.   Let U be a uniform random variable on $[0, 1]$, and let $V = 1/U$. To find the density of V, we first find the cdf:

$$F_V(v) = P(V \leq v)$$

$$= P\left(\frac{1}{U} \leq v\right)$$

$$= P\left(U \geq \frac{1}{v}\right)$$

$$= 1 - \frac{1}{v}$$

This expression is valid for $v \geq 1$; for $v < 1$, $F_V(v) = 0$. We can now find the density by differentiation:

$$f_V(v) = \frac{1}{v^2}, \quad 1 \leq v < \infty$$ □

Looking back over these examples, we see that we have gone through the same basic steps in each case: first finding the cdf of the transformed variable, then differentiating to find the density, and then specifying in what region the result holds. These same steps can be used to prove the following general result.

PROPOSITION B.   Let X be a continuous random variable with density $f(x)$ and let $Y = g(X)$ where g is a differentiable, strictly monotonic function on some interval I. Suppose that $f(x) = 0$ if x is not in I. Then Y has the density function

$$f_Y(y) = f_X(g^{-1}(y)) \left| \frac{d}{dy} g^{-1}(y) \right|$$

for y such that $y = g(x)$ for some x, and $f_Y(y) = 0$ if $y \neq g(x)$ for any x in I. Here g^{-1} is the inverse function of g; that is, $g^{-1}(y) = x$ if $y = g(x)$.

For any specific problem, it is usually easier to proceed from scratch than to decipher the notation and apply the proposition.

We conclude this section by developing some results relating the uniform distribution to other continuous distributions. Throughout, we consider a random variable X, with density f and cdf F, where F is strictly increasing on some interval I, $F = 0$ to the left of I, and $F = 1$ to the right of I. I may be a bounded interval or an unbounded interval such as the whole real line. $F^{-1}(x)$ is then well-defined for $x \in I$.

PROPOSITION C.   Let $Z = F(X)$; then Z has a uniform distribution on $[0, 1]$.

Proof.

$$P(Z \le z) = P(F(X) \le z) = P(X \le F^{-1}(z)) = F(F^{-1}(z)) = z$$

This is the uniform cdf.   ☐

PROPOSITION D.   Let U be uniform on $[0, 1]$, and let $X = F^{-1}(U)$. Then the cdf of X is F.

Proof.

$$P(X \le x) = P(F^{-1}(U) \le x) = P(U \le F(x)) = F(x)$$ ☐

This last proposition is quite useful in generating pseudorandom numbers with a given cdf F. Many computer packages have routines for generating pseudorandom numbers that are uniform on $[0, 1]$. These numbers are called pseudorandom because they are generated according to some rule or algorithm and thus are not "really" random; nonetheless, we hope that they behave like random numbers in many respects. Proposition D tells us that to generate random variables with cdf F, we just apply F^{-1} to uniform random numbers. This is quite practical as long as F^{-1} can be calculated easily.

EXAMPLE D.   Suppose that, as part of a simulation study, we want to generate random variables from an exponential distribution. The cdf is $F(t) = 1 - e^{-\lambda t}$. F^{-1} can be found by solving $x = 1 - e^{-\lambda t}$ for t:

$$e^{-\lambda t} = 1 - x$$

$$-\lambda t = \log(1 - x)$$

$$t = -\log(1 - x)/\lambda$$

Thus, if U is uniform on $[0, 1]$, then $T = -\log(1 - U)/\lambda$ is an exponential random variable with parameter λ. This can be simplified slightly by noting that

$V = 1 - U$ is also uniform on $[0, 1]$ since

$$P(V \leq v) = P(1 - U \leq v) = P(U \geq 1 - v) = 1 - (1 - v) = v \qquad \square$$

2.4 Concluding Remarks

This chapter introduced the concept of a random variable, one of the fundamental ideas of probability theory. A fully rigorous discussion of random variables requires a background in measure theory and is typically covered in a graduate-level course in probability. The development here is sufficient for the needs of this course.

Discrete and continuous random variables have been defined, and it should be mentioned that more general random variables can also be defined and are useful on occasion. In particular, it makes sense to consider random variables that have both a discrete and a continuous component. For example, the lifetime of a transistor might be 0 with some probability $p > 0$ if it does not function at all; if it does function, the lifetime could be modeled as a continuous random variable.

2.5 Problems

1. Suppose that X is a discrete random variable with $P(X = 0) = .25$, $P(X = 1) = .125$, $P(X = 2) = .125$, and $P(X = 3) = .5$. Graph the frequency function and the cumulative distribution function of X.
2. An experiment consists of throwing a fair coin four times. Find the frequency function and the cumulative distributive function of the following random variables: (a) the number of heads before the first tail, (b) the number of heads following the first tail, (c) the number of heads minus the number of tails, and (d) the number of tails times the number of heads.
3. The following table shows the cumulative distribution function of a discrete random variable. Find the frequency function.

k	$F(k)$
0	0
1	.1
2	.3
3	.7
4	.8
5	1.0

4. Let A and B be events, and let I_A and I_B be the associated indicator random variables. Show that

$$I_{A \cap B} = I_A I_B$$

and

$$I_{A \cup B} = \max(I_A, I_B)$$

5. For what values of p is a two-out-of-three majority decoder better than transmission of the message once?

6. Appending three extra bits to a four-bit word in a particular way (a Hamming code) allows detection and correction of up to one error in any of the bits. If each bit has probability .05 of being changed during communication, and the bits are changed independently of each other, what is the probability that the word is correctly received (that is, zero or one bit is in error)? How does this probability compare to the probability that the word will be transmitted correctly with no check bits, in which case all four bits would have to be transmitted correctly for the word to be correct?

7. Consider the binomial distribution with n trials and probability p of success on each trial. For what value of k is $P(X = k)$ maximized? This value is called the **mode** of the distribution. (*Hint:* Consider the ratio of successive terms.)

8. Which is more likely: 9 heads in 10 tosses of a fair coin or 18 heads in 20 tosses?

9. A multiple-choice test consists of 20 items, each with four choices. A student is able to eliminate one of the choices on each question as incorrect and chooses randomly from the remaining three choices. A passing grade is 12 items or more correct.
   (a) What is the probability that the student passes?
   (b) Answer the question in part (a) again, assuming that the student can eliminate two of the choices on each question.

10. Two boys play basketball in the following way. They take turns shooting and stop when a basket is made. Player A goes first and has probability p_1 of making a basket on any throw. Player B, who shoots second, has probability p_2 of making a basket. The outcomes of the successive trials are assumed to be independent.
   (a) Find the frequency function for the total number of attempts.
   (b) What is the probability that player A wins?

11. Two teams, A and B, play a series of games. If team A has probability .4 of winning each game, is it to its advantage to play the best three out of five games or the best four out of seven? Assume the outcomes of successive games are independent.

12. Show that if n approaches ∞ and r/n approaches p and m is fixed, the hypergeometric frequency function tends to the binomial frequency function. Give a heuristic argument for why this is true.

13. Suppose that in a sequence of independent Bernoulli trials, the number of failures up to the first success is counted. What is the frequency function for this random variable?

14. Continuing with Problem 13, find the frequency function for the number of failures up to the rth success.

15. Find an expression for the cumulative distribution function of a geometric random variable.

16. If X is a geometric random variable with $p = .5$, for what value of k is $P(X \leq k) \approx .99$?

17. If X is a geometric random variable, show that

$$P(X > n + k - 1 | X > n - 1) = P(X > k)$$

In light of the construction of a geometric distribution from a sequence of independent Bernoulli trials, how can this be interpreted so that it is "obvious"?

18. Three identical fair coins are thrown simultaneously until all three show the same face. What is the probability that they are thrown more than three times?

19. In a sequence of independent trials with probability p of success, what is the probability that there are r successes before the kth failure?

20. (Banach Match Problem) A pipe smoker carries one box of matches in his left pocket and one in his right. Initially, each box contains n matches. If he needs a match, the smoker is equally likely to choose either pocket. What is the frequency function for the number of matches in the other box when he first discovers that one box is empty?

21. The probability of being dealt a royal straight flush (ace, king, queen, jack, and ten of the same unit) in poker is about 1.3×10^{-8}. Suppose that an avid poker player sees 100 hands a week, 52 weeks a year, for 20 years.
    (a) What is the probability that she never sees a royal straight flush dealt?
    (b) What is the probability that she sees two royal straight flushes dealt?

22. The university administration assures a mathematician that he has only 1 chance in 10,000 of being trapped in a much maligned elevator in the mathematics building. If he goes to work 5 days a week, 52 weeks a year, for 10 years, and always rides the elevator up to his office when he first arrives, what is the probability that he will never be trapped in the elevator on his way up? What is the probability that he will be trapped once? Twice? Assume that the outcomes on all the days are mutually independent (a dubious assumption in practice).

23. Suppose that a rare disease has an incidence of 1 in 1000. Assuming that members of the population are affected independently, find the probability of k cases in a population of 100,000 for $k = 0, 1, 2, \ldots$.

24. Suppose that in a city the number of suicides can be approximated by a Poisson process with $\lambda = .33$ per month.
    (a) Find the probability of k suicides in a year for $k = 0, 1, 2, \ldots$. What is the most probable number of suicides?
    (b) What is the probability of two suicides in one week?

25. Phone calls are received at a certain residence as a Poisson process with parameter $\lambda = 2$ per hour.

(a) If Diane takes a 10-minute shower, what is the probability that the phone rings during that time?

(b) How long can her shower be if she wishes the probability of receiving no phone calls to be at most .5?

26. For what value of k is the Poisson frequency function with parameter λ maximized? (*Hint:* Consider the ratio of consecutive terms.)

27. Let $F(x) = 1 - \exp(-\alpha x^\beta)$ for $x \geq 0$, $\alpha > 0$, $\beta > 0$, and $F(x) = 0$ for $x < 0$. Show that F is a cdf, and find the corresponding density.

28. Let $f(x) = (1 + \alpha x)/2$ for $-1 \leq x \leq 1$ and $f(x) = 0$ otherwise, where $-1 \leq \alpha \leq 1$. Show that f is a density, and find the corresponding cdf. Find the quartiles and the median of the distribution in terms of α.

29. A line segment of length 1 is cut once at random. What is the probability that the longer piece is more than twice the length of the shorter piece?

30. If f and g are densities, show that $\alpha f + (1 - \alpha)g$ is a density, where $0 \leq \alpha \leq 1$.

31. The Cauchy cumulative distribution function is

$$F(x) = \tfrac{1}{2} + \tfrac{1}{\pi}\arctan(x), \quad -\infty < x < \infty$$

(a) Show that this is a cdf.

(b) Find the density function.

(c) Find x such that $P(X > x) = .1$.

32. Suppose that X has the density function $f(x) = cx^2$ for $0 \leq x \leq 1$ and $f(x) = 0$ otherwise.

(a) Find c.

(b) Find the cdf.

(c) What is $P(.1 \leq X < .5)$?

33. Find the upper and lower quartiles of the exponential distribution.

34. Find the probability density for the distance from an event to its nearest neighbor for a Poisson process in the plane.

35. Find the probability density for the distance from an event to its nearest neighbor for a Poisson process in three-dimensional space.

36. Let T be an exponential random variable with parameter λ. Let X be a discrete random variable defined as $X = k$ if $k \leq T < k + 1, k = 0, 1, \ldots$. Find the frequency function of X.

37. Suppose that the lifetime of an electronic component follows an exponential distribution with $\lambda = .1$.

(a) Find the probability that the lifetime is less than 10.

(b) Find the probability that the lifetime is between 5 and 15.

(c) Find t such that the probability that the lifetime is greater than t is .01.

38. T is an exponential random variable, and $P(T < 1) = .05$. What is λ?

39. If $\alpha > 1$, show that the gamma density has a maximum at $(\alpha - 1)/\lambda$.

40. Show that the gamma density integrates to 1.

41. The gamma function is a generalized factorial function.

(a) Show that $\Gamma(1) = 1$.

(b) Show that $\Gamma(x + 1) = x\Gamma(x)$. (*Hint:* Use integration by parts.)

(c) Conclude that $\Gamma(n) = (n - 1)!$, for $n = 1, 2, 3, \ldots$.

(d) Use the fact that $\Gamma(\tfrac{1}{2}) = \sqrt{\pi}$ to show that, if n is an odd integer,

$$\Gamma\left(\frac{n}{2}\right) = \frac{\sqrt{\pi}\,(n - 1)!}{2^{n-1}\left(\dfrac{n - 1}{2}\right)!}$$

42. Show by a change of variables that

$$\Gamma(x) = 2\int_0^\infty t^{2x-1} e^{-t^2}\, dt$$

$$= \int_0^\infty e^{xt} e^{-e^{-t}}\, dt$$

43. Let X be a normal random variable with $\mu = 5$ and $\sigma = 10$. Find (a) $P(X > 10)$, (b) $P(-20 < X < 15)$, and (c) the value of x such that $P(X > x) = .05$.

44. If $X \sim N(\mu, \sigma^2)$, show that $P(|X - \mu| \le .675\sigma) = .5$.

45. If $X \sim N(\mu, \sigma^2)$, find the value of c in terms of μ and σ such that $P(-c \le X \le c) = .95$.

46. If $X \sim N(0, \sigma^2)$, find the density of $Y = |X|$.

47. If $X \sim N(\mu, \sigma^2)$ and $Y = aX + b$, where $a < 0$, show that $Y \sim N(a\mu + b, a^2\sigma^2)$.

48. If U is uniform on $[0, 1]$, find the density function of $\sqrt{U}$.

49. If U is uniform on $[-1, 1]$, find the density function of U^2.

50. Find the density function of $Y = e^Z$, where $Z \sim N(\mu, \sigma^2)$. This is called the **lognormal density**, since $\log Y$ is normally distributed.

51. Find the density of cX when X follows a gamma distribution. Show that only λ is affected by such a transformation, which justifies calling λ a scale parameter.

52. Show that if X has a density function f_X and $Y = aX + b$, then

$$f_Y(y) = \frac{1}{|a|} f_X\left(\frac{x - b}{a}\right)$$

53. Suppose that Θ follows a uniform distribution on the interval $[-\pi/2, \pi/2]$. Find the cdf and density of $\tan \Theta$.

54. A particle of mass m has a random velocity, V, which is normally distributed with parameters $\mu = 0$ and σ. Find the density function of the kinetic energy, $E = \tfrac{1}{2}mV^2$.

55. How could random variables with the following density function be generated from a uniform random number generator?

$$f(x) = \frac{1 + \alpha x}{2}, \quad -1 \le x \le 1, \quad -1 \le \alpha \le 1$$

56. Let $f(x) = \alpha x^{-\alpha-1}$ for $x \ge 1$ and $f(x) = 0$ otherwise, where α is a positive parameter. Show how to generate random variables from this density from a uniform random number generator.

57. The Weibull cumulative distribution function is

$$F(x) = 1 - e^{-(x/\alpha)^\beta}, \quad x \ge 0$$

(a) Find the density function
(b) Show that if W follows a Weibull distribution, then $X = (W/\alpha)^\beta$ follows an exponential distribution.
(c) How could Weibull random variables be generated from a uniform random number generator?

58. If the radius of a circle is an exponential random variable, find the density function of the area.

59. If the radius of a sphere is an exponential random variable, find the density function of the volume.

60. Let U be a uniform random variable. Find the density function of $V = U^{-\alpha}$, $\alpha > 0$. Compare the rates of decrease of the tails of the densities as a function of α. Does the comparison make sense intuitively?

61. This problem shows one way to generate discrete random variables from a uniform random number generator. Suppose that F is the cdf of an integer-valued random variable; let U be uniform on $[0, 1]$. Define a random variable $Y = k$ if $F(k - 1) < U \le F(k)$. Show that Y has cdf F. Apply this result to show how to generate geometric random variables from uniform random variables.

62. One of the most commonly used (but not one of the best) methods of generating pseudorandom numbers is the linear congruential method, which works as follows. Let x_0 be an initial number (the "seed"). The sequence is generated recursively as

$$x_n = (ax_{n-1} + c) \bmod m$$

(a) Choose values of a, c, and m, and try this out. Do the sequences "look" random?
(b) Making good choices of a, c, and m involves both art and theory. The following are some values that have been proposed: (1) $a = 69069, c = 0$, $m = 2^{31}$; (2) $a = 65539$, $c = 0$, $m = 2^{31}$. The latter is an infamous generator called RANDU. Try out these schemes, and examine the results.

3

Joint Distributions

3.1 Introduction

This chapter is concerned with the joint probability structure of two or more random variables defined on the same probability space. Joint distributions arise naturally in many applications, of which the following are illustrative:

- In ecological studies, counts, modeled as random variables, of several species are often made. One species is often the prey of another; clearly, the number of predators will be related to the number of prey.
- The joint probability distribution of the x, y, and z components of wind velocity can be experimentally measured in studies of atmospheric turbulence.
- The joint distribution of the values of various physiological variables in a population of patients is often of interest in medical studies.
- A model for the joint distribution of age and length in a population of fish can be used to estimate the age distribution from the length distribution. The age distribution is relevant to the setting of reasonable harvesting policies.

In the case of two random variables, X and Y, their joint behavior is determined by the cumulative distribution function

$$F(x, y) = P(X \leq x, Y \leq y)$$

whether X and Y are continuous or discrete. The cdf gives the probability that the point (X, Y) belongs to a semiinfinite rectangle in the plane, as shown in Figure 3-1. The probability that (X, Y) belongs to a given rectangle is, from Figure 3-2,

$$P(x_1 < X \leq x_2, y_1 < Y \leq y_2) = F(x_2, y_2) - F(x_2, y_1) - F(x_1, y_2)$$
$$+ F(x_1, y_1)$$

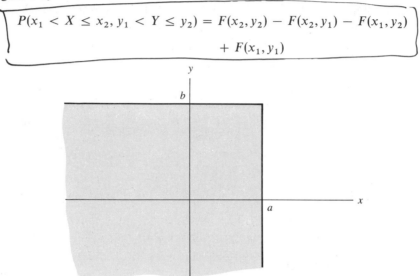

Figure 3-1. $F(a, b)$ gives the probability of the shaded rectangle.

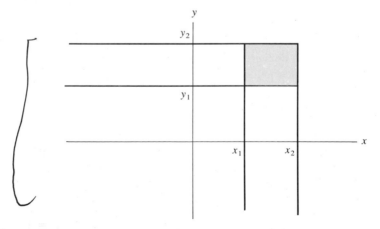

Figure 3-2.   The probability of the shaded rectangle can be found by subtracting from the probability of the (semiinfinite) rectangle having the upper-right corner (x_2, y_2) the probabilities of the (x_1, y_2) and (x_2, y_1) rectangles, and then adding back in the probability of the (x_1, y_1) rectangle.

The probability that (X, Y) belongs to a set A, for a large enough class of sets for practical purposes, can be determined by taking limits of intersections and unions of rectangles. In general, if $X_1, \ldots, X_n$ are jointly distributed random variables, their joint cdf is

$$F(x_1, x_2, \ldots, x_n) = P(X_1 \leq x_1, X_2 \leq x_2, \ldots, X_n \leq x_n)$$

Two- and higher-dimensional versions of density functions and frequency functions exist. We will start with a detailed description of such functions for the discrete case, since it is the easier one to understand.

3.2 Discrete Random Variables

Suppose that X and Y are discrete random variables defined on the same probability space and that they take on values $x_1, x_2, \ldots$ and $y_1, y_2, \ldots$, respectively. Their **joint frequency function**, or joint probability mass function, $p(x, y)$, is simply

$$p(x_i, y_j) = P(X = x_i, Y = y_j)$$

A simple example will illustrate these ideas. A fair coin is tossed three times; let X denote the number of heads on the first toss and Y the total number of heads. From the sample space, which is

$$\Omega = \{hhh, hht, hth, htt, thh, tht, tth, ttt\}$$

we see that the joint frequency function of X and Y is as given in the following table:

		y		
x	0	1	2	3
0	$\frac{1}{8}$	$\frac{2}{8}$	$\frac{1}{8}$	0
1	0	$\frac{1}{8}$	$\frac{2}{8}$	$\frac{1}{8}$

Thus, for example, $p(0, 2) = P(X = 0, Y = 2) = \frac{1}{8}$.

Suppose that we wish to find the frequency function of Y from the joint frequency function. This is straightforward:

$$p_Y(0) = P(Y = 0)$$

$$= P(Y = 0, X = 0) + P(Y = 0, X = 1)$$

$$= \tfrac{1}{8} + 0$$

$$= \tfrac{1}{8}$$

$$p_Y(1) = P(Y = 1)$$

$$= P(Y = 1, X = 0) + P(Y = 1, X = 1)$$

$$= \tfrac{3}{8}$$

In general, to find the frequency function of Y, we simply sum down the appropriate column of the table. For this reason, p_Y is called the **marginal frequency function** of Y. Similarly, summing across the rows gives

$$p_X(x) = \sum_i p(x, y_i)$$

which is the marginal frequency function of X.

The case for several random variables is analogous. If $X_1, \ldots, X_m$ are discrete random variables defined on the same probability space, their joint frequency function is

$$p(x_1, \ldots, x_m) = P(X_1 = x_1, \ldots, X_m = x_m)$$

The marginal frequency function of X_1, for example, is

$$p_{X_1}(x_1) = \sum_{x_2 \cdots x_m} p(x_1, x_2, \ldots, x_m)$$

The two-dimensional marginal frequency function of X_1 and X_2, for example, is

$$p_{X_1 X_2}(x_1, x_2) = \sum_{x_3 \cdots x_m} p(x_1, x_2, \ldots, x_m)$$

EXAMPLE A.   (Multinomial Distribution) The multinomial distribution, an important generalization of the binomial distribution, arises in the following way. Suppose that each of n independent trials can result in one of r types of outcomes and that on each trial the probabilities of the r outcomes are $p_1, p_2, \ldots, p_r$. Let N_i be the total number of outcomes of type i in the n trials, $i = 1, \ldots, r$. To calculate the joint frequency function, we observe that any particular sequence of trials giving rise to $N_1 = n_1, N_2 = n_2, \ldots, N_r = n_r$ occurs with probability $p_1^{n_1} p_2^{n_2} \cdots p_r^{n_r}$. From Proposition C in Section 1.4.2, there are $n!/n_1! n_2! \cdots n_r!$ such sequences, and thus the joint frequency function is

$$p(n_1, \ldots, n_r) = \binom{n}{n_1 \cdots n_r} p_1^{n_1} p_2^{n_2} \cdots p_r^{n_r}$$

The marginal distribution of any particular N_i can be obtained by summing the joint frequency function over the other n_j. This formidable algebraic task can be avoided, however, by noting that N_i can be interpreted as the number of successes in n trials, each of which has probability p_i of success and $1 - p_i$ of failure. Therefore, N_i is a binomial random variable, and

$$p_{N_i}(n_i) = \binom{n}{n_i} p_i^{n_i}(1 - p_i)^{n - n_i}$$

The multinomial distribution is applicable in considering the probabilistic properties of a **histogram**. As a concrete example, suppose that 100 independent observations are taken from a uniform distribution on $[0, 1]$, that the interval $[0, 1]$ is partitioned into 10 equal bins, and that the counts $n_1, \ldots, n_{10}$ in each of the 10 bins are recorded and graphed as the heights of vertical bars above the respective bins. The joint distribution of the heights is multinomial with $n = 100$ and $p_i = .1, i = 1, \ldots, 10$. Figure 3-3 shows four histograms constructed in this manner from a pseudorandom number generator; the figure illustrates the sort of random fluctuations that can be expected in histograms. □

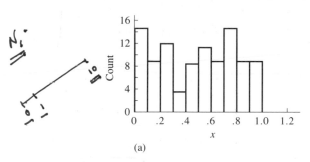

(a)

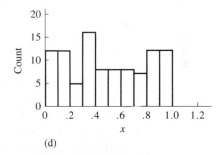

(b)

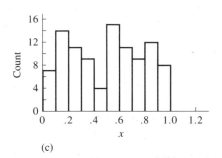

(c)

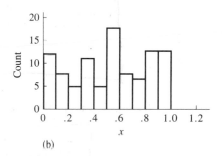

(d)

Figure 3-3. Four histograms, each formed from 100 independent uniform random numbers.

3.3   Continuous Random Variables

Suppose that X and Y are continuous random variables with a joint cdf, $F(x, y)$. Their **joint density function** is a piecewise continuous function of two variables, $f(x, y)$, such that for any "reasonable" two-dimensional set A

$$P((X, Y) \in A) = \iint_A f(x, y) \, dy \, dx$$

In particular,

$$F(x, y) = \int_{-\infty}^{x} \int_{-\infty}^{y} f(u, v)\, dv\, du$$

From the fundamental theorem of multivariable calculus, it follows that

$$f(x, y) = \frac{\partial^2}{\partial x \partial y} F(x, y)$$

wherever the derivative is defined.

For small δ_x and δ_y, if f is continuous at (x, y),

$$P(x \leq X \leq x + \delta_x, y \leq Y \leq y + \delta_y) = \int_{x}^{x+\delta_x} \int_{y}^{y+\delta_y} f(u, v)\, dv\, du$$

$$\approx f(x, y)\delta_x \delta_y$$

Thus, the probability that (X, Y) is in a small neighborhood of (x, y) is proportional to $f(x, y)$. Differential notation is sometimes useful:

$$P(x \leq X \leq x + dx, y \leq Y \leq y + dy) = f(x, y)\, dx\, dy$$

The **marginal cdf** of X, or F_X, is

$$F_X(x) = P(X \leq x)$$

$$= \lim_{y \to \infty} F(x, y)$$

$$= \int_{-\infty}^{x} \int_{-\infty}^{\infty} f(u, y)\, dy\, du$$

From this, it follows that the density function of X alone, known as the **marginal density** of X, is

$$f_X(x) = F_X'(x) = \int_{-\infty}^{\infty} f(x, y)\, dy$$

In the discrete case, the marginal frequency function was found by summing the joint frequency function over the other variable; in the continuous case, it is found by integration.

For several jointly continuous random variables, we can make the obvious generalizations. The joint density function is a function of several variables, and the marginal density functions are found by integration. There are marginal density functions of various dimensions. Suppose that X, Y, and Z are jointly continuous random variables with density function $f(x, y, z)$. The one-dimensional marginal distribution of X is

$$f_X(x) = \int_{-\infty}^{\infty} \int_{-\infty}^{\infty} f(x, y, z)\, dy\, dz$$

and the two-dimensional marginal distribution of X and Y is

$$f_{XY}(x, y) = \int_{-\infty}^{\infty} f(x, y, z)\, dz$$

EXAMPLE A.   (Farlie–Morgenstern Family) If $F(x)$ and $G(y)$ are one-dimensional cdf's, it can be shown that, for any α for which $|\alpha| \le 1$,

$$H(x, y) = F(x)G(y)\{1 + \alpha[1 - F(x)][1 - G(y)]\}$$

is a bivariate distribution function. Since $F(\infty) = G(\infty) = 1$, the marginal distributions are

$$H(x, \infty) = F(x)$$
$$H(\infty, y) = G(y)$$

In this way, an infinite number of different bivariate distributions with given marginals can be constructed.

As an example, we will construct bivariate distributions with marginals that are uniform on $[0, 1]$. First, with $\alpha = -1$, we have

$$H(x, y) = xy[1 - (1 - x)(1 - y)]$$
$$= x^2 y + y^2 x - x^2 y^2, \quad 0 \le x \le 1, 0 \le y \le 1$$

The bivariate density is

$$h(x, y) = \frac{\partial^2}{\partial x \partial y} H(x, y)$$
$$= 2x + 2y - 4xy, \quad 0 \le x \le 1, 0 \le y \le 1$$

The density is shown in Figure 3-4. Perhaps you can imagine integrating over y (pushing all the mass onto the x axis) to produce a marginal uniform density for x.

Next, if $\alpha = 1$,

$$H(x, y) = xy[1 + (1 - x)(1 - y)]$$
$$= 2xy - x^2 y - y^2 x - x^2 y^2, \quad 0 \le x \le 1, 0 \le y \le 1$$

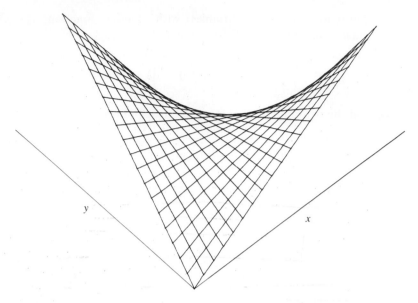

Figure 3-4.   The joint density $h(x, y) = 2x + 2y - 4xy$, where $0 \leq x \leq 1$ and $0 \leq y \leq 1$, which has uniform marginal densities.

The density is

$$h(x, y) = 2 - 2x - 2y + 4xy$$

This density is shown in Figure 3-5.

We have just constructed two different bivariate distributions, both of which have uniform marginals.   □

EXAMPLE B.   Consider the following joint density:

$$f(x, y) = \begin{cases} \lambda^2 e^{-\lambda y}, & 0 \leq x \leq y \\ 0, & \text{elsewhere} \end{cases}$$

To find the marginal distributions, it is helpful to draw a picture showing where the density is nonzero to aid in determining the limits of integration (see Figure 3-6). We then have

$$f_X(x) = \int_x^\infty \lambda^2 e^{-\lambda y} \, dy = \lambda e^{-\lambda x}, \quad x \geq 0$$

The marginal distribution of X is exponential. The marginal distribution of Y is

$$f_Y(y) = \int_0^y \lambda^2 e^{-\lambda y} \, dx = \lambda^2 y e^{-\lambda y}, \quad y \geq 0$$

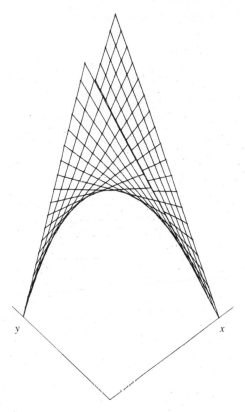

Figure 3-5. The joint density $h(x, y) = 2 - 2x - 2y + 4xy$, where $0 \le x \le 1$ and $0 \le y \le 1$, which has uniform marginal densities.

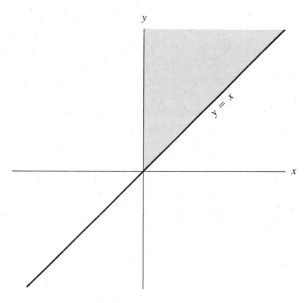

Figure 3-6. The joint density of Example B is nonzero over the shaded region of the plane.

The marginal distribution of Y is a gamma distribution. Note that we would not be able to reconstruct the joint distribution of X and Y from their marginal distributions.   □

In some applications, it is useful to analyze distributions that are uniform over some region of space. For example, in the plane, the random point (X, Y) is uniform over a region, R, if for any $A \subset R$,

$$P((X, Y) \in A) = \frac{|A|}{|R|}$$

where $|\ |$ denotes area.

EXAMPLE C.   A point is chosen randomly in a disk of radius 1. Since the area of the disk is π,

$$f(x, y) = \begin{cases} \dfrac{1}{\pi}, & \text{if } x^2 + y^2 \le 1 \\ \\ 0, & \text{otherwise} \end{cases}$$

We can easily calculate the distribution of R, the distance of the point from the origin. If the point lies in a disk of radius r, $R \le r$. Since this disk has area πr^2,

$$F_R(r) = P(R \le r) = \frac{\pi r^2}{\pi} = r^2$$

The density function of R is thus $f_R(r) = 2r, 0 \le r \le 1$.

Let us now find the marginal density of the x coordinate of the random point:

$$f_X(x) = \int_{-\infty}^{\infty} f(x, y)\, dy$$

$$= \frac{1}{\pi} \int_{-\sqrt{1-x^2}}^{\sqrt{1-x^2}} dy$$

$$= \frac{2}{\pi} \sqrt{1 - x^2}, \quad -1 \le x \le 1$$

Note that we chose the limits of integration carefully; outside these limits the joint density is zero. By symmetry, the marginal density of Y is

$$f_Y(y) = \frac{2}{\pi} \sqrt{1 - y^2}, \quad -1 \le y \le 1 \qquad\qquad □$$

EXAMPLE D.   (Bivariate Normal Density) The bivariate normal density is given by the complicated expression

$$f(x, y) = \frac{1}{2\pi\sigma_X\sigma_Y\sqrt{1 - \rho^2}} \exp\left(-\frac{1}{2(1 - \rho^2)}\left[\frac{(x - \mu_X)^2}{\sigma_X^2} + \frac{(y - \mu_Y)^2}{\sigma_Y^2}\right.\right.$$

$$\left.\left. - \frac{2\rho(x - \mu_X)(y - \mu_Y)}{\sigma_X\sigma_Y}\right]\right)$$

One of the earliest uses of this bivariate density was as a model for the joint distribution of the heights of fathers and sons. The density depends on five parameters:

$$-\infty < \mu_X < \infty \qquad -\infty < \mu_Y < \infty$$

$$\sigma_X > 0 \qquad \sigma_Y > 0$$

$$-1 < \rho < 1$$

The contour lines of the density are the lines in the xy plane on which the joint density is constant. From the equation above, we see that $f(x, y)$ is constant if

$$\frac{(x - \mu_X)^2}{\sigma_X^2} + \frac{(y - \mu_Y)^2}{\sigma_Y^2} - \frac{2\rho(x - \mu_X)(y - \mu_Y)}{\sigma_X\sigma_Y} = \text{constant}$$

The locus of such points is an ellipse centered at (μ_X, μ_Y). If $\rho = 0$, the axes of the ellipse are parallel to the x and y axes, and if $\rho \neq 0$, they are tilted. Figure 3-7 shows several bivariate normal densities, and Figure 3-8 shows the corresponding elliptical contours.

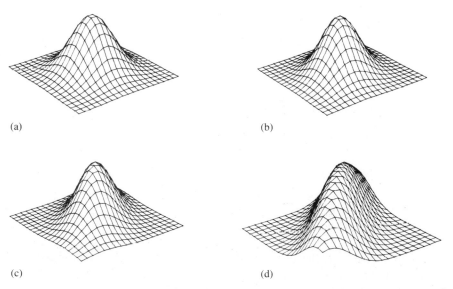

(a)

(b)

(c)

(d)

Figure 3-7.   Bivariate normal densities with $\mu_X = \mu_Y = 0$ and $\sigma_X = \sigma_Y = 1$ and (a) $\rho = 0$, (b) $\rho = .3$, (c) $\rho = .6$, (d) $\rho = .9$.

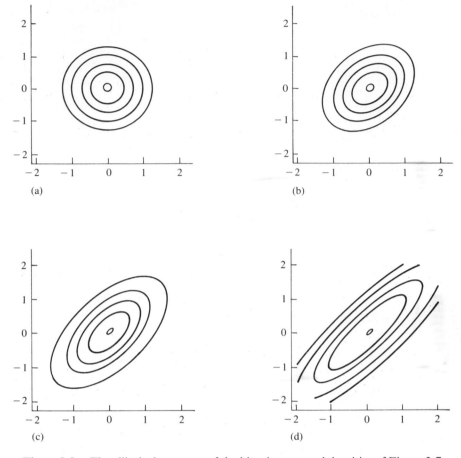

Figure 3-8.   The elliptical contours of the bivariate normal densities of Figure 3-7.

The marginal distributions of X and Y are $N(\mu_X, \sigma_X^2)$ and $N(\mu_Y, \sigma_Y^2)$, respectively, as we will now demonstrate. The marginal density of X is

$$f_X(x) = \int_{-\infty}^{\infty} f_{XY}(x, y)\, dy$$

Making the changes of variables $u = (x - \mu_X)/\sigma_X$ and $v = (y - \mu_Y)/\sigma_Y$ gives us

$$f_X(x) = \frac{1}{2\pi\sigma_X\sqrt{1 - \rho^2}} \int_{-\infty}^{\infty} \exp\left[-\frac{1}{2(1 - \rho^2)}(u^2 + v^2 - 2\rho uv) \right] dv$$

To evaluate this integral, we use the technique of "completing the square." Using the identity

$$u^2 + v^2 - 2\rho uv = (v - \rho u)^2 + u^2(1 - \rho^2)$$

we have

$$f_X(x) = \frac{1}{2\pi\sigma_X\sqrt{1-\rho^2}} e^{-u^2/2} \int_{-\infty}^{\infty} \exp\left[-\frac{1}{2(1-\rho^2)}(v-\rho u)^2\right] dv$$

Finally, recognizing the integral as that of a normal density with mean ρu and variance $(1-\rho^2)$, we obtain

$$f_X(x) = \frac{1}{\sigma_X\sqrt{2\pi}} e^{-(1/2)[(x-\mu_X)^2/\sigma_X^2]}$$

which is a normal density, as was to be shown. $\square$

3.4 Independent Random Variables

DEFINITION.   Random variables $X_1, X_2, \ldots, X_n$ are said to be independent if their joint cdf factors into the product of their marginal cdf's:

$$F(x_1, x_2, \ldots, x_n) = F_{X_1}(x_1)F_{X_2}(x_2)\cdots F_{X_n}(x_n)$$

for all $x_1, x_2, \ldots, x_n$.

   This definition holds for both continuous and discrete random variables. For discrete random variables, it is equivalent to state that their joint frequency function factors; for continuous random variables, it is equivalent to state that their joint density function factors. To see why this is true, consider the case of two jointly continuous random variables, X and Y. If they are independent, then

$$F(x, y) = F_X(x)F_Y(y)$$

and taking the second mixed partial derivative makes it clear that the density function factors. On the other hand, if the density function factors, then the joint cdf can be expressed as a product:

$$F(x, y) = \int_{-\infty}^{x} \int_{-\infty}^{y} f_X(u)f_Y(v)\, dv\, du$$

$$= \left[\int_{-\infty}^{x} f_X(u)\, du\right]\left[\int_{-\infty}^{y} f_Y(v)\, dv\right]$$

$$= F_X(x)F_Y(y)$$

It can be shown that the definition implies that if X and Y are independent, then

$$P(X \in A, Y \in B) = P(X \in A)P(Y \in B)$$

It can also be shown that if g and h are functions, then $Z = g(X)$ and $W = h(Y)$ are independent as well. A sketch of an argument goes like this (the details are beyond the level of this course): We wish to find $P(Z \le z, W \le w)$. Let $A(z)$ be the set of x such that $g(x) \le z$, and let $B(w)$ be the set of y such that $h(y) \le w$. Then

$$P(Z \le z, W \le w) = P(X \in A(z), Y \in B(w))$$

$$= P(X \in A(z))P(Y \in B(w))$$

$$= P(Z \le z)P(W \le w)$$

EXAMPLE A. (Farlie–Morgenstern Family) From Example A in Section 3.3, we see that X and Y are independent only if $\alpha = 0$, since only in this case does the joint cdf H factor into the product of the marginals F and G.   □

EXAMPLE B. If X and Y follow a bivariate normal distribution and $\rho = 0$, their joint density factors into the product of two normal densities, and therefore X and Y are independent.   □

EXAMPLE C. Suppose that a node in a communications network has the property that if two packets of information arrive within time τ of each other they "collide" and then have to be retransmitted. If the times of arrival of the two packets are independent and uniform on $[0, T]$, what is the probability that they collide?

The times of arrival of two packets, T_1 and T_2, are independent and uniform on $[0, T]$, so their joint density is the product of the marginals, or

$$f(t_1, t_2) = \frac{1}{T^2}$$

for t_1 and t_2 in the square with sides $[0, T]$. Therefore, (T_1, T_2) is uniformly distributed over the square. The probability that the two packets collide is proportional to the area of the shaded strip in Figure 3-9. It is easy to see that the area of the unshaded portion is $(T - \tau)^2$, from which it follows that the probability that the packets collide is $1 - (1 - \tau/T)^2$.   □

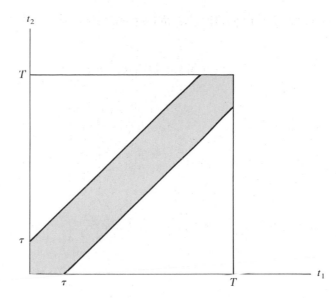

Figure 3-9. The probability that the two packets collide is proportional to the area of the shaded region.

3.5 Conditional Distributions

3.5.1 The Discrete Case

If X and Y are jointly distributed discrete random variables, the conditional probability that $X = x_i$ given that $Y = y_j$ is, if $p_Y(y_j) > 0$,

$$P(X = x_i | Y = y_j) = \frac{P(X = x_i, Y = y_j)}{P(Y = y_j)}$$

$$= \frac{p_{XY}(x_i, y_j)}{p_Y(y_j)}$$

This probability is defined to be zero if $p_Y(y_j) = 0$. We will denote this conditional probability by $p_{X|Y}(x|y)$. Note that this function of x is a genuine frequency function since it is nonnegative and sums to 1.

EXAMPLE A.   We return to the simple discrete distribution considered in Section 3.2, reproducing the table of values for convenience here:

	y			
x	0	1	2	3
0	$\frac{1}{8}$	$\frac{2}{8}$	$\frac{1}{8}$	0
1	0	$\frac{1}{8}$	$\frac{2}{8}$	$\frac{1}{8}$

The conditional frequency function of X given $Y = 1$ is

$$p_{X|Y}(0|1) = \frac{\frac{2}{8}}{\frac{3}{8}} = \frac{2}{3}$$

$$p_{X|Y}(1|1) = \frac{\frac{1}{8}}{\frac{3}{8}} = \frac{1}{3} \qquad\qquad \square$$

The definition of the conditional frequency function given above can be re-expressed as

$$p_{XY}(x, y) = p_{X|Y}(x|y)p_Y(y)$$

which follows from the multiplication principle of chapter 1. This useful equation gives a relationship between the joint and conditional frequency functions. Summing both sides over all values of y, we have an extremely useful application of the law of total probability:

$$p_X(x) = \sum_y p_{X|Y}(x|y)p_Y(y)$$

EXAMPLE B. Suppose that a particle counter is imperfect and detects each incoming particle with probability p and independently of any other particles. If the distribution of the number of incoming particles in a unit of time is a Poisson distribution with parameter λ, what is the distribution of the number of counted particles?

Let N denote the true number of particles and X the counted number. From the statement of the problem, the conditional distribution of X given $N = n$ is binomial, with n trials and probability p of success. By the law of total probability,

$$P(X = k) = \sum_{n=0}^{\infty} P(N = n)P(X = k|N = n)$$

$$= \sum_{n=k}^{\infty} \frac{\lambda^n e^{-\lambda}}{n!} \binom{n}{k} p^k (1 - p)^{n-k}$$

$$= \frac{(\lambda p)^k}{k!} e^{-\lambda} \sum_{n=k}^{\infty} \lambda^{n-k} \frac{(1 - p)^{n-k}}{(n - k)!}$$

$$= \frac{(\lambda p)^k}{k!} e^{-\lambda} \sum_{j=0}^{\infty} \frac{\lambda^j (1 - p)^j}{j!}$$

$$= \frac{(\lambda p)^k}{k!} e^{-\lambda} e^{\lambda(1-p)}$$

$$= \frac{(\lambda p)^k}{k!} e^{-\lambda p}$$

We see that the distribution of X is a Poisson distribution with parameter λp. This model arises in other applications as well. For example, N might denote the number of traffic accidents in a given time period, with each accident being fatal or nonfatal; X would then be the number of fatal accidents. □

3.5.2 The Continuous Case

In analogy with the definition in the preceding section, if X and Y are jointly continuous random variables, the **conditional density** of Y given X is defined to be

$$f_{Y|X}(y|x) = \frac{f_{XY}(x, y)}{f_X(x)}$$

if $0 < f_X(x) < \infty$ and 0, otherwise. This definition is in accord with the result to which a differential argument would lead. We would define $f_{Y|X}(y|x)\,dy$ as $P(y \le Y \le y + dy | x \le X \le x + dx)$ and calculate

$$P(y \le Y \le y + dy | x \le X \le x + dx) = \frac{f_{XY}(x, y)\,dx\,dy}{f_X(x)\,dx} = \frac{f_{XY}(x, y)}{f_X(x)}\,dy$$

The joint distribution can be expressed in terms of the marginal and conditional distributions as follows:

$$f_{XY}(x, y) = f_{Y|X}(x|y)f_X(x)$$

Integrating both sides over x allows the marginal distribution of Y to be expressed as

$$f_Y(y) = \int_{-\infty}^{\infty} f_{Y|X}(y|x)f_X(x)\,dx$$

which is the law of total probability for the continuous case.

EXAMPLE A. In Example B in Section 3.3, we saw that

$$f_{XY}(x, y) = \lambda^2 e^{-\lambda y}, \quad 0 \le x \le y$$

$$f_X(x) = \lambda e^{-\lambda x}, \quad x \ge 0$$

$$f_Y(y) = \lambda^2 y e^{-\lambda y}, \quad y \ge 0$$

Let us find the conditional densities.
  First,

$$f_{Y|X}(y|x) = \frac{\lambda^2 e^{-\lambda y}}{\lambda e^{-\lambda x}} = \lambda e^{-\lambda(y-x)}, \quad y \ge x$$

The conditional density of Y given $X = x$ is exponential on the interval $[x, \infty)$.

Expressing the joint density as

$$f_{XY}(x, y) = f_{Y|X}(y|x)f_X(x)$$

we see that we could generate X and Y according to f_{XY} in the following way: First, generate X as an exponential random variable (f_X), and then generate Y as another exponential random variable ($f_{Y|X}$) on the interval $[x, \infty)$. From this representation, we see that Y may be interpreted as the sum of two independent exponential random variables and that the distribution of this sum is a gamma one, a fact that we will derive later by a different method.

Now,

$$f_{X|Y}(x|y) = \frac{\lambda^2 e^{-\lambda y}}{\lambda^2 y e^{-\lambda y}} = \frac{1}{y}, \quad 0 \le x \le y$$

The conditional density of X given $Y = y$ is uniform on the interval $[0, y]$. Finally, expressing the joint density as

$$f_{XY}(x, y) = f_{X|Y}(x|y)f_Y(y)$$

we see that alternatively we could generate X and Y according to the density f_{XY} by first generating Y from a gamma density and then generating X uniformly on $[0, y]$. Another interpretation of this result is that, conditional on the sum of two independent exponential random variables, the first is uniformly distributed. $\square$

EXAMPLE B.   (Stereology) In metallography and other applications of quantitative microscopy, aspects of three-dimensional structure are deduced from studying two-dimensional cross sections. Concepts of probability and statistics play an important role (DeHoff and Rhines, 1968). In particular, the following problem arises. Spherical particles are dispersed in a medium (grains in a metal, for example); the density function of the radii of the spheres can be denoted as $f_R(r)$. When the medium is sliced, two-dimensional, circular cross sections of the spheres are observed; the density function of the radii of these circles can be denoted as $f_X(x)$. How are these density functions related?

To derive the relationship, we assume that the cross-sectioning plane is chosen at random, fix $R = r$, and find the conditional density $f_{X|R}(x|r)$. As shown in Figure 3-10, let H denote the distance from the center of the sphere to the planar cross section. By our assumption, H is uniformly distributed on $[0, r]$, and $X = \sqrt{r^2 - H^2}$. We can thus find the conditional distribution of X given $R = r$:

$$\begin{aligned}
F_{X|R}(x|r) &= P(X \le x) \\
&= P(\sqrt{r^2 - H^2} \le x) \\
&= P(H \ge \sqrt{r^2 - x^2}) \\
&= 1 - \frac{\sqrt{r^2 - x^2}}{r}, \quad 0 \le x \le r
\end{aligned}$$

Differentiating, we find

$$f_{X|R}(x|r) = \frac{x}{r\sqrt{r^2 - x^2}}, \quad 0 \le x \le r$$

The marginal distribution of X is, from the law of total probability,

$$f_X(x) = \int_{-\infty}^{\infty} f_{X|R}(x|r)f_R(r)\,dr$$

$$= \int_x^{\infty} \frac{x}{r\sqrt{r^2 - x^2}} f_R(r)\,dr$$

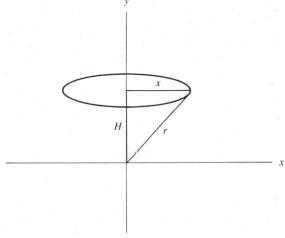

Figure 3-10.   A plane slices a sphere of radius r at a distance H from its center, producing a circle of radius x.

[The limits of integration are x and ∞ since for $r \le x$, $f_{X|R}(x|r) = 0$.] This equation is called Abel's Equation. In practice, the marginal density f_X can be approximated by making measurements of the radii of cross-sectional circles. Then the problem becomes that of trying to solve for an approximation to f_R, since it is the distribution of spherical radii that is of real interest.   □

EXAMPLE C.   (Bivariate Normal Density) The conditional density of Y given X is the ratio of the bivariate normal density to a univariate normal density. After some messy algebra, this ratio simplifies to

$$f_{Y|X}(y, x) = \frac{1}{\sigma_Y\sqrt{2\pi(1 - \rho^2)}} \exp\left(-\frac{1}{2}\frac{\left[y - \mu_Y - \rho\frac{\sigma_Y}{\sigma_X}(x - \mu_X)\right]^2}{\sigma_Y^2(1 - \rho^2)}\right)$$

This is a normal density with mean $\mu_Y + \rho(x - \mu_X)\sigma_Y/\sigma_X$ and variance $\sigma_Y^2(1 - \rho^2)$. The conditional distribution of Y given X is a univariate normal distribution.

In Example B in Section 2.2.3, the distribution of the velocity of a turbulent wind flow was shown to be approximately normally distributed. Van Atta and Chen (1968) also measured the joint distribution of the velocity at a point at two different times, t and $t + \tau$. Figure 3-11 shows the measured conditional distributions of the velocity, v_2, at time $t + \tau$, given various values of v_1. There is a

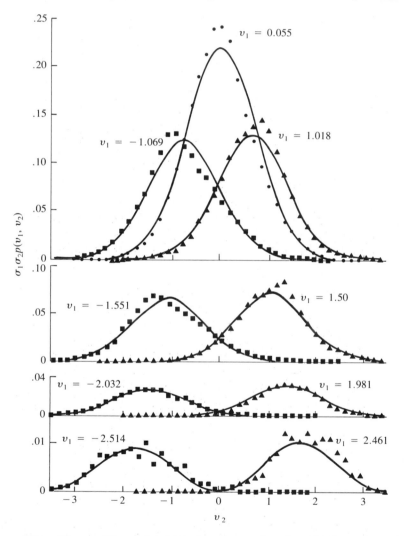

Figure 3-11. The conditional densities of v_2 given v_1 for selected values of v_1, where v_1 and v_2 are components of the velocity of a turbulent wind flow at different times. The solid lines are the conditional densities according to a normal fit, and the circles and squares are empirical values determined from 409,600 observations.

systematic departure from the normal distribution. Therefore, it appears that, even though the velocity is normally distriubted, the joint distribution of v_1 and v_2 is not bivariate normal. This should not be totally unexpected, since the relation of v_1 and v_2 must conform to equations of motion and continuity, which may not permit a joint normal distribution.   □

Example C illustrates that even when two random variables are marginally normally distributed, they need not be jointly normally distributed.

EXAMPLE D.   (Rejection Method) The rejection method is commonly used to generate random variables from a density function, especially when the inverse of the cdf cannot be found in closed form and therefore the inverse cdf method, Proposition D in Section 2.3, cannot be used. Suppose that f is a density function that is nonzero on an interval $[a, b]$ and zero outside the interval (a and b may be infinite). Let $M(x)$ be a function such that $M(x) \geq f(x)$ on $[a, b]$, and let

$$m(x) = \frac{M(x)}{\int_a^b M(x)\, dx}$$

be a probability density function. As we will see, the idea is to choose M so that it is easy to generate random variables from m. If $[a, b]$ is finite, m can be chosen to be the uniform distribution on $[a, b]$. The algorithm is as follows:

Step 1: Generate T with the density m.
Step 2: Generate U, uniform on $[0, 1]$ and independent of T. If $M(T) \times U \leq f(T)$, then let $X = T$ (accept T). Otherwise, go to Step 1 (reject T).

See Figure 3-12.

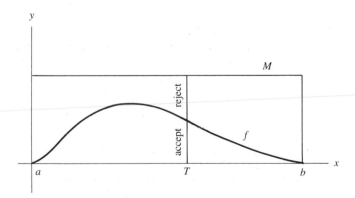

Figure 3-12.   Illustration of the rejection method.

We must check that the density function of the random variable X thus obtained is in fact f:

$$P(x \leq X \leq x + dx)$$

$$= P(x \leq T \leq x + dx | \text{accept})$$

$$= \frac{P(x \leq T \leq x + dx)P(UM(T) \leq f(T) | x \leq T \leq x + dx)}{P(\text{accept})}$$

$$= \frac{P(x \leq T \leq x + dx)P(U \leq f(x)/M(x))}{P(\text{accept})}$$

The numerator of the last expression is

$$\frac{m(x)\,dx\,f(x)}{M(x)} = \frac{f(x)\,dx}{\int_a^b M(x)\,dx}$$

and the denominator is, from the law of total probability,

$$P(\text{accept}) = P(U \leq f(T)/M(T))$$

$$= \int_a^b \frac{f(t)}{M(t)} m(t)\,dt$$

$$= \frac{1}{\int_a^b M(t)\,dt}$$

where the last two steps follow from the definition of m and since f integrates to 1. Finally, we see that the numerator over the denominator is $f(x)\,dx$. □

In order for the rejection method to be computationally efficient, the algorithm should lead to acceptance with high probability; otherwise, many rejection steps may have to be looped through for each acceptance.

3.6 Functions of Jointly Distributed Random Variables

The distribution of a function of a single random variable was developed in Section 2.3. In this section, that development is extended to several random variables, but first some important special cases are considered.

3.6.1 Sums and Quotients

Suppose that X and Y are discrete random variables taking values on the positive and negative integers and having the joint frequency function $f(x, y)$, and let $Z = X + Y$. To find the frequency function of Z, we note that $Z = z$ whenever

$X = x$ and $Y = z - x$, where x is an integer. The probability that $Z = z$ is thus the sum over all x of the joint probabilities, or

$$f_Z(z) = \sum_{x=-\infty}^{\infty} f(x, z - x)$$

If X and Y are independent, so $f(x, y) = f_X(x)f_Y(y)$, then

$$f_Z(z) = \sum_{x=-\infty}^{\infty} f_X(x)f_Y(z - x)$$

This sum is called the **convolution** of the sequences f_X and f_Y.

The continuous case is very similar. Supposing that X and Y are continuous random variables, we first find the cdf of Z and then differentiate to find the density. Since $Z \leq z$ whenever the point (X, Y) is in the shaded region R_z shown in Figure 3-13, we have

$$F_Z(z) = \iint_{R_z} f(x, y)\,dx\,dy$$

$$= \int_{-\infty}^{\infty} \int_{-\infty}^{z-x} f(x, y)\,dy\,dx$$

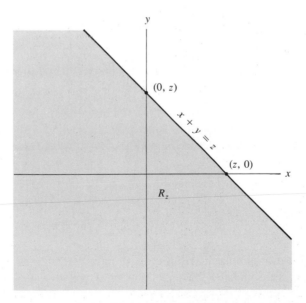

Figure 3-13. $X + Y \leq z$ whenever (X, Y) is in the shaded region R_z.

In the inner integral, we make the change of variables $y = v - x$ to obtain

$$F_Z(z) = \int_{-\infty}^{\infty} \int_{-\infty}^{z} f(x, v - x) \, dv \, dx$$

$$= \int_{-\infty}^{z} \int_{-\infty}^{\infty} f(x, v - x) \, dv \, dx$$

Differentiating, we have, if $\int_{-\infty}^{\infty} f(x, z - x) \, dx$ is continuous,

$$f_Z(z) = \int_{-\infty}^{\infty} f(x, z - x) \, dx$$

which is the obvious analogue of the result for the discrete case.

If X and Y are independent,

$$f_Z(z) = \int_{-\infty}^{\infty} f_X(x) f_Y(z - x) \, dx$$

This integral is called the **convolution** of the functions f_X and f_Y.

EXAMPLE A.   Suppose that the lifetime of a component is exponentially distributed and that an identical and independent backup component is available. The system operates as long as one of the components is functional; therefore, the distribution of the life of the system is that of the sum of two independent exponential random variables. Let T_1 and T_2 be independent exponentials with parameter λ, and let $S = T_1 + T_2$.

$$f_S(s) = \int_0^s \lambda e^{-\lambda t} \lambda e^{-\lambda(s-t)} \, dt$$

It is important to note the limits of integration. Beyond these limits, one of the two component densities is zero. When dealing with densities that are nonzero only on some subset of the real line, we must always be careful. Continuing, we have

$$f_S(s) = \lambda^2 \int_0^s e^{-\lambda s} \, dt$$

$$= \lambda^2 s e^{-\lambda s}$$

This is a gamma distribution with parameters 2 and λ (compare with Example B in Section 3.5.2).   □

Let us next consider the quotient of two continuous random variables. The derivation is very similar to that for the sum of such variables, given above: We first find the cdf and then differentiate to find the density. Suppose that X and Y are continuous with joint density function f and that $Z = Y/X$. Then $F_Z(z) = P(Z \le z)$ is the probability of the set of (x, y) such that $y/x \le z$. If $x > 0$, this is the set $y \le xz$; if $x < 0$, it is the set $y \ge xz$. Thus,

$$F_Z(z) = \int_{-\infty}^{0} \int_{xz}^{\infty} f(x, y)\,dy\,dx + \int_{0}^{\infty} \int_{-\infty}^{xz} f(x, y)\,dy\,dx$$

To remove the dependence of the inner integrals on x, we make the change of variables $y = xv$ in the inner integrals and obtain

$$f_Z(z) = \int_{-\infty}^{0} \int_{z}^{-\infty} xf(x, xv)\,dv\,dx + \int_{0}^{\infty} \int_{-\infty}^{z} xf(x, xv)\,dv\,dx$$

$$= \int_{-\infty}^{0} \int_{-\infty}^{z} (-x)f(x, xv)\,dv\,dx + \int_{0}^{\infty} \int_{-\infty}^{z} xf(x, xv)\,dv\,dx$$

$$= \int_{-\infty}^{z} \int_{-\infty}^{\infty} |x|f(x, xv)\,dx\,dv$$

Finally, differentiating (again under an assumption of continuity), we find

$$f_Z(z) = \int_{-\infty}^{\infty} |x|f(x, xz)\,dx$$

In particular, if X and Y are independent,

$$f_Z(z) = \int_{-\infty}^{\infty} |x|f_X(x)f_Y(xz)\,dx$$

EXAMPLE B.   Suppose that X and Y are independent standard normal random variables and that $Z = Y/X$. We then have

$$f_Z(z) = \int_{-\infty}^{\infty} \frac{|x|}{2\pi} e^{-x^2/2} e^{-x^2 y^2/2}\,dx$$

From the symmetry of the integrand about zero,

$$f_Z(z) = \frac{1}{\pi} \int_{0}^{\infty} x e^{-x^2((z^2+1)/2)}\,dx$$

To simplify this, we make the change of variables $u = x^2$ to obtain

$$f_Z(z) = \frac{1}{2\pi} \int_{0}^{\infty} e^{-u((z^2+1)/2)}\,du$$

Next, using the fact that $\int_0^\infty \lambda \exp(-\lambda x)\,dx = 1$, we get

$$f_Z(z) = \frac{1}{\pi(z^2 + 1)}, \quad -\infty < z < \infty$$

This density is called the **Cauchy density**. Like the standard normal density, the Cauchy density is symmetric about zero and "bell-shaped," but the tails of the Cauchy tend to zero very slowly compared to the tails of the normal. This can be interpreted as being due to a substantial probability that X in the quotient Y/X is near zero.   ☐

Example B indicates one method of generating Cauchy random variables—we can generate independent standard normal random variables and form their quotient. The next section shows how to generate standard normals.

3.6.2   The General Case

The following example illustrates the concepts that are important to the general case of functions of several random variables and is also interesting in its own right.

EXAMPLE A.   Suppose that X and Y are independent standard normal random variables, which means that their joint distribution is the standard bivariate normal distribution, or

$$f_{XY}(x, y) = \frac{1}{2\pi} e^{-(x^2/2)-(y^2/2)}$$

We wish to change to polar coordinates and then reexpress the density in this new coordinate system:

$$R = \sqrt{X^2 + Y^2}$$

$$\Theta = \begin{cases} \tan^{-1}\left(\dfrac{Y}{X}\right), & \text{if } X > 0 \\[2ex] \tan^{-1}\left(\dfrac{Y}{X}\right) + \pi, & \text{if } X < 0 \\[2ex] \dfrac{\pi}{2}\,\text{sgn}(Y), & \text{if } X = 0,\, Y \neq 0 \\[2ex] 0, & \text{if } X = 0,\, Y = 0 \end{cases}$$

The inverse transformation is

$$X = R \cos \Theta$$

$$Y = R \sin \Theta$$

The joint density of R and Θ is

$$f_{R\Theta}(r, \theta) \, dr \, d\theta = P(r \leq R \leq r + dr, \theta \leq \Theta \leq \theta + d\theta)$$

This probability is equal to the area of the shaded patch in Figure 3-14 times $f_{XY}[x(r, \theta), y(r, \theta)]$. The area in question is clearly $r \, dr \, d\theta$, so

$$P(r \leq R \leq r + dr, \theta \leq \Theta \leq \theta + d\theta) = f_{XY}(r \cos \theta, r \sin \theta)r \, dr \, d\theta$$

and

$$f_{R\Theta}(r, \theta) = rf_{XY}(r \cos \theta, r \sin \theta)$$

Thus,

$$f_{R\Theta}(r, \theta) = \frac{r}{2\pi} e^{-(r^2 \cos^2 \theta)/2 - (r^2 \sin^2 \theta)/2}$$

$$= \frac{1}{2\pi} r e^{-r^2/2}$$

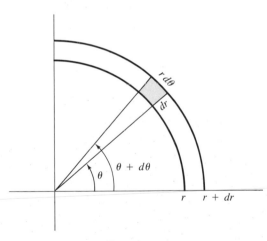

Figure 3-14.   The area of the shaded patch is $r \, dr \, d\theta$.

From this, we see that R and Θ are independent random variables, that Θ is uniform on $[-\pi/2, 3\pi/2]$ or, equivalently, on $[0, 2\pi]$, and that R has the density

$$f_R(r) = re^{-r^2/2}, \quad r \geq 0$$

which is called the **Rayleigh density**.

An interesting relationship can be found by changing variables again, letting $T = R^2$. Using the standard techniques for finding the density of a function of a single random variable, we obtain

$$f_T(t) = \tfrac{1}{2} e^{-t/2}$$

This is an exponential distribution with parameter $\tfrac{1}{2}$. Since R and Θ are independent, so are T and Θ, and the joint density of the latter pair is

$$f_{T\Theta}(t, \theta) = \frac{1}{2\pi} \left(\frac{1}{2} \right) e^{-t/2}$$

We have thus arrived at a characterization of the standard bivariate normal distribution: Θ is uniform on $[0, 2\pi]$, and R^2 is exponential with parameter $\tfrac{1}{2}$. (Also, from Example B in Section 3.6.1, $\tan \Theta$ follows a Cauchy distribution.)

These relationships can be used to construct an algorithm for generating standard normal random variables, which is quite useful since Φ, the cdf, and Φ^{-1} cannot be expressed in closed form. First, generate U_1 and U_2, which are independent and uniform on $[0, 1]$. Then $-2 \log U_1$ is exponential with parameter $\tfrac{1}{2}$, and $2\pi U_2$ is uniform on $[0, 2\pi]$. It follows that

$$X = \sqrt{-2 \log U_1} \cos(2\pi U_2)$$

and

$$Y = \sqrt{-2 \log U_2} \sin(2\pi U_2)$$

are independent standard normal random variables. This method of generating normally distributed random variables is sometimes called the **polar method**. □

For the general case, suppose that X and Y are jointly distributed continuous random variables, that X and Y are mapped onto U and V by the transformation

$$u = g_1(x, y)$$

$$v = g_2(x, y)$$

and that the transformation can be inverted to obtain

$$x = h_1(u, v)$$

$$y = h_2(u, v)$$

Assume that g_1 and g_2 have continuous partial derivatives and that the Jacobian

$$J(x, y) = \det \begin{bmatrix} \dfrac{\partial g_1}{\partial x} & \dfrac{\partial g_1}{\partial y} \\[2mm] \dfrac{\partial g_2}{\partial x} & \dfrac{\partial g_2}{\partial y} \end{bmatrix} = \left(\dfrac{\partial g_1}{\partial x}\right)\left(\dfrac{\partial g_2}{\partial y}\right) - \left(\dfrac{\partial g_2}{\partial x}\right)\left(\dfrac{\partial g_1}{\partial y}\right) \neq 0$$

for all x and y. This leads directly to the following result.

PROPOSITION A. Under the assumptions stated just above, the joint density of U and V is

$$f_{UV}(u, v) = f_{XY}(h_1(u, v), h_2(u, v))|J^{-1}(h_1(u, v), h_2(u, v))|$$

We will not prove Proposition A here. It follows from the formula established in advanced calculus for a change of variables in multiple integrals. The essential elements of the proof follow the discussion in Example A.

EXAMPLE B. To illustrate the formalism, let us redo Example A. The roles of u and v are played by r and θ:

$$r = \sqrt{x^2 + y^2}$$

$$\theta = \tan^{-1}\left(\frac{y}{x}\right)$$

The inverse transformation is

$$x = r\cos\theta$$

$$y = r\sin\theta$$

After some algebra, we obtain the partial derivatives:

$$\frac{\partial r}{\partial x} = \frac{x}{\sqrt{x^2 + y^2}} \qquad \frac{\partial r}{\partial y} = \frac{y}{\sqrt{x^2 + y^2}}$$

$$\frac{\partial \theta}{\partial x} = \frac{-y}{x^2 + y^2} \qquad \frac{\partial \theta}{\partial y} = \frac{x}{x^2 + y^2}$$

The Jacobian is the determinant of the matrix of these expressions, or

$$J(x, y) = \frac{1}{\sqrt{x^2 + y^2}} = \frac{1}{r}$$

Proposition A therefore says that

$$f_{R\Theta}(r, \theta) = rf_{XY}(r\cos\theta, r\sin\theta)$$

which is the same as the result we obtained by a direct argument in Example A.   □

Proposition A extends readily to transformations of more than two random variables. If $X_1, \ldots, X_n$ have the joint density function $f_{X_1 \cdots X_n}$ and

$$Y_i = g_i(X_1, \ldots, X_n), \quad i = 1, \ldots, n$$

$$X_i = h_i(Y_1, \ldots, Y_n), \quad i = 1, \ldots, n$$

and if $J(x_1, \ldots, x_n)$ is the determinant of the matrix with the ij entry $\partial g_i/\partial x_j$, then the joint density of $Y_1, \ldots, Y_n$ is

$$f_{Y_1 \cdots Y_n}(y_1, \ldots, y_n) = f_{X_1 \cdots X_n}(x_1, \ldots, x_n)|J^{-1}(x_1, \ldots, x_n)|$$

wherein each x_i is expressed in terms of the y's; $x_i = h_i(y_1, \ldots, y_n)$.

3.7   Extrema and Order Statistics

This section is concerned with ordering a collection of independent continuous random variables. In particular, let us assume that $X_1, X_2, \ldots, X_n$ are independent random variables with the common cdf F and density f. Let U denote the maximum of the X_i and V the minimum. The cdf's of U and V, and therefore their densities, can be found by a simple trick.

First, we note that $U \leq u$ if and only if $X_i \leq u$ for all i. Thus,

$$F_U(u) = P(U \leq u)$$

$$= P(X_1 \leq u)P(X_2 \leq u) \cdots P(X_n \leq u)$$

$$= [F(u)]^n$$

Differentiating, we find the density,

$$f_U(u) = nf(u)[F(u)]^{n-1}$$

Similarly, $V \geq v$ if and only if $X_i \geq v$ for all i. Thus,

$$1 - F_V(v) = [1 - F(v)]^n$$

and

$$F_V(v) = 1 - [1 - F(v)]^n$$

The density function of V is therefore

$$f_V(v) = nf(v)[1 - F(v)]^{n-1}$$

EXAMPLE A. Suppose that n system components are connected in series, which means that the system fails if any one of them fails, and that the lifetimes of the components, $T_1, \ldots, T_n$, are independent random variables that are exponentially distributed with parameter λ. The random variable that represents the length of time the system operates is V, which is the minimum of the T_i and has the density

$$f_V(v) = n\lambda e^{-\lambda v}(e^{-\lambda v})^{n-1}$$

$$= n\lambda e^{-n\lambda v}$$

We see that V is exponentially distributed with parameter $n\lambda$. □

EXAMPLE B. Suppose that a system has components as described in Example A but connected in parallel, which means that the system fails only when they all fail. The system's lifetime is thus the maximum of n exponential random variables and has the density

$$f_U(u) = n\lambda e^{-\lambda u}(1 - e^{-\lambda u})^{n-1}$$

By expanding the last term using the binomial theorem, we see that this density is a weighted sum of exponential terms rather than a simple exponential density. □

We will now derive the above results once more, by the differential technique, and generalize them. To find $f_U(u)$, we observe that $u \leq U \leq u + du$ if one of the n X_i falls in the interval $(u, u + du)$ and the other $n - 1$ X_i fall to the left of u. The probability of any particular such arrangement is $[F(u)]^{n-1}f(u)\,du$, and since there are n such arrangements,

$$f_U(u) = n[F(u)]^{n-1}f(u)$$

Now we again assume that $X_1, \ldots, X_n$ are independent continuous random variables with density $f(x)$. We sort the X_i and denote by $X_{(1)} < X_{(2)} < \cdots < X_{(n)}$ the **order statistics**. Note that X_1 is not necessarily equal to $X_{(1)}$ (in fact, this equality holds with probability n^{-1}). Thus, $X_{(n)}$ is the maximum, and $X_{(1)}$ is the minimum. If n is odd, say, $n = 2m + 1$, then $X_{(m+1)}$ is called the **median** of the X_i.

THEOREM A. The density of $X_{(k)}$, the kth order statistic, is

$$f_k(x) = \frac{n!}{(k-1)!(n-k)!}\,f(x)F^{k-1}(x)[1 - F(x)]^{n-k}$$

Proof. We will use a differential argument to derive this result heuristically. (The alternative approach of first deriving the cdf and then differentiating is developed in Problem 58 at the end of this chapter.) The event $x \leq X_{(k)} \leq x + dx$ occurs if $k - 1$ observations are less than x, one observation is in the interval $[x, x + dx]$, and $n - k$ observations are greater than $x + dx$. The probability of any particular arrangement of this type is $f(x)F^{k-1}(x)[1 - F(x)]^{n-k}$, and, by the multinomial theorem, there are $n!/(k - 1)!1!(n - k)!$ such arrangements, which completes the argument. $\qquad\square$

EXAMPLE C.   For the case where the X_i are uniform on $[0, 1]$, the density of the kth order statistic reduces to

$$\frac{n!}{(k - 1)!(n - k)!} x^{k-1}(1 - x)^{n-k}, \quad 0 \leq x \leq 1$$

An interesting by-product of this result is that, since the density integrates to 1,

$$\int_0^1 x^{k-1}(1 - x)^{n-k}\, dx = \frac{(k - 1)!(n - k)!}{n!} \qquad\square$$

Joint distributions of order statistics can also be worked out. For example, to find the joint density of the minimum and maximum, we note that $x \leq X_{(1)} \leq x + dx$ and $y \leq X_{(n)} \leq y + dy$ if one X_i falls in $[x, x + dx]$, one falls in $[y, y + dy]$, and $n - 2$ fall in $[x, y]$. There are $n(n - 1)$ ways to choose the minimum and maximum, and thus $U = X_{(1)}$ and $V = X_{(n)}$ have the joint density

$$f(u, v) = n(n - 1)f(v)f(u)[F(u) - F(v)]^{n-2}, \quad v \leq u$$

For example, for the uniform case,

$$f(u, v) = n(n - 1)(u - v)^{n-2}, \quad 1 \geq u \geq v \geq 0$$

The **range** of $X_{(1)}, \ldots, X_{(n)}$ is $R = X_{(n)} - X_{(1)}$. Using the same kind of analysis we used in Section 3.6.1 to derive the distribution of a sum, we find

$$f_R(r) = \int_{-\infty}^{\infty} f(v + r, v)\, dv$$

EXAMPLE D.   Find the distribution of the range, $U - V$, for the uniform case.

The integrand is $f(v + r, v) = n(n - 1)r^{n-2}$ for $0 \leq v \leq v + r \leq 1$ or, equivalently, $0 \leq v \leq 1 - r$. Thus,

$$f_R(r) = \int_0^{1-r} n(n - 1)r^{n-2}\, dv = n(n - 1)r^{n-2}(1 - r)$$

The corresponding cdf is

$$F_R(r) = nr^{n-1} - (n-1)r^n \qquad \square$$

EXAMPLE E. (Tolerance Interval) If a large number of independent random variables having the common density function f are observed, it seems intuitively likely that most of the probability mass of the density $f(x)$ is contained in the interval $(X_{(1)}, X_{(n)})$ and unlikely that a future observation will lie outside this interval. In fact, very precise statements can be made. For example, the amount of the probability mass in the interval is $F(X_{(n)}) - F(X_{(1)})$, a random variable that we will denote by Q. From Proposition C of Section 2.3, the distribution of $F(X_i)$ is uniform; therefore, the distribution of Q is the distribution of $U_{(n)} - U_{(1)}$, which is the range of n independent uniform random variables. Thus, $P(Q > \alpha)$, the probability that more than $100\alpha\%$ of the probability mass is contained in the range, is

$$P(Q > \alpha) = 1 - n\alpha^{n-1} + (n-1)\alpha^n$$

For example, if $n = 100$ and $\alpha = .95$, this probability is .96. In words, this means that the probability is .96 that the range of 100 independent random variables covers 95% or more of the probability mass, or, with probability .96, 95% of all further observations from the same distribution will fall between the minimum and maximum. This statement does not depend on the actual form of the distribution. $\qquad \square$

3.8 Problems

1. The joint frequency function of two discrete random variables, X and Y, is given in the following table:

y	x 1	2	3	4
1	.10	.05	.02	.02
2	.05	.20	.05	.02
3	.02	.05	.20	.04
4	.02	.02	.04	.10

   (a) Find the marginal frequency functions of X and Y.
   (b) Find the conditional frequency function of X given $Y = 1$ and of Y given $X = 1$.
2. An urn contains p black balls, q white balls, and r red balls, and n balls are chosen without replacement.
   (a) Find the joint distribution of the numbers of black, white, and red balls in the sample.

(b) Find the joint distribution of the numbers of black and white balls in the sample.

(c) Find the marginal distribution of the number of white balls in the sample.

3. Three players play ten rounds of a game, and each player has probability $\frac{1}{3}$ of winning each round. Find the joint distribution of the numbers of games won by each of the three players.

4. A sieve is made of a square mesh of wires. Each wire has diameter d, and the holes in the mesh are squares whose side length is w. A spherical particle of radius r is dropped on the mesh. What is the probability that it passes through? What is the probability that it fails to pass through if it is dropped n times? (Calculations such as these are relevant to the theory of sieving for analyzing the size distribution of particulate matter.)

5. (Buffon's Needle Problem) A needle of length L is dropped randomly on a plane ruled with parallel lines that are a distance D apart, where $D \geq L$. Show that the probability that the needle comes to rest crossing a line is $2L/\pi D$. Explain how this gives a mechanical means of estimating the value of π.

6. A point is chosen randomly in the interior of an ellipse:

$$\frac{x^2}{a^2} + \frac{y^2}{b^2} = 1$$

Find the marginal densities of the x and y coordinates of the point.

7. Find the joint and marginal densities corresponding to the cdf

$$F(x, y) = (1 - e^{-\alpha x})(1 - e^{-\beta y}), \quad x \geq 0, y \geq 0$$

8. Let X and Y have the joint density

$$f(x, y) = \tfrac{6}{7}(x + y)^2, \quad 0 \leq x \leq 1, 0 \leq y \leq 1$$

(a) By integrating over the appropriate regions, find (i) $P(X > Y)$, (ii) $P(X + Y \leq 1)$, (iii) $P(X \leq \frac{1}{2})$.

(b) Find the marginal densities of X and Y.

(c) Find the two conditional densities.

9. Suppose that (X, Y) is uniformly distributed over the region defined by $0 \leq y \leq 1 - x^2$ and $-1 \leq x \leq 1$.

(a) Find the marginal densities of X and Y.

(b) Find the two conditional densities.

10. A point is uniformly distributed in a unit sphere in three dimensions.

(a) Find the marginal densities of the x, y, and z coordinates.

(b) Find the joint density of the x and y coordinates.

(c) Find the density of the xy coordinates conditional on $Z = 0$.

11. Let U_1, U_2, and U_3 be independent random variables uniform on $[0, 1]$. Find the probability that the roots of the quadratic $U_1 x^2 + U_2 x + U_3$ are real.

12. Let

$$f(x, y) = c(x^2 - y^2)e^{-x}, \quad 0 \le x < \infty, -x \le y < x$$

(a) Find c.
(b) Find the marginal densities.
(c) Find the conditional densities.

13. A fair coin is thrown once; if it lands heads up, it is thrown a second time. Find the frequency function of the total number of heads.

14. Suppose that

$$f(x, y) = xe^{-x(y+1)}, \quad 0 \le x < \infty, 0 \le y < \infty$$

(a) Find the marginal densities of X and Y.
(b) Find the conditional densities of X and Y.

15. Suppose that two components have independent exponentially distributed lifetimes, T_1 and T_2, with parameters α and β, respectively. Find (a) $P(T_1 > T_2)$ and (b) $P(T_1 > 2T_2)$.

16. If X_1 is uniform on $[0, 1]$, and, conditional on X_1, X_2 is uniform on $[0, X_1]$, find the joint and marginal distributions of X_1 and X_2.

17. Find the conditional densities of X and Y as given in Example B in Section 3.3.

18. Consider a Poisson process on the real line, and denote by $N(t_1, t_2)$ the number of events in the interval (t_1, t_2). If $t_0 < t_1 < t_2$, find the conditional distribution of $N(t_0, t_1)$ given that $N(t_0, t_2) = n$. (*Hint:* Use the fact that the numbers of events in disjoint subsets are independent.)

19. Suppose that, conditional on N, X has a binomial distribution with N trials and probability p of success, and that N is a binomial random variable with m trials and probability r of success. Find the unconditional distribution of X.

20. Let P have a uniform distribution on $[0, 1]$, and, conditional on $P = p$, let X have a Bernoulli distribution with parameter p. Find the conditional distribution of P given X.

21. Let X have the density function f, and let $Y = X$ with probability $\frac{1}{2}$ and $Y = -X$ with probability $\frac{1}{2}$. Show that the density of Y is symmetric about zero—that is, $f_Y(y) = f_Y(-y)$.

22. Spherical particles whose radii have the density function $f_R(r)$ are dropped on a mesh as in Problem 4. Find an expression for the density function of the particles that pass through.

23. Prove that X and Y are independent if and only if $f_{X|Y}(x|y) = f_X(x)$ for all x and y.

24. Let $f(x) = (1 + \alpha x)/2$, for $-1 \le x \le 1$ and $-1 \le \alpha \le 1$.
(a) Describe an algorithm to generate random variables from this density using the rejection method.
(b) Write a computer program to do so, and test it out.

25. Let $f(x) = 6x^2(1 - x)^2$, for $1 \le x \le 1$.

(a) Describe an algorithm to generate random variables from this density using the rejection method. In what proportion of the trials will the acceptance step be taken?

(b) Write a computer program to do so, and test it out.

26. Show that the number of iterations necessary to generate a random variable using the rejection method is a geometric random variable, and evaluate the parameter of the geometric frequency function. Show that in order to keep the number of iterations small, $M(x)$ should be chosen to be close to $f(x)$.

27. Show that the following method of generating discrete random variables works (D. R. Fredkin). Suppose, for concreteness, that X takes on values 0, 1, 2, ... with probabilities $p_0, p_1, p_2, \ldots$. Let U be a uniform random variable. If $U < p_0$, return $X = 0$. If not, replace U by $U - p_0$, and if the new U is less than p_1, return $X = 1$. If not, decrement U by p_1, compare U to p_2, etc.

28. (a) Let T be an exponential random variable with parameter λ, let W be random variable, which is ± 1 with probability $\frac{1}{2}$ each, and let $X = WT$. Show that the density of X is

$$f_X(x) = \frac{\lambda}{2} e^{-\lambda|x|}$$

which is called the **double exponential density**.

(b) Show that for some constant, c,

$$\frac{1}{\sqrt{2\pi}} e^{-x^2/2} \le c e^{-|x|}$$

Use this result and that of part (a) to show how to use the rejection method to generate random variables from a standard normal density.

29. Let U_1 and U_2 be independent and uniform on $[0, 1]$. Find and sketch the density function of $S = U_1 + U_2$.

30. Let N_1 and N_2 be independent random variables following Poisson distributions with parameters λ_1 and λ_2. Show that the distribution of $N = N_1 + N_2$ is Poisson with parameter $\lambda_1 + \lambda_2$.

31. For a Poisson distribution, suppose that events are independently labeled A and B with probabilities $p_A + p_B = 1$. If the parameter of the Poisson distribution is λ, show that the number of events labeled A follows a Poisson distribution with parameter $p_A \lambda$.

32. Let X and Y be jointly continuous random variables. Find an expression for the density of $Z = X - Y$.

33. Let X and Y be independent standard normal random variables. Find the density of $Z = X + Y$, and show that Z is normally distributed as well. (*Hint:* Use the technique of completing the square to help in evaluating the integral.)

34. Let T_1 and T_2 be independent exponentials with parameters λ_1 and λ_2. Find the density function of $T_1 + T_2$.

35. Find the density function of $X + Y$, where X and Y have a joint density as given in Example B in Section 3.3.

36. Suppose that X and Y are independent discrete random variables and each assumes the values 0, 1, and 2 with probability $\frac{1}{3}$ each. Find the frequency function of $X + Y$.

37. Let X and Y have the joint density function $f(x, y)$, and let $Z = XY$. Show that the density function of Z is

$$f_Z(z) = \int_{-\infty}^{\infty} f\left(x, \frac{z}{y}\right) \frac{1}{|y|} dy$$

38. Find the density of the quotient of two independent uniform random variables.

39. Consider forming a random rectangle in two ways. Let U_1, U_2, and U_3 be independent random variables uniform on $[0, 1]$. One rectangle has sides U_1 and U_2, and the other is a square with sides U_3. Find the probability that the area of the square is greater than the area of the other rectangle.

40. Let X, Y, and Z be independent $N(0, \sigma^2)$. Let Θ, Φ, and R be the corresponding random variables that are the spherical coordinates of (X, Y, Z):

$$x = r \sin \phi \cos \theta$$

$$y = r \sin \phi \sin \theta$$

$$z = r \cos \phi$$

$$0 \le \phi \le \pi, \quad 0 \le \theta \le 2\pi$$

Find the joint and marginal densities of Θ, Φ, and R. (Hint: $dx\, dy\, dz = r^2 \sin \phi \, dr \, d\theta \, d\phi$.)

41. A point is generated on a unit disk in the following way: The radius, R, is uniform on $[0, 1]$, and the angle Θ is uniform on $[0, 2\pi]$ and is independent of R.
    (a) Find the joint density of $X = R \cos \Theta$ and $Y = R \sin \Theta$.
    (b) Find the marginal densities of X and Y.
    (c) Is the density uniform over the disk? If not, modify the method to produce a uniform density.

42. If X and Y are independent exponential random variables, find the joint density of the polar coordinates R and Θ of the point (X, Y). Are R and Θ independent?

43. Let X and Y be jointly continuous random variables. Find an expression for the joint density of $U = a + bX$ and $V = c + dY$.

44. If X and Y are independent standard normal random variables, find $P(X^2 + Y^2 \le 1)$.

45. Let X and Y be jointly continuous random variables.
    (a) Develop an expression for the joint density of $X + Y$ and $X - Y$.

(b) Develop an expression for the joint density of XY and Y/X.

(c) Specialize the expressions from parts (a) and (b) to the case where X and Y are independent.

46. Find the joint density of $X + Y$ and X/Y, where X and Y are independent exponential random variables with parameter λ. Show that $X + Y$ and X/Y are independent.

47. Suppose that a system's components are connected in series and have life-times that are independent exponential random variables with parameters λ_i. Show that the lifetime of the system is exponential with parameter $\sum \lambda_i$.

48. Each component of the following system (Figure 3-15) has an exponentially distributed lifetime with parameter λ. Find the cdf and the density of the system's lifetime.

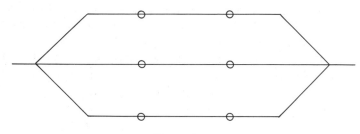

Figure 3-15

49. A card contains n chips and has an error-correcting mechanism such that the card still functions if a single chip fails but does not function if two or more chips fail. If each chip has a lifetime that is an independent exponential with parameter λ, find the density function of the card's lifetime.

50. Suppose that a queue has n servers and that the length of time to complete a job is an exponential random variable. If a job is at the top of the queue and will be handled by the next available server, what is the distribution of the waiting time until service? What is the distribution of the waiting time until service of the next job in the queue?

51. Find the density of the minimum of n independent Weibull random variables, each of which has the density

$$f(t) = \alpha\beta t^{\beta-1} e^{-\alpha t^{\beta}}$$

52. If five numbers are chosen at random in the interval $[0, 1]$, what is the probability that they all lie in the middle half of the interval?

53. Let $X_1, \ldots, X_n$ be independent random variables, each with the density function f. Find an expression for the probability that the interval $(-\infty, X_{(n)}]$ encompasses at least $100\gamma\%$ of the probability mass of f.

54. Show that the joint cdf of $X_{(1)}$ and $X_{(n)}$ is

$$F(x, y) = F^n(y) - [F(y) - F(x)]^n, \quad x \leq y$$

55. If $X_1, \ldots, X_n$ are independent random variables, each with the density function f, show that the joint density of $X_{(1)}, \ldots, X_{(n)}$ is

$$n!f(x_1)f(x_2)\cdots f(x_n), \quad x_1 < x_2 < \cdots < x_n$$

56. Let U_1, U_2, and U_3 be independent uniform random variables.
    (a) Find the joint density of $U_{(1)}$, $U_{(2)}$, and $U_{(3)}$.
    (b) The locations of three gas stations are independently and randomly placed along a mile of highway. What is the probability that no two gas stations are less than $\frac{1}{3}$ mile apart?

57. Use the differential method to find the joint density of $X_{(i)}$ and $X_{(j)}$, where $i < j$.

58. Prove Theorem A of Section 3.7 by finding the cdf of $X_{(k)}$ and differentiating. (*Hint:* $X_{(k)} \leq x$ if and only if k or more of the X_i are less than or equal to x. The number of X_i less than or equal to x is a binomial random variable.)

59. Find the density of $U_{(k)} - U_{(k-1)}$ if the U_i, $i = 1, \ldots, n$, are independent uniform random variables. This is the density of the spacing between adjacent points chosen uniformly in the interval $[0, 1]$.

60. Show that

$$\int_0^1 \int_0^y (y - x)^n \, dx \, dy = \frac{1}{(n+1)(n+2)}$$

61. If T_1 and T_2 are independent exponential random variables, find the density function of $R = T_{(2)} - T_{(1)}$.

62. Let $U_1, \ldots, U_n$ be independent uniform random variables, and let V be uniform and independent of the U_i.
    (a) Find $P(V \leq U_{(n)})$.
    (b) Find $P(U_{(1)} < V < U_{(n)})$.

63. Do both parts of Problem 62 again, assuming that the U_i and V have the density function f and the cdf F, with F^{-1} uniquely defined. [*Hint:* $F^{-1}(U_i)$ has a uniform distribution.]

4

Expected Values

4.1 The Expected Value of a Random Variable

The concept of the expected value of a random variable parallels the notion of a weighted average. The possible values of the random variable are weighted by their probabilities, as specified in the following definition.

DEFINITION. If X is a discrete random variable with frequency function $p(x)$, the expected value of X, denoted by $E(X)$, is

$$E(X) = \sum_i x_i p(x_i)$$

provided that $\sum_i |x_i| p(x_i) < \infty$. If the sum diverges, the expectation is undefined.

$E(X)$ is also referred to as the **mean** of X and is often denoted by μ or μ_X. You may find it helpful to think of the expected value of X as the center of mass of the frequency function. Imagine placing the masses $p(x_i)$ at the points x_i on a beam; the balance point of the beam is the expected value of X.

EXAMPLE A.   (Roulette) A roulette wheel has the numbers 1 through 36, as well as 0 and 00. If you bet \$1 that an odd number comes up, you win or lose \$1 according to whether or not that event occurs. If X denotes your net gain, $X = 1$ with probability $\frac{18}{38}$ and $X = -1$ with probability $\frac{20}{38}$. The expected value of X is

$$E(X) = 1 \times \tfrac{18}{38} + (-1) \times \tfrac{20}{38} = -\tfrac{1}{19}$$

Thus, your expected loss is about \$.05. In chapter 5, it will be shown that this coincides in the limit with the actual average loss per game if you play a long sequence of independent games.   □

EXAMPLE B.   (Expectation of a Geometric Random Variable) Suppose that items produced in a plant are defective with probability p independently of each other. Items are inspected one by one until a defective item is found. On the average, how many items must be inspected?

The number of items inspected, X, is a geometric random variable, with $P(X = k) = q^{k-1}p$, where $q = 1 - p$. Therefore,

$$E(X) = \sum_{k=1}^{\infty} kpq^{k-1} = p \sum_{k=1}^{\infty} kq^{k-1}$$

We use a trick to calculate the sum. Since $kq^{k-1} = \dfrac{d}{dq} q^k$,

$$E(X) = p \frac{d}{dq} \sum_{k=1}^{\infty} q^k$$

$$= p \frac{d}{dq} \frac{q}{1 - q}$$

$$= \frac{p}{(1 - q)^2}$$

$$= \frac{1}{p}$$

It can be shown that the interchange of differentiation and summation is justified. Thus, for example, if 10% of the items are defective, an average of ten items must be examined in order to find one that is defective, as might have been guessed.   □

EXAMPLE C.   (Poisson Distribution) The expected value of a Poisson random variable is

$$E(X) = \sum_{k=0}^{\infty} \frac{k\lambda^k}{k!} e^{-\lambda}$$

$$= \lambda \sum_{k=1}^{\infty} \frac{\lambda^{k-1}}{(k - 1)!} e^{-\lambda}$$

$$= \lambda \sum_{j=0}^{\infty} \frac{\lambda^j}{j!} e^{-\lambda}$$

Since $\sum_{j=0}^{\infty}(\lambda^j/j!) = e^{\lambda}$, we have

$$E(X) = \lambda$$

The parameter λ of the Poisson distribution can thus be interpreted as the average count. $\square$

EXAMPLE D. (St. Petersburg Paradox) A gambler has the following strategy for playing a sequence of games: He starts off betting \$1; if he loses, he doubles his bet; and he continues to double his bet until he finally wins. To analyze this scheme, suppose that the game is fair and that he wins or loses the amount he bets. At trial 0, he bets \$1; if he loses, he bets \$2 at trial 1; and if he has not won by the kth trial, he bets $(\$2)^k$. When he finally wins, he will be \$1 ahead, which can be checked by going through the scheme for the first few values of k. Let X denote the amount of money bet on the very last game (the game he wins). Since the probability that k losses are followed by one win is $2^{-(k+1)}$,

$$P(X = 2^k) = \frac{1}{2^{k+1}}$$

and

$$E(X) = \sum_{n=0}^{\infty} np(n)$$

$$= \sum_{k=0}^{\infty} 2^k \frac{1}{2^{k+1}} = \infty$$

Formally, $E(X)$ is not defined. Practically, the analysis shows that a flaw in this scheme is that it does not take into account the enormous amount of capital required. $\square$

The definition of expectation for a continuous random variable is a fairly obvious extension of the discrete case—summation is replaced by integration.

DEFINITION. If X is a continuous random variable with density $f(x)$, then

$$E(X) = \int_{-\infty}^{\infty} xf(x)\,dx$$

provided that $\int |x|f(x)\,dx < \infty$. If the integral diverges, the expectation is undefined.

Again $E(X)$ can be regarded as the center of mass of the density. We next consider some examples.

EXAMPLE E. (Gamma Density) If X follows a gamma density with parameters α and λ,

$$E(X) = \int_0^\infty \frac{\lambda^\alpha}{\Gamma(\alpha)} x^\alpha e^{-\lambda x}\, dx$$

This integral is easy to evaluate once we realize that $\lambda^{\alpha+1} x^\alpha e^{-\lambda x}/\Gamma(\alpha + 1)$ is a gamma density and therefore integrates to 1. We have

$$\int_0^\infty x^\alpha e^{-\lambda x}\, dx = \frac{\Gamma(\alpha + 1)}{\lambda^{\alpha+1}}$$

from which it follows that

$$E(X) = \frac{\lambda^\alpha}{\Gamma(\alpha)} \left[\frac{\Gamma(\alpha + 1)}{\lambda^{\alpha+1}} \right]$$

Finally, using the relation $\Gamma(\alpha + 1) = \alpha\Gamma(\alpha)$, we find

$$E(X) = \frac{\alpha}{\lambda}$$

For the exponential density, $\alpha = 1$, so $E(X) - 1/\lambda$. This may be contrasted to the median of the exponential density, which was found in Section 2.2.1 to be $\log 2/\lambda$. The mean and the median can both be interpreted as "typical" values of X, but they measure different attributes of the probability distribution. □

EXAMPLE F. (Normal Distribution) From the definition of the expectation, we have

$$E(X) = \frac{1}{\sigma\sqrt{2\pi}} \int_{-\infty}^\infty x e^{-(x-\mu)^2/2\sigma^2}\, dx$$

Making the change of variables $z = x - \mu$ changes the above equation to

$$E(X) = \frac{1}{\sigma\sqrt{2\pi}} \int_{-\infty}^\infty z e^{-z^2/2\sigma^2}\, dz + \frac{\mu}{\sigma\sqrt{2\pi}} \int_{-\infty}^\infty e^{-z^2/2\sigma^2}\, dz$$

The first integral is 0 since the contributions from $z < 0$ cancel those from $z > 0$, and the second integral is μ since the normal density integrates to 1. Thus,

$$E(X) = \mu$$

The parameter μ of the normal density is the expectation, or mean value. We could have made the derivation much shorter by claiming that it was "obvious" that since the center of symmetry of the density is μ, the expectation must be μ.   □

EXAMPLE G.   (Cauchy Density) Recall that the Cauchy density is

$$f(x) = \frac{1}{\pi}\left(\frac{1}{1+x^2}\right), \quad -\infty < x < \infty$$

The density is symmetric about zero, so it would seem that $E(X) = 0$. However,

$$\int_{-\infty}^{\infty} \frac{|x|}{1+x^2} = \infty$$

Therefore, the expectation does not exist. The reason that it fails to exist is, of course, that the density decreases so slowly that very large values of X can occur with substantial probability.   □

The expected value can be interpreted as a long-run average. In chapter 5, it will be shown that if $E(X)$ exists and if $X_1, X_2, \ldots$ is a sequence of independent random variables with the same distribution as X and if $S_n = \sum_{i=1}^{n} X_i$, then, as $n \to \infty$,

$$\frac{S_n}{n} \to E(X)$$

This statement will be made more precise in chapter 5. For now, a simple empirical demonstration will be sufficient.

EXAMPLE H.   Using a pseudorandom number generator, a sequence $X_1, X_2, \ldots$ of independent standard normal random variables was generated, as well as a sequence $Y_1, Y_2, \ldots$ of independent Cauchy random variables. Figure 4-1 shows the graphs of

$$G(n) = \frac{1}{n}\sum_{i=1}^{n} X_i$$

and

$$C(n) = \frac{1}{n}\sum_{i=1}^{n} Y_i$$

for $n = 1, 2, \ldots, 500$. Note how $G(n)$ appears to be tending to a limit, whereas $C(n)$ does not.   □

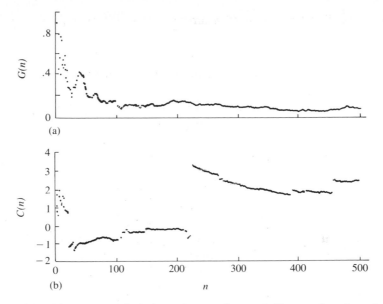

Figure 4-1. The average of n independent random variables as a function of n for (a) normal random variables and (b) Cauchy random variables.

4.1.1 Expectations of Functions of Random Variables

We often need to find $E[g(X)]$, where X is a random variable and g is a fixed function. The straightforward way to do this would seem to be the following: Let $Y = g(X)$, find the density or frequency function of Y, say, f_Y, and then compute $E(Y)$ from the definition. It turns out, fortunately, that the process is much more simple than that.

THEOREM A.   Suppose that $Y = g(X)$.
(a) If X is discrete with frequency function $p(x)$, then

$$E(Y) = \sum_x g(x)p(x)$$

provided that $\sum |g(x)| p(x) < \infty$.
(b) If X is continuous with density function $f(x)$, then

$$E(Y) = \int_{-\infty}^{\infty} g(x)f(x)\,dx$$

provided that $\int |g(x)| f(x)\,dx < \infty$.

Proof.   We will prove this result for the discrete case. The basic argument is the same for the continuous case, but to make that proof rigorous requires some advanced theory of integration. By definition,

$$E(Y) = \sum_i y_i p_Y(y_i)$$

Let A_i denote the set of x's mapped to y_i by g; that is, $x \in A_i$ if $g(x) = y_i$. Then

$$p_Y(y_i) = \sum_{x \in A_i} p(x)$$

and

$$
\begin{aligned}
E(Y) &= \sum_i y_i \sum_{x \in A_i} p(x) \\
&= \sum_i \sum_{x \in A_i} y_i p(x) \\
&= \sum_i \sum_{x \in A_i} g(x) p(x) \\
&= \sum_x g(x) p(x)
\end{aligned}
$$

This last step follows because the A_i are disjoint and every x belongs to some A_i. $\square$

It is worth pointing out explicitly that $E[g(X)] \neq g[E(X)]$; that is, the average value of the function is not equal to the function of the average value. Suppose, for example, that X takes on values 1 and 2, each with probability $\frac{1}{2}$, so $E(X) = \frac{3}{2}$. Let $Y = 1/X$. Then $E(Y)$ is clearly $1 \times .5 + .5 \times .5 = .75$, but $1/E(X) = \frac{2}{3}$.

EXAMPLE A.   The magnitude of the velocity of a gas molecule follows Maxwell's distribution:

$$f_V(v) = \frac{\sqrt{2/\pi}}{\sigma^3} v^2 e^{-v^2/2\sigma^2}, \quad 0 \le v < \infty$$

where the parameter σ depends on the temperature of the gas. Suppose that we wish to find the average kinetic energy, $X = \frac{1}{2}mV^2$. To do this from the definition, we would first have to find the density function of X and then calculate the expected value. But Theorem A allows us instead to calculate

$$
\begin{aligned}
E(X) &= \int_{-\infty}^{\infty} \tfrac{1}{2}mv^2 f_V(v)\, dv \\
&= \frac{m}{2} \frac{\sqrt{2/\pi}}{\sigma^3} \int_0^{\infty} v^4 e^{-v^2/2\sigma^2}\, dv
\end{aligned}
$$

which is much simpler. To evaluate the integral, we make the change of variables $u = v^2/2\sigma^2$ to reduce it to

$$\frac{2m\sigma^2}{\sqrt{\pi}} \int_0^\infty u^{3/2} e^{-u} \, du = \frac{2m\sigma^2}{\sqrt{\pi}} \Gamma\left(\frac{5}{2}\right)$$

Finally, using the facts $\Gamma(\frac{1}{2}) = \sqrt{\pi}$ and $\Gamma(\alpha + 1) = \alpha \Gamma(\alpha)$, we have

$$E(X) = \tfrac{3}{2} m\sigma^2 \qquad \square$$

Now suppose that $Y - g(X_1, \ldots, X_n)$, where the X_i have a joint distribution, and that we wish to find $E(Y)$. We do not have to find the density or frequency function of Y, which again could be a formidable task.

THEOREM B. Suppose that $X_1, \ldots, X_n$ are jointly distributed random variables and $Y = g(X_1, \ldots, X_n)$.
(a) If the X_i are discrete with frequency function $p(x_1, \ldots, x_n)$, then

$$E(Y) = \sum_{x_1, \ldots, x_n} g(x_1, \ldots, x_n) p(x_1, \ldots, x_n)$$

provided that $\sum_{x_1, \ldots, x_n} |g(x_1, \ldots, x_n)| p(x_1, \ldots, x_n) < \infty$.
(b) If the X_i are continuous with joint density function $f(x_1, \ldots, x_n)$, then

$$E(Y) = \int \int \cdots \int g(x_1, \ldots, x_n) f(x_1, \ldots, x_n) \, dx_1 \cdots dx_n$$

provided that the integral with $|g|$ in place of g converges.

Proof.   The proof is similar to that of Theorem A. $\square$

EXAMPLE B.   A stick of unit length is broken randomly in two places. What is the average length of the middle piece?
    We interpret this question to mean that the locations of the two break points are independent uniform random variables U_1 and U_2. Therefore, we need to compute $E|U_1 - U_2|$. Theorem B tells us that we do not need to find the density function of $|U_1 - U_2|$ and that

$$E|U_1 - U_2| = \int_0^1 \int_0^1 |u_1 - u_2| \, du_1 \, du_2$$

$$= \int_0^1 \int_0^{u_1} (u_1 - u_2) \, du_2 \, du_1 + \int_0^1 \int_{u_1}^1 (u_2 - u_1) \, du_2 \, du_1$$

With some care, we find the expectation to be $\tfrac{1}{3}$. This is in accord with the intuitive argument that the smaller of U_1 and U_2 should be $\frac{1}{3}$ on the average and the larger should be $\frac{2}{3}$ on the average, which means that the difference should be $\frac{1}{3}$.   ∎

We note the following immediate consequence of Theorem B.

COROLLARY A.   If X and Y are independent random variables and g and h are fixed functions, then $E[g(X)h(Y)] = \{E[g(X)]\}\{E[h(Y)]\}$, provided that the expectations on the right-hand side exist.

In particular, if X and Y are independent, $E(XY) = E(X)E(Y)$. The proof of this corollary is left to Problem 27 of the end-of-chapter problems.

4.1.2 Expectations of Linear Combinations of Random Variables

One of the most useful properties of the expectation is that it is a linear operation, as is stated by the following theorem.

THEOREM A.   If $X_1, \ldots, X_n$ are jointly distributed random variables with expectations $E(X_i)$ and Y is a linear function of the X_i, $Y = a + \sum_{i=1}^{n} b_i X_i$, then

$$E(Y) = a + \sum_{i=1}^{n} b_i E(X_i)$$

Proof.   We will prove this for the continuous case. The proof in the discrete case is parallel and is left to Problem 22 at the end of this chapter. For notational simplicity, we take $n = 2$. From Theorem B of Section 4.1.1, we have

$$E(Y) = \int \int (a + b_1 x_1 + b_2 x_2) f(x_1, x_2)\, dx_1\, dx_2$$

$$= a \int \int f(x_1, x_2)\, dx_1\, dx_2 + b_1 \int \int x_1 f(x_1, x_2)\, dx_1\, dx_2$$

$$+ b_2 \int \int x_2 f(x_1, x_2)\, dx_1\, dx_2$$

The first double integral of the last expression is merely the integral of the bivariate density, which is equal to 1. The second double integral can be evaluated as follows:

$$\int \int x_1 f(x_1, x_2)\, dx_1\, dx_2 = \int x_1 \left[\int f(x_1, x_2)\, dx_2 \right] dx_1$$

$$= \int x_1 f_{X_1}(x_1)\, dx_1$$

$$= E(X_1)$$

A similar evaluation for the third double integral brings us to

$$E(Y) = a + b_1 E(X_1) + b_2 E(X_2)$$

This proves the theorem once we check that the expectation is well-defined, or that

$$\int\int |a + b_1 x_1 + b_2 x_2| f(x_1, x_2)\, dx_1\, dx_2 < \infty$$

This can be verified using the inequality

$$|a + b_1 x_1 + b_2 x_2| \le |a| + |b_1||x_1| + |b_2||x_2|$$

and the assumption that the $E(X_i)$ exist.   □

Theorem A is extremely useful. We can illustrate its utility with several examples.

EXAMPLE A.   Suppose that we wish to find the expectation of a binomial random variable, Y. From the binomial frequency function,

$$E(Y) = \sum_{k=0}^{n} \binom{n}{k} k p^k (1 - p)^{n-k}$$

It is not immediately obvious how to evaluate this sum. We can, however, represent Y as the sum of Bernoulli random variables, X_i, which equal 1 or 0 depending on whether there is success or failure on the ith trial,

$$Y = \sum_{k=1}^{n} X_i$$

Since $E(X_i) = 0 \times (1 - p) + 1 \times p = p$, we have immediately that $E(Y) = np$.
    □

EXAMPLE B.   (Coupon Collection) Suppose that you collect coupons, that there are n distinct types of coupons, and that on each trial you are equally likely to get a coupon of any of the types. How many trials would you expect to go through until you had a complete set of coupons? (This might be a model for collecting baseball cards or for certain grocery store promotions.)
    The solution of this problem is greatly simplified by representing the number of trials as a sum. Let X_1 be the number of trials up to and including the trial on which the first coupon is collected: $X_1 = 1$. Let X_2 be the number of trials from that point up to and including the trial on which the next coupon different from the first is obtained; let X_3 be the number of trials from that point up to

and including the trial on which the third distinct coupon is collected; etc., up to X_n. Then the total number of trials, X, is the sum of the X_i, $i = 1, 2, \ldots, n$.

We now find the distribution of X_r. At this point, $r - 1$ of n coupons have been collected, so on each trial the probability of success is $(n - r + 1)/n$. Therefore, X_r is a geometric random variable, with $E(X_r) = n/(n - r + 1)$ (see Example B in Section 4.1). Thus,

$$E(X) = \sum_{r=1}^{n} E(X_r)$$

$$= \frac{n}{n} + \frac{n}{n-1} + \frac{n}{n-2} + \cdots + \frac{n}{1}$$

$$= n \sum_{r=1}^{n} \frac{1}{r}$$

For example, if there are 10 types of coupons, the expected number of trials necessary to obtain at least one of each kind is 29.3.

Finally, we note the following famous approximation:

$$\sum_{r=1}^{n} \frac{1}{r} = \log n + \gamma + \varepsilon_n$$

where log is the natural log or $\log_e$ (unless otherwise specified, log means natural log throughout this text), γ is Euler's Constant, $\gamma = .57 \ldots$, and ε_n approaches zero as n approaches infinity. Using this approximation for $n = 10$, we find that the approximate expected number of trials is 28.8. Generally, we see that the number of trials grows at the rate $n \log n$, or slightly faster than n. $\quad\square$

EXAMPLE C. (Group Testing) Suppose that a large number, n, of blood samples are to be screened for a relatively rare disease. If each sample is assayed individually, n tests will be required. On the other hand, if each sample is divided in half and one of the halves is put into a pool with all the other halves, the pooled lot can be tested. Then, provided that the test method is sensitive enough, if this test is negative, no further assays are necessary and only one test has to be performed. If the test on the pooled blood is positive, each reserved half-sample can be tested individually. In this case, a total of $n + 1$ tests will be required. It is therefore plausible, assuming that the disease is rare, that some savings can be achieved through this pooling procedure.

To analyze this more quantitatively, let us first generalize the scheme and suppose that the n samples are first grouped into m groups of k samples each, or $n = mk$. Each group is then tested; if a group tests positively, each individual in the group is tested. If X_i is the number of tests run on the ith group, the total number of tests run is $N = \sum_{i=1}^{m} X_i$, and the expected total number of tests is

$$E(N) = \sum_{i=1}^{m} E(X_i)$$

Let us find $E(X_i)$. If the probability of a negative on any individual sample is p, then the X_i take on the value 1 with probability p^k or the value $k + 1$ with probability $1 - p^k$. Thus,

$$E(X_i) = p^k + (k + 1)(1 - p^k)$$
$$= k + 1 - kp^k$$

We now have

$$E(N) = m(k + 1) - mkp^k = n\left(1 + \frac{1}{k} - p^k\right)$$

Recalling that n tests are necessary with no pooling, we see that the factor $(1 + 1/k - p^k)$ is the average number of samples used in group testing as a proportion of n. Figure 4-2 shows this proportion as a function of k for $p = .99$. From the figure, we see that for group testing with a group size of about 10, only 20% of the number of tests used with the straightforward method are needed on the average. ☐

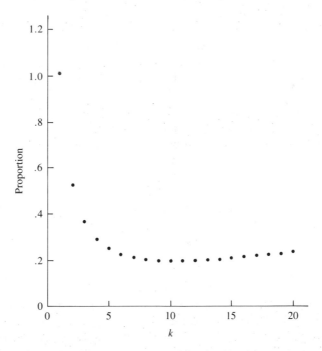

Figure 4-2. The proportion of n in the average number of samples tested using group testing as a function of k.

4.2   Variance and Standard Deviation

The expected value of a random variable is its average value and can be viewed as an indication of the central value of the density or frequency function. The expected value is therefore sometimes referred to as a **location parameter**. The median of a distribution is also a location parameter, one that does not necessarily equal the mean. This section introduces another parameter, the **standard deviation** of a random variable, which is an indication of how dispersed the probability distribution is about its center, of how spread out on the average are the values of the random variable about its expectation. We first define the **variance** of a random variable and then define the standard deviation in terms of the variance.

DEFINITION.   If X is a random variable with expected value $E(X)$, the variance of X is

$$\text{Var}(X) = E\{[X - E(X)]^2\}$$

provided that the expectation exists. The standard deviation of X is the square root of the variance.

The variance is often denoted by σ^2 and the standard deviation by σ. From the above definition, the variance of X is the average value of the squared deviation of X from its mean. If X has units of meters, for example, the variance has units of meters squared, and the standard deviation has units of meters. Although we are often interested ultimately in the standard deviation rather than the variance, it is usually easier to find the variance first and then take the square root.

The variance of a random variable changes in a simple way under linear transformations.

THEOREM A.   If $\text{Var}(X)$ exists and $Y = a + bX$, then $\text{Var}(Y) = b^2 \text{Var}(X)$.

Proof.   Since $E(Y) = a + bE(X)$,

$$
\begin{aligned}
E[(Y - E(Y))^2] &= E\{[bX - bE(X)]^2\} \\
&= E\{b^2[X - E(X)]^2\} \\
&= b^2 E\{[X - E(X)]^2\} \\
&= b^2 \text{Var}(X) \qquad\qquad \square
\end{aligned}
$$

This result seems reasonable once you realize that the addition of a constant does not affect the variance, since the variance is a measure of the spread around a center and the center has merely been shifted.

The standard deviation transforms in a natural way: $\sigma_Y = |b|\sigma_X$. Thus, if the units of measurement are changed from meters to centimeters, for example, the standard deviation is simply multiplied by 100.

EXAMPLE A. (Bernoulli Distribution) If X has a Bernoulli distribution—that is, X takes on values 0 and 1 with probability $1 - p$ and p, respectively—then we have seen (Example A in Section 4.1.2) that $E(X) = p$. By the definition of variance,

$$\text{Var}(X) = (0 - p)^2 \times (1 - p) + (1 - p)^2 \times p$$

$$= p^2 - p^3 + p - 2p^2 + p^3$$

$$= p(1 - p)$$

Note that the expression $p(1 - p)$ is a quadratic with a maximum at $p = \frac{1}{2}$. If p is 0 or 1, the variance is 0, which makes sense since the probability distribution is concentrated at a single point and the random variable is not variable at all. The distribution is most dispersed when $p = \frac{1}{2}$. □

EXAMPLE B. (Normal Distribution) We have seen that $E(X) = \mu$. Then

$$\text{Var}(X) = E[(X - \mu)^2] = \frac{1}{\sigma\sqrt{2\pi}} \int_{-\infty}^{\infty} (x - \mu)^2 e^{-(x-\mu)^2/2\sigma^2} \, dx$$

Making the change of variables $z = (x - \mu)/\sigma$ changes the right-hand side to

$$\frac{\sigma^2}{\sqrt{2\pi}} \int_{-\infty}^{\infty} z^2 e^{-z^2/2} \, dz$$

Finally, making the change of variables $u = z^2/2$ reduces the integral to a gamma function, and we find that $\text{Var}(X) = \sigma^2$. □

The following theorem gives an alternative way of calculating the variance.

THEOREM B. The variance of X, if it exists, may also be calculated as follows:

$$\text{Var}(X) = E(X^2) - [E(X)]^2$$

Proof. Denote $E(X)$ by μ.

$$\text{Var}(X) = E[(X - \mu)^2]$$

$$= E(X^2 - 2\mu X + \mu^2)$$

By the linearity of the expectation, this becomes

$$\text{Var}(X) = E(X^2) - 2\mu E(X) + \mu^2$$
$$= E(X^2) - 2\mu^2 + \mu^2$$
$$= E(X^2) - \mu^2$$

as was to be shown.   □

According to Theorem B, the variance of X can be found in two steps: First find $E(X)$, and then find $E(X^2)$.

EXAMPLE C.   (Uniform Distribution) Let us apply Theorem B to find the variance of a random variable that is uniform on $[0, 1]$. We know that $E(X) = \frac{1}{2}$; next we need to find $E(X^2)$:

$$E(X^2) = \int_0^1 x^2 \, dx = \frac{1}{3}$$

We thus have

$$\text{Var}(X) = \frac{1}{3} - \left(\frac{1}{2}\right)^2 = \frac{1}{12}$$ □

It was stated earlier that the variance or standard deviation of a random variable gives an indication as to how spread out its possible values are. A famous inequality, **Chebyshev's Inequality**, lends a quantitative aspect to this indication.

THEOREM C.   (Chebyshev's Inequality) Let X be a random variable with mean μ and variance σ^2. Then, for any $t > 0$,

$$P(|X - \mu| > t) \leq \frac{\sigma^2}{t^2}$$

Proof.   We will prove this for the continuous case. The discrete case is entirely analogous. Let $R = \{x : |x - \mu| > t\}$. Then

$$P(|X - \mu| > t) = \int_R f(x) \, dx$$

If $x \in R$,

$$\frac{|x - \mu|^2}{t^2} \geq 1$$

Thus,

$$\int_R f(x) \, dx \leq \int_R \frac{(x - \mu)^2}{t^2} f(x) \, dx \leq \int_{-\infty}^{\infty} \frac{(x - \mu)^2}{t^2} f(x) \, dx = \frac{\sigma^2}{t^2} \qquad \square$$

Theorem C says that if σ^2 is very small, there is a high probability that X will not deviate much from μ. For another interpretation, we can set $t = k\sigma$ so that the inequality becomes

$$P(|X - \mu| \geq k\sigma) \leq \frac{1}{k^2}$$

For example, the probability that X is more than 4σ away from μ is less than or equal to $\frac{1}{16}$. These results hold for any random variable with any distribution. In particular cases, the bounds are often much narrower. For example, if X is normally distributed, we find from tables of the normal distribution that $P(|X - \mu| > 2\sigma) = .05$ (compared to $\frac{1}{4}$ obtained from Chebyshev's Inequality).

Chebyshev's Inequality has the following consequence.

COROLLARY A.   If $\text{Var}(X) = 0$ then $P(X = \mu) = 1$.

Proof.   We will give a proof by contradiction. Suppose that $P(X = \mu) < 1$. Then, for some $\varepsilon > 0$, $P(|X - \mu| \geq \varepsilon) > 0$. However, by Chebyshev's Inequality, for any $\varepsilon > 0$,

$$P(|X - \mu| \geq \varepsilon) = 0 \qquad \qquad \square$$

4.2.1   A Model for Measurement Error

Values of physical constants are not precisely known but must be determined by experimental procedures. Such seemingly simple operations as weighting an object, determining a voltage, or measuring an interval of time are actually quite complicated when all the details and possible sources of error are taken into account. The National Bureau of Standards in the United States and similar agencies in other countries are charged with developing and maintaining measurement standards. Such agencies employ probabilists and statisticians as well as physical scientists in this endeavor.

A distinction is usually made between random and systematic measurement errors. A sequence of repeated independent measurements made with no deliberate change in the apparatus or experimental procedure may not yield identical values, and the uncontrollable fluctuations are often modeled as random. At the same time, there may be errors that have the same effect on every measurement; equipment may be slightly out of calibration, for example, or there may be errors associated with the theory underlying the method of measurement. If the true value of the quantity being measured is denoted by x_0 the measurement, X, is modeled as

$$X = x_0 + \beta + \varepsilon$$

where β is the constant, or systematic, error and ε is the random component of the error; ε is a random variable with $E(\varepsilon) = 0$ and $\text{Var}(\varepsilon) = \sigma^2$. We then have

$$E(X) = x_0 + \beta$$

and

$$\text{Var}(X) = \sigma^2$$

β is often called the **bias** of the measurement procedure. The two factors affecting the size of the error are the bias and the size of the variance, σ^2. A perfect measurement would have $\beta = 0$ and $\sigma^2 = 0$.

EXAMPLE A.   (Measurement of the Gravity Constant) This and the next example are taken from an interesting and readable paper by Youden (1972), a statistician at the National Bureau of Standards. Measurement of the acceleration due to gravity at Ottawa was done 32 times with each of two different methods (Preston–Thomas et al., 1960). The results are displayed as histograms in Figure 4-3. There is clearly some systematic difference between the two methods as well as some

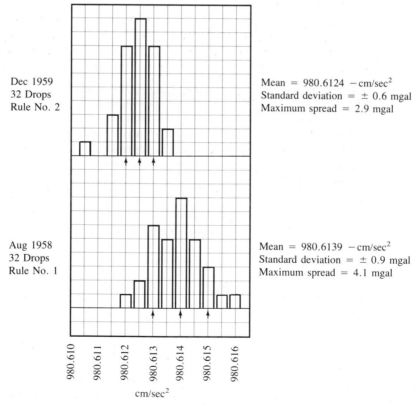

Figure 4-3.   Histograms of two sets of measurements of the acceleration due to gravity.

variation within each method. It appears that the two biases are unequal and that both standard deviations are positive. □

An overall measure of the size of the measurement error that is often used is the **mean squared error**, which is defined as

$$MSE = E[(X - x_0)^2]$$

The mean squared error can be decomposed into contributions from the bias and the variance.

THEOREM A. $MSE = \beta^2 + \sigma^2$.

Proof. From Theorem B of Section 4.2,

$$E[(X - x_0)^2] = \text{Var}(X - x_0) + [E(X - x_0)]^2$$
$$= \text{Var}(X) + \beta^2$$
$$= \sigma^2 + \beta^2 \qquad\qquad □$$

Measurements are often reported in the form 102 ± 1.6, for example. Although it is not always clear what precisely is meant by such notation, 102 is the experimentally determined value and 1.6 is some measure of the error. It is often claimed or hoped that β is negligible relative to σ, and in that case 1.6 represents σ or some multiple of σ. In the graphical presentation of experimentally obtained data, error bars, usually of width σ or some multiple of σ, are placed around measured values. In some cases, efforts are made to bound the magnitude of β, and the bound is incorporated into the error bars in some fashion.

EXAMPLE B. (Measurement of the Speed of Light) Figure 4-4, taken from McNish (1962) and discussed by Youden (1972), shows 24 independent determinations of the speed of light, c, with error bars. The right column of the figure contains codes for the experimental methods used to obtain the measurements; for example, G denotes a method called the geodimeter method. The range of values for c is about 3.5 km/sec, and many of the errors are less than .5 km/sec. Examination of the figure makes it clear that the error bars are too small and that the spread of values cannot be accounted for by different experimental techniques alone—the geodimeter method produced both the smallest and the next to largest value for c. Youden remarks, "Surely the evidence suggests that individual investigators are unable to set realistic limits of error to their reported values." He goes on to suggest that the differences are largely due to calibration errors for equipment. □

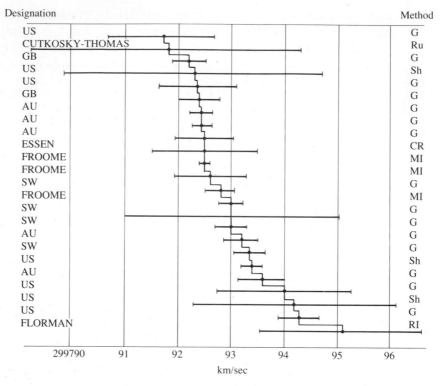

Figure 4-4. A plot of 24 independent determinations of the speed of light with the reported error bars. The investigator or country is listed in the left column, and the experimental method is coded in the right column.

4.3 Covariance and Correlation

The variance of a random variable is a measure of its variability, and the **covariance** of two random variables is a measure of their joint variability, or their degree of association. After defining covariance, we will develop some of its properties and discuss a measure of association called correlation, which is defined in terms of covariance. You may find this material somewhat formal and abstract at first, but as you use them, covariance, correlation, and their properties will begin to seem natural and familiar.

DEFINITION. If X and Y are jointly distributed random variables with expectations μ_X and μ_Y, respectively, the covariance of X and Y is

$$\text{Cov}(X, Y) = E[(X - \mu_X)(Y - \mu_Y)]$$

provided that the expectation exists.

The covariance is the average value of the product of the deviation of X from its mean and the deviation of Y from its mean. If the random variables are positively associated—that is, when X is larger than its mean, Y tends to be larger than its mean as well—the covariance will be positive. If the association is negative—that is, when X is larger than its mean, Y tends to be smaller than its mean—the covariance is negative. These statements will be expanded in the discussion of correlation.

By expanding the product and using the linearity of the expectation, we obtain an alternative expression for the covariance, paralleling Theorem B of Section 4.2:

$$\text{Cov}(X, Y) = E(XY - X\mu_Y - Y\mu_X + \mu_X\mu_Y)$$

$$= E(XY) - E(X)\mu_Y - E(Y)\mu_X + \mu_X\mu_Y$$

$$= E(XY) - E(X)E(Y)$$

In particular, if X and Y are independent, then $E(XY) = E(X)E(Y)$ and $\text{Cov}(X, Y) = 0$ (but the converse is not true).

EXAMPLE A.   Let us return to the bivariate uniform distributions of Example A in Section 3.3. First, note that since the marginal distributions are uniform, $E(X) = E(Y) = \frac{1}{2}$. For the case for which $\alpha = -1$, by ordinary calculus,

$$E(XY) = \int_0^1 \int_0^1 xy(2x + 2y - 4xy)\,dx\,dy = \frac{2}{9}$$

Thus,

$$\text{Cov}(X, Y) = \frac{2}{9} - (\frac{1}{2})\frac{1}{2} = -\frac{1}{36}$$

The covariance is negative, indicating a negative relationship between X and Y. In fact, from Figure 3-4, we see that if X is less than its mean, $\frac{1}{2}$, then Y tends to be larger than its mean, and vice versa. A similar analysis shows that when $\alpha = 1$, $\text{Cov}(X, Y) = \frac{1}{36}$. ◻

We will now develop an expression for the covariance of linear combinations of random variables, proceeding in a number of small steps. First, since $E(a + X) = a + E(X)$,

$$\text{Cov}(a + X, Y) = E\{[a + X - E(a + X)][Y - E(Y)]\}$$

$$= E\{[X - E(X)][Y - E(Y)]\}$$

$$= \text{Cov}(X, Y)$$

Next, since $E(aX) = aE(X)$,

$$\text{Cov}(aX, bY) = E\{[aX - aE(X)][bY - bE(Y)]\}$$
$$= E\{ab[X - E(X)][Y - E(Y)]\}$$
$$= abE\{[X - E(X)][Y - E(Y)]\}$$
$$= ab\,\text{Cov}(X, Y)$$

Next, we consider $\text{Cov}(X, Y + Z)$:

$$\text{Cov}(X, Y + Z) = E([X - E(X)]\{[Y - E(Y)] + [Z - E(Z)]\})$$
$$= E\{[X - E(X)][Y - E(Y)] + [X - E(X)][Z - E(Z)]\}$$
$$= E\{[X - E(X)][Y - E(Y)]\}$$
$$+ E\{[X - E(X)][Z - E(Z)]\}$$
$$= \text{Cov}(X, Y) + \text{Cov}(X, Z)$$

We can now put these results together to find $\text{Cov}(aW + bX, cY + dZ)$:

$$\text{Cov}(aW + bX, cY + dZ) = \text{Cov}(aW + bX, cY) + \text{Cov}(aW + bX, dZ)$$
$$= \text{Cov}(aW, cY) + \text{Cov}(bX, cY) + \text{Cov}(aW, dZ)$$
$$+ \text{Cov}(bX, dZ)$$
$$= ac\,\text{Cov}(W, Y) + bc\,\text{Cov}(X, Y) + ad\,\text{Cov}(W, Z)$$
$$+ bd\,\text{Cov}(X, Z)$$

In general, the same kind of argument gives the important bilinear property of covariance.

THEOREM A. Suppose that $U = a + \sum_{i=1}^{n} b_i X_i$ and $V = c + \sum_{j=1}^{m} d_j Y_j$. Then

$$\text{Cov}(U, V) = \sum_{i=1}^{n} \sum_{j=1}^{m} b_i d_j \text{Cov}(X_i, Y_j)$$

This theorem has many applications. In particular, since $\text{Var}(X) = \text{Cov}(X, X)$,

$$\text{Var}(X + Y) = \text{Cov}(X + Y, X + Y)$$
$$= \text{Var}(X) + \text{Var}(Y) + 2\,\text{Cov}(X, Y)$$

More generally, we have the following result for the variance of a linear combination of random variables.

COROLLARY A. $\text{Var}(a + \sum_{i=1}^{n} b_i X_i) = \sum_{i=1}^{n} \sum_{j=1}^{n} b_i b_j \text{Cov}(X_i, X_j)$.

If the X_i are independent, then $\text{Cov}(X_i, X_j) = 0$, and we have another corollary.

COROLLARY B. $\text{Var}(\sum_{i=1}^{n} X_i) = \sum_{i=1}^{n} \text{Var}(X_i)$, if the X_i are independent.

Corollary B is very useful. Note that $E(\sum X_i) = \sum E(X_i)$ whether or not the X_i are independent, but it is generally *not* the case that $\text{Var}(\sum X_i) = \sum \text{Var}(X_i)$.

EXAMPLE B.   Finding the variance of a binomial random variable from the definition of variance and the frequency function of the binomial distribution is not easy (try it). But expressing a binomial random variable as a sum of independent Bernoulli random variables makes the computation of the variance trivial. Specifically, if Y is a binomial random variable, it can be expressed as $Y = X_1 + X_2 + \cdots + X_n$, where the X_i are independent Bernoulli random variables with $P(X_i = 1) = p$. We saw earlier (Example A in Section 4.2) that $\text{Var}(X_i) = p(1 - p)$, from which it follows from Corollary B that $\text{Var}(Y) = np(1 - p)$.   □

EXAMPLE C.   Suppose that a fair coin is tossed n times. Let X be the proportion of heads in the n tosses. If Y is the number of heads in the n tosses, then Y has a binomial distribution with n trials and probability $\frac{1}{2}$ of success, and $X = Y/n$. From Example B and from Theorem B of Section 4.2, $\text{Var}(X) = 1/4n$. Note that $\text{Var}(X)$ approaches zero as n approaches infinity; in chapter 6, we will see that this implies that X approaches $\frac{1}{2}$.   □

The **correlation coefficient** is defined in terms of the covariance.

DEFINITION.   If X and Y are jointly distributed random variables and the variances and covariances of both X and Y exist and the variances are nonzero, then the correlation of X and Y, denoted by ρ, is

$$\rho = \frac{\text{Cov}(X, Y)}{\sqrt{\text{Var}(X)\,\text{Var}(Y)}}$$

Note that because of the way the ratio is formed, the correlation is a dimensionless quantity (it has no units, such as inches, since the units in the numerator and denominator cancel). From the properties of the variance and covariance that we have established, it follows easily that if X and Y are both subjected to linear transformations (such as changing their units from inches to meters), the correlation coefficient does not change. Since it does not depend on the units of measurement, ρ is in many cases a more useful measure of association than is the covariance.

EXAMPLE D.   Let us return to the bivariate uniform distribution of Example A. Since X and Y are marginally uniform, $\text{Var}(X) = \text{Var}(Y) = \frac{1}{12}$. In the one case, we found

$$\text{Cov}(X, Y) = -\tfrac{1}{36}, \text{ so}$$

$$\rho = -\tfrac{1}{36} \times 12 = -\tfrac{1}{3}$$

In the other case, the covariance was $\tfrac{1}{36}$, so the correlation is $\tfrac{1}{3}$. $\qquad\square$

The following notation and relationship are often useful. The standard deviations of X and Y are denoted by σ_X and σ_Y and their covariance by σ_{XY}. We thus have

$$\rho = \frac{\sigma_{XY}}{\sigma_X \sigma_Y}$$

and

$$\sigma_{XY} = \rho \sigma_X \sigma_Y$$

The following theorem states some further properties of ρ:

THEOREM B. $-1 \le \rho \le 1$. Furthermore, $\rho = \pm 1$ if and only if $P(Y = a + bX) = 1$ for some constants a and b.

Proof. Since the variance of a random variable is nonnegative,

$$0 \le \text{Var}\left(\frac{X}{\sigma_X} + \frac{Y}{\sigma_Y}\right)$$

$$= \text{Var}\left(\frac{X}{\sigma_X}\right) + \text{Var}\left(\frac{Y}{\sigma_Y}\right) + 2\,\text{Cov}\left(\frac{X}{\sigma_X}, \frac{Y}{\sigma_Y}\right)$$

$$= \frac{\text{Var}(X)}{\sigma_X^2} + \frac{\text{Var}(Y)}{\sigma_Y^2} + \frac{2\,\text{Cov}(X, Y)}{\sigma_X \sigma_Y}$$

$$= 2(1 + \rho)$$

From this, we see that $\rho \ge -1$. Similarly,

$$0 \le \text{Var}\left(\frac{X}{\sigma_X^2} - \frac{Y}{\sigma_Y^2}\right) = 2(1 - \rho)$$

implies that $\rho \le 1$. Suppose that $\rho = 1$. Then

$$\text{Var}\left(\frac{X}{\sigma_X^2} - \frac{Y}{\sigma_Y^2}\right) = 0$$

which by Corollary A of Section 4.2 implies that

$$P\left(\frac{X}{\sigma_X} - \frac{Y}{\sigma_Y} = c\right) = 1$$

for some constant, c. This is equivalent to $P(Y = a + bX) = 1$ for some a and b. A similar argument holds for $\rho = -1$. $\qquad\square$

EXAMPLE E.   (Bivariate Normal Distribution) We will show that the covariance of X and Y when they follow a bivariate normal distribution is $\rho\sigma_X\sigma_Y$, which means that ρ is the correlation coefficient. The covariance is

$$\text{Cov}(X, Y) = \int_{-\infty}^{\infty} \int_{-\infty}^{\infty} (x - \mu_X)(y - \mu_Y) f(x, y) \, dx \, dy$$

Making the changes of variables $u = (x - \mu_X)/\sigma_X$ and $v = (y - \mu_Y)/\sigma_Y$ changes the right-hand side to

$$\frac{\sigma_X\sigma_Y}{2\pi\sqrt{1 - \rho^2}} \int_{-\infty}^{\infty} \int_{-\infty}^{\infty} uv \exp\left[-\frac{1}{2(1 - \rho^2)}(u^2 + v^2 - 2\rho uv) \right] du \, dv$$

As in Example D in Section 3.3, we use the technique of completing the square to rewrite this expression as

$$\frac{\sigma_X\sigma_Y}{2\pi\sqrt{1 - \rho^2}} \int_{-\infty}^{\infty} v \exp(-v^2/2) \left(\int_{-\infty}^{\infty} u \exp\left[-\frac{1}{2(1 - \rho^2)}(u - \rho v)^2 \right] du \right) dv$$

The inner integral is the mean of a $N[\rho v, (1 - \rho^2)]$ random variable, missing only the normalizing factor, and we thus have

$$\text{Cov}(X, Y) = \frac{\rho\sigma_X\sigma_Y}{\sqrt{2\pi}} \int_{-\infty}^{\infty} v^2 e^{-v^2/2} \, dv = \rho\sigma_X\sigma_Y$$

as was to be shown. $\qquad\square$

The correlation coefficient ρ measures the strength of the linear relationship between X and Y (compare with Figure 3-7). Correlation also affects the appearance of a scatterplot, which is constructed by generating n independent pairs (X_i, Y_i), where $i = 1, \ldots, n$, and plotting the points. Figure 4-5 shows scatterplots of 100 pairs of pseudorandom bivariate normal random variables for various values of ρ. Note that the clouds of points are roughly elliptical in shape.

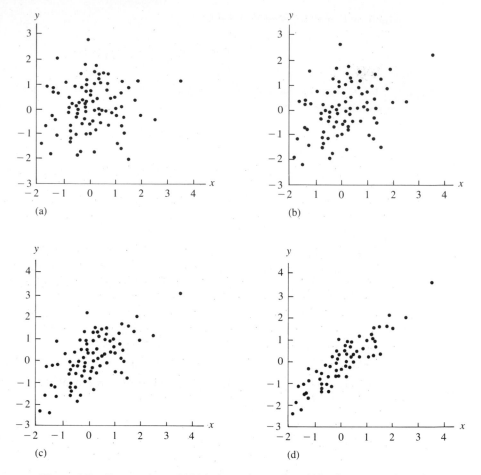

Figure 4-5. Scatterplots of 100 independent pairs of bivariate normal random variables, (a) $\rho = 0$, (b) $\rho = .3$, (c) $\rho = .6$, (d) $\rho = .9$.

4.4   Conditional Expectation and Prediction

4.4.1   Definitions and Examples

In Section 3.5, conditional frequency functions and density functions were defined. We noted that these had the properties of ordinary frequency and density functions. In particular, associated with a conditional distribution is a conditional mean. Suppose that Y and X are discrete random variables and that the conditional frequency function of Y is $p_{Y|X}(y|x)$. The **conditional expectation** of Y given $X = x$ is

$$E(Y|X = x) = \sum_{y} y p_{Y|X}(y|x)$$

For the continuous case, we have

$$E(Y|X = x) = \int y f_{Y|X}(y|x) \, dy$$

More generally, the conditional expectation of a function $h(Y)$ is

$$E[h(Y)|X = x] = \int h(y) f_{Y|X}(y|x) \, dy$$

in the continuous case. A similar equation holds in the discrete case.

EXAMPLE A. Consider a Poisson process on $[0, 1]$ with mean λ, and let N be the number of points in $[0, 1]$. For $p < 1$, let X be the number of points in $[0, p]$. Find the conditional distribution and conditional mean of X given N.

We first find the joint distribution: $P(X = x, N = n)$, which is the probability of x events in $[0, p]$ and $n - x$ events in $[p, 1]$. From the assumption of a Poisson process, the counts in the two intervals are independent Poisson random variables with parameters $p\lambda$ and $(1 - p)\lambda$, so

$$p_{XN}(x, n) = \frac{(p\lambda)^x e^{-p\lambda}}{x!} \frac{[(1 - p)\lambda]^{n-x} e^{-(1-p)\lambda}}{(n - x)!}$$

The marginal distribution of N is Poisson, so the conditional frequency function of X is, after some algebra,

$$p_{X|N}(x|n) = \frac{p_{XN}(x, n)}{p_N(n)}$$

$$= \frac{n!}{x!(n - x)!} p^x (1 - p)^{n-x}$$

This is the binomial distribution with parameters n and p. The conditional expectation is thus np. $\square$

EXAMPLE B. (Bivariate Normal Distribution) From Example C in Section 3.5.2, if Y and X follow a bivariate normal distribution, the conditional density of Y given X is

$$f_{Y|X}(y|x) = \frac{1}{\sigma_Y \sqrt{2\pi(1 - \rho^2)}} \exp\left(-\frac{1}{2} \frac{\left[y - \mu_Y - \rho \frac{\sigma_Y}{\sigma_X}(x - \mu_X) \right]^2}{\sigma_Y^2(1 - \rho^2)} \right)$$

This is a normal density with mean $\mu_Y + \rho(x - \mu_X)\sigma_Y/\sigma_X$ and variance $\sigma_Y^2(1 - \rho^2)$. The former is the conditional mean and the latter the conditional variance of Y given X.

Note that the conditional mean is a linear function of X and that as $|\rho|$ increases, the conditional variance decreases; both of these facts are suggested by the elliptical contours of the joint density. To see this more exactly, consider the case in which $\sigma_X = \sigma_Y = 1$ and $\mu_X = \mu_Y = 0$. The contours then are ellipses satisfying

$$x^2 - 2\rho xy + y^2 = \text{constant}$$

The major and minor axes of such an ellipse are at $45°$ and $135°$. But the conditional expectation of Y given $X = x$ is the line $y = \rho x$; this line does not lie along the major axis of the ellipse. $\qquad\square$

Since the conditional expectation of Y given $X = x$ is defined for every value of x, the conditional expectation is a function of X and hence is a random variable, which we write as $E(Y|X)$. Provided that the appropriate sums or integrals converge, this random variable has an expectation and a variance. Its expectation is $E[E(Y|X)]$; for this expression, note that since $E(Y|X)$ is a random variable that is a function of X, the outer expectation is taken with respect to the distribution of X.

THEOREM A. $E(Y) = E[E(Y|X)]$.

Proof.   We will prove this for the discrete case. The continuous case is proved similarly. We need to show that

$$E(Y) = \sum_x E(Y|X = x)p_X(x)$$

where

$$E(Y|X = x) = \sum_y y p_{Y|X}(y|x)$$

This follows since interchanging the order of summation gives us

$$\sum_x E(Y|X = x)p_X(x) = \sum_y y \sum_x p_{Y|X}(y|x)p_X(x)$$

(it can be shown that this interchange can be made). From the law of total probability, we have

$$p_Y(y) = \sum_x p_{Y|X}(y|x)p_X(x)$$

Therefore,

$$\sum_y y \sum_x p_{Y|X}(y|x)p_X(x) = \sum_y y p_Y(y) = E(Y) \qquad\square$$

Theorem A gives what might be called a law of total expectation: The expectation of a random variable Y can be calculated by weighting the conditional expectations appropriately and summing or integrating.

EXAMPLE C. Suppose that in a system a component and a backup unit both have mean lifetimes equal to μ. If the component fails, the system automatically substitutes the backup unit, but there is probability p that something will go wrong and it will fail to do so. Let T be the total lifetime, and let $X = 1$ if the substitution of the backup takes place successfully, and $X = 0$ if it does not. Then

$$E(T|X = 1) = 2\mu$$

$$E(T|X = 0) = \mu$$

Thus,

$$E(T) = E(T|X = 1)P(X = 1) + E(T|X = 0)P(X = 0) = \mu(2 - p) \qquad \square$$

EXAMPLE D. (Random Sums) Consider sums of the type

$$T = \sum_{i=1}^{N} X_i$$

where N is a random variable with a finite expectation and the X_i are random variables that are independent of N and have the common mean $E(X)$. Such sums arise in a variety of applications. An insurance company might receive N claims in a given period of time, and the amounts of the individual claims might be modeled as random variables $X_1, X_2, \ldots$. The random variable N could denote the number of customers entering a store and X_i the expenditure of the ith customer, or N could denote the number of jobs in a single-server queue and X_i the service time for the ith job. For this last case, T is the time to serve all the jobs in the queue. According to the Theorem A,

$$E(T) = E[E(T|N)]$$

Since $E(T|N = n) = nE(X)$, $E(T|N) = NE(X)$ and thus

$$E(T) = E[NE(X)] = E(N)E(X)$$

This agrees with the intuitive guess that the average time to complete N jobs, where N is random, is the average value of N times the average amount of time to complete a job. $\qquad \square$

We have seen that the expectation of the random variable $E(Y|X)$ is $E(Y)$. We now find its variance.

THEOREM B. $\text{Var}(Y) = \text{Var}[E(Y|X)] + E[\text{Var}(Y|X)]$.

Proof.   We will explain what is meant by the notation in the course of the proof. First,

$$\text{Var}(Y|X = x) = E(Y^2|X = x) - [E(Y|X = x)]^2$$

which is defined for all values of x. Thus, just as we defined $E(Y|X)$ to be a random variable by letting X be random, we can define $\text{Var}(Y|X)$ as a random variable. In particular, $\text{Var}(Y|X)$ has the expectation $E[\text{Var}(Y|X)]$. Since

$$\text{Var}(Y|X) = E(Y^2|X) - [E(Y|X)]^2$$

$$E[\text{Var}(Y|X)] = E[E(Y^2|X)] - E\{[E(Y|X)]^2\}$$

Also,

$$\text{Var}[E(Y|X)] = E\{[E(Y|X)]^2\} - \{E[E(Y|X)]\}^2$$

The final piece that we need is

$$\text{Var}(Y) = E(Y^2) - [E(Y)]^2$$

$$= E[E(Y^2|X)] - \{E[E(Y|X)]\}^2$$

by the law of total expectation. Now we can put all the pieces together:

$$\text{Var}(Y) = E[E(Y^2|X)] - \{E[E(Y|X)]\}^2$$

$$= E[E(Y^2|X)] - E\{[E(Y|X)]^2\} + E\{[E(Y|X)]^2\} - \{E[E(Y|X)]\}^2$$

$$= E[\text{Var}(Y|X)] + \text{Var}[E(Y|X)] \qquad \square$$

EXAMPLE E.   (Random Sums) Let us continue Example D, but with the additional assumptions that the X_i are independent random variables with the same mean, $E(X)$, and the same variance, $\text{Var}(X)$, and that $\text{Var}(N) < \infty$. According to Theorem B,

$$\text{Var}(T) = E[\text{Var}(T|N)] + \text{Var}[E(T|N)]$$

Since $E(T|N) = NE(X)$,

$$\text{Var}[E(T|N)] = [E(X)]^2 \text{Var}(N)$$

Also since $\text{Var}(T|N = n) = \text{Var}(\sum_{i=1}^n X_i) = n \text{Var}(X)$,

$$\text{Var}(T|N) = N \text{Var}(X)$$

and

$$E[\text{Var}(T|N)] = E(N)\,\text{Var}(X)$$

We thus have

$$\text{Var}(T) = [E(X)]^2\,\text{Var}(N) + E(N)\,\text{Var}(X)$$

If N is fixed, say, $N = n$, then $\text{Var}(T) = n\,\text{Var}(X)$. Thus, we see from the equation just above that extra variability occurs in T because N is random.   □

4.4.2 Prediction

This section treats the problem of predicting the value of one random variable from another. We might wish, for example, to measure the value of some physical quantity, such as pressure, using an instrument. The actual pressures to be measured are unknown and variable, so we might model them as values of a random variable, Y. Assume that measurements are to be taken by some instrument that produces a response, X, related to Y in some fashion but corrupted by random noise as well; X might represent current flow, for example. Y and X have some joint distribution, and we wish to predict the actual pressure, Y, from the instrument response, X.

As another example, in forestry, the volume of a tree (in board-feet) is sometimes estimated from its diameter. For a whole forest, it is reasonable to model diameter (X) and volume (Y) as random variables with some joint distribution, and then the only problem is that of predicting Y from X.

Let us first consider a relatively trivial situation: the problem of predicting Y by means of a constant value, c. If we wish to choose the "best" value of c, we need some measure of the effectiveness of a prediction. One that is amenable to mathematical analysis and that is widely used is the mean squared error:

$$\text{MSE} = E[(Y - c)^2]$$

This is the average squared error of prediction, the averaging being done with respect to the distribution of Y. The problem then becomes finding the value of c that minimizes the mean squared error. To solve this problem, we denote $E(Y)$ by μ (see Theorem A of Section 4.2.1) and observe that

$$E[(Y - c)^2] = \text{Var}(Y - c) + [E(Y - c)]^2$$
$$= \text{Var}(Y) + (\mu - c)^2$$

The first term of the last expression does not depend on c, and the second term is minimized for $c = \mu$, which is the optimal choice of c.

Now let us consider predicting Y by some function $h(X)$ in order to minimize $\text{MSE} = E\{[Y - h(X)]^2\}$. From Theorem A of Section 4.4.1, the right-hand side can be expressed as

$$E\{[Y - h(X)]^2\} = E(E\{[Y - h(X)]^2|X\})$$

For every x, the inner expectation is minimized by setting $h(x)$ equal to the constant $E(Y|X = x)$, from the result of the preceding paragraph. We thus have that the minimizing function $h(X)$ is

$$h(X) = E(Y|X)$$

EXAMPLE A.   For the bivariate normal distribution, we found that

$$E(Y|X) = \mu_Y + \rho\frac{\sigma_Y}{\sigma_X}(X - \mu_X)$$

This linear function of X is thus the minimum mean squared error predictor of Y from X.   □

A practical limitation of the optimal prediction scheme is that its implementation depends on knowing the joint distribution of Y and X in order to find $E(Y|X)$, and often this information is not available, not even approximately. For this reason, we can try to attain the more modest goal of finding the optimal *linear* predictor of Y. (In Example A, it turned out that the best predictor was linear, but this is not generally the case.) That is, rather than finding the best function h among all functions, we try to find the best function of the form $h(x) = \alpha + \beta x$. This merely requires optimizing over the two parameters α and β. Now

$$E[(Y - \alpha - \beta X)^2] = \mathrm{Var}(Y - \alpha - \beta X) + [E(Y - \alpha - \beta X)]^2$$

$$= \mathrm{Var}(Y - \beta X) + [E(Y - \alpha - \beta X)]^2$$

The first term of the last expression does not depend on α, so α can be chosen so as to minimize the second term. To do this, note that

$$E(Y - \alpha - \beta X) = \mu_Y - \alpha - \beta\mu_X$$

and that the right-hand side is zero, and hence its square is minimized, if

$$\alpha = \mu_Y - \beta\mu_X$$

As for the first term,

$$\mathrm{Var}(Y - \beta X) = \sigma_Y^2 + \beta^2\sigma_X^2 - 2\beta\sigma_{XY}$$

where $\sigma_{XY} = \text{Cov}(X, Y)$. This is a quadratic function of β, and the minimum is found by setting the derivative with respect to β equal to zero, which yields

$$\beta = \frac{\sigma_{XY}}{\sigma_X^2}$$

With this choice of α and β, the mean squared error predictor becomes

$$\text{Var}(Y - \beta X) = \sigma_Y^2 + \frac{\sigma_{XY}^2}{\sigma_X^4}\sigma_X^2 - 2\frac{\sigma_{XY}}{\sigma_X^2}\sigma_{XY}$$

$$= \sigma_Y^2 - \frac{\sigma_{XY}^2}{\sigma_X^2}$$

$$= \sigma_Y^2 - \rho^2\sigma_Y^2$$

$$= \sigma_Y^2(1 - \rho^2)$$

Note that the optimal linear predictor depends on the joint distribution of X and Y only through their means, variances, and covariance. Thus, in practice, it is generally easier to construct the optimal linear predictor or an approximation to it than to construct the general optimal predictor $E(Y|X)$. Second, note that the form of the optimal linear predictor is the same as that of $E(Y|X)$ for the bivariate normal distribution. Third, note that the mean squared error predictor depends only on σ_Y and ρ and that it is small if ρ is close to $+1$ or -1. Here we see again, from a different point of view, that the correlation coefficient is a measure of the strength of the linear relationship between X and Y.

4.5   The Moment-Generating Function

This section develops and applies some of the properties of the moment-generating fuction. It turns out, despite its unlikely appearance, to be a very useful tool that can dramatically simplify certain calculations.

The **moment-generating function (mgf)** of a random variable X is $M(t) = E(e^{tX})$ if the expectation is defined. In the discrete case,

$$M(t) = \sum_x e^{tx}p(x)$$

and in the continuous case,

$$M(t) = \int_{-\infty}^{\infty} e^{tx}f(x)\,dx$$

The expectation, and hence the moment-generating function, may or may not exist for any particular value of t. In the continuous case, the existence of the expectation depends on how rapidly the tails of the density decrease; for example, since the tails of the Cauchy density die down at the rate x^{-2}, the expectation does not exist for any t and the moment-generating function is undefined. Since the tails of the normal density die down at the rate e^{-t^2}, the integral converges for all t.

PROPERTY A. If the moment-generating function exists for t in an open interval containing zero, it uniquely determines the probability distribution.

We cannot prove this important property here—its proof depends on properties of the Laplace transform. Note that Property A says that if two random variables have the same mgf in an open interval containing zero, they have the same distribution. For some problems, we can find the mgf and then deduce the unique probability distribution corresponding to it.

The rth moment of a random variable is $E(X^r)$ if the expectation exists. We have already encountered the first and second moments earlier in this chapter, that is, $E(X)$ and $E(X^2)$. Central moments rather than ordinary moments are often used: The rth central moment is $E\{[X - E(X)]^r\}$. The variance is the second central moment and is a measure of dispersion about the mean. The third central moment, called the **skewness**, is used as a measure of the asymmetry of a density or a frequency function about its mean; if a density is symmetric about its mean, the skewness is zero (see Problem 62 at the end of this chapter). As its name implies, the moment-generating function has something to do with moments. To see this, consider the continuous case:

$$M(t) = \int_{-\infty}^{\infty} e^{tx} f(x)\, dx$$

The derivative of $M(t)$ is

$$M'(t) = \frac{d}{dt} \int_{-\infty}^{\infty} e^{tx} f(x)\, dx$$

It can be shown that differentiation and integration can be interchanged, so that

$$M'(t) = \int_{-\infty}^{\infty} x e^{tx} f(x)\, dx$$

and

$$M'(0) = \int_{-\infty}^{\infty} x f(x)\, dx = E(X)$$

Differentiating r times, we find

$$M^{(r)}(0) = E(X^r)$$

It can further be argued that if the moment-generating function exists in an interval containing zero, then so do all the moments. We thus have the following property.

PROPERTY B.   If the moment-generating function exists in an open interval containing zero, then $M^{(r)}(0) = E(X^r)$.

In order to find the moments of a random variable from the definition of expectation, we must sum a series or carry out an integration. The utility of Property B is that, if the mgf can be found, the process of integration or summation, which may be difficult, can be replaced by the process of differentiation, which is mechanical. We can illustrate these concepts using some familiar distributions.

EXAMPLE A.   (Poisson Distribution) By definition,

$$M(t) = \sum_{k=0}^{\infty} e^{tk} \frac{\lambda^k}{k!} e^{-\lambda}$$

$$= \sum_{k=0}^{\infty} \frac{(\lambda e^t)^k}{k!} e^{-\lambda}$$

$$= e^{-\lambda} e^{\lambda e^t}$$

$$= e^{\lambda(e^t - 1)}$$

The sum converges for all t. Differentiating, we have

$$M'(t) = \lambda e^t e^{\lambda(e^t - 1)}$$

$$M''(t) = \lambda e^t e^{\lambda(e^t - 1)} + \lambda^2 e^{2t} e^{\lambda(e^t - 1)}$$

Evaluating these derivatives at $t = 0$, we find

$$E(X) = \lambda$$

$$E(X^2) = \lambda^2 + \lambda$$

from which it follows that

$$\text{Var}(X) = E(X^2) - [E(X)]^2 = \lambda$$

We have found that the mean and the variance of a Poisson distribution are equal.   □

EXAMPLE B.   (Gamma Distribution) The mgf of a gamma distribution is

$$M(t) = \int_0^\infty e^{tx} \frac{\lambda^\alpha}{\Gamma(\alpha)} x^{\alpha-1} e^{-\lambda x} \, dx$$

$$= \frac{\lambda^\alpha}{\Gamma(\alpha)} \int_0^\infty x^{\alpha-1} e^{x(t-\lambda)} \, dx$$

The latter integral converges for $t < \lambda$ and can be evaluated by relating it to the gamma density having parameters α and $\lambda - t$. We thus obtain

$$M(t) = \frac{\lambda^\alpha}{\Gamma(\alpha)} \left(\frac{\Gamma(\alpha)}{(\lambda - t)^\alpha} \right) = \left(\frac{\lambda}{\lambda - t} \right)^\alpha$$

Differentiating, we find

$$M'(0) = E(X) = \frac{\alpha}{\lambda}$$

$$M''(0) = E(X^2) = \frac{\alpha(\alpha + 1)}{\lambda^2}$$

From these equations, we find that

$$\text{Var}(X) = E(X^2) - [E(X)]^2$$

$$= \frac{\alpha(\alpha + 1)}{\lambda^2} - \frac{\alpha^2}{\lambda^2}$$

$$= \frac{\alpha}{\lambda^2} \qquad\qquad \square$$

EXAMPLE C.   (Standard Normal Distribution) For the standard normal distribution, we have

$$M(t) = \frac{1}{\sqrt{2\pi}} \int_{-\infty}^\infty e^{tx} e^{-x^2/2} \, dx$$

The integral converges for all t and can be evaluated using the technique of completing the square. Since

$$\frac{x^2}{2} - tx = \frac{1}{2}(x^2 - 2tx + t^2) - \frac{t^2}{2}$$

$$= \frac{1}{2}(x - t)^2 - \frac{t^2}{2}$$

therefore

$$M(t) = \frac{e^{t^2/2}}{\sqrt{2\pi}} \int_{-\infty}^{\infty} e^{-(x-t)^2/2} \, dx$$

Making the change of variables $u = x - t$ and using the fact that the standard normal density integrates to 1, we find that

$$M(t) = e^{t^2/2}$$

From this result, we easily see that $E(X) = 0$ and $\text{Var}(X) = 1$.   □

Let us continue with the development of the properties of the moment-generating function.

PROPERTY C.   If X has the mgf $M_X(t)$ and $Y = a + bX$, then Y has the mgf $M_Y(t) = e^{at}M_X(bt)$.

Proof.

$$M_Y(t) = E(e^{tY})$$
$$= E(e^{at+btX})$$
$$= E(e^{at}e^{btX})$$
$$= e^{at}E(e^{btX})$$
$$= e^{at}M_X(bt)$$ □

EXAMPLE D.   (General Normal Distribution) If Y follows a general normal distribution with parameters μ and σ, the distribution of Y is the same as that of $\mu + \sigma X$, where X follows a standard normal distribution. Thus, from Example C and Property C,

$$M_Y(t) = e^{\mu t}M_X(\sigma t) = e^{\mu t}e^{\sigma^2 t^2/2}$$ □

PROPERTY D.   If X and Y are independent random variables with mgf's M_X and M_Y and $Z = X + Y$, then $M_Z(t) = M_X(t)M_Y(t)$ on the common interval where both mgf's exist.

Proof.

$$M_Z(t) = E(e^{tZ})$$
$$= E(e^{tX+tY})$$
$$= E(e^{tX}e^{tY})$$

From the assumption of independence,

$$M_Z(t) = E(e^{tX})E(e^{tY})$$

$$= M_X(t)M_Y(t) \qquad \square$$

By induction, Property D can be extended to sums of several independent random variables. This is one of the most useful properties of the moment-generating function. The next three examples show how it can be used to easily derive results that would take a lot more work to achieve without recourse to the mgf.

EXAMPLE E.   The sum of independent Poisson random variables is a Poisson random variable: If X is Poisson with λ and Y is Poisson with μ, then $X + Y$ is Poisson with $\lambda + \mu$, since

$$e^{\lambda(e^t-1)}e^{\mu(e^t-1)} = e^{(\lambda+\mu)(e^t-1)} \qquad \square$$

EXAMPLE F.   If X follows a gamma distribution with parameters α_1 and λ and Y follows a gamma distribution with parameters α_2 and λ, the mgf of $X + Y$ is

$$\left(\frac{\lambda}{\lambda - t}\right)^{\alpha_1}\left(\frac{\lambda}{\lambda - t}\right)^{\alpha_2} = \left(\frac{\lambda}{\lambda - t}\right)^{\alpha_1+\alpha_2}$$

where $t < \lambda$. The right-hand expression is the mgf of a gamma distribution with parameters λ and $\alpha_1 + \alpha_2$. It follows from this that the sum of n independent exponential random variables with parameter λ follows a gamma distribution with parameters n and λ. Thus, the time between n consecutive events of a Poisson process in time follows a gamma distribution. Assuming that the service times in a queue are independent exponential random variables, the length of time to serve n customers follows a gamma distribution. $\qquad \square$

EXAMPLE G.   If $X \sim N(\mu, \sigma^2)$ and, independent of X, $Y \sim N(\nu, \tau^2)$, then the mgf of $X + Y$ is

$$e^{\mu t}e^{t^2\sigma^2/2}e^{\nu t}e^{t^2\tau^2/2} = e^{(\mu+\nu)t}e^{t^2(\sigma^2+\tau^2)/2}$$

which is the mgf of a normal distribution with mean $\mu + \nu$ and variance $\sigma^2 + \tau^2$. The sum of independent normal random variables is thus normal. $\qquad \square$

The preceding three examples are atypical. In general, if two independent random variables follow some type of distribution, it is not necessarily true that their sum follows the same type of distribution. For example, the sum of two gamma random variables having different values for the parameter λ does not follow a gamma distribution, as can be easily seen from the mgf.

We now apply moment-generating functions to random sums of the type introduced in Section 4.4.1. Suppose that

$$S = \sum_{i=1}^{N} X_i$$

where the X_i are independent and have the same mgf, M_X, and where N has the mgf M_N and is independent of the X_i. By conditioning, we have

$$M_S(t) = E[E(e^{tS}|N)]$$

Given $N = n$, $M_S(t) = [M_X(t)]^n$ from Property D. We thus have

$$M_S(t) = E[M_X(t)^N]$$
$$= E(e^{N \log M_X(t)})$$
$$= M_N[\log M_X(t)]$$

(We must carefully note the values of t for which this is defined.)

EXAMPLE H. (Compound Poisson Distribution) This example presents a model that occurs for certain chain reactions, or "cascade" processes. When a single primary electron, having been accelerated in an electrical field, hits a plate, several secondary electrons are produced. In a multistage multiplying tube, each of these secondary electrons hits another plate and thereby produces a number of tertiary electrons. The process can continue through several stages in this manner. Woodward (1948) considered models of this type in which the number of electrons produced by the impact of a single electron on the plate is random and, in particular, in which the number of secondary electrons follows a Poisson distribution. The number of electrons produced at the third stage is described by a random sum of the type described just above, where N is the number of secondary electrons and X_i is the number of electrons produced by the ith secondary electron. Suppose that the X_i are independent Poisson random variables with parameter λ and that N is a Poisson random variable with parameter μ. According to the result above, the mgf of S, the total number of particles, is

$$M_S(t) = \exp[\mu(e^{\lambda(e^t-1)} - 1)] \qquad \square$$

If X and Y have a joint distribution, their joint moment-generating function is defined as

$$M_{XY}(s, t) = E(e^{sX+tY})$$

which is a function of two variables, s and t. If the joint mgf is defined on an open set containing the origin, it uniquely determines the joint distribution. The mgf of the marginal distribution of X alone is

$$M_X(s) = M_{XY}(s, 0)$$

and similarly for Y. It can be shown that X and Y are independent if and only if their joint mgf factors into the product of the mgf's of the marginal distributions. $E(XY)$ and other higher-order joint moments can be obtained from the joint mgf. Analogous properties hold for the joint mgf of several random variables.

The major limitation of the mgf is that it may not exist. The **characteristic function** of a random variable X is defined to be

$$\phi(t) = E(e^{itX})$$

where $i = \sqrt{-1}$. In the continuous case,

$$\phi(t) = \int_{-\infty}^{\infty} e^{itx} f(x)\, dx$$

This integral converges for all values of t, since $|e^{itx}| \leq 1$. The characteristic function is thus defined for all distributions. Its properties are similar to those of the mgf: Moments can be obtained by differentiation, the characteristic function changes simply under linear transformations, and the characteristic function of a sum of independent random variables is the product of their characteristic functions. But applying the characteristic function to problem solving requires some familiarity with the techniques of complex variables.

4.6 Approximate Methods

In many applications, only the first two moments of a random variable, and not the entire probability distribution, are known, and even these may only be known approximately. We will see in chapter 5 that repeated independent observations of a random variable allow reliable estimates to be made of its mean and variance. Suppose that we know the expectation and the variance of a random variable X, but not the entire distribution and that we are interested in the mean and variance of $Y = g(X)$ for some fixed function g. For example, we might be able to measure X and determine its mean and variance, but really be interested in Y, which is related to X in a known way. We might wish to know $\mathrm{Var}(Y)$, at least approximately, in order to assess the accuracy of the indirect measurement process. From the results given in this chapter, we cannot in general find $E(Y) = \mu_Y$ and $\mathrm{Var}(Y) = \sigma_Y^2$ from $E(X) = \mu_X$ and $\mathrm{Var}(X) = \sigma_X^2$, unless the function g is linear. However, if g is nearly linear in a range in which X has high probability, it can be approximated by a linear function and approximate moments of Y can be found.

In proceeding as just described, we follow a tack often taken in applied mathematics: When confronted with a nonlinear problem that we cannot solve, we linearize. In probability and statistics, this method is called **propagation of error**, or the **δ method**. Linearization is carried out through a Taylor Series expansion of g about μ_X. To the first order,

$$Y = g(X) \approx g(\mu_X) + (X - \mu_X)g'(\mu_X)$$

We have expressed Y as approximately equal to a linear function of X. Since we know how to find expectations and variances of linear functions, we have

$$\mu_Y \approx g(\mu_X)$$

$$\sigma_Y^2 \approx \sigma_X^2[g'(\mu_X)]^2$$

We know that in general $E(Y) \neq g(E(X))$, as given by the approximation. In fact, we can carry out the Taylor Series expansion to the second order to get an improved approximation of μ_Y:

$$Y = g(X) \approx g(\mu_X) + (X - \mu_X)g'(\mu_X) + \tfrac{1}{2}(X - \mu_X)^2 g''(\mu_X)$$

Taking the expectation of the right-hand side, we have, since $E(X - \mu_X) = 0$,

$$E(Y) \approx g(\mu_X) + \tfrac{1}{2}\sigma_X^2 g''(\mu_X)$$

How good such approximations are depends on how nonlinear g is in a neighborhood of μ_X and on the size of σ_X. From Chebyshev's Inequality, we know that X is unlikely to be many standard deviations away from μ_X; if g can be reasonably well approximated in this range by a linear function, the approximations for the moments will be reasonable as well.

EXAMPLE A.   The relation of voltage, current, and resistance is $V = IR$. Suppose that the voltage is held constant at a value V_0 across a medium whose resistance fluctuates randomly as a result, say, of random fluctuations at the molecular level. The current therefore also varies randomly. Suppose that it can be determined experimentally to have mean $\mu_I \neq 0$ and variance and σ_I^2. We wish to find the mean and variance of the resistance, R, and since we do not know the distribution of I, we must resort to an approximation. We have

$$R = g(I) = \frac{V_0}{I}$$

$$g'(\mu_I) = -\frac{V_0}{\mu_I^2}$$

$$g''(\mu_I) = \frac{2V_0}{\mu_I^3}$$

Thus,

$$\mu_R \approx \frac{V_0}{\mu_I} + \frac{V_0}{\mu_I^3}\sigma_I^2$$

$$\sigma_R^2 \approx \frac{V_0^2}{\mu_I^4}\sigma_I^2$$

We see that the variability of R depends on both the mean level of I and the variance of I. This makes sense, since if I is quite small, small variations in I will result in large variations in $R = V_0/I$, whereas if I is large, small variations will not affect R as much. The second-order correction factor for μ_R also depends on μ_I and is large if μ_I is small. In fact, when I is near zero, the function $g(I) = V_0/I$ is quite nonlinear, and the linearization is not a good approximation. □

EXAMPLE B. This example examines the accuracy of the approximations using a simple test case. We choose the function $g(x) = \sqrt{x}$ and consider two cases: X uniform on $[0, 1]$, and X uniform on $[1, 2]$. The graph of $g(x)$ in Figure 4-6 shows that g is more nearly linear in the latter case, so we would expect the approximations to work better there.

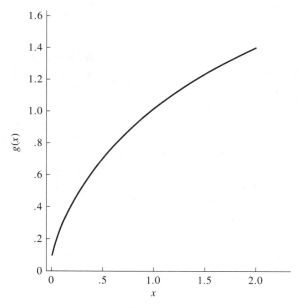

Figure 4-6. The function $g(x) = \sqrt{x}$ is more nearly linear over the interval $[1, 2]$ than over the interval $[0, 1]$.

First, if X is uniform on $[0, 1]$, we find that the density function of $Y = \sqrt{X}$ is $f_Y(y) = 2y$, for $0 \le y \le 1$. We are therefore able to compute the exact mean and variance of Y and compare them to the approximations. By simple calculation, we find that $E(Y) = .667$, $\text{Var}(Y) = .055$, and $\sigma_Y = .236$. Using the approximation method, we first calculate

$$g'(x) = \tfrac{1}{2}x^{-1/2}$$

$$g''(x) = -\tfrac{1}{4}x^{-3/2}$$

Since X is uniform on $[0, 1]$, $\mu_X = \tfrac{1}{2}$, and evaluating the derivatives at this value gives us

$$g'(\mu_X) = \frac{\sqrt{2}}{2}$$

$$g''(\mu_X) = -\frac{\sqrt{2}}{2}$$

We know that $\text{Var}(X) = \tfrac{1}{12}$ for a random variable uniform on $[0, 1]$, so the approximations are

$$E(Y) \approx \sqrt{\frac{1}{2}} - \frac{1}{2}\left(\frac{\sqrt{2}}{12 \times 2}\right) = .678$$

$$\text{Var}(Y) \approx \tfrac{1}{2} \times \tfrac{1}{12} = .042$$

$$\sigma_Y \approx .204$$

The approximation to the mean is .678, and compared to the actual value of .667, it is off by about 1.6%. The approximation to the standard deviation is .204, and compared to the actual value of .236, it is off by 13%.

Now let us consider the case in which X is uniform on $[1, 2]$. We find that $Y = \sqrt{X}$ has the density function $f_Y(y) = 2y$, for $1 \le y \le \sqrt{2}$, from which the mean of Y can be calculated to be 1.219. The variance and standard deviation are .0142 and .119, respectively. To compare these to the approximations we note that $\mu_X = \tfrac{3}{2}$ and $\text{Var}(X) = \tfrac{1}{12}$ (the random variable uniform on $[1, 2]$ can be obtained by adding the constant 1 to a random variable uniform on $[0, 1]$; compare with Corollary A in Section 4.2). We find

$$g'(\mu_X) = .408$$

$$g''(\mu_X) = -.136$$

so the approximations are

$$E(Y) \approx \sqrt{\frac{3}{2}} - \frac{1}{2}\left(\frac{.136}{12}\right) = 1.219$$

$$\text{Var}(Y) \approx \frac{.408^2}{12} = .0138$$

$$\sigma_Y \approx .118$$

These values are much closer to the actual values than are the approximations for the first case.　□

Suppose that we have $Z = g(X, Y)$, a function of two variables. We can again carry out Taylor Series expansions to approximate the mean and variance of Z. To the first order, letting μ denote the point (μ_X, μ_Y),

$$Z = g(X, Y) \approx g(\mu) + (X - \mu_X)\frac{\partial g(\mu)}{\partial x} + (Y - \mu_Y)\frac{\partial g(\mu)}{\partial y}$$

The notation $\partial g(\mu)/\partial x$ means that the derivative is evaluated at the point μ. Here Z is expressed as approximately equal to a linear function of X and Y, and the mean and variance of this linear function are easily calculated to be

$$E(Z) \approx g(\mu)$$

and

$$\text{Var}(Z) \approx \sigma_X^2\left(\frac{\partial g(\mu)}{\partial x}\right)^2 + \sigma_Y^2\left(\frac{\partial g(\mu)}{\partial y}\right)^2 + 2\sigma_{XY}\left(\frac{\partial g(\mu)}{\partial x}\right)\left(\frac{\partial g(\mu)}{\partial y}\right)$$

(For the latter calculation, see Corollary A in Section 4.3.) As is the case with a single variable, a second-order expansion can be used to obtain an improved estimate of $E(Z)$:

$$Z = g(X, Y) \approx g(\mu) + (X - \mu_X)\frac{\partial g(\mu)}{\partial x} + (Y - \mu_Y)\frac{\partial g(\mu)}{\partial y}$$

$$+ \frac{1}{2}(X - \mu_X)^2\frac{\partial^2 g(\mu)}{\partial x^2} + \frac{1}{2}(Y - \mu_Y)^2\frac{\partial^2 g(\mu)}{\partial y^2}$$

$$+ (X - \mu_X)(Y - \mu_Y)\frac{\partial^2 g(\mu)}{\partial x \partial y}$$

Taking expectations term by term on the right-hand side yields

$$E(Z) \approx g(\mu) + \frac{1}{2}\sigma_X^2\frac{\partial^2 g(\mu)}{\partial x^2} + \frac{1}{2}\sigma_Y^2\frac{\partial^2 g(\mu)}{\partial y^2} + \sigma_{XY}\frac{\partial^2 g(\mu)}{\partial x \partial y}$$

The general case of a function of n variables can be worked out similarly; the basic concepts are illustrated by the two-variable case.

EXAMPLE C.　(Expectation and Variance of a Ratio) Let us consider the case where $Z = Y/X$, which arises frequently in practice. For example, a chemist might measure the concentrations of two substances, both with some measurement error that is indicated by their standard deviations, and then report the relative concentrations in the form of a ratio. What is the approximate standard deviation of the ratio, Z?

Using the method of propagation of error derived above, for $g(x, y) = y/x$, we have

$$\frac{\partial g}{\partial x} = \frac{-y}{x^2} \qquad \frac{\partial g}{\partial y} = \frac{1}{x}$$

$$\frac{\partial^2 g}{\partial x^2} = \frac{2y}{x^3} \qquad \frac{\partial^2 g}{\partial y^2} = 0$$

$$\frac{\partial^2 g}{\partial x \partial y} = \frac{-1}{x^2}$$

Evaluating these derivatives at (μ_X, μ_Y) and using the result above, we find, if $\mu_X \neq 0$,

$$E(Z) \approx \frac{\mu_Y}{\mu_X} + \sigma_X^2 \frac{\mu_Y}{\mu_X^3} - \frac{\sigma_{XY}}{\mu_X^2}$$

$$= \frac{\mu_Y}{\mu_X} + \frac{1}{\mu_X^2}\left(\sigma_X^2 \frac{\mu_Y}{\mu_X} - \rho\sigma_X\sigma_Y\right)$$

From this equation, we see that the difference between $E(Z)$ and μ_Y/μ_X depends on several factors. If σ_X and σ_Y are small—that is, if the two concentrations are measured quite accurately—the difference is small. If μ_X is small, the difference is relatively large. Finally, correlation between X and Y affects the difference.

We now consider the variance. Again using the result above and evaluating the partial derivatives at (μ_X, μ_Y), we find

$$\text{Var}(Z) \approx \sigma_X^2 \frac{\mu_Y^2}{\mu_X^4} + \frac{\sigma_Y^2}{\mu_X^2} - 2\sigma_{XY}\frac{\mu_Y}{\mu_X^3}$$

$$= \frac{1}{\mu_X^2}\left(\sigma_X^2 \frac{\mu_Y^2}{\mu_X^2} + \sigma_Y^2 - 2\rho\sigma_X\sigma_Y \frac{\mu_Y}{\mu_X}\right)$$

From this equation, we see that the ratio is quite variable when μ_X is small, paralleling the results in Example A, and that correlation between X and Y, if of the same sign as μ_Y/μ_X, decreases $\text{Var}(Z)$.   □

4.7 Problems

1. Show that if a random variable is bounded—that is, $|X| < M < \infty$—then $E(X)$ exists.

2. If X is a discrete uniform random variable—that is, $P(X = k) = 1/n$ for $k = 1, 2, \ldots, n$—find $E(X)$ and $\text{Var}(X)$.

3. Find $E(X)$ and $\text{Var}(X)$ for Problem 3 in chapter 2.

4. Let X have the cdf $F(x) = 1 - x^{-\alpha}, x \geq 1$.
   (a) Find $E(X)$ for those values of α for which $E(X)$ exists.
   (b) Find $\text{Var}(X)$ for those values of α for which it exists.

5. Let X have the density

$$f(x) = \frac{1 + \alpha x}{2}, \quad -1 \leq x \leq 1, \quad -1 \leq \alpha \leq 1$$

   Find $E(X)$ and $\text{Var}(X)$.

6. Show that if X is a discrete random variable, taking values on the positive integers, then $E(X) = \sum_{k=1}^{n} P(X > k)$. Apply this result to find the expected value of a geometric random variable.

7. This is a simplified inventory problem. Suppose that it costs c dollars to stock an item, and that the item sells for s dollars. Suppose that the number of items that will be asked for by customers is a random variable with the frequency function $p(k)$. Find a rule for the number of items that should be stocked in order to maximize the expected income. (*Hint:* Consider the difference of successive terms.)

8. A list of n items is arranged in random order; to find a requested item, they are searched sequentially until the desired item is found. What is the expected number of items that must be searched through, assuming that each item is equally likely to be the one requested? (Questions of this nature arise in the design of computer algorithms.)

9. Referring to Problem 8, suppose that the items are not equally likely to be requested but have known probabilities $p_1, p_2, \ldots, p_n$. Suggest an alternative searching procedure that will decrease the average number of items that must be searched through, and show that in fact it does so.

10. If X is a continuous random variable with a density that is symmetric about some point, ξ, show that $E(X) = \xi$, provided that $E(X)$ exists.

11. If X is a nonnegative continuous random variable, show that

$$E(X) = \int_0^\infty [1 - F(x)]\, dx$$

   Apply this result to find the mean of the exponential distribution.

12. Let X be a continuous random variable with the density function

$$f(x) = 2x, \quad 0 \leq x \leq 1$$

   (a) Find $E(X)$.
   (b) Find $E(X^2)$ and $\text{Var}(X)$.

13. Suppose that two lotteries each have n possible numbers and the same payoff. In terms of expected gain, is it better to buy two tickets from one of the lotteries or one from each?

14. Suppose that $E(X) = \mu$ and $\mathrm{Var}(X) = \sigma^2$. Let $Z = (X - \mu)/\sigma$. Show that $E(Z) = 0$ and $\mathrm{Var}(Z) = 1$.

15. Find (a) the expectation and (b) the variance of the kth order statistic of a sample of n independent random variables uniform on $[0, 1]$. The density function is given in Example C in Section 3.7.

16. If $U_1, \ldots, U_n$ are independent uniform random variables, find $E(U_{(n)} - U_{(1)})$.

17. Find $E(U_{(k)} - U_{(k-1)})$, where the $U_{(i)}$ are as in Problem 16.

18. A stick of unit length is broken into two pieces. Find the expected ratio of the length of the longer piece to the length of the shorter piece.

19. A random square has a side length that is a uniform random variable. Find the expected area of the square.

20. A random rectangle has sides the lengths of which are independent uniform random variables. Find the expected area of the rectangle, and compare this result to that of Problem 19.

21. Repeat Problems 19 and 20 assuming that the distribution of the lengths is exponential.

22. Prove Theorem A of Section 4.1.2 for the discrete case.

23. If X_1 and X_2 are independent random variables following a gamma distribution with parameters α and λ, find $E(R^2)$, where $R^2 = X_1^2 + X_2^2$.

24. Referring to Example B in Section 4.1.2, what is the expected number of coupons needed to collect r different types, where $r < n$?

25. If n men throw their hats into a pile and each man takes a hat at random, what is the expected number of matches? (*Hint:* Express the number as a sum.)

26. Suppose that n enemy aircraft are shot at simulataneously by m gunners, that each gunner selects an aircraft to shoot at independently of the other gunners, and that each gunner hits the selected aircraft with probability p. Find the expected number of aircraft hit by the gunners.

27. Prove Corollary A of Section 4.1.1.

28. Find $E[1/(X + 1)]$, where X is a Poisson random variable.

29. Prove Chebyshev's Inequality in the discrete case.

30. Let X be uniform on $[0, 1]$, and let $Y = \sqrt{X}$. Find $E(Y)$ by (a) finding the density of Y and then finding the expectation and (b) using Theorem A of Section 4.1.1.

31. Find the mean of a negative binomial random variable. (*Hint:* Express the random variable as a sum.)

32. Consider the following scheme for group testing. The original lot of samples is divided into two groups, and each of the subgroups is tested as a whole. If either subgroup tests positive, it is divided in two, and the procedure is repeated. If any of the groups thus obtained tests positive, test every member of that group. Find the expected number of tests performed, and compare it to the number performed with no grouping and with the scheme described in Example C in Section 4.1.2.

33. For what values of p is the group testing of Example C in Section 4.1.2 inferior to testing every individual?

34. Let X be an exponential random variable with standard deviation σ. Find $P(|X - E(X)| > k\sigma)$ for $k = 2, 3, 4$, and compare the results to the bounds from Chebyshev's Inequality.

35. Show that $\text{Var}(X - Y) = \text{Var}(X) + \text{Var}(Y) - 2\,\text{Cov}(X, Y)$.

36. If X and Y are independent random variables with equal variances, find $\text{Cov}(X + Y, X - Y)$.

37. Find the covariance and the correlation of N_i and N_j, where $N_1, N_2, \ldots, N_r$ are multinomial random variables. (*Hint:* Express them as sums.)

38. If $U = a + bX$ and $V = c + dY$, show that $\rho_{UV} = \rho_{XY}$.

39. If X and Y are independent random variables and $Z = Y - X$, find expressions for the covariance and the correlation of X and Z in terms of the variances of X and Y.

40. Let U and V be independent random variables with means μ and variances σ^2. Let $Z = \alpha U + V\sqrt{1 - \alpha^2}$. Find $E(Z)$ and ρ_{UZ}.

41. Two independent measurements, X and Y, are taken of a quantity μ. $E(X) = E(Y) = \mu$, but σ_X and σ_Y are unequal. The two measurements are combined by means of a weighted average to give

$$Z = \alpha X + (1 - \alpha)Y$$

where α is a scalar and $0 \le \alpha \le 1$.
(a) Show that $E(Z) = \mu$.
(b) Find α in terms of σ_X and σ_Y to minimize $\text{Var}(Z)$.

42. Suppose that X_i, where $i = 1, \ldots, n$, are independent random variables with $E(X_i) = \mu$ and $\text{Var}(X_i) = \sigma^2$. Let $\bar{X} = n^{-1}\sum_{i=1}^{n} X_i$. Show that $E(\bar{X}) = \mu$ and $\text{Var}(\bar{X}) = \sigma^2/n$.

43. Show that $\text{Cov}(X, Y) \le \sqrt{\text{Var}(X)\,\text{Var}(Y)}$.

44. Let X, Y, and Z be uncorrelated random variables with variances σ_X^2, σ_Y^2, and σ_Z^2, respectively. Let

$$U = Z + X$$
$$V = Z + Y$$

Find $\text{Cov}(U, V)$ and ρ_{UV}.

45. Let $T = \sum_{k=1}^{n} kX_k$, where the X_k are independent random variables with means μ and variances σ^2. Find $E(T)$ and $\text{Var}(T)$.

46. Let $S = \sum_{k=1}^{n} X_k$, where the X_k are as in Problem 45. Find the covariance and the correlation of S and T.

47. If X and Y are independent random variables, find $\text{Var}(XY)$ in terms of the means and variances of X and Y.

48. A function is measured at two points with some error (for example, the position of an object is measured at two times). Let

$$X_1 = f(x) + \varepsilon_1$$
$$X_2 = f(x + h) + \varepsilon_2$$

where ε_1 and ε_2 are independent random variables with mean zero and variance σ^2. Since the derivative of f is

$$\lim_{h \to 0} \frac{f(x + h) - f(x)}{h}$$

it is estimated by

$$Z = \frac{X_2 - X_1}{h}$$

(a) Find $E(Z)$ and $Var(Z)$. What is the effect of choosing a value of h that is very small, as is suggested by the definition of the derivative?

(b) Find an approximation to the mean squared error of Z using a Taylor Series expansion. Can you find the value of h that minimizes the mean squared error?

(c) Suppose that f is measured at three points with some error. How could you construct an estimate of the second derivative of f, and what are the mean and the variance of your estimate?

49. Let (X, Y) be a random point uniformly distributed on a unit disk. Show that $Cov(X, Y) = 0$, but that X and Y are not independent.

50. Let Y have a density that is symmetric about zero, and let $X = SY$, where S is an independent random variable taking on the values $+1$ and -1 with probability $\frac{1}{2}$ each. Show that $Cov(X, Y) = 0$, but that X and Y are not independent.

51. In Section 3.7, the joint density of the minimum and maximum of n independent uniform random variables was found. In the case $n = 2$, this amounts to X and Y, the minimum and maximum, respectively, of two independent random variables uniform on $[0, 1]$, is

$$f(x, y) = 2, \quad 0 \le x \le y$$

(a) Find the covariance and the correlation of X and Y. Does the sign of the correlation make sense intuitively?

(b) Find $E(X|Y)$ and $E(Y|X)$. Do these results make sense intuitively?

(c) Find the mean squared error of the best predictor of Y in terms of X.

52. Show that $E[Var(Y|X)] \le Var(Y)$.

53. Suppose that a bivariate normal distribution has $\mu_X = \mu_Y = 0$ and $\sigma_X = \sigma_Y = 1$. Sketch the contours of the density and the lines $E(Y|X)$ and $E(X|Y)$ for $\rho = 0, .5$, and $.9$.

54. If X and Y are independent, show that $E(X|Y = y) = E(X)$.

55. Let X be a binomial random variable representing the number of successes in n independent Bernoulli trials. Let Y be the number of successes in the first m trials, where $m < n$. Find the conditional frequency function of Y given $X = x$ and the conditional mean.

56. An item is present in a list of n items with probability p; if it is present, its position in the list is uniformly distributed. A computer program searches through the list sequentially. Find the expected number of items searched through before the program terminates.

57. A fair coin is tossed n times, and the number of heads, N, is counted. The coin is then tossed N more times. Find the expected total number of heads generated by this process.

58. The number of offspring of an organism is a discrete random variable with mean μ and variance σ^2. Each of its offspring reproduces in the same manner. Find the expected number of offspring in the third generation and its variance.

59. Let T be an exponential random variable, and conditional on T, let U be uniform on $[0, T]$. Find the unconditional mean and variance of U.

60. Let the point (X, Y) be uniformly distributed over the half disk $x^2 + y^2 \leq 1$, where $y \geq 0$. If you observe X, what is the best prediction for Y? If you observe Y, what is the best prediction for X? For both questions, "best" means having the minimum mean squared error.

61. Let X and Y have the joint density

$$f(x, y) = e^{-y}, \quad 0 \leq x \leq y$$

Find $E(Y|X)$ and $E(X|Y)$.

62. Show that if a density is symmetric about zero, its skewness is zero.

63. Find the moment-generating function of a Bernoulli random variable, and use it to find the mean, variance, and third moment.

64. Use the result of Problem 63 to find the mgf of a binomial random variable and its mean and variance.

65. Show that if X_i follows a binomial distribution with n_i trials and probability of success $p_i = p$, where $i = 1, \ldots, n$ and the X_i are independent, then $\sum_{i=1}^{n} X_i$ follows a binomial distribution.

66. Referring to Problem 65, show that if the p_i are unequal, the sum does not follow a binomial distribution.

67. Find the mgf of a geometric random variable, and use it to find the mean and the variance.

68. Use the result of Problem 67 to find the mgf of a negative binomial random variable and its mean and variance.

69. Under what conditions is the sum of independent negative binomial random variables also negative binomial?

70. Let X and Y be independent random variables, and let α and β be scalars. Find an expression for the mgf of $Z = \alpha X + \beta Y$ in terms of the mgf's of X and Y.

71. Let $X_1, X_2, \ldots, X_n$ be independent normal random variables with means μ_i and variances σ_i^2. Show that $Y = \sum_{i=1}^{n} \alpha_i X_i$, where the α_i are scalars, is normally distributed, and find its mean and variance.

72. Assuming that $X \sim N(0, \sigma^2)$, use the mgf to show that the odd moments are zero and the even moments are

$$\mu_{2n} = \frac{(2n)! \sigma^{2n}}{2^n (n!)}$$

73. Use the mgf to show that if X follows an exponential distribution, cX does also.

74. Suppose that Θ is a random variable that follows a gamma distribution with parameters λ and α, where α is an integer, and suppose that, conditional on Θ, X follows a Poisson distribution with parameter Θ. Find the unconditional distribution of X. (*Hint:* Find the mgf by using iterated conditional expectations.)

75. Find the distribution of a geometric sum of exponential random variables by using moment-generating functions.

76. If X is a nonnegative integer-valued random variable, the **probability-generating function** of X is defined to be

$$G(s) = \sum_{k=0}^{\infty} s^k p_k$$

where $p_k = P(X = k)$.

(a) Show that

$$p_k = \frac{1}{k!} \frac{d^k}{ds^k} G(s) \Big|_{s=0}$$

(b) Show that

$$\frac{dG}{ds} \Big|_{s=1} = E(X)$$

$$\frac{d^2 G}{ds^2} \Big|_{s=1} = E[X(X + 1)]$$

(c) Express the probability-generating function in terms of the moment-generating function.

(d) Find the probability-generating function of the Poisson distribution.

77. Show that if X and Y are independent, their joint moment-generating function factors.

78. Show how to find $E(XY)$ from the joint moment-generating function of X and Y.

79. Use moment-generating functions to show that if X and Y are independent, then

$$\text{Var}(aX + bY) = a^2 \text{Var}(X) + b^2 \text{Var}(Y)$$

80. Find the mean and variance of the compound Poisson distribution (Example H in Section 4.5).

81. Find expressions for the approximate mean and variance of $Y = g(X)$ for (a) $g(x) = \sqrt{x}$, (b) $g(x) = \log x$, and (c) $g(x) = \sin^{-1} x$.

82. If X is uniform on $[10, 20]$, find the approximate and exact mean and variance of $Y = 1/X$, and compare them.

83. Find the approximate mean and variance of $Y = \sqrt{X}$, where X is a random variable following a Poisson distribution.

84. Two sides, x_0 and y_0, of a right triangle are independently measured as X and Y, where $E(X) = x_0$ and $E(Y) = y_0$ and $\text{Var}(X) = \text{Var}(Y) = \sigma^2$. The angle between the two sides is then determined as

$$\Theta = \arctan\left(\frac{Y}{X}\right)$$

Find the approximate mean and variance of Θ.

5

Limit Theorems

5.1   Introduction

This chapter is principally concerned with the limiting behavior of the sum of independent random variables as the number of summands becomes large. The results presented here are both intrinsically interesting and useful in statistics. Many commonly computed statistical quantities, such as averages, can be represented as sums.

5.2   The Law of Large Numbers

It is commonly believed that if a fair coin is tossed many times and the proportion of heads is calculated, that proportion will be close to $\frac{1}{2}$. The law of large numbers is a mathematical formulation of this belief. The successive tosses of the coin are modeled as independent random trials. The random variable X_i takes on the value 0 or 1 according to whether the ith trial results in a tail or a head, and the proportion of heads in n trials is

$$\bar{X}_n = \frac{1}{n} \sum_{i=1}^{n} X_i$$

The law of large numbers states that $\bar{X}_n$ approaches $\frac{1}{2}$ in a sense that is specified by the following theorem.

THEOREM A. (Law of Large Numbers) Let $X_1, X_2, \ldots, X_i \ldots$ be a sequence of independent random variables with $E(X_i) = \mu$ and $\text{Var}(X_i) = \sigma^2$. Let $\bar{X}_n = n^{-1} \sum_{i=1}^{n} X_i$. Then, for any $\varepsilon > 0$,

$$P(|\bar{X}_n - \mu| > \varepsilon) \to 0 \quad \text{as } n \to \infty$$

Proof.   We first find $E(\bar{X}_n)$ and $\text{Var}(\bar{X}_n)$:

$$E(\bar{X}_n) = \frac{1}{n} \sum_{i=1}^{n} E(X_i) = \mu$$

Since the X_i are independent,

$$\text{Var}(\bar{X}_n) = \frac{1}{n^2} \sum_{i=1}^{n} \text{Var}(X_i) = \frac{\sigma^2}{n}$$

The desired result now follows immediately from Chebyshev's Inequality, which states that

$$P(|\bar{X}_n - \mu| > \varepsilon) \leq \frac{\text{Var}(\bar{X}_n)}{\varepsilon^2} = \frac{\sigma^2}{n\varepsilon^2} \to 0, \quad \text{as } n \to \infty \qquad \square$$

If a sequence of random variables, $\{Z_n\}$, is such that $P(|Z_n - \alpha| > \varepsilon)$ approaches zero as n approaches infinity, for any $\varepsilon > 0$ and where α is some scalar, then Z_n is said to **converge in probability to** α. There is another mode of convergence, called strong convergence or almost sure convergence, which asserts a bit more. Z_n is said to **converge almost surely to** α if for every $\varepsilon > 0, |Z_n - \alpha| > \varepsilon$ only a finite number of times with probability 1; that is, beyond some point in the sequence, the difference is always less than ε, but where that point is random. The version of the law of large numbers stated and proved above asserts that $\bar{X}_n$ converges to μ in probability. This version is usually called the weak law of large numbers. Under the same assumptions, a strong law of large numbers, which asserts that $\bar{X}_n$ converges almost surely to μ, can also be proved, but we will not do so.

We now consider some examples that illustrate the utility of the law of large numbers.

EXAMPLE A.   (Monte Carlo Integration) Suppose that we wish to calculate

$$I(f) = \int_0^1 f(x)\,dx$$

where the integration cannot be done by elementary means or evaluated using tables of integrals. The most common approach is to use a numerical method in which the integral is approximated by a sum; various schemes and computer packages exist for doing this. Another method, called the **Monte Carlo method**, works in the following way. Generate independent uniform random variables on $[0, 1]$—that is, $X_1, X_2, \ldots, X_n$—and compute

$$\bar{f} = \frac{1}{n} \sum_{i=1}^n f(X_i)$$

By the law of large numbers, this should be close to $E[f(X)]$, which is simply

$$E[f(X)] = \int_0^1 f(x)\,dx = I(f)$$

This simple scheme can be easily modified in order to change the range of integration and in other ways. Compared to the standard numerical methods, it is not especially efficient in one dimension, but becomes increasingly efficient as the dimensionality of the integral grows.

As a concrete example, let us consider the evaluation of

$$I(f) = \frac{1}{\sqrt{2\pi}} \int_0^1 e^{-x^2/2}\,dx$$

The integral is that of the standard normal density, which cannot be evaluated in closed form. From the table of the normal distribution (Table 2 in Appendix B), an accurate numerical approximation is $I(f) = .3413$. If 1000 points, $X_1, \ldots, X_{1000}$, uniformly distributed over the interval $0 \le x \le 1$, are generated using a pseudorandom number generator, the integral is then approximated by

$$\hat{I}(f) = \frac{1}{1000}\left(\frac{1}{\sqrt{2\pi}}\right) \sum_{i=1}^{1000} e^{-X_i^2/2}$$

which produced for one realization of the X_i the value .3417.   □

EXAMPLE B.   (Repeated Measurements) Suppose that repeated unbiased measurements, $X_1, \ldots, X_n$, of a quantity are made. If n is large, the law of large numbers allows us to hope that $\bar{X}$ will be close to the true value, μ, of the quantity, but how close $\bar{X}$ is not only depends on μ but on the variance of the measurement error, σ^2,

Fortunately, σ^2 can be estimated and therefore

$$\text{Var}(\bar{X}) = \frac{\sigma^2}{n}$$

can be estimated from the data to assess the precision of $\bar{X}$. First, note that $n^{-1}\sum_{i=1}^{n} X_i^2$ converges to $E(X^2)$, from the law of large numbers. Second, it can be shown that if Z_n converges to α in probability and g is a continuous function, then

$$g(Z_n) \rightarrow g(\alpha)$$

which implies that

$$\bar{X}^2 \rightarrow [E(X)]^2$$

Finally, since $n^{-1}\sum_{i=1}^{n} X_i^2$ converges to $E(X^2)$ and $\bar{X}^2$ converges to $[E(X)]^2$, with a little additional argument it can be shown that

$$\frac{1}{n}\sum_{i=1}^{n} X_i^2 - \bar{X}^2 \rightarrow E(X^2) - [E(X)]^2 = \text{Var}(X)$$

More generally, it follows from the law of large numbers that the sample moments, $n^{-1}\sum_{i=1}^{n} X_i^r$, converge in probability to the moments of X, $E(X^r)$.   □

EXAMPLE C.   A muscle or nerve cell membrane contains a very large number of channels; when open, these channels allow ions to pass through. Individual channels seem to open and close randomly, and it is often assumed that in an equilibrium situation the channels open and close independently of each other and that only a very small fraction are open at any one time. Suppose then that the probability that a channel is open is p, a very small number, that there are m channels in all, and that the amount of current flowing through an individual channel is c. The number of channels open at any particular time is N, a binomial random variable with m trials and probability p of success on each trial. The total amount of current is $S = cN$ and can be measured. We then have

$$E(S) = cE(N) = cmp$$

$$\text{Var}(S) = c^2 mp(1 - p)$$

and

$$\frac{\text{Var}(S)}{E(S)} = c(1 - p) \approx c$$

since p is small. Thus, through independent measurements, $S_1, \ldots, S_n$, we can estimate $E(S)$ and $\text{Var}(S)$ and therefore c, the amount of current flowing through a single channel. $\square$

5.3   Convergence in Distribution and the Central Limit Theorem

In applications, we often want to find $P(a < X < b)$ when we do not know the cdf of X precisely; it is sometimes possible to do this by approximating F_X. The approximation is often arrived at by some sort of limiting argument. The most famous limit theorem in probability theory is the central limit theorem, which is the main topic of this section. Before discussing the central limit theorem, we develop some introductory terminology, theory, and examples.

DEFINITION.   Let $X_1, X_2, \ldots$ be a sequence of random variables with cumulative distribution functions $F_1, F_2, \ldots$, and let X be a random variable with distribution function F. We say that X_n converges in distribution to X if

$$\lim_{n \to \infty} F_n(x) = F(x)$$

at every point at which F is continuous.

Moment-generating functions are often useful for establishing the convergence of distribution functions. We know from Property A of Section 4.5 that a distribution function F_n is uniquely determined by its mgf, M_n. The following theorem, which we give without proof, states that this unique determination holds for limits as well.

THEOREM A.   (Continuity Theorem) Let F_n be a sequence of cumulative distribution functions with the corresponding moment-generating function M_n. Let F be a cumulative distribution function with the moment-generating function M. If $M_n(t) \to M(t)$ for all t in an open interval containing zero, then $F_n(x) \to F(x)$ at all continuity points of F.

EXAMPLE A.   We will show that the Poisson distribution can be approximated by the normal distribution for large values of λ. This is suggested by examining Figure 2-6, which shows that as λ increases, the probability mass function of the Poisson distribution becomes more symmetric and bell-shaped.

Let $\lambda_1, \lambda_2, \ldots$ be an increasing sequence with $\lambda_n \to \infty$, and let $\{X_n\}$ be a sequence of Poisson random variables with the corresponding parameters. We know that $E(X_n) = \text{Var}(X_n) = \lambda_n$. If we wish to approximate the Poisson distribution function by a normal distribution function, the normal must have the same mean and variance as the Poisson does. In addition, if we wish to prove a limiting result, we run into the difficulty that the mean and variance are tending to infinity. This difficulty is dealt with by **standardizing** the random variables—that is, by letting

$$Z_n = \frac{X_n - \lambda}{\sqrt{\lambda}}$$

We then have $E(Z_n) = 0$ and $\text{Var}(Z_n) = 1$, and we will show that the mgf of Z_n converges to the mgf of the standard normal distribution.

The mgf of X_n is

$$M_{X_n}(t) = e^{\lambda_n(e^t - 1)}$$

By Property C of Section 4.5, the mgf of Z_n is

$$M_{Z_n}(t) = e^{-t\sqrt{\lambda_n}} M_{X_n}\left(\frac{t}{\sqrt{\lambda_n}}\right)$$

$$= e^{-t\sqrt{\lambda_n}} e^{\lambda_n(e^{t/\sqrt{\lambda_n}} - 1)}$$

It will be easier to work with the log of this expression. Using the power series expansion for e^x, we have

$$\log M_{Z_n}(t) = -t\sqrt{\lambda_n} + \lambda_n\left[\frac{t}{\sqrt{\lambda_n}} + \frac{1}{2}\left(\frac{t}{\sqrt{\lambda_n}}\right)^2 + \frac{1}{3!}\left(\frac{t}{\sqrt{\lambda_n}}\right)^3 + \cdots\right]$$

From this, we see that

$$\lim_{n \to \infty} \log M_{Z_n}(t) = \frac{t^2}{2}$$

or

$$\lim_{n \to \infty} M_{Z_n}(t) = e^{t^2/2}$$

The last expression is the mgf of the standard normal distribution.

We have shown that a standardized Poisson random variable converges in distribution to a standard normal variable as λ approaches infinity. Practically, we wish to use this limiting result as a basis for an approximation for large but finite values of λ_n. How adequate the approximation is for $\lambda = 100$, say, is a matter for theoretical and/or empirical investigation. It turns out that the approximation is increasingly good for large values of λ and that λ does not have to be all that large (see the end-of-chapter problems).   □

The next example shows how the approximation of the Poisson distribution can be applied in a specific case.

EXAMPLE B.   A certain type of particle is emitted at a rate of 900 per hour. What is the probability that more than 950 particles will be emitted in a given hour if the counts form a Poisson process?

Let X be a Poisson random variable with mean 900. We find $P(X > 950)$ by standardizing:

$$P(X > 950) = P\left(\frac{X - 900}{\sqrt{900}} > \frac{950 - 900}{\sqrt{900}}\right)$$

$$\approx 1 - \Phi(\tfrac{5}{3})$$

$$= .05$$

where Φ is the standard normal cdf.   □

We now turn to the central limit theorem, which is concerned with a limiting property of sums of random variables. If $X_1, X_2, \ldots$ is a sequence of independent random variables with mean μ and variance σ^2 and if

$$S_n = \sum_{i=1}^{n} X_i$$

we know from the law of large numbers that S_n/n converges to μ in probability. This followed from the fact that

$$\text{Var}\left(\frac{S_n}{n}\right) = \frac{1}{n^2} \text{Var}(S_n) = \frac{\sigma^2}{n} \to 0$$

The central limit theorem is concerned not with the fact that the ratio S_n/n converges to μ but with how it fluctuates around μ. To analyze these fluctuations, we have to examine them on the correct scale, which is again obtained by standardizing:

$$Z_n = \frac{S_n - n\mu}{\sigma\sqrt{n}}$$

It is easily verified that Z_n has mean 0 and variance 1. The central limit theorem states that the distribution of Z_n converges to the standard normal distribution.

THEOREM B.   (Central Limit Theorem) Let $X_1, X_2, \ldots$ be a sequence of independent random variables having mean 0 and variance σ^2 and the common distribution function F and moment-generating function M defined in a neighborhood of zero. Let

$$S_n = \sum_{i=1}^{n} X_i$$

Then

$$\lim_{n\to\infty} P\left(\frac{S_n}{\sigma\sqrt{n}} \le x\right) = \Phi(x), \quad -\infty < x < \infty$$

Proof. Let $Z_n = S_n/\sigma\sqrt{n}$. We will show that the mgf of Z_n tends to the mgf of the standard normal distribution. Since S_n is a sum of independent random variables,

$$M_{S_n}(t) = [M(t)]^n$$

and

$$M_{Z_n}(t) = \left[M\left(\frac{t}{\sigma\sqrt{n}}\right)\right]^n$$

$M(s)$ has a Taylor Series expansion about zero:

$$M(s) = M(0) + sM'(0) + \tfrac{1}{2}s^2 M''(0) + \varepsilon_s$$

where $\varepsilon_s/s^2 \to 0$ as $s \to 0$. Since $E(X) = 0$, $M'(0) = 0$, and $M''(0) = \sigma^2$. As $n \to \infty$, $t/\sigma\sqrt{n} \to 0$, and

$$M\left(\frac{t}{\sigma\sqrt{n}}\right) = 1 + \frac{1}{2}\sigma^2\left(\frac{t}{\sigma\sqrt{n}}\right)^2 + \varepsilon_n$$

where $\varepsilon_n/(t^2/n\sigma^2) \to 0$ as $n \to \infty$. We thus have

$$M_{Z_n}(t) = \left(1 + \frac{t^2}{2n} + \varepsilon_n\right)^n$$

It can be shown that if $a_n \to a$, then

$$\lim_{n\to\infty}\left(1 + \frac{a_n}{n}\right)^n = e^a$$

From this result, it follows that

$$M_{Z_n}(t) \to e^{t^2/2} \quad \text{as } n \to \infty$$

where $\exp(t^2/2)$ is the mgf of the standard normal distribution, as was to be shown. ☐

Theorem B is one of the simplest versions of the central limit theorem; there are many central limit theorems of various degrees of abstraction and generality. We have proved Theorem B under the assumption that the moment-generating functions exist, which is a rather strong assumption. By using characteristic functions instead, we could modify the proof so that it would only be necessary that first and second moments exist. Further generalizations weaken the assumption that the X_i have the same distribution and apply to linear combinations of independent random variables. Central limit theorems have also been proved that weaken the independence assumption and allow the X_i to be dependent but not "too" dependent. Central limit theorems are still an active area of research in probability theory.

For practical purposes, especially for statistics, the limiting result in itself is not of primary interest. Statisticians are more interested in its use as an approximation with finite values of n. It is impossible to give a concise and definitive statement of how good the approximation is, but some general guidelines are available, and examining special cases can give insight. How fast the approximation becomes good depends on the distribution of the summands, the X_i. If the distribution is fairly symmetric and has tails that die off rapidly, the approximation becomes good for relatively small values of n. If the distribution is very skewed or if the tails die down very slowly, a larger value of n is needed for a good approximation. The following examples deal with two special cases.

EXAMPLE C.   Since the uniform distribution on $[0, 1]$ has mean $\frac{1}{2}$ and variance $\frac{1}{12}$, the sum of 12 uniform random variables, minus 6, has mean 0 and variance 1. The distribution of this sum is quite close to normal; in fact, before better algorithms were developed, it was commonly used in computers for generating normal random variables from uniform ones. It is possible to compare the real and approximate distributions analytically, but we will content ourselves with a simple demonstration. Figure 5-1 shows a histogram of 1000 such sums with a

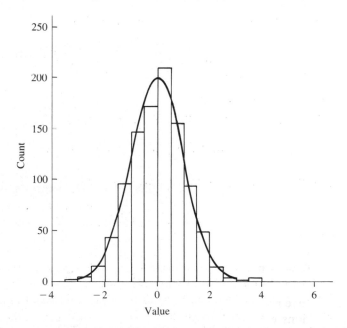

Figure 5-1.   A histogram of 1000 values, each of which is the sum of 12 uniform $[-\frac{1}{2}, \frac{1}{2}]$ pseudorandom variables, with an approximating standard normal density.

superimposed normal density function. The fit is surprisingly good, especially considering that 12 is not usually regarded as a large value of n.   □

EXAMPLE D.   The sum of n independent exponential random variables with parameter $\lambda = 1$ follows a gamma distribution with $\lambda = 1$ and $\alpha = n$. The exponential density is quite skewed; therefore, a good approximation of a standardized gamma by a standardized normal would not be expected for small n. Figure 5-2 shows the cdf's of the standard normal and standardized gamma distributions for increasing values of n. Note how the approximation improves as n increases.

□

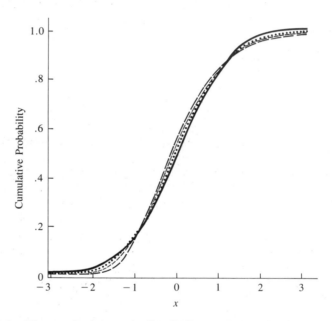

Figure 5-2.   The standard normal cdf (solid line) and the cdf's of standardized gamma distributions with $\alpha = 5$ (long dashes), $\alpha = 10$ (short dashes), and $\alpha = 30$ (dots).

Let us now consider some applications of the central limit theorem.

EXAMPLE E.   (Measurement Error) Suppose that $X_1, \ldots, X_n$ are repeated, independent unbiased measurements of a quantity, μ, and that $\text{Var}(X_i) = \sigma^2$. The average of the measurements, $\overline{X}$, is used as an estimate of μ. The law of large numbers tells us that $\overline{X}$ converges to μ in probability, so we can hope that $\overline{X}$ is close to μ if n is large. Chebyshev's Inequality allows us to bound the probability of an error of a given size, but the central limit theorem gives a much sharper approximation to the actual error. Suppose that we wish to find $P(|\overline{X} - \mu| < c)$ for some constant c. To use the central limit theorem to approximate this probability, we first standardize, using $E(\overline{X}) = \mu$ and $\text{Var}(\overline{X}) = \sigma^2/n$:

$$P(|\bar{X} - \mu| < c) = P(-c < \bar{X} - \mu < c)$$

$$= P\left(\frac{-c}{\sigma/\sqrt{n}} < \frac{\bar{X} - \mu}{\sigma/\sqrt{n}} < \frac{c}{\sigma/\sqrt{n}}\right)$$

$$\approx \Phi\left(\frac{c\sqrt{n}}{\sigma}\right) - \Phi\left(-\frac{c\sqrt{n}}{\sigma}\right)$$

For example, suppose that 16 measurements are taken with $\sigma = 1$. The probability that the average deviates from μ by less than .5 is

$$P(|\bar{X} - \mu| < .5) = \Phi(.5 \times 4) - \Phi(-.5 \times 4) = .954$$

This sort of reasoning can be turned around. That is, given c and γ, n can be found such that

$$P(|\bar{X} - \mu| < c) \geq \gamma \qquad \Box$$

EXAMPLE F.   (Normal Approximation to the Binomial Distribution) Since a binomial random variable is the sum of independent Bernoulli random variables, its distribution can be approximated by a normal distribution. The approximation is best when the binomial distribution is symmetric—that is, when $p = \frac{1}{2}$. A frequently used rule of thumb is that the approximation is reasonable when $np > 5$ and $n(1 - p) > 5$. The approximation is especially useful for large values of n, for which tables are not readily available.

Suppose that a coin is tossed 100 times and lands heads up 60 times. Should we be surprised and perhaps doubt that the coin is fair?

To answer this question, we note that if the coin is fair, the number of heads, X, is a binomial random variable with 100 trials and probability of success $\frac{1}{2}$, and $E(X) = 50$ and $\text{Var}(X) = 25$. We could calculate $P(X = 60)$, which would be a small number. But since there are so many possible outcomes, $P(X = 50)$ is also a small number, so this calculation would not really answer the question. Instead, we calculate the probability of a deviation as extreme as or more extreme than 60 if the coin is fair; that is, we calculate $P(X \geq 60)$. To approximate this probability from the normal distribution, we standardize:

$$P(X \geq 60) = P\left(\frac{X - 50}{5} \geq \frac{60 - 50}{5}\right)$$

$$\approx 1 - \Phi(2)$$

$$= .0228$$

The probability is rather small, so the fairness of the coin is called into question.
$\Box$

EXAMPLE G.   (Particle Size Distribution) The distribution of the sizes of grains of particulate matter is often found to be quite skewed, with a slowly decreasing right tail. A distribution called the lognormal is sometimes fit to such a distribution, and X is said to follow a lognormal distribution if $\log X$ has a normal distribution. The central limit theorem gives a theoretical rationale for the use of the lognormal distribution in some situations.

Suppose that a particle of initial size y_0 is subjected to repeated impacts, that on each impact a proportion, X_i, of the particle is broken off, and that the X_i are modeled as independent random variables having the same distribution. After the first impact, the size of the particle is $Y_1 = X_1 y_0$; after the second impact, the size is $Y_2 = X_2 X_1 y_0$; and after the nth impact, the size is

$$Y_n = X_n X_{n-1} \cdots X_2 X_1 y_0$$

Then

$$\log Y_n = \log y_0 + \sum_{i=1}^{n} \log X_i$$

and the central limit theorem applies to $\log Y_n$.   ☐

5.4   Problems

1. Let $X_1, X_2, \ldots$ be a sequence of independent random variables with $E(X_i) = \mu$ and $\mathrm{Var}(X_i) = \sigma_i^2$. Show that if $n^{-1} \sum_{i=1}^{n} \sigma_i^2 \to 0$, then $\bar{X} \to \mu$ in probability.
2. Let X_i be as in Problem 1 but with $E(X_i) = \mu_i$ and $n^{-1} \sum_{i=1}^{n} \mu_i \to \mu$. Show that $\bar{X} \to \mu$ in probability.
3. Using moment-generating functions, show that as $n \to \infty$, $p \to 0$, and $np \to \lambda$, the binomial distribution with parameters n and p tends to the Poisson distribution.
4. Using moment-generating functions, show that as $\alpha \to \infty$ the gamma distribution with parameters α and λ, properly standardized, tends to the standard normal distribution.
5. Show that if $X_n \to c$ in probability and if g is a continuous function, then $g(X_n) \to g(c)$ in probability.
6. Compare the Poisson cdf and the normal approximation for (a) $\lambda = 10$, (b) $\lambda = 20$, and (c) $\lambda = 40$.
7. Compare the binomial cdf and the normal approximation for (a) $n = 20$ and $p = .2$, and (b) $n = 40$ and $p = .5$.
8. A six-sided die is rolled 100 times. Using the normal approximation, find the probability that the face showing a six turns up between 15 and 20 times. Find the probability that the sum of the face values of the 100 trials is less than 300.

9. A skeptic gives the following argument to show that there must be a flaw in the central limit theorem: "We know that the sum of independent Poisson random variables follows a Poisson distribution with a parameter that is the sum of the parameters of the summands. In particular, if n independent Poisson random variables, each with parameter n^{-1}, are summed, the sum has a Poisson distribution with parameter 1. The central limit theorem says that as n approaches infinity the distribution of the sum tends to a normal distribution, but the Poisson with parameter 1 is not the normal." What do you think of this argument?

10. The central limit theorem can be used to analyze round-off error. Suppose that the round-off error is represented as a uniform random variable on $[-\frac{1}{2}, \frac{1}{2}]$. If 100 numbers are added, find the probability that the round-off error exceeds (a) 1, (b) 2, and (c) 5.

11. Suppose that you bet $5 on each of a sequence of 50 independent fair games. What is the probability that you will lose more than $75?

12. Suppose that $X_1, \ldots, X_{20}$ are independent random variables with density functions

$$f(x) = 2x, \quad 0 \le x \le 1$$

Let $S = X_1 + \cdots + X_{20}$. Use the central limit theorem to approximate $P(S \le 10)$.

13. Suppose that a measurement has mean μ and variance $\sigma^2 = 25$. Let $\bar{X}$ be the average of n such independent measurements. How large should n be so that $P(|\bar{X} - \mu| < 1) = .95$?

14. Suppose that a company ships packages that are variable in weight, with an average weight of 15 lb and a standard deviation of 10. Assuming that the packages come from a large number of different customers so that it is reasonable to model their weights as independent random variables, find the probability that 100 packages will have a total weight exceeding 17,000 lb.

15. Use the Monte Carlo method with $n = 100$ and $n = 1000$ to estimate $\int_0^1 \cos(2\pi x)\, dx$. Compare the estimates to the exact answer.

16. Prove that if $a_n \to a$, then $(1 + a_n/n)^n \to e^a$.

17. Let f_n be a sequence of frequency functions with $f_n(x) = \frac{1}{2}$ if $x = \pm(\frac{1}{2})^n$ and $f_n(x) = 0$ otherwise. Show that $\lim f_n(x) = 0$ for all x, which means that the frequency functions do not converge to a frequency function, but that there exists a cdf F such that $\lim F_n(x) = F(x)$.

18. In addition to limit theorems that deal with sums, there are limit theorems that deal with extreme values such as maxima or minima. Here is an example. Let $U_1, \ldots, U_n$ be independent uniform random variables on $[0, 1]$, and let $U_{(n)}$ be the maximum. Find the cdf of $U_{(n)}$ and a standardized $U_{(n)}$, and show that the cdf of the standardized variable tends to a limiting value.

6

Distributions Derived from the Normal Distribution

6.1 Introduction

This chapter assembles some results concerning three probability distributions derived from the normal distribution—the χ^2, t, and F distributions. These distributions occur in many statistical problems.

6.2 χ^2, t, and F Distributions

DEFINITION. If Z is a standard normal random variable, the distribution of $U = Z^2$ is called the chi-square distribution with 1 degree of freedom.

We have already encountered the chi-square distribution in Section 2.3, where we saw that it is a special case of the gamma distribution with parameters $\frac{1}{2}$ and $\frac{1}{2}$. The chi-square distribution with 1 degree of freedom is denoted χ_1^2. It is useful to note that if $X \sim N(\mu, \sigma^2)$, then $(X - \mu)/\sigma \sim N(0, 1)$, and therefore $[(X - \mu)/\sigma]^2 \sim \chi_1^2$.

DEFINITION.   If $U_1, U_2, \ldots, U_n$ are independent chi-square random variables with 1 degree of freedom, the distribution of $V = U_1 + U_2 + \cdots + U_n$ is called the chi-square distribution with n degrees of freedom and is denoted by χ_n^2.

From Example F in Section 4.5, we know that the sum of independent gamma random variables that have the same value of λ follows a gamma distribution, and therefore the chi-square distribution with n degrees of freedom is a gamma distribution with $\alpha = n/2$ and $\lambda = \frac{1}{2}$. Its density is

$$f(v) = \frac{1}{2^{n/2}\Gamma(n/2)} e^{n/2-1} e^{-v/2}, \quad v \geq 0$$

Its moment-generating function is

$$M(t) = (1 - 2t)^{-n/2}$$

Also, $E(V) = n$ and $\text{Var}(V) = 2n$. To indicate that V follows a chi-square distribution with n degrees of freedom, we write $V \sim \chi_n^2$. A notable consequence of the definition of the chi-square distribution is that if U and V are independent and $U \sim \chi_n^2$ and $V \sim \chi_m^2$, then $U + V \sim \chi_{m+n}^2$.

We now turn to the t distribution.

DEFINITION.   If $Z \sim N(0, 1)$ and $U \sim \chi_n^2$ and Z and U are independent, then the distribution of $Z/\sqrt{U/n}$ is called the t distribution with n degrees of freedom.

PROPOSITION A.   The density function of the t distribution with n degrees of freedom is

$$f(t) = \frac{\Gamma[(n + 1)/2]}{\sqrt{n\pi}\,\Gamma(n/2)} \left(1 + \frac{t^2}{n} \right)^{-(n+1)/2}$$

Proof.   This is proved by a standard method. The density function of $\sqrt{U/n}$ is straightforward to obtain, and the density function of the quotient of two independent random variables was derived in Section 3.6.1. The details of the proof are left as an end-of-chapter problem.   □

From the density function of Proposition A, the t distribution is symmetric about zero. As the number of degrees of freedom approaches infinity, the t distribution tends to the standard normal distribution; in fact, for more than 20 or 30 degrees of freedom, the distributions are very close. Figure 6-1 shows several t densities. Note that the tails become lighter as the degrees of freedom increase.

DEFINITION.   Let U and V be independent chi-square random variables with m and n degrees of freedom, respectively. The distribution of

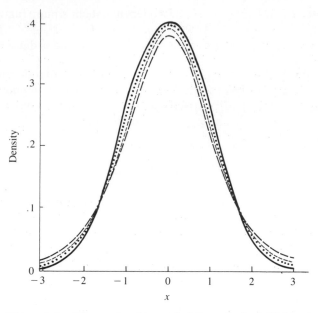

Figure 6-1. Three t densities with 5 (long dashes), 10 (short dashes), and 30 (dots) degrees of freedom and the standard normal density (solid line).

$$W = \frac{U/m}{V/n}$$

is called the F distribution with m and n degrees of freedom and is denoted by $F_{m,n}$.

PROPOSITION B. The density function of W is given by

$$f(w) = \frac{\Gamma[(m+n)/2]}{\Gamma(m/2)\Gamma(n/2)} \left(\frac{m}{n}\right)^{m/2} w^{m/2-1} \left(1 + \frac{m}{n}w\right)^{-(m+n)/2}, \quad w \geq 0$$

Proof. W is the ratio of two independent random variables, and its density follows from the results given in Section 3.6.1. □

It can be shown that, for $n > 2$, $E(W)$ exists and equals $n/(n-2)$. From the definitions of the t and F distributions, it is clear that the square of a t_n random variable follows an $F_{1,n}$ distribution (see Problem 6 at the end of this chapter).

6.3 The Sample Mean and the Sample Variance

Let $X_1, \ldots, X_n$ be independent $N(\mu, \sigma^2)$ random variables; we sometimes refer to them as a **sample** from a normal distribution. In this section, we will find the joint and marginal distributions of

$$\bar{X} = \frac{1}{n} \sum_{i=1}^{n} X_i$$

$$S^2 = \frac{1}{n-1} \sum_{i=1}^{n} (X_i - \bar{X})^2$$

These are called the sample mean and the sample variance, respectively. First note that since $\bar{X}$ is a linear combination of independent normal random variables, it is normally distributed with

$$E(\bar{X}) = \mu$$

$$\text{Var}(\bar{X}) = \frac{\sigma^2}{n}$$

As a preliminary to showing that $\bar{X}$ and S^2 are independently distributed, we establish the following theorem.

THEOREM A. The random variable $\bar{X}$ and the vector of random variables $(X_1 - \bar{X}, X_2 - \bar{X}, \ldots, X_n - \bar{X})$ are independent.

Proof. At the level of this course, it is difficult to give a proof that provides sufficient insight into why this result is true; a rigorous proof essentially depends on geometric properties of the multivariate normal distribution, which this book does not cover. We present a proof based on moment-generating functions; in particular, we will show that the joint moment-generating function

$$M(s, t_1, \ldots, t_n) = E\{\exp[s\bar{X} + t_1(X_1 - \bar{X}) + \cdots + t_n(X_n - \bar{X})]\}$$

factors into the product of two moment-generating functions—one of $\bar{X}$ and the other of the vector. The factoring implies (Section 4.5) that the random variables are independent of each other and is accomplished through some algebraic trickery. First, we observe that since

$$\sum_{i=1}^{n} t_i(X_i - \bar{X}) = \sum_{i=1}^{n} t_i X_i - n\bar{X}\bar{t}$$

then

$$s\bar{X} + \sum_{i=1}^{n} t_i(X_i - \bar{X}) = \sum_{i=1}^{n} \left[\frac{s}{n} + (t_i - \bar{t})\right] X_i$$

$$= \sum_{i=1}^{n} a_i X_i$$

where

$$a_i = \frac{s}{n} + (t_i - \bar{t})$$

Furthermore, we observe that

$$\sum_{i=1}^{n} a_i = s$$

$$\sum_{i=1}^{n} a_i^2 = \frac{s^2}{n} + \sum_{i=1}^{n} (t_i - \bar{t})^2$$

Now we have

$$M(s, t_1, \ldots, t_n) = M_{X_1 \cdots X_n}(a_1, \ldots, a_n)$$

and since the X_i are independent normal random variables, we have

$$M(s, t_1, \ldots, t_n) = \prod_{i=1}^{n} M_{X_i}(a_i)$$

$$= \prod_{i=1}^{n} \exp\left(\mu a_i + \frac{\sigma^2}{2} a_i^2\right)$$

$$= \exp\left(\mu \sum_{i=1}^{n} a_i + \frac{\sigma^2}{2} \sum_{i=1}^{n} a_i^2\right)$$

$$= \exp\left[\mu s + \frac{\sigma^2}{2}\left(\frac{s^2}{n}\right) + \frac{\sigma^2}{2} \sum_{i=1}^{n} (t_i - \bar{t})^2\right]$$

$$= \exp\left(\mu s + \frac{\sigma^2}{2n} s^2\right) \exp\left[\frac{\sigma^2}{2} \sum_{i=1}^{n} (t_i - \bar{t})^2\right]$$

The first factor is the mgf of $\bar{X}$. Since the mgf of the vector $(X_1 - \bar{X}, \ldots, X_n - \bar{X})$ can be obtained by setting $s = 0$ in M, the second factor is this mgf.   ☐

COROLLARY A. $\bar{X}$ and S^2 are independently distributed.

Proof.   This follows immediately since S^2 is a function of the vector $(X_i - \bar{X}, \ldots, X_n - \bar{X})$, which is independent of $\bar{X}$.   ☐

The next theorem gives the marginal distribution of S^2.

THEOREM B.   The distribution of $(n - 1)S^2/\sigma^2$ is the chi-square distribution with $n - 1$ degrees of freedom.

Proof.   We first note that

$$\frac{1}{\sigma^2} \sum_{i=1}^{n} (X_i - \mu)^2 = \sum_{i=1}^{n} \left(\frac{X_i - \mu}{\sigma}\right)^2 \sim \chi_n^2$$

Also,

$$\frac{1}{\sigma^2} \sum_{i=1}^{n} (X_i - \mu)^2 = \frac{1}{\sigma^2} \sum_{i=1}^{n} [(X_i - \bar{X}) + (\bar{X} - \mu)]^2$$

Expanding the square and using the fact that $\sum_{i=1}^{n} (X_i - \bar{X}) = 0$, we obtain

$$\frac{1}{\sigma^2} \sum_{i=1}^{n} (X_i - \bar{X})^2 + \left(\frac{\bar{X} - \mu}{\sigma/\sqrt{n}}\right)^2$$

This is a relation of the form $W = U + V$. Since U and V are independent by Corollary A, $M_W(t) = M_U(t)M_V(t)$. Since W and V both follow chi-square distributions,

$$M_U(t) = \frac{M_W(t)}{M_V(t)}$$

$$= \frac{(1 - 2t)^{-n/2}}{(1 - 2t)^{-1/2}}$$

$$= (1 - 2t)^{-(n-1)/2}$$

The last expression is the mgf of a random variable with a χ_{n-1}^2 distribution. □

One final result concludes this chapter's collection.

COROLLARY B.   Let $\bar{X}$ and S^2 be as given at the beginning of this section. Then

$$\frac{\bar{X} - \mu}{S/\sqrt{n}} \sim t_{n-1}$$

Proof.   We simply express the given ratio in a different form:

$$\frac{\bar{X} - \mu}{S/\sqrt{n}} = \frac{\left(\dfrac{\bar{X} - \mu}{\sigma/\sqrt{n}}\right)}{\sqrt{s^2/\sigma^2}}$$

The latter is the ratio of an $N(0, 1)$ random variable to the square root of an independent random variable with a χ^2_{n-1} distribution divided by its degrees of freedom. □

6.4   Problems

1. Prove Proposition A of Section 5.2.
2. Prove Proposition B of Section 5.2.
3. Let X be the average of a sample of 16 independent normal random variables with mean 0 and variance 1. Determine c such that

$$P(|\bar{X}| < c) = .5$$

4. If T follows a t_7 distribution, find t_0 such that (a) $P(|T| < t_0) = .9$ and (b) $P(T > t_0) = .05$.
5. Show that if $X \sim F_{n,m}$, then $X^{-1} \sim F_{m,n}$.
6. Show that if $T \sim t_n$, then $T^2 \sim F_{1,n}$.
7. Show that the Cauchy distribution and the t distribution with 1 degree of freedom are the same.
8. Show that if X and Y are independent exponential random variables with $\lambda = 1$, then X/Y follows an F distribution. Also, identify the degrees of freedom.
9. Find the mean and variance of S^2, where S^2 is as in Section 5.3.
10. Show how to use the chi-square distribution to calculate $P(a < S^2/\sigma^2 < b)$.
11. Let $X_1, \ldots, X_n$ be a sample from an $N(\mu_X, \sigma^2)$ distribution and $Y_1, \ldots, Y_m$ be an independent sample from an $N(\mu_Y, \sigma^2)$ distribution. Show how to use the F distribution to find $P(S_X^2/S_Y^2 > c)$.

7

Survey Sampling

7.1 Introduction

Resting on the probabilistic foundations of the preceding chapters, this chapter marks the beginning of our study of statistics by introducing the subject of survey sampling. As well as being of considerable intrinsic interest, the development of the elementary theory of survey sampling serves to introduce several concepts and techniques that will recur and be amplified in later chapters.

Sample surveys are used to obtain information about a large population by examining only a small fraction of that population. Sampling techniques have been used in many fields, such as the following:

- Governments survey human populations; for example, the U.S. government conducts health surveys and census surveys.
- Sampling techniques have been extensively employed in agriculture to estimate such quantities as the total acreage of wheat in a state by surveying a sample of farms.
- The Interstate Commerce Commission has carried out sampling studies of rail and highway traffic. In one such study, records of shipments of household

goods by motor carriers were sampled to evaluate the accuracy of preshipment estimates of charges, claims for damages, and other variables.
- In the practice of quality control, the output of a manufacturing process may be sampled in order to examine the items for defects.
- During audits of the financial records of large companies, sampling techniques may be used when examination of the entire set of records is impractical.

The sampling techniques discussed here are probabilistic in nature—each member of the population has a specified probability of being included in the sample, and the actual composition of the sample is random. Such techniques differ markedly from the type of sampling scheme in which particular population members are included in the sample because the investigator thinks they are typical in some way. Such a scheme may be effective in some situations, but there is no way mathematically to guarantee its unbiasedness (a term that will be precisely defined later) or to estimate the magnitude of any error committed, such as that arising from estimating the population mean by the sample mean. We will see that using a random sampling technique means that estimates can be guaranteed to be unbiased and probabilistic bounds on errors can be calculated. Among the advantages of using random sampling are the following:

- The selection of sample units at random is a guard against investigator biases, even unconscious ones.
- A small sample costs far less and is much faster to survey than a complete enumeration.
- The results from a small sample may actually be more accurate than those from a complete enumeration. The quality of the data in a small sample can be more easily monitored and controlled, and a complete enumeration may require a much larger, and therefore perhaps more poorly trained, staff.
- Random sampling techniques provide for the calculation of an estimate of the error due to sampling.
- In designing a sample, it is frequently possible to determine the sample size necessary to obtain a prescribed error level.

Several interesting papers written for a general audience are in Tanur et al. (1972). For example, a paper by Hansen discusses methods used by the U.S. Bureau of the Census.

7.2 Population Parameters

This section defines those numerical characteristics, or parameters, of the population that we will estimate from a sample. We will assume that the population is of size N and that associated with each member of the population is a numerical value of interest. These numerical values will be denoted by $x_1, x_2, \ldots, x_N$. The variable x_i may be a numerical variable such as age or weight, or it may take on the value 1 or 0 to denote the presence or absence of some characteristic. We will refer to the latter situation as the dichotomous case.

EXAMPLE A.   This is the first of many examples in this chapter dealing with a study by Herkson (1976). The population consists of $N = 393$ short-stay hospitals. We will let x_i denote the number of patients discharged from the ith hospital during January 1968. A histogram of the population values is shown in Figure 7-1. The histogram was constructed in the following way: The numbers of hospitals that discharged 0–200, 201–400, . . . , 2801–3000 patients were graphed as horizontal lines above the respective intervals. For example, the figure indicates that about 40 hospitals discharged from 601 to 800 patients. The histogram is a convenient graphical representation of the distribution of the values in the population, being more quickly assimilated than would a list of 393 values.   □

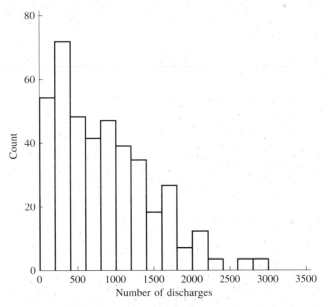

Figure 7-1.   Histogram of the numbers of patients discharged during January 1968 from 393 short-stay hospitals.

We will be particularly interested in the **population mean**, or average,

$$\mu = \frac{1}{N} \sum_{i=1}^{N} x_i$$

For the population of 393 hospitals, the mean number of discharges is 814.6. Note the location of this value in Figure 7-1. In the dichotomous case, where the presence or absence of a characteristic is to be determined, μ equals the proportion, p, of individuals in the population having the particular characteristic.

The **population total** is

$$\tau = \sum_{i=1}^{N} x_i = N\mu$$

The total number of people discharged from the population of hospitals is $\tau = 320{,}138$. In the dichotomous case, the population total is the total number of members of the population possessing the characteristic of interest.

We will also need to consider the **population variance**,

$$\sigma^2 = \frac{1}{N} \sum_{i=1}^{N} (x_i - \mu)^2$$

A useful identity can be obtained by expanding the square in this equation:

$$\sigma^2 = \frac{1}{N} \left(\sum_{i=1}^{N} x_i^2 - 2\mu \sum_{i=1}^{N} x_i + N\mu^2 \right)$$

$$= \frac{1}{N} \left(\sum_{i=1}^{N} x_i^2 - 2N\mu^2 + N\mu^2 \right)$$

$$= \frac{1}{N} \sum_{i=1}^{N} x_i^2 - \mu^2$$

In the dichotomous case, the population variance reduces to $p(1 - p)$:

$$\sigma^2 = \frac{1}{N} \sum_{i=1}^{N} x_i^2 - \mu^2$$

$$= p - p^2$$

$$= p(1 - p)$$

Here we used the fact that since each x_i is 0 or 1, each x_i^2 is also 0 or 1.

The **population standard deviation** is the square root of the population variance and is used as a measure of how spread out, dispersed, or scattered the individual values are. The standard deviation is given in the same units (for example, inches) as are the population values, whereas the variance is given in those units squared. The variance of the discharges is 347,766, and the standard deviation is 589.7; examination of the histogram in Figure 7-1 makes it clear that the latter number is the more reasonable description of the spread of the population values.

7.3   Simple Random Sampling

The most elementary form of sampling is **simple random sampling** (s.r.s.): each particular sample of size n has the same probability of occurrence; that is, each of the $\binom{N}{n}$ possible samples of size n taken without replacement has the same probability. We assume that sampling is done without replacement so that each member of the population will appear in the sample at most once. The actual composition of the sample is usually determined by using a table of random numbers or a random number generator on a computer. Conceptually, we can regard the population members as balls in an urn, a specified number of which are selected for inclusion in the sample at random and without replacement.

Since the composition of the sample is random, the sample mean is random. An analysis of the accuracy with which the sample mean approximates the population mean must therefore be probabilistic in nature. In this section, we will derive some statistical properties of the sample mean.

7.3.1   The Expectation and Variance of the Sample Mean

We will denote the sample size by n (n is less than N) and the values of the sample members by $X_1, X_2, \ldots, X_n$. It is important to realize that each X_i is a random variable. In particular, X_i is not the same as x_i: X_i is the value of the ith member of the sample, and x_i is that of the ith member of the population.

The joint distribution of the X_i is determined by that of the x_i. Let us denote the distinct values of x_i by $\zeta_1, \ldots, \zeta_m$, where n_1 of the x_i equal ζ_1, n_2 equal $\zeta_2, \ldots$, and n_m equal ζ_m. (In the dichotomous case, $\zeta_1 = 1$ and $\zeta_2 = 0$, or vice versa.) Since each member of the population is equally likely to be in the sample,

$$P(X_i = \zeta_j) = \frac{n_j}{N}$$

We will consider the **sample mean**,

$$\bar{X} = \frac{1}{n} \sum_{i=1}^{n} X_i$$

as an estimate of the population mean. As an estimate of the population total, we will consider

$$T = N\bar{X}$$

Properties of T will follow readily from those of $\bar{X}$. Since each X_i is a random variable, so is the sample mean; its probability distribution is called its **sampling distribution**. In general, any numerical value, or statistic, computed from a random sample is a random variable and has an associated sampling distribution. The sampling distribution of $\bar{X}$ determines how accurately $\bar{X}$ estimates μ; roughly speaking, the more tightly the sampling distribution is centered around μ, the better the estimate.

EXAMPLE A.   To illustrate the concept of a sampling distribution, let us look again at the population of 393 hospitals. In practice, of course, the population would not be known, and only one sample would be drawn. For pedagogical purposes here, we can consider the sampling distribution of the sample mean from this known population. Suppose, for example, that we want to find the sampling distribution of the mean of a sample of size 16. In principle, we could form all $\binom{393}{16}$ samples and compute the mean of each one—this would give the sampling distribution. But since the number of such samples is of the order 10^{33}, this is clearly not practical. We will thus employ a technique known as **simulation**. We can estimate the sampling distribution of the mean of a sample of size n by drawing many samples of size n, computing the mean of each sample, and then forming a histogram of the collection of sample means. Figure 7-2 shows the results of such a simulation for sample sizes of 8, 16, 32, and 64 with 500 replications for each sample size. Three features of Figure 7-2 are noteworthy:

1. All the histograms are centered about the population mean, 814.6.
2. As the sample size increases, the histograms become less spread out.
3. Although the shape of the histogram of population values (Figure 7-1) is not symmetric about the mean, the histograms in Figure 7-2 are more nearly so.

These features will be explained quantitatively in later sections of this chapter.   □

As a measure of the center of the sampling distribution, we will use $E(\bar{X})$. As a measure of the dispersion of the sampling distribution about this center, we will use the standard deviation of $\bar{X}$. The key results that will be obtained below are that the sampling distribution is centered at μ and that its spread is inversely proportional to the square root of the sample size, n. We first show that the sampling distribution is centered at μ.

THEOREM A.   With simple random sampling, $E(\bar{X}) = \mu$.

Proof.   Since

$$E(X_i) = \sum_{j=1}^{m} \zeta_j P(X_i = \zeta_j) = \frac{1}{N} \sum_{j=1}^{m} n_j \zeta_j = \mu$$

it follows from Theorem A in Section 4.1.2 that

$$E(\bar{X}) = \frac{1}{n} \sum_{i=1}^{n} E(X_i) = \mu \qquad\qquad □$$

From Theorem A, we have the following corollary.

COROLLARY A.   With simple random sampling, $E(T) = \tau$.

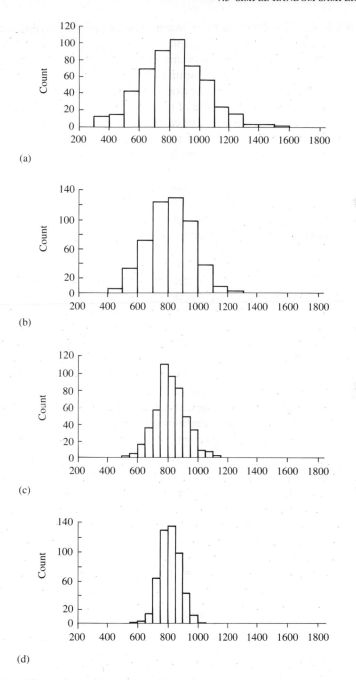

Figure 7-2. Histograms of the values of the mean number of discharges in 500 simple random samples from the population of 393 hospitals. Sample sizes: (a) $n = 8$, (b) $n = 16$, (c) $n = 32$, (d) $n = 64$.

Proof.

$$E(T) = E(N\bar{X})$$

$$= NE(\bar{X})$$

$$= N\mu$$

$$= \tau \qquad \qquad \square$$

In the dichotomous case, $\mu = p$, and $\bar{X}$ is the proportion of the sample that possesses the characteristic of interest. In this case, $\bar{X}$ will be denoted by $\hat{p}$. We have shown that $E(\hat{p}) = p$.

$\bar{X}$ and T are said to be **unbiased** estimates of the population mean and total. We say that an estimate is unbiased if its expectation equals the quantity we wish to estimate—that is, the estimate is correct "on the average." Section 4.2.1 introduced the concepts of bias and variance in the context of a model of measurement error, and these concepts are also relevant in this new context. In chapter 4, it was shown that

$$\text{Mean squared error} = \text{variance} + \text{bias}^2$$

Since $\bar{X}$ and T are unbiased, their mean squared errors are equal to their variances.

We now derive the variance and thus the standard deviation, also called the **standard error**, of the sampling distribution of $\bar{X}$. This derivation is considerably more complicated than the preceding one. The crucial ingredient is the following lemma.

LEMMA A.   With simple random sampling,

$$\text{Cov}(X_i, X_j) = \begin{cases} \sigma^2 & \text{if } i = j \\ -\sigma^2/(N - 1), & \text{if } i \neq j \end{cases}$$

Proof.   The case in which $i = j$ is simple:

$$\text{Cov}(X_i, X_i) = \text{Var}(X_i)$$

$$= E(X_i^2) - [E(X_i)]^2$$

$$= \frac{1}{N} \sum_{j=1}^{m} n_j \zeta_j^2 - \mu^2$$

$$= \sigma^2$$

The case in which $i \neq j$ is more complicated. X_i and X_j are not independent random variables, and their covariance is nonzero since the sampling is done without replacement. We have

$$\text{Cov}(X_i, X_j) = E(X_i X_j) - E(X_i)E(X_j)$$

and

$$E(X_i X_j) = \sum_{k=1}^{m} \sum_{l=1}^{m} \zeta_k \zeta_l P(X_i = \zeta_k \text{ and } X_j = \zeta_l)$$

$$= \sum_{k=1}^{m} \zeta_k P(X_i = \zeta_k) \sum_{l=1}^{m} \zeta_l P(X_j = \zeta_l | X_i = \zeta_k)$$

from the formula for conditional probability. Now,

$$P(X_j = \zeta_l | X_i = \zeta_k) = \begin{cases} n_l/(N-1), & \text{if } k \neq l \\ (n_l - 1)/(N-1), & \text{if } k = l \end{cases}$$

If we express

$$\frac{n_k - 1}{N - 1} \zeta_k = \frac{n_k}{N - 1} \zeta_k - \frac{\zeta_k}{N - 1}$$

the expression for $E(X_i X_j)$ becomes

$$\sum_{l=1}^{m} \zeta_k \frac{n_k}{N} \left(\sum_{l=1}^{m} \zeta_l \frac{n_l}{N - 1} - \frac{\zeta_k}{N - 1} \right) = \frac{1}{N(N-1)} \left(\tau^2 - \sum_{k=1}^{m} \zeta_k^2 n_k \right)$$

$$= \frac{\tau^2}{N(N-1)} - \frac{1}{N(N-1)} \sum_{k=1}^{m} \zeta_k^2 n_k$$

$$= \frac{N\mu^2}{N-1} - \frac{1}{N-1}(\mu^2 + \sigma^2)$$

$$= \mu^2 - \frac{\sigma^2}{N-1}$$

Finally, subtracting $E(X_i)E(X_j) = \mu^2$ from the last equation, we have

$$\text{Cov}(X_i, X_j) = -\frac{\sigma^2}{N-1}$$

for $i \neq j$. □

(A slicker, but less straightforward, proof of Lemma A is outlined in Problem 3 at the end of this chapter.) This lemma shows that X_i and X_j are not independent of each other for $i \neq j$, but that the covariance is very small for large values of N. We are now able to derive the following theorem.

THEOREM B.   With simple random sampling,

$$\text{Var}(\bar{X}) = \frac{\sigma^2}{n}\left(\frac{N-n}{N-1}\right)$$

$$= \frac{\sigma^2}{n}\left(1 - \frac{n-1}{N-1}\right)$$

Proof.   From Corollary B of Section 4.3,

$$\text{Var}(\bar{X}) = \frac{1}{n^2}\sum_{i=1}^{n}\sum_{j=1}^{n}\text{Cov}(X_i, X_j)$$

$$= \frac{1}{n^2}\sum_{i=1}^{n}\text{Var}(X_i) + \frac{1}{n^2}\sum_{i=1}^{n}\sum_{j\neq i}\text{Cov}(X_i, X_j)$$

$$= \frac{\sigma^2}{n} - \frac{1}{n^2}n(n-1)\frac{\sigma^2}{N-1}$$

After some algebra, this gives the desired result. $\square$

The ratio n/N is called the **sampling fraction**. The factor

$$\left(1 - \frac{n-1}{N-1}\right)$$

in Theorem B is called the **finite population correction**. Frequently, the sampling fraction is very small, and then the standard deviation of $\bar{X}$ is

$$\sigma_{\bar{X}} \approx \frac{\sigma}{\sqrt{n}}$$

We see that, apart from the usually small finite population correction, the spread of the sampling distribution and therefore the precision of $\bar{X}$ are determined by the sample size (n) and not by the population size (N). As will be made more explicit later, the appropriate measure of the precision of the sample mean is its standard error, which is inversely proportional to the square root of the sample size. Thus, in order to double the accuracy, the sample size must be quadrupled. (You might examine Figure 7-2 with this in mind.) The other factor that determines the accuracy of the sample mean is the population standard deviation, σ. If σ is small, the population values are not very dispersed and a small sample will be fairly accurate. But if the values are widely dispersed, a much larger sample will be required in order to attain the same accuracy.

The precision of the estimate of the population total does depend on the population size, N.

COROLLARY B.   With simple random sampling,

$$\text{Var}(T) = N^2 \left(\frac{\sigma^2}{n} \right) \frac{N - n}{N - 1}$$

Proof.   Since $T = N\bar{X}$,

$$\text{Var}(T) = N^2 \, \text{Var}(\bar{X}) \qquad \qquad \square$$

7.3.2   Estimation of the Population Variance

The precision of the sample mean is determined by its variance, the formula for which involves the sample size and the (usually unknown) population variance. In this section, we derive an estimate of the population variance, which we will later use to assess the precision of the sample mean. Since the population variance is the average squared deviation from the population mean, estimating it by the average squared deviation from the sample mean seems natural:

$$\hat{\sigma}^2 = \frac{1}{n} \sum_{i=1}^{n} (X_i - \bar{X})^2$$

The following theorem shows that this estimate is biased.

THEOREM A.   With simple random sampling,

$$E(\hat{\sigma}^2) = \sigma^2 \left(\frac{n - 1}{n} \right) \frac{N}{N - 1}$$

Proof.   We can express $\hat{\sigma}^2$ as follows:

$$\hat{\sigma}^2 = \frac{1}{n} \sum_{i=1}^{n} X_i^2 - \bar{X}^2$$

Thus,

$$E(\hat{\sigma}^2) = \frac{1}{n} \sum_{i=1}^{n} E(X_i^2) + E(\bar{X}^2)$$

Now, we know that

$$E(X_i^2) = \text{Var}(X_i) + [E(X_i)]^2$$

$$= \sigma^2 + \mu^2$$

Similarly, from Theorems A and B of Section 7.3.1,

$$E(\bar{X}^2) = \text{Var}(\bar{X}) + [E(\bar{X})]^2$$

$$= \frac{\sigma^2}{n}\left(1 - \frac{n-1}{N-1}\right) + \mu^2$$

Substituting these expressions for $E(X_i^2)$ and $E(\bar{X}^2)$ in the equation for $E(\hat{\sigma}^2)$ above gives the desired result. $\square$

The estimate of the population variance is slightly biased; if the population is large relative to n, the dominant bias is due to the term $(n-1)/n$. From Theorem A, we see that an unbiased estimate of σ^2 may be obtained by multiplying $\hat{\sigma}^2$ by the factor $n(N-1)/(n-1)N$. We also have the following corollary.

COROLLARY A. An unbiased estimate of $\text{Var}(\bar{X})$ is

$$s_{\bar{X}}^2 = \frac{\hat{\sigma}^2}{n}\left(\frac{n}{n-1}\right)\left(\frac{N-1}{N}\right)\left(\frac{N-n}{N-1}\right)$$

$$= \frac{s^2}{n}\left(1 - \frac{n}{N}\right)$$

where

$$s^2 = \frac{1}{n-1}\sum_{i=1}^{n}(X_i - \bar{X})^2$$

Proof. Since

$$\text{Var}(\bar{X}) = \frac{\sigma^2}{n}\left(\frac{N-n}{N-1}\right)$$

an unbiased estimate of $\text{Var}(\bar{X})$ may be obtained by substituting in an unbiased estimate of σ^2. Algebra then yields the desired result. $\square$

Similarly, an unbiased estimate of the variance of T, the estimator of the population total, is

$$s_T^2 = N^2 s_{\bar{X}}^2$$

For the dichotomous case, in which each X_i is 0 or 1, note that

$$\frac{1}{n}\sum_{i=1}^{n}(X_i - \bar{X})^2 = \frac{1}{n}\sum_{i=1}^{n}X_i^2 - \bar{X}^2$$

$$= \hat{p}(1 - \hat{p})$$

Therefore,

$$s^2 = \frac{n}{n-1} \hat{p}(1 - \hat{p})$$

Thus, as a special case of Corollary A, we have the following corollary.

COROLLARY B. An unbiased estimate of $\text{Var}(\hat{p})$ is

$$s_{\hat{p}}^2 = \frac{\hat{p}(1 - \hat{p})}{n-1}\left(1 - \frac{n}{N}\right)$$

In many cases, the sampling fraction, n/N, is small and may be neglected. Furthermore, it often makes little difference whether $n - 1$ or n is used as the divisor.

The quantities $s_{\bar{X}}$, s_T, and $s_{\hat{p}}$ are called **estimated standard errors**.

EXAMPLE A. A simple random sample of 50 of the 393 hospitals was taken. From this sample, $\bar{X} = 938.5$ (recall that, in fact, $\mu = 814.6$) and $s = 614.53$ ($\sigma = 590$). An estimate of the variance of $\bar{X}$ is

$$s_{\bar{X}}^2 = \frac{s^2}{n}\left(1 - \frac{n}{N}\right) = 6592$$

The estimated standard error of $\bar{X}$ is

$$s_{\bar{X}} = 81.19$$

This estimated standard error gives a rough idea of how accurate the value of $\bar{X}$ is; in this case, we see that the magnitude of the error is of the order 80, as opposed to 8 or 800, say. In the next section, the error estimates will be sharpened. □

EXAMPLE B. The estimate of the total number of discharges in the population of hospitals is

$$T = N\bar{X} = 368,831$$

Recall that the true value of the population total is 320,139. The estimated standard error of T is

$$s_T = Ns_{\bar{X}} = 31,908$$

Again, this estimated standard error can be used as a rough gauge of the estimation error. □

EXAMPLE C.   Let p be the proportion of hospitals that had fewer than 1000 discharges—that is, $p = .654$. In the sample, 26 of 50 hospitals had fewer than 1000 discharges, so

$$\hat{p} = \frac{26}{50} = .52$$

The variance of $\hat{p}$ is estimated by

$$s_{\hat{p}}^2 = \frac{\hat{p}(1 - \hat{p})}{n - 1}\left(1 - \frac{n}{N}\right) = .0045$$

Thus, the estimated standard error of $\hat{p}$ is

$$s_{\hat{p}} = .067$$

Crudely, this tells us that the error of $\hat{p}$ is in the second or first decimal place—that we are probably not so fortunate as to have an error only in the third decimal place. $\square$

7.3.3   The Normal Approximation to the Sampling Distribution of $\bar{X}$

We have found the mean and the standard deviation of the sampling distribution of $\bar{X}$. Ideally, we would like to know the sampling distribution, since it would tell us everything we could hope to know about the accuracy of the estimate. Without knowledge of the population itself, however, we cannot determine the sampling distribution. In this section, we will use the central limit theorem to deduce an approximation to the sampling distribution—the normal, or Gaussian, distribution. This approximation will be used to find probabilistic bounds for the estimation error.

In Section 5.3, we considered a sequence of independent and identically distributed (i.i.d.) random variables, $X_1, X_2, \ldots$ having the common mean and variance μ and σ^2. The sample mean of $X_1, X_2, \ldots, X_n$ is

$$\bar{X}_n = \frac{1}{n}\sum_{i=1}^{n} X_i$$

This sample mean has the properties

$$E(\bar{X}_n) = \mu$$

and

$$\mathrm{Var}(\bar{X}_n) = \frac{\sigma^2}{n}$$

The central limit theorem says that, for a fixed number z,

$$P\left(\frac{\bar{X}_n - \mu}{\sigma/\sqrt{n}} \le z\right) \to \Phi(z) \quad \text{as } n \to \infty$$

where Φ is the cumulative distribution function of the standard normal distribution. Using a more compact and suggestive notation, we have

$$P\left(\frac{\bar{X}_n - \mu}{\sigma_{\bar{X}_n}} \le z\right) \to \Phi(z)$$

The context of survey sampling is not exactly like that of the central limit theorem as stated above—as we have seen, in sampling the X_i are not independent of each other, and it makes no sense to have n tend to infinity while N remains fixed. But other central limit theorems have been proved that are appropriate to the sampling context. These show that if n is large, but still small relative to N, then $\bar{X}_n$, the mean of a simple random sample, is approximately normally distributed.

To demonstrate the use of the central limit theorem, we will apply it to approximate $P(|X - \mu| \le \delta)$, the probability that the error made in estimating μ by $\bar{X}$ is less than some constant δ.

$$P(|\bar{X} - \mu| \le \delta) = P(-\delta \le \bar{X} - \mu \le \delta)$$

$$= P\left(-\frac{\delta}{\sigma_{\bar{X}}} \le \frac{\bar{X} - \mu}{\sigma_{\bar{X}}} \le \frac{\delta}{\sigma_{\bar{X}}}\right)$$

$$\approx \Phi\left(\frac{\delta}{\sigma_{\bar{X}}}\right) - \Phi\left(-\frac{\delta}{\sigma_{\bar{X}}}\right)$$

$$= 2\Phi\left(\frac{\delta}{\sigma_{\bar{X}}}\right) - 1$$

since $\Phi(-z) = 1 - \Phi(z)$, from the symmetry of the standard normal distribution about zero.

EXAMPLE A. Let us again consider the population of 393 hospitals. The standard deviation of the mean of a sample of size $n = 64$ is, using the finite population correction,

$$\sigma_{\bar{X}} = \frac{\sigma}{\sqrt{n}} \sqrt{1 - \frac{n-1}{N-1}}$$

$$= \frac{589.7}{8} \sqrt{1 - \frac{63}{392}}$$

$$= 67.5$$

The approximation using the central limit theorem gives the probability that the sample mean differs from the population mean by more than 100 (in absolute value) as

$$1 - \left[2\Phi\left(\frac{100}{67.5}\right) - 1 \right] = 2(1 - .931)$$

$$= .138$$

or about 14%. In fact, among the 500 samples of size 64 in Example A in Section 7.3.1, 82, or 16.4%, differed by more than 100 from the population mean. Similarly, the central limit theorem approximation gives .026 as the probability of deviations of more than 150 from the population mean. In the simulation in Example A in Section 7.3.1, 11 of 500, or 2.2%, differed by more than 150. If we are not too finicky, the central limit theorem gives us reasonable approximations.   □

EXAMPLE B.   For a sample of size 50, the standard error of the sample mean number of discharges is

$$\sigma_{\bar{X}} = 78$$

For the particular sample of size 50 discussed in Example A in Section 7.3.2, we found $\bar{X} = 938.35$, so $\bar{X} - \mu = 123.9$. We now calculate an approximation of the probability of an error this large or larger:

$$P(|\bar{X} - \mu| \geq 123.9) = 1 - P(|\bar{X} - \mu| < 123.9)$$

$$\approx 1 - \left[2\Phi\left(\frac{123.9}{78}\right) - 1 \right]$$

$$= 2 - 2\Phi(1.59)$$

$$= .12$$

Thus, we can expect an error this large or larger to occur about 12% of the time.   □

EXAMPLE C.   In Example C in Section 7.3.2, we found from the sample of size 50 an estimate $\hat{p} = .52$ of the proportion of hospitals that discharged more than 1000 patients; in fact, the actual proportion in the population is .65. Thus, $|\hat{p} - p| = .13$. What is the probability that the estimate will be off by an amount this large or larger?

We have

$$\sigma_{\hat{p}} = \sqrt{\frac{p(1 - p)}{n}} \sqrt{1 - \frac{n - 1}{N - 1}}$$

$$= .068 \times .94 = .064$$

We can therefore calculate

$$P(|p - \hat{p}| > .13) = 1 - P(|p - \hat{p}| \leq .13)$$

$$= 1 - P\left(\frac{|p - \hat{p}|}{\sigma_{\hat{p}}} \leq \frac{.13}{\sigma_{\hat{p}}}\right)$$

$$\approx 2[1 - \Phi(2.03)]$$

$$= .04$$

We see that the sample was rather "unlucky"—an error this large or larger would only occur about 4% of the time. $\square$

We can now derive a **confidence interval** for the population mean, μ. A confidence interval for a population parameter, θ, is a random interval, calculated from the sample, that contains θ with some specified probability. For example, a 95% confidence interval for μ is a random interval that contains μ with probability .95; if we were to take many random samples and form a confidence interval from each one, about 95% of these intervals would contain μ. If the coverage probability is $1 - \alpha$, the interval is called a $100(1 - \alpha)\%$ confidence interval. Confidence intervals are frequently used in conjunction with point estimates to convey information about the uncertainty of the estimates.

For $0 \leq \alpha \leq 1$, let $z(\alpha)$ be that number such that the area under the standard normal density function to the right of $z(\alpha)$ is α (Figure 7-3). Note that the symmetry of the standard normal density function about zero implies that $z(1 - \alpha) = -z(\alpha)$. If Z follows a standard normal distribution, then, by definition of $z(\alpha)$,

$$P(-z(\alpha/2) \leq Z \leq z(\alpha/2)) = 1 - \alpha$$

From the central limit theorem, $(\bar{X} - \mu)/\sigma_{\bar{X}}$ has approximately a standard normal distribution, so

$$P\left(-z(\alpha/2) \leq \frac{\bar{X} - \mu}{\sigma_{\bar{X}}} \leq z(\alpha/2)\right) \approx 1 - \alpha$$

Elementary manipulation of the inequalities gives

$$P(\bar{X} - z(\alpha/2)\sigma_{\bar{X}} \leq \mu \leq \bar{X} + z(\alpha/2)\sigma_{\bar{X}}) \approx 1 - \alpha$$

That is, the probability that μ lies in the confidence interval $\bar{X} \pm z(\alpha/2)\sigma_{\bar{X}}$ is approximately $1 - \alpha$. It is important to understand that this interval is random and that the above equation states that the probability that this random interval covers μ is $1 - \alpha$. In practice, α is assigned a small value, such as .1, .05, or .01, so that the probability that the interval covers μ will be large. Also, since the population variance is typically not known, $s_{\bar{X}}$ is substituted for $\sigma_{\bar{X}}$. For large samples, it can be shown that the effect of this substitution is practically negligible. It is impossible to give a precise answer to the question "How large is large?" As a rule of thumb, a value of n greater than 25 or 30 is usually adequate.

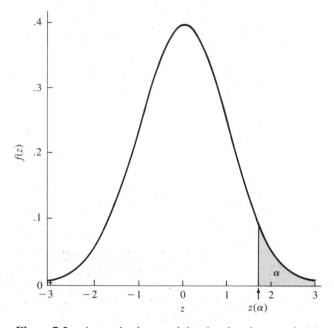

Figure 7-3. A standard normal density showing α and $z(\alpha)$.

The following example illustrates the procedure for calculating confidence intervals.

EXAMPLE D.   A particular area contains 8000 condominium units. In a survey of the occupants, a simple random sample of size 100 yields the information that the average number of motor vehicles per unit is 1.6, with a sample standard deviation of .8. The estimated standard error of $\bar{X}$ is thus

$$s_{\bar{X}} = \frac{s}{\sqrt{n}}\sqrt{1 - \frac{n}{N}}$$

$$= \frac{.8}{10}\sqrt{1 - \frac{100}{8000}}$$

$$= .08$$

Note that the finite population correction makes almost no difference. Since

$\hat{p}(1-\hat{p})$ S x

$z(.025) = 1.96$, a 95% confidence interval for the population average is $\bar{x} \pm 1.96s_{\bar{x}}$, or $(1.44, 1.76)$.

An estimate of the total number of motor vehicles is $T = 8000(1.6) = 12,800$. The standard error of T is

$$s_T = Ns_{\bar{x}} = 640$$

A 95% confidence interval for the total number of motor vehicles is $T \pm 1.96s_T$, or $(11,546, 14,054)$.

In the same survey, 12% of the respondents said they planned to sell their condos within the next year; $\hat{p} = .12$ is an estimate of the population proportion p. The estimated standard error is

$$s_{\hat{p}} = \sqrt{\frac{\hat{p}(1 - \hat{p})}{n - 1}} \sqrt{1 - \frac{100}{8000}} = .03$$

A 95% confidence interval for p is $\hat{p} \pm 1.96s_{\hat{p}}$, or $(.06, .18)$.

The total number of owners planning to sell is estimated as $T = N\hat{p} = 960$. The estimated standard error of T is $s_T = Ns_{\hat{p}} = 240$. A 95% confidence interval for the number in the population planning to sell is $T \pm 1.96S_T$, or $(490, 1430)$. □

The width of a confidence interval is determined by the sample size n and the population standard deviation σ. If σ is known approximately, perhaps from earlier samples of the population, n can be chosen so as to obtain a confidence interval close to some desired length. Such analysis is usually an important aspect of planning the design of a sample survey.

EXAMPLE E.   The interval for the total number of owners planning to sell in Example D might be considered too wide for practical purposes; reducing its width would require a larger sample size. Suppose that an interval with a half-width of 200 is desired. Neglecting the finite population correction, the half-width is

$$1.96s_T = 1.96N\sqrt{\frac{\hat{p}(1 - \hat{p})}{n}} = \frac{5095}{\sqrt{n}}$$

Setting the last expression equal to 200 and solving for n yields $n = 649$ as the necessary sample size. □

7.4   Estimation of a Ratio

The foundations of the theory of survey sampling have been laid in the preceding sections on simple random sampling. This and the next section build on that foundation, developing some advanced topics in survey sampling.

In this section, we consider the estimation of a ratio. Suppose that for each member of a population, two values, x and y, may be measured. The ratio of interest is

$$r = \frac{\sum_{i=1}^{N} y_i}{\sum_{i=1}^{N} x_i} = \frac{\mu_y}{\mu_x}$$

Ratios arise frequently in sample surveys; for example, if households are sampled, the following ratios might be calculated:

- If y is the number of unemployed males aged 20–30 in a household and x is the number of males aged 20–30 in a household, then r is the proportion of unemployed males aged 20–30.
- If y is weekly food expenditure and x is number of inhabitants, then r is weekly food cost per inhabitant.
- If y is the number of motor vehicles and x is the number of inhabitants of driving age, then r is the number of motor vehicles per inhabitant of driving age.

In a survey of farms, y might be the acres of wheat planted and x the total acreage. In an inventory audit, y might be the audited value of an item and x the book value.

In this section, we first consider directly the problem of estimating a ratio. Later, we will use the estimation of a ratio as a technique for estimating μ_y. We will produce a new estimate, the ratio estimate, which we will compare to the ordinary estimate, $\bar{Y}$.

Before continuing, we note the elementary but sometimes overlooked fact that

$$r \neq \frac{1}{N} \sum_{i=1}^{N} \frac{y_i}{x_i}$$

Suppose that a sample is drawn consisting of the pairs (X_i, Y_i); the natural estimate of r is $R = \bar{Y}/\bar{X}$. We wish to derive expressions for $E(R)$ and $\text{Var}(R)$, but since R is a nonlinear function of the random variables $\bar{X}$ and $\bar{Y}$, we cannot do this in closed form. We will therefore employ the approximate methods of Section 4.6.

In order to calculate the approximate variance of R, we need to know $\text{Var}(\bar{X})$, $\text{Var}(\bar{Y})$, and $\text{Cov}(\bar{X}, \bar{Y})$. The first two quantities we know from Theorem B of Section 7.3.1. For the last quantity, we define the **population covariance** of x and y to be

$$\sigma_{xy} = \frac{1}{N} \sum_{i=1}^{N} (x_i - \mu_x)(y_i - \mu_y)$$

It can then be shown, in a manner entirely analogous to the proof of Theorem B in Section 7.3.1, that

$$\text{Cov}(\bar{X}, \bar{Y}) = \frac{\sigma_{xy}}{n}\left(1 - \frac{n-1}{N-1}\right)$$

From Example C in Section 4.6, we have the following theorem.

THEOREM A.   With simple random sampling, the approximate variance of $R = \bar{Y}/\bar{X}$ is

$$\text{Var}(R) \approx \frac{1}{\mu_x^2}(r\sigma_{\bar{X}}^2 + \sigma_{\bar{Y}}^2 - 2r\sigma_{\bar{X}\bar{Y}})$$

$$= \frac{1}{n}\left(1 - \frac{n-1}{N-1}\right)\frac{1}{\mu_x^2}(r^2\sigma_x^2 + \sigma_y^2 - 2r\sigma_{xy})$$

The **population correlation coefficient** is defined as

$$\rho = \frac{\sigma_{xy}}{\sigma_x \sigma_y}$$

and is used as a measure of the strength of the linear relationship between the x and y values in the population. It can be shown that $-1 \le \rho \le 1$; large values of ρ indicate a strong positive relationship between x and y, and small values indicate a strong negative relationship. The equation in Theorem A can be expressed in terms of the population correlation coefficient as follows:

$$\text{Var}(R) \approx \frac{1}{n}\left(1 - \frac{n-1}{N-1}\right)\frac{1}{\mu_x^2}(r^2\sigma_x^2 + \sigma_y^2 - 2r\rho\sigma_x\sigma_y)$$

From this equation, we see that strong correlation of the same sign as r decreases the variance. We also note that the variance is affected by the size of μ_x—if μ_x is small, the variance is large, essentially because small values of $\bar{X}$ in the ratio $R = \bar{Y}/\bar{X}$ cause R to fluctuate wildly.

We now consider the approximate expectation of R. From Example C in Section 4.6 and the calculations above, we have the following theorem.

THEOREM B.   With simple random sampling, the expectation of R is given approximately by

$$E(R) \approx r + \frac{1}{n}\left(1 - \frac{n-1}{N-1}\right)\frac{1}{\mu_x^2}(r\sigma_x^2 - \rho\sigma_x\sigma_y)$$

From the equation in Theorem B, we see that strong correlation of the same sign as r decreases the bias and that the bias is large if μ_x is small. Furthermore, note that the bias is of the order $1/n$, so its contribution to the mean squared error is of the order $1/n^2$. In comparison, the contribution of the variance is of the order $1/n$. Therefore, for large samples, the bias is negligible compared to the standard error of the estimate.

For large samples, truncating the Taylor Series after the linear term provides a good approximation, since the deviations $\bar{X} - \mu_X$ and $\bar{Y} - \mu_Y$ are likely to be small. To this order of approximation, R is expressed as a linear combination of $\bar{X}$ and $\bar{Y}$, and an argument based on the central limit theorem can be used to show that R is approximately normally distributed. Approximate confidence intervals can thus be formed for r by using the normal distribution.

In order to estimate the standard error of R, we substitute R for r in the formula of Theorem A. The x and y population variances are estimated by s_x^2 and s_y^2. The population covariance is estimated by

$$s_{xy} = \frac{1}{n-1} \sum_{i=1}^{n} (X_i - \bar{X})(Y_i - \bar{Y})$$

$$= \frac{1}{n-1} \left(\sum_{i=1}^{n} X_i Y_i - n\bar{X}\bar{Y} \right)$$

(as can be seen by expanding the product), and the population correlation is estimated by

$$\hat{\rho} = \frac{s_{xy}}{s_x s_y}$$

The estimated variance of R is thus

$$s_R^2 = \frac{1}{n}\left(1 - \frac{n-1}{N-1}\right)\frac{1}{\bar{X}^2}(R^2 s_x^2 + s_y^2 - 2R s_{xy})$$

An approximate $100(1 - \alpha)\%$ confidence interval for r is $R \pm z(\alpha/2)s_R$.

EXAMPLE A.   Suppose that 100 people who recently bought houses are surveyed, and the monthly mortgage payment and gross income of each buyer are determined. Let y denote the mortgage payment and x the gross income. Suppose that

$$\bar{X} = \$3100 \qquad \bar{Y} = \$868$$

$$s_y = \$250 \qquad s_x = \$1200$$

$$\hat{\rho} = .85 \qquad R = .28$$

Neglecting the finite population correction, the estimated standard error of R is

$$s_R = \frac{1}{10}\left(\frac{1}{3100}\right)\sqrt{.28^2 \times 1200^2 + 250^2 - 2 \times .28 \times .85 \times 250 \times 1200}$$

$$= .006$$

An approximate 95% confidence interval for r is $.28 \pm (1.96) \times (.006)$, or $.28 \pm .012$. Note that the high correlation between x and y causes the standard error of R to be small. We can use the observed values for the variances, covariances, and means to gauge the order of magnitude of the bias by substituting them in place of the population parameters in the formula of Theorem B. Doing so, and again neglecting the finite population correction, gives the value .00025 for the bias. Note that the large value of $\bar{X}$ and the large positive correlation coefficient cause the bias to be small.   □

Ratios may also be used as tools for estimating population means and totals. To illustrate the concept, we return to the example of hospital discharges. For this population, the number of beds in each hospital is also known; let us denote the number of beds in the ith hospital by x_i and the number of discharges by y_i. Suppose that all the x_i are known, perhaps from an earlier enumeration, before a sample has been taken to estimate the number of discharges, and that we would like to take advantage of this information. One way to do this is to form a **ratio estimate** of μ_y:

$$\bar{Y}_R = \frac{\mu_x}{\bar{X}} \bar{Y} = \mu_x R$$

where $\bar{X}$ is the average number of beds and $\bar{Y}$ is the average number of discharges in the sample. The idea is fairly simple: We expect x_i and y_i to be closely related in the population, since a hospital with a large number of beds should tend to have a large number of discharges. This is borne out by Figure 7-4, a scatterplot

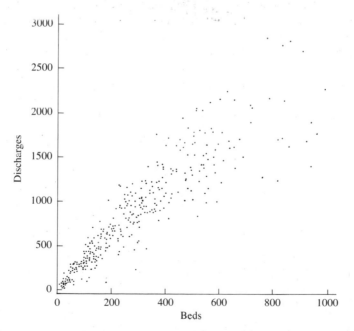

Figure 7-4. Scatterplot of the number of discharges versus the number of beds for the 393 hospitals.

of the number of discharges versus the number of beds. If $\bar{X} < \mu_x$, the sample underestimates the number of beds and probably the number of discharges as well; multiplying $\bar{Y}$ by $\mu_x/\bar{X}$ increases $\bar{Y}$ to $\bar{Y}_R$.

To see how the above ratio estimate works in practice, it was simulated from 500 samples of size 64. The histogram of the results is shown in Figure 7-5 along with the histogram of the means of 500 simple random samples of size 64. The comparison shows dramatically how effective the ratio estimate is at reducing variability.

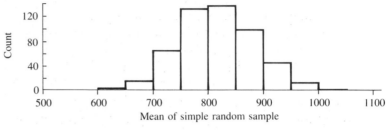

(a)

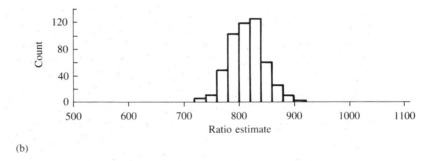

(b)

Figure 7-5.   (a) A histogram of the means of 500 simple random samples of size 64 from the population of discharges; (b) a histogram of the values of 500 ratio estimates of the mean number of discharges from samples of size 64.

Two more examples will illustrate the scope of the ratio estimation method.

EXAMPLE B.   Suppose that we wish to estimate the total number of unemployed males aged 20–30 from a sample of households and that we know τ_x, the total number of males aged 20–30, from census data. The ratio estimate is

$$T_R = \tau_x \frac{\bar{Y}}{\bar{X}}$$

where $\bar{Y}$ is the average number of unemployed males aged 20–30 per household in the sample, and $\bar{X}$ is the sample average number of males aged 20–30 per household. ∎

EXAMPLE C. A sample of items in an inventory is taken to estimate the total value of the inventory. Let Y_i be the audited value of the ith sample item, and let X_i be its book value. We assume that τ_x, the total book value of the inventory, is known, and we estimate the total audited value by

$$T_R = \tau_x \frac{\bar{Y}}{\bar{X}}$$

∎

We will now analyze the observed success of the ratio estimate. From Theorem A, and using that $\text{Var}(\bar{Y}_R) = \mu_x^2 \, \text{Var}(R)$, we have the following corollary.

COROLLARY A. The approximate variance of the ratio estimate of μ_y is

$$\text{Var}(\bar{Y}_R) \approx \frac{1}{n}\left(1 - \frac{n-1}{N-1}\right)(r^2\sigma_x^2 + \sigma_y^2 - 2r\rho\sigma_x\sigma_y)$$

Similarly, from Theorem B, we have another corollary.

COROLLARY B. The approximate bias of the ratio estimate of μ_y is

$$E(\bar{Y}_R) - \mu_Y \approx \frac{1}{n}\left(1 - \frac{n-1}{N-1}\right)\frac{1}{\mu_x}(r\sigma_x^2 - \rho\sigma_x\sigma_y)$$

In the following, the finite population correction is neglected for simplicity. Since the variance of the ordinary estimate $\bar{Y}$ is

$$\text{Var}(\bar{Y}) = \frac{\sigma_y^2}{n}$$

the ratio estimate has a smaller variance if

$$r^2\sigma_x^2 - 2\rho\sigma_x\sigma_y < 0$$

or

$$2\rho\sigma_y > r\sigma_x$$

Letting $C_x = \sigma_x/\mu_x$ and $C_y = \sigma_y/\mu_y$, this last inequality is equivalent to

$$\rho > \frac{1}{2}\left(\frac{C_x}{C_y}\right)$$

C_x and C_y are called **coefficients of variation** and give the standard deviation as a proportion of the mean. Coefficients of variation are often more meaningful than standard deviations. For example, a standard deviation of 10 means one thing if the true value of the quantity being measured is 100 and something entirely different if the true value is 10,000.

EXAMPLE D.   For the population of 393 hospitals, we have

$$\mu_x = 274.8 \qquad \sigma_x = 213.2$$

$$\mu_y = 814.6 \qquad \sigma_y = 589.7$$

$$r = 2.96 \qquad \rho = .91$$

Thus,

$$\text{Var}(\bar{Y}_R) \approx \frac{1}{n}(2.96^2 \times 213.2^2 + 589.7^2 - 2 \times 2.96 \times .91 \times 213.2 \times 598.7)$$

$$= \frac{68,697.4}{n}$$

and

$$\sigma_{\bar{Y}_R} \approx \frac{262.1}{\sqrt{n}}$$

Including the finite population correction, the linearized approximation predicts that, with $n = 64$,

$$\sigma_{\bar{Y}_R} = \tfrac{1}{8}(262.1)\sqrt{1 - \frac{63}{392}} = 31.0$$

The actual standard deviation of the 500 sample values displayed in Figure 7-5 is 29.9, which is remarkably close. The mean of the 500 values is 816.2, compared to the population mean of 814.6; the slight apparent bias is consistent with Corollary B.

The standard deviation of $\bar{Y}$ from a simple random sample is, neglecting the finite population correction,

$$\sigma_{\bar{Y}} = \frac{589.7}{\sqrt{n}}$$

The ratio estimate achieved a substantially smaller standard deviation.

The following is another way of interpreting this comparison. If a simple random sample of size n_1 is taken, the variance of the estimate is $\text{Var}(\bar{Y}) = 589.7^2/n_1$. A ratio estimate from a sample of size n_2 will have the same variance if

$$\frac{262.1^2}{n_2} = \frac{589.7^2}{n_1}$$

or

$$n_2 = n_1 \left(\frac{262.1}{589.7}\right)^2 = .1975 n_1$$

Thus, in this case, we can obtain the same precision from a ratio estimate *using a sample about 80% smaller* than the simple random sample. Note that this comparison neglects the bias of the ratio estimate, which is justifiable in this case because the bias is quite small. Here is a case in which a biased estimate performs substantially better than an unbiased estimate, the bias being quite small and the reduction in variance being quite large. ☐

The formation of an approximate confidence interval around a ratio estimate of a mean is fairly straightforward. The standard error of $\bar{Y}_R$ is estimated by substituting estimates for the unknown quantities in the expression for $\text{Var}(\bar{Y}_R)$ in Corollary A. Since the values of x are known, σ_x^2 and the denominator of r are known; σ_y^2 and ρ must be estimated from the sample.

7.5 Stratified Random Sampling

7.5.1 Introduction and Notation

In stratified random sampling, the population is partitioned into subpopulations, or **strata**, which are then independently sampled. The results from the strata are then combined to estimate population parameters, such as the mean.

Following are some examples that suggest the range of situations in which stratification is natural:

- In auditing transactions, the transactions may be grouped into strata on the basis of their nominal values. For example, high-value, medium-value, and low-value strata might be formed.
- In samples of human populations, geographical areas often form natural strata.
- In a study of records of shipments of household goods by motor carriers, the carriers were grouped into three strata: large carriers, medium carriers, and small carriers.

Stratified samples are used for a variety of reasons. We are often interested in obtaining information about each of a number of natural subpopulations in addition to information about the population as a whole. The subpopulations might be defined by geographical areas or age groups. In an industrial application in which the population consists of items produced by a manufacturing process, relevant subpopulations might consist of items produced during different shifts or from different lots of raw material. The use of a stratified random sample guarantees a prescribed number of observations from each subpopulation, whereas the use of a simple random sample can result in underrepresentation of some subpopulations. A second reason for using stratification is that, as will be shown below, the stratified sample mean can be considerably more precise than the mean of a simple random sample, especially if the population members within each stratum are relatively homogeneous and if there is considerable variation between strata.

In the next section, properties of the stratified estimated are derived. Since a simple random sample is taken within each stratum, the results will follow easily from the derivations of earlier sections. The section after that takes up the problem of how to allocate the total number of observations, n, among the various strata. Comparisons will be made of the efficiencies of different allocation schemes and also of the precisions of these allocation schemes relative to that of a simple random sample of the same total size.

7.5.2   Properties of Stratified Estimates

We will denote by N_l, where $l = 1, \ldots, L$, the population sizes in the L strata; also, $N_1 + N_2 + \cdots + N_L = N$, the total population size. The population mean and variance of the lth stratum are denoted by μ_l and σ_l^2. The overall population mean can be expressed in terms of the μ_l as follows:

$$\mu = \frac{1}{N} \sum_{l=1}^{L} \sum_{i=1}^{N_l} x_{il}$$

$$= \frac{1}{N} \sum_{l=1}^{L} N_l \mu_l$$

$$= \sum_{l=1}^{L} W_l \mu_l$$

where x_{il} denotes the ith population value in the lth stratum and $W_l = N_l/N$ is the fraction of the population contained in the lth stratum.

Within each stratum, a simple random sample of size n_l is taken. The sample mean in stratum l is denoted by

$$\bar{X}_l = \frac{1}{n_l} \sum_{i=1}^{n_l} X_{il}$$

Here X_{il} denotes the ith observation in the lth stratum. By analogy with the above relationship between the overall population mean and the population means of the various strata, the obvious estimate of μ is

$$\bar{X}_s = \sum_{l=1}^{L} \frac{N_l \bar{X}_l}{N}$$

$$= \sum_{l=1}^{L} W_l \bar{X}_l$$

THEOREM A. The stratified estimate, $\bar{X}_s$, of the population mean is unbiased.

Proof.

$$E(\bar{X}_s) = \sum_{l=1}^{L} W_l E(\bar{X}_l)$$

$$= \frac{1}{N} \sum_{l=1}^{L} N_l \mu_l$$

$$= \mu \qquad \qquad \square$$

Since we assume that the samples from different strata are independent of one another and that within each stratum a simple random sample is taken, the variance of $\bar{X}_s$ can be easily calculated.

THEOREM B. The variance of the stratified sample mean is given by

$$\text{Var}(\bar{X}_s) = \sum_{l=1}^{L} W_l^2 \left(\frac{1}{n_l}\right)\left(1 - \frac{n_l - 1}{N_l - 1}\right)\sigma_l^2$$

Proof. Since the $\bar{X}_l$ are independent,

$$\text{Var}(\bar{X}_s) = \sum_{l=1}^{L} W_l^2 \, \text{Var}(\bar{X}_l)$$

From Theorem B of Section 7.3.1, we have

$$\text{Var}(\bar{X}_l) = \frac{1}{n_l}\left(1 - \frac{n_l - 1}{N_l - 1}\right)\sigma_l^2$$

Therefore, the desired result follows. $\qquad \square$

If the sampling fractions within all strata are small,

$$\text{Var}(\bar{X}_s) \approx \sum_{l=1}^{L} \frac{W_l^2 \sigma_l^2}{n_l}$$

EXAMPLE A.   We again consider the population of hospitals. As we did in the discussion of ratio estimates, we assume that the number of beds in each hospital is known but that the number of discharges is not. We will try to make use of this knowledge by stratifying the hospitals according to the number of beds. Let stratum A consist of the 98 smallest hospitals, stratum B of the 98 next larger, stratum C of the 98 next larger, and stratum D of the 99 largest. The following table shows the results of this stratification of hospitals by size:

Stratum	N_l	W_l	μ_l	σ_l
A	98	.249	182.9	103.4
B	98	.249	526.5	204.8
C	98	.249	956.3	243.5
D	99	.251	1591.2	419.2

Suppose that we use a sample of total size n and let

$$n_1 = n_2 = n_3 = n_4 = \frac{n}{4}$$

so that we have equal sample sizes in each stratum. Then, from Theorem B, neglecting the finite population corrections and using the numerical values in the above table, we have

$$\text{Var}(\bar{X}_s) = \sum_{l=1}^{4} \frac{W_l^2 \sigma_l^2}{n_l}$$

$$= \frac{4}{n} \sum_{l=1}^{4} W_l^2 \sigma_l^2$$

$$= \frac{72,137.9}{n}$$

and

$$\sigma_{\bar{X}_s} = \frac{268.6}{\sqrt{n}}$$

The standard deviation of the mean of a simple random sample is

$$\sigma_{\bar{X}} = \frac{587.7}{\sqrt{n}}$$

Comparing the two standard deviations, we see that a tremendous gain in precision has resulted from the stratification. The ratio of the variances is .20; thus a stratified estimate based on a total sample size of $n/5$ is as precise as a simple random sample of size n. The reduction in variance due to stratification is comparable to that achieved by using a ratio estimate (Example D in Section 7.4). In later parts of this section, we will look more analytically at why the stratification done here produced such dramatic improvement.   □

Let us next consider the stratified estimate of the population total, $T_s = N\bar{X}_s$. From Theorem B, we have the following corollary.

COROLLARY A.   The expectation and variance of the stratified estimate of the population total are

$$E(T_s) = \tau$$

and

$$Var(T_s) = N^2 Var(\bar{X}_s)$$

$$= \sum_{l=1}^{L} N_l^2 \left(\frac{1}{n_l}\right)\left(1 - \frac{n_l - 1}{N_l - 1}\right)\sigma_l^2$$

In order to estimate the standard errors of $\bar{X}_s$ and T_s, the variances of the individual strata must be separately estimated and substituted into the formulae above. The estimate of σ_l^2 is given by

$$s_l^2 = \frac{1}{n_l - 1} \sum_{i=1}^{n_l} (X_{il} - \bar{X}_l)^2$$

$Var(\bar{X}_s)$ is estimated by

$$s_{\bar{X}_s}^2 = \sum_{l=1}^{L} W_l^2 \left(\frac{1}{n_l}\right)\left(1 - \frac{n_l}{N_l}\right)s_l^2$$

EXAMPLE B.   A sample of size 10 was drawn from each of the four strata of hospitals described in Example A, yielding the following:

$$\bar{X}_1 = 240.6 \qquad s_1^2 = 6827.6$$

$$\bar{X}_2 = 507.4 \qquad s_2^2 = 23{,}790.7$$

$$\bar{X}_3 = 865.1 \qquad s_3^2 = 42{,}573.0$$

$$\bar{X}_4 = 1716.5 \qquad s_4^2 = 152{,}099.6$$

Therefore, $\bar{X}_s = 832.5$. The variance of the stratified sample mean is estimated by

$$s_{\bar{X}_s}^2 = \frac{1}{10} \sum_{l=1}^{4} W_l^2 \left(1 - \frac{n_l - 1}{N_l - 1}\right)s_l^2$$

$$= 1281.0$$

Thus,

$$s_{\bar{X}_s} = 35.8$$

An approximate 95% confidence interval for the population mean number of discharges is $\bar{X}_s \pm 1.96s_{\bar{X}_s}$, or (762.4, 902.7).

The total number of discharges is estimated by $T_s = 393\bar{X}_s = 327{,}172$. The standard error of T_s is estimated by $s_{T_s} = 393s_{\bar{X}_s} = 14{,}069$. An approximate 95% confidence interval for the population total is $T_s \pm 1.96s_{T_s}$, or (299,596, 354,748).   □

7.5.3   Methods of Allocation

In Section 7.5.2, it was shown that, neglecting the finite population correction,

$$\text{Var}(\bar{X}_s) = \sum_{l=1}^{L} \frac{W_l^2 \sigma_l^2}{n_l}$$

If the resources of a survey allow only a total of n units to be sampled, the question arises of how to choose $n_1, \ldots, n_L$ to minimize $\text{Var}(\bar{X}_s)$ subject to the constraint $n_1 + \cdots + n_L = n$.

For the sake of simplicity, the calculations in this section ignore the finite population correction within each stratum. The analysis may be extended to include these corrections, but the cost is some additional algebra. More complete results are contained in Cochran (1977).

THEOREM A.   The sample sizes $n_1, \ldots, n_L$ that minimize $\text{Var}(\bar{X}_s)$ subject to the constraint $n_1 + \cdots + n_L = n$ are given by

$$n_l = n \frac{W_l \sigma_l}{\sum_{k=1}^{L} W_k \sigma_k}$$

where $l = 1, \ldots, L$.

Proof.   We introduce a Lagrange multiplier, and we must then minimize

$$L(n_1, \ldots, n_L, \lambda) = \sum_{l=1}^{L} \frac{W_l^2 \sigma_l^2}{n_l} + \lambda\left(\sum_{l=1}^{L} n_l - n\right)$$

For $l = 1, \ldots, L$, we have

$$\frac{\partial L}{\partial n_l} = -\frac{W_l^2 \sigma_l^2}{n_l^2} + \lambda$$

Setting these partial derivatives equal to zero, we have the system of equations

$$n_l = \frac{W_l \sigma_l}{\sqrt{\lambda}}$$

where $l = 1, \ldots, L$. To determine λ, we first sum these equations over l:

$$n = \frac{1}{\sqrt{\lambda}} \sum_{l=1}^{L} W_l \sigma_l$$

Thus,

$$\frac{1}{\sqrt{\lambda}} = \frac{n}{\sum_{l=1}^{L} W_l \sigma_l}$$

and

$$n_l = n \frac{W_l \sigma_l}{\sum_{l=1}^{L} W_l \sigma_l}$$

which proves the theorem. □

This theorem shows that those strata for which $W_l \sigma_l$ is large should be sampled heavily. This makes sense intuitively. If W_l is large, the stratum contains a large fraction of the population; if σ_l is large, the population values in the stratum are quite variable, and in order to obtain a good determination of the stratum's mean, a relatively large sample size must be used. This optimal allocation scheme is called **Neyman allocation**.

Substituting the optimal values of n_l as given in Theorem A into the equation for $\text{Var}(\bar{X}_s)$ given in Theorem B in Section 7.5.2 gives us the following corollary.

COROLLARY A.   Denoting by $\bar{X}_{so}$, the stratified estimate using the optimal allocations as given in Theorem A and neglecting the finite population correction,

$$\text{Var}(\bar{X}_{so}) = \frac{\left(\sum\limits_{l=1}^{L} W_l \sigma_l \right)^2}{n}$$

EXAMPLE A.   For the population of hospitals, the weights for optimal allocation, $W_l \sigma_l / \sum W_l \sigma_l$, are, from the table of Example A of Section 7.5.2,

	Stratum			
	A	B	C	D
Weight	.108	.208	.249	.432

Note that, because of its larger standard deviation, stratum D is sampled four times as heavily as stratum A.   □

The optimal allocations depend on the individual variances of the strata, which generally will not be known. Furthermore, if a survey measures several attributes for each population member, it is usually impossible to find an allocation that is simultaneously optimal for each of those variables. A simple and popular alternative method of allocation is to use the same sampling fraction in each stratum,

$$\frac{n_1}{N_1} = \frac{n_2}{N_2} = \cdots = \frac{n_L}{N_L}$$

which holds if

$$n_l = n\frac{N_l}{N} = nW_l$$

for $l = 1, \ldots, L$. This method is called **proportional allocation**. The estimate of the population mean based on proportional allocation is

$$\bar{X}_{sp} = \sum_{l=1}^{L} W_l \bar{X}_l$$

$$= \sum_{l=1}^{L} W_l \frac{1}{n_l} \sum_{i=1}^{n_l} X_{il}$$

$$= \frac{1}{n} \sum_{l=1}^{L} \sum_{i=1}^{n_l} X_{il}$$

since $W_l/n_l = 1/n$. This estimate is simply the unweighted mean of the sample values.

THEOREM B.   With stratified sampling based on proportional allocation, ignoring the finite population correction,

$$\mathrm{Var}(\bar{X}_{sp}) = \frac{1}{n} \sum_{l=1}^{L} W_l \sigma_l^2$$

Proof.   From Theorem B of Section 7.5.2, we have

$$\mathrm{Var}(\bar{X}_{sp}) = \sum_{l=1}^{L} \frac{W_l^2}{nW_l} \sigma_l^2$$

which simplifies to the desired result.   □

We now compare $\text{Var}(\bar{X}_{sp})$ and $\text{Var}(\bar{X}_{so})$ in order to discover the circumstances under which optimal allocation is substantially better than proportional allocation.

THEOREM C. With stratified random sampling, the difference between the variance of the estimate of the population mean based on proportional allocation and the variance of that estimate based on optimal allocation is, ignoring the finite population correction,

$$\text{Var}(\bar{X}_{sp}) - \text{Var}(\bar{X}_{so}) = \frac{1}{n} \sum_{l=1}^{L} W_l(\sigma_l - \bar{\sigma})^2$$

where

$$\bar{\sigma} = \sum_{l=1}^{L} W_l \sigma_l$$

Proof.

$$\text{Var}(\bar{X}_{sp}) - \text{Var}(\bar{X}_{so}) = \frac{1}{n}\left[\sum_{l=1}^{L} W_l \sigma_l^2 - \left(\sum_{l=1}^{L} W_l \sigma_l \right)^2 \right]$$

The term within the large brackets equals $\sum_{l=1}^{L} W_l(\sigma_l - \bar{\sigma})^2$, which may be verified by expanding the square and collecting terms. $\qquad\square$

According to Theorem C, if the variances of the strata are all the same, proportional allocation yields the same results as optimal allocation. The more variable these variances are, the better it is to use optimal allocation.

EXAMPLE B. Let us calculate how much better optimal allocation is than proportional allocation for the population of hospitals. From Theorem C and Corollary A, we have

$$\text{Var}(\bar{X}_{sp}) = \text{Var}(\bar{X}_{so}) + \frac{1}{n} \sum W_l(\sigma_l - \bar{\sigma})^2$$

Therefore,

$$\frac{\text{Var}(\bar{X}_{sp})}{\text{Var}(\bar{X}_{so})} = 1 + \frac{\frac{1}{n} \sum W_l(\sigma_l - \bar{\sigma})^2}{\text{Var}(\bar{X}_{so})}$$

$$= 1 + \frac{\sum W_l(\sigma_l - \bar{\sigma})^2}{(\sum W_l \sigma_l)^2}$$

$$= 1 + .218$$

Thus, under proportional allocation, the variance of the mean is about 20% larger than it is under optimal allocation. $\qquad\square$

We can also compare the variance under simple random sampling with the variance under proportional allocation. The variance under simple random sampling is, neglecting the finite population correction,

$$\text{Var}(\bar{X}) = \frac{\sigma^2}{n}$$

In order to compare this equation with that for the variance under proportional allocation, we need a relationship between the overall population variance, σ^2, and the strata variances, σ_l^2. The overall population variance may be expressed as

$$\sigma^2 = \frac{1}{N} \sum_{l=1}^{L} \sum_{i=1}^{N_l} (x_{il} - \mu)^2$$

Also,

$$(x_{il} - \mu)^2 = [(x_{il} - \mu_l) + (\mu_l - \mu)]^2$$
$$= (x_{il} - \mu_l)^2 + 2(x_{il} - \mu_l)(\mu_l - \mu) + (\mu_l - \mu)^2$$

When both sides of this last equation are summed over l, the middle term on the right-hand side becomes zero, so we have

$$\sum_{i=1}^{N_l} (x_{il} - \mu)^2 = \sum_{i=1}^{N_l} (x_{il} - \mu_l)^2 + N_l(\mu_l - \mu)^2$$
$$= N_l \sigma_l^2 + N_l(\mu_l - \mu)^2$$

Dividing both sides by N and summing over l, we have

$$\sigma^2 = \sum_{l=1}^{L} W_l \sigma_l^2 + \sum_{l=1}^{L} W_l(\mu_l - \mu)^2$$

Substituting this expression for σ^2 into $\text{Var}(\bar{X}) = \sigma^2/n$ and using the formula for $\text{Var}(\bar{X}_{sp})$ given in Theorem B completes a proof of the following theorem.

THEOREM D. The difference between the variance of the mean of a simple random sample and the variance of the mean of a stratified random sample based on proportional allocation is, neglecting the finite population correction,

$$\text{Var}(\bar{X}) - \text{Var}(\bar{X}_{sp}) = \frac{1}{n} \sum_{l=1}^{L} W_l(\mu_l - \mu)^2$$

Thus, stratified random sampling with proportional allocation always gives a smaller variance than does simple random sampling, providing that the finite population correction is ignored. Comparing the equations for the variances under simple random sampling, proportional allocation, and optimal allocation, we see that stratification with proportional allocation is better than simple random sampling if the strata means are quite variable and that stratification with optimal allocation is even better than stratification with proportional allocation if the strata standard deviations are variable.

EXAMPLE C.   We calculate the improvement that would result from using stratification with proportional allocation rather than simple random sampling for the population of hospitals. From Theorems C and D, we have

$$\frac{\text{Var}(\bar{X}_{srs})}{\text{Var}(\bar{X}_{sp})} = 1 + \frac{\sum W_l(\mu_l - \bar{\mu})^2}{\sum W_l\sigma_l^2}$$

$$= 1 + 3.83$$

As is frequently the case, the gain from using stratification with proportional allocation rather than simple random sampling is much greater than the gain from using optimal allocation rather than proportional allocation. Furthermore, proportional allocation only requires knowledge of the sizes of the strata, whereas optimal allocation requires knowledge of the standard deviations of the strata, and such knowledge is usually unavailable.   ☐

Typically, stratified random sampling can result in substantial increases in precision for populations containing values that vary greatly in size. For example, a population of transactions, a sample of which is to be audited for errors, might contain transactions in the hundreds of thousands of dollars and transactions in the hundreds of dollars. If such a population were divided into several strata according to the dollar amounts of the transactions, there might well be considerable variation in the mean transaction errors between the strata, since there may be rather large errors on large transactions and small errors on small transactions. The variability of the errors might also be larger in the strata as well.

We have not addressed the question of how many strata to form and how to define the strata. In order to construct the optimal number of strata, the population values themselves, which are of course unknown, would have to be used. Stratification must therefore be done on the basis of some related variable that is known (such as transaction amount in the preceding paragraph) or on the results of earlier samples. In practice, it usually turns out that such relationships are not strong enough to make it worthwhile constructing more than a few strata.

7.6   Concluding Remarks

This chapter introduced survey sampling. It first covered the most elementary method of probability sampling—simple random sampling. The theory of this method underlies the theory of more complex sampling techniques. Stratified sampling was also introduced and shown to increase the precision of estimates substantially in many cases.

Several concepts and techniques introduced here recur throughout statistics: the concept of a random estimate of a population parameter, such as the population mean; bias; the standard error of an estimate; confidence intervals based on the central limit theorem; and linearization, or propagation of error.

The theory and technique of survey sampling go far beyond the material in this introduction. One method that deserves mention because of its widespread use is **systematic sampling**. The population members are given in a list. If, say, a 10% sample is desired, every tenth member of the list is sampled starting from some random point among the first ten. If the list is in totally random order, this method is similar to simple random sampling. If, however, there is some correlation or relationship between successive members, the method is more similar to stratified sampling. The clear danger of this method is that there may be some periodic structure in the list, in which case bias can ensue.

Another commonly used method is **cluster sampling**. In sampling residential households, a survey might choose blocks randomly and then either sample every dwelling on each chosen block or further subsample the dwellings. Since one would expect dwellings within a single block to be relatively homogeneous, this method can be less precise than a simple random sample of the same size.

We have developed a mathematical model for survey sampling and have deduced consequences of that model, including probabilistic error bounds for the estimates. As is always the case, reality never quite matches the mathematical model. The basic assumptions of the model are that every population member appears in the sample with a specified probability and that an exact measurement or response is obtained from every sample member. The first assumption may well be in error; there may be real difficulties in obtaining an exact enumeration of all the population members. Furthermore, the serious problem of nonresponse haunts all surveys. For example, it is easier to contact families with children than to contact childless couples or singles. Finkner (1950) reported for a survey of fruit growers that larger growers were more likely to respond to a mailed questionnaire than were small growers. If the respondents and nonrespondents differ with respect to the characteristics being measured by the survey, serious bias can result. In a very readable article, Williams (1978) gives an illustration of how large this bias can be, even with a very high response rate. In a survey of unemployment, 98% of the employed and 95% of the unemployed responded; there was, nevertheless, a 4% relative bias incurred in the estimate of the unemployment rate. Response levels of 60–70% are common in surveys; therefore, the bias may clearly be much larger than the standard error of the estimate.

The *Literary Digest* poll of 1936, which predicted a 57% to 43% victory for Republican Alfred Landon over incumbent president Franklin Roosevelt, is one of the most famous of flawed surveys. Questionnaires were mailed to about 10 million voters, who were selected from lists such as telephone books and club memberships, and approximately 2.4 million of the questionnaires were returned.

There were two intrinsic problems: (1) nonresponse—those who did not respond may have voted differently from those who did—and (2) selection bias—even if all 10 million voters had responded, they would not have constituted a random sample; those in lower socioeconomic classes (who were more likely to vote for Roosevelt) were less likely to have telephone service or belong to clubs and thus less likely to be included in the sample than were wealthier voters.

Substantial random and systematic errors can also be present in the measurements themselves, especially when surveys seek responses from human populations. Responses can be influenced by the wording of the questions and/or by the personality characteristics of interviewers. The interesting paper by Hansen included in Tanur et al. (1972) reports some efforts of the U.S. Bureau of the Census in investigating these sorts of problems.

7.7   Problems

1. Consider a population consisting of five values—1, 2, 2, 4, and 8. Calculate the sampling distribution of the mean of a sample of size 2 by generating all possible such samples. Find the mean and variance of the sampling distribution, and compare the results to Theorems A and B in Section 7.3.1.

2. For a random sample of size n from a population of size N, consider the following as an estimate of μ:

$$\bar{X}_c = \sum_{i=1}^{n} c_i X_i$$

where the c_i are fixed numbers and $X_1, \ldots, X_n$ is the sample.
   (a) Find a condition on the c_i such that the estimate is unbiased.
   (b) Show that the choice of c_i that minimizes the variances of the estimate subject to this condition is $c_i = 1/n$, where $i = 1, \ldots, n$.

3. Here is an alternative proof of Lemma A in Section 7.3.1. Consider a random permutation $Y_1, Y_2, \ldots, Y_N$ of $x_1, x_2, \ldots, x_N$. Argue that the joint distribution of any subcollection, $Y_{i_1}, \ldots, Y_{i_n}$, of the Y_i is the same as that of a simple random sample, $X_1, \ldots, X_n$. In particular,

$$\text{Var}(Y_i) = \text{Var}(X_i) = \sigma^2$$

and

$$\text{Cov}(Y_i, Y_j) = \text{Cov}(X_k, X_l) = \gamma$$

if $i \neq j$ and $k \neq l$. Since $Y_1 + Y_2 + \cdots + Y_N = \tau$,

$$\text{Var}\left(\sum_{i=1}^{n} Y_i\right) = 0$$

(Why?) Express $\text{Var}(\sum_{i=1}^{n} Y_i)$ in terms of σ^2 and the unknown covariance, γ. Solve for γ, and conclude that

$$\gamma = -\frac{\sigma^2}{N-1}$$

for $i \neq j$.

4. In surveys, it is difficult to obtain accurate answers to sensitive questions such as "Have you ever used heroin?" or "Have you ever cheated on an exam?" Warner (1965) introduced the method of **randomized response** to deal with such situations. A respondent spins an arrow on a wheel or draws a ball from an urn containing balls of two colors to determine which of two statements to respond to: (1) "I have characteristic A," or (2) "I do not have characteristic A." The interviewer does not know which statement is being responded to, but merely records a yes or a no. The hope is that an interviewee is more likely to answer truthfully if he or she realizes that the interviewer does not know which statement is being responded to. Let R be the proportion of a sample answering "yes." Let p be the probability that statement 1 is responded to (p is known from the structure of the randomizing device), and let q be the proportion of the population that has characteristic A. Let r be the probability that a respondent answers "yes."
   (a) Show that $r = (2p - 1)q + (1 - p)$. [*Hint*: $P(\text{yes}) = P(\text{yes given question 1})P(\text{question 1}) + P(\text{yes given question 2})P(\text{question 2}).$]
   (b) If r were known, how could q be determined?
   (c) Show that $E(R) = r$, and propose an estimate, Q, for q. Show that the estimate is unbiased.
   (d) Ignoring the finite population correction, show that

$$\text{Var}(R) = \frac{r(1-r)}{n}$$

   where n is the sample size.
   (e) Find an expression for $\text{Var}(Q)$.
5. A variation of the method described in Problem 4 has been proposed. Instead of responding to statement 2, the respondent answers an unrelated question for which the probability of a "yes" response is known, for example, "Were you born in June?"
   (a) Propose an estimate of q for this method.
   (b) Show that the estimate is unbiased.
   (c) Obtain an expression for the variance of the estimate.
6. Compare the accuracies of the methods of Problems 4 and 5 by comparing their standard deviations. You may do this by substituting some plausible numerical values for p and q.
7. Referring to Example D in Section 7.3.3, how large should the sample be in order that the 95% confidence interval for the total number of owners planning to sell will have a width of 500?

8. Referring again to Example D in Section 7.3.3, suppose that a survey is done of another condominium project of 12,000 units. The sample size is 200, and the proportion planning to sell in this sample is .18.
   (a) What is the standard error of this estimate? Give a 90% confidence interval.
   (b) Suppose we use the notation $\hat{p}_1 = .12$ and $\hat{p}_2 = .18$ to refer to the proportions in the two samples. Let $\hat{d} = \hat{p}_1 - \hat{p}_2$ be an estimate of the difference, d, of the two population proportions p_1 and p_2. Using the fact that $\hat{p}_1$ and $\hat{p}_2$ are independent random variables, find expressions for the variance and standard error of $\hat{d}$.
   (c) Since $\hat{p}_1$ and $\hat{p}_2$ are approximately normally distributed, so is $\hat{d}$. Use this fact to construct 99%, 95%, and 90% confidence intervals for d. Is there clear evidence that p_1 is really less than p_2?
9. Two populations are independently surveyed using simple random samples of size n, and two proportions, p_1 and p_2, are estimated. It is expected that both population proportions are close to .5. What should the sample size be so that the standard error of the difference, $\hat{p}_1 - \hat{p}_2$, will be less than .02?
10. In a survey of a very large population, the incidences of two health problems are to be estimated from the same sample. It is expected that the first problem will affect about 3% of the population and the second about 40%. Ignore the finite population correction in answering the following questions.
    (a) How large should the sample be in order for the standard errors of both estimates to be less than .01? What are the actual standard errors for this sample size?
    (b) Suppose that instead of imposing the same limit on both standard errors, the investigator wants the standard error to be less than 10% of the true value in each case. What should the sample size be?
11. A simple random sample of a population of size 2000 yields the following 25 values:

104	109	111	109	87
86	80	119	88	122
91	103	99	108	96
104	98	98	83	107
79	87	94	92	97

   (a) Calculate an unbiased estimate of the population mean.
   (b) Calculate unbiased estimates of the population variance and $\text{Var}(\bar{X})$.
   (c) Give approximate 99% confidence intervals for the population mean and total.
12. With simple random sampling, is $\bar{X}^2$ an unbiased estimate of μ^2? If not, what is the bias?
13. Two surveys were independently conducted to estimate a population mean, μ. Denote the estimates and their standard errors by X_1 and X_2 and $\sigma_{\bar{X}_1}$ and $\sigma_{\bar{X}_2}$. Assume that $\bar{X}_1$ and $\bar{X}_2$ are unbiased. For some α and β, the two

estimates can be combined to give a better estimator:

$$X = \alpha \bar{X}_1 + \beta \bar{X}_2$$

(a) Find the conditions on α and β that make the combined estimate unbiased.

(b) What choice of α and β minimizes the variances, subject to the condition of unbiasedness?

14. Let $X_1, \ldots, X_n$ be a simple random sample. Show that $\dfrac{1}{n}\sum_{i=1}^{n} X_i^3$ is an unbiased estimate of $\dfrac{1}{N}\sum_{i=1}^{N} x_i^3$.

15. Suppose that of a population of N items, k are defective in some way. For example, the items might be documents, a small proportion of which are fraudulent. How large should a sample be so that with a specified probability it will contain at least one of the defective items? For example, if $N = 10,000$, $k = 50$, and $p = .95$, what should the sample size be? Such calculations are useful in planning sample sizes for acceptance sampling.

16. This problem presents an algorithm for drawing a simple random sample from a population in a sequential manner. The members of the population are considered for inclusion in the sample one at a time in some prespecified order (for example, the order in which they are listed). The ith member of the population is included in the sample with probability

$$\frac{n - n_i}{N - i + 1}$$

where n_i is the number of population members already in the sample before the ith member is examined. Show that the sample selected in this way is in fact a simple random sample; that is, show that every possible sample occurs with probability

$$\frac{1}{\binom{N}{n}}$$

17. In accounting and auditing, the following sampling method is sometimes used to estimate a population total. In estimating the value of an inventory, suppose that a book value exists for each item and is readily accessible. For each item in the sample, the difference D between the book value and the audited value is determined. The net inventory value is estimated by the sum of the book values of the population and $N\bar{D}$, where N is the population size.

(a) Show that the estimate is unbiased.

(b) Find an expression for the variance of the estimate.

(c) Compare the expression obtained in part (b) to the variance of the usual estimate, which is the product of N and the average audited value. Under what circumstances would the proposed method be more accurate?

(d) How could a ratio estimate be employed in this situation? Would there be any advantage or disadvantage to using a ratio estimate rather than the proposed method?

18. Show that the population correlation coefficient is less than or equal to 1 in absolute value.

19. Suppose that for Example D in Section 7.3.3, the average number of occupants per condominium unit in the sample is 2.2 with a sample standard deviation of .7 and that the sample correlation coefficient between the number of occupants and the number of motor vehicles is .85. Estimate the population ratio of the number of motor vehicles per occupant and its standard error. Find an approximate 95% confidence interval for the estimate.

20. Hartley and Ross (1954) derived the following exact bound on the relative size of the bias and standard error of a ratio estimate:

$$\frac{|E(R) - r|}{\sigma_R} \le \frac{\sigma_{\bar{X}}}{\mu_x} = \frac{\sigma_x}{\mu_x} \sqrt{\frac{1}{n}\left(1 - \frac{n-1}{N-1}\right)}$$

(a) Derive this bound from the relation

$$\text{Cov}(R, X) = E\left(\frac{\bar{Y}}{\bar{X}}\bar{X}\right) - E\left(\frac{\bar{Y}}{\bar{X}}\right)E(\bar{X})$$

(b) Apply the bound to Problem 19 using sample estimates in place of the given population parameters.

21. This problem introduces a technique called the "jackknife," originally proposed by Quenouille (1956) for reducing bias. Many nonlinear estimates, including the ratio estimator, have the property that

$$E(\hat{\theta}) = \theta + \frac{b_1}{n} + \frac{b_2}{n^2} + \cdots$$

where $\hat{\theta}$ is an estimate of θ. The jackknife forms an estimate $\hat{\theta}_J$, which has a leading bias term of the order n^{-2} rather than n^{-1}. Thus, for sufficiently large n, the bias of $\hat{\theta}_J$ is substantially smaller than that of $\hat{\theta}$. The technique involves splitting the sample into several subsamples, computing the estimate for each subsample, and then combining the several estimates. The sample is split into p groups of size m, where $n = mp$. For $j = 1, \ldots, p$, the estimate $\hat{\theta}_j$ is calculated from the $m(p-1)$ observations left after the jth group has been deleted. From the expression above,

$$E(\hat{\theta}_j) = \theta + \frac{b_1}{m(p-1)} + \frac{b_2}{[m(p-1)]^2} + \cdots$$

Now, p "pseudovalues" are defined:

$$V_j = p\hat{\theta} - (p-1)\hat{\theta}_j$$

The jackknife estimate, $\hat{\theta}_J$, is defined as the average of the pseudovalues:

$$\hat{\theta}_J = \frac{1}{p}\sum_{j=1}^{p} V_j$$

Show that the bias of $\hat{\theta}_J$ is of the order n^{-2}.

22. A population consists of three strata with $N_1 = N_2 = 1000$ and $N_3 = 500$. A stratified random sample with 10 observations in each stratum yields the following data:

Stratum 1	94	99	106	106	101	102	122	104	97	97
Stratum 2	183	183	179	211	178	179	192	192	201	177
Stratum 3	343	302	286	317	289	284	357	288	314	276

Estimate the population mean and total and give 90% confidence intervals.

23. The following table (Cochran, 1977) shows the stratification of all farms in a county by farm size and the mean and standard deviation of the number of acres of corn in each stratum.

Farm Size	N_l	μ_l	σ_l
0–40	394	5.4	8.3
41–80	461	16.3	13.3
81–120	391	24.3	15.1
121–160	334	34.5	19.8
161–200	169	42.1	24.5
201–240	113	50.1	26.0
241–	148	63.8	35.2

(a) For a sample size of 100 farms, compute the sample sizes from each stratum for proportional and optimal allocation, and compare them.

(b) Calculate the variances of the sample mean for each allocation and compare them to each other and to the variance of an estimate formed from simple random sampling.

24. (a) Suppose that the cost of a survey is $C = C_0 + C_1 n$, where C_0 is a startup cost and C_1 is the cost per observation. For a given cost C, find the allocation $n_1, \ldots, n_L$ to L strata that is optimal in the sense that it minimizes the variance of the estimate of the population mean subject to the cost constraint.

(b) Suppose that the cost of an observation varies from stratum to stratum—in some strata the observations might be relatively cheap and in others relatively expensive. The cost of a survey with an allocation $n_1, \ldots, n_L$ is

$$C = C_0 + \sum_{l=1}^{L} C_l n_l$$

For a fixed total cost C, what choice of $n_1, \ldots, n_L$ minimizes the variance?

(c) Assuming that the cost function is as given in part (b), for a fixed variance, find n_l to minimize cost.

25. Consider stratifying the population of Problem 1 into two strata: $(1, 2, 2)$ and $(4, 8)$. Assuming that one observation is taken from each stratum, find the sampling distribution of the estimate of the population mean and the mean and standard deviation of the sampling distribution. Compare to Theorems A and B in Section 7.5.2 and the results of Problem 1.

26. (Computer Exercise) Construct a population consisting of the integers from 1 to 100. Simulate the sampling distribution of the sample mean of a sample of size 12 by drawing 100 samples of size 12 and making a histogram of the results.

27. (Computer Exercise) Continuing with Problem 26, divide the population into two strata of equal size, allocate 6 observations per stratum, and simulate the distribution of the stratified estimate of the population mean. Do the same thing with four strata. Compare the results to each other and to the results of Problem 26.

28. A population consists of two strata, H and L, of sizes 100,000 and 500,000 and standard deviations 20 and 12, respectively. A stratified sample of size 100 is to be taken.
   (a) Find the optimal allocation for estimating the population mean.
   (b) Find the optimal allocation for estimating the difference of the means of the strata, $\mu_H - \mu_L$.

29. The value of a population mean increases linearly through time: $\mu(t) = \alpha + \beta t$. Independent simple random samples of size n are taken at times $t = 1, 2$, and 3.
   (a) Find conditions on w_1, w_2, and w_3 such that

$$\hat{\beta} = w_1 \bar{X}_1 + w_2 \bar{X}_2 + w_3 \bar{X}_3$$

   is an unbiased estimate of the rate of change, β. Here $\bar{X}_i$ denotes the sample mean at time t_i.
   (b) What values of the w_i minimize the variance subject to the constraint that the estimate is unbiased?

30. The value of an inventory is to be estimated by sampling. The items are stratified by book value in the following way:

Stratum	N_i	μ_i	σ_i
$1000–	70	3000	1250
$200–1000	500	500	100
$1–200	10,000	90	30

(a) What should the relative sampling fraction in each stratum be for proportional and for optimal allocation?

(b) How do the variances under each type of allocation compare to each other and to the variance under simple random sampling?

31. The following table gives values for breast cancer mortality from 1950 to 1960 (y) and the adult white female population in 1960 (x) for 301 counties in North Carolina, South Carolina, and Georgia.

(a) Make a histogram of the population values for cancer mortality.

(b) What are the population mean and total cancer mortality? What are the population variance and standard deviation?

(c) Simulate the sampling distribution of the mean of a sample of 25 observations of cancer mortality.

(d) Draw a simple random sample of size 25 and use it to estimate the mean and total cancer mortality.

(e) Estimate the population variance and standard deviation from the sample of part (d).

(f) Form 95% confidence intervals for the population mean and total from the sample of part (d). Do the intervals cover the population values?

(g) Repeat parts (d) through (f) for a sample of size 100.

(h) Suppose that the size of the total population of each county is known and that this information is used to improve the cancer mortality estimates by forming a ratio estimator. Do you think this will be effective? Why or why not?

(i) Simulate the sampling distribution of ratio estimators of mean cancer mortality based on a simple random sample of size 25. Compare this result to that of part (c).

(j) Draw a simple random sample of size 25 and estimate the population mean and total cancer mortality by calculating ratio estimates. How do these estimates compare to those formed in the usual way in part (d) from the same data?

(k) Form confidence intervals about the estimates obtained in part (j).

(l) Stratify the counties into 4 strata by population size. Randomly sample 6 observations from each stratum and form estimates of the population mean and total mortality.

(m) Stratify the counties into 4 strata by population size. What are the sampling fractions for proportional allocation and optimal allocation? Compare the variances of the estimates of the population mean obtained using simple random sampling, proportional allocation, and optimal allocation.

(n) How much better than those in part (m) will the estimates of the population mean be if 8, 16, 32, or 64 strata are used instead?

Breast cancer mortality and population

y	x	y	x	y	x
1	445	0	559	3	677
1	950	5	976	5	1096
5	1236	6	1285	3	1291
9	1438	7	1479	4	1536
4	1696	7	1792	7	1795
3	1847	8	1933	8	1959
7	2091	8	2099	5	2104
11	2172	9	2174	13	2183
4	2236	4	2245	8	2261
10	2404	4	2419	11	2462
11	2511	14	2591	6	2624
9	2736	13	2747	18	2782
12	2894	12	2906	14	2929
7	3112	9	3118	11	3185
11	3314	4	3316	13	3401
11	3488	12	3511	4	3549
15	3654	15	3680	12	3683
21	3800	16	3802	13	3832
20	4093	21	4149	15	4162
22	4329	14	4331	16	4399
16	4681	28	4737	11	4784
27	4967	17	5041	20	5051
24	5124	27	5156	25	5167
15	5773	22	5932	21	5983
25	6035	26	6074	17	6134
15	6445	33	6624	24	6841
21	6916	32	6934	23	6978
33	7115	20	7256	19	7288
36	7407	26	7408	33	7503
37	7910	20	7917	28	7957
39	8249	29	8289	22	8313
34	8493	35	8531	21	8773
51	9225	30	9243	32	9435
60	9605	19	9841	29	9994
31	10303	35	10416	27	10461
39	10890	41	11105	61	11622
43	12608	45	12775	46	12915
64	13407	64	13647	66	13870
36	14620	28	14816	59	14952
48	15204	37	16161	72	16239
51	16925	71	17027	60	17201
52	17742	65	18482	77	18731
53	19906	58	20065	75	20140
69	20969	41	21353	73	21757
92	24296	60	24351	63	24692
90	25715	111	26245	103	26408
83	28477	90	29254	97	29422
142	35112	105	35876	145	36307
104	47672	179	49126	152	53464
246	62398	236	62652	250	62931
360	88456				

8

Estimation of Parameters and Fitting of Probability Distributions

8.1 Introduction

This chapter introduces the theory of parameter estimation, one of the major themes of mathematical statistics. The results that are developed are applied to the problem of fitting probability laws to data. Many families of probability laws depend on a small number of parameters; for example, the Poisson family depends on the parameter λ (the mean number of counts), and the Gaussian family depends on two parameters, μ and σ. Unless the values of parameters are known in advance, they must be estimated from data in order to fit the probability law. Some general methods of parameter estimation will be discussed in Section 8.3.

After parameter values have been chosen, the model should be compared to the actual data to see if the fit is reasonable; chapter 9 is concerned with measures and tests of goodness of fit, but some aspects are introduced in this chapter.

In order to introduce and illustrate some of the ideas and to provide a concrete basis for later theoretical discussions, we will first consider a classical example— the fitting of a Poisson distribution to radioactive decay. The concepts introduced in this example will be elaborated on in this and the next chapter.

8.2   Fitting the Poisson Distribution to Emissions of Alpha Particles

Records of emissions of alpha particles from radioactive sources show that the number of emissions per unit of time is not constant but fluctuates in a seemingly random fashion. If the underlying rate of emission is constant over the period of observation (which will be the case if the half-life is much longer than the time period of observation) and if the particles come from a very large number of independent sources (atoms), the Poisson model seems appropriate. For this reason, the Poisson distribution is frequently used as a model for radioactive decay. You should recall that the Poisson distribution as a model for random counts in space or time rests on three assumptions: 1) the underlying rate at which the events occur is constant in space or time, 2) events in disjoint intervals of space or time occur independently, and 3) there are no multiple events.

Berkson (1966) conducted a careful analysis of data obtained from the National Bureau of Standards. The source of the alpha particles was americium 241. The experimenters recorded 10,220 times between successive emissions in 20 groups of 511 each. The observed mean emission rate (total number of emissions divided by total time) was .8392 emissions per second. The clock used to record the times was accurate to .0002 second.

There are many possible ways to test whether a Poission process is a plausible model. For example, we could test whether the time intervals between events follow an exponential distribution, whether the intervals are independent of one another, or whether the counts in a time interval of a specified length follow a Poisson distribution. Here we will focus on the counts in 1207 time intervals of length 10 seconds. The first two columns of the following table summarize the data. These columns are interpreted as follows: In 18 of the 1207 intervals, there were 0, 1, or 2 counts; in 28 of the intervals, there were 3 counts; ...; in 5 of the intervals, there were 17 or more counts.

n	Observed	Expected	χ^2
0–2	18	12.2	2.76
3	28	27.0	.04
4	56	56.5	.01
5	105	94.9	1.07
6	126	132.7	.34
7	146	159.1	1.08
8	164	166.9	.05
9	161	155.6	.19
10	123	130.6	.44
11	101	99.7	.02
12	74	69.7	.27
13	53	45.0	1.42
14	23	27.0	.59
15	15	15.1	.00
16	9	7.9	.57
17+	5	7.1	.57
	1207	1207	8.99

Let us consider modeling the 1207 values as 1207 independent realizations of Poisson random variables. Since the average number of emissions is .8392 per second, the average number in a 10-second interval is 8.392, which we view as an estimate of λ, the parameter of the Poisson distribution. (A more systematic discussion of parameter estimation is the subject of Section 8.3.)

Now consider the 16 cells into which the counts are grouped. Under the hypothesized model, the probability that a random count falls in any one of the cells may be calculated from the Poisson probability law. If X is a Poisson random variable, then $P(X = k)$ is given by

$$\pi_k = \frac{\lambda^k e^{-\lambda}}{k!}, \quad k = 0, 1, 2, \ldots$$

The probability that an observation falls in the first cell (0, 1, or 2 counts) is

$$p_1 = \pi_0 + \pi_1 + \pi_2$$

The probability that an observation falls in the second cell is $p_2 = \pi_3$. The probability that an observation falls in the sixteenth cell is

$$p_{16} = \sum_{k=17}^{\infty} \pi_k$$

Under the assumption that $X_1, \ldots, X_{1207}$ are independent Poisson random variables, the number of observations out of 1207 falling in a given cell follows a binomial distribution with a mean, or expected value, of $1207 p_k$, and the joint distribution of the counts in all the cells is multinomial with $n = 1207$ and probabilities $p_1, p_2, \ldots, p_{16}$. The third column of the table above gives the expected number of counts in each cell; for example, since $p_4 = .0786$, the expected count in the corresponding cell is $1207 \times .0786 = 94.9$.

To assess the goodness of fit of the Poisson model, we must compare the observed and expected counts. Of course, even if the model were correct, we would not expect the observed and expected counts to be in perfect agreement; indeed, agreement that was too close might make us suspicious that the data were somehow fudged. Casual examination of the table may lead us to say that the fit does not look unreasonable, but we have no yardstick with which to measure goodness of fit, and it might be difficult to convince a skeptic that the fit is good. To take into account the random fluctuations of the observed counts about their expectations, we use the following basic idea: We concoct a measure of the distance or discrepancy between the observed values (O) and the expected values (E), say, $D(O, E)$. Note that $D(O, E)$ is a random variable and has an associated sampling distribution. Suppose that we know, at least approximately, the sampling distribution of $D(O, E)$ for the case in which the model is actually correct, and that schematically it looks like Figure 8-1. Large values of D are relatively unlikely if the model is correct, so the larger the value of D, the stronger

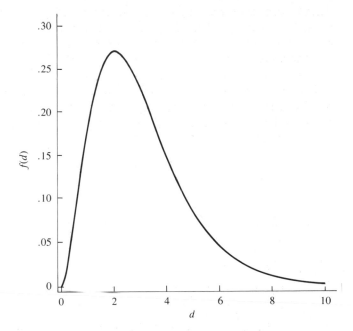

Figure 8-1.   Schematic graph of the probability density of $D(O, E)$.

the evidence against the model. If we observe $D = d$, we will use as a measure of evidence against the model

$$p^* = P(D \geq d|\text{model is correct})$$

where p^* is called a **p-value**, or a **significance probability**. [We use $P(D \geq d)$ rather than $P(D = d)$ since if D is a continuous random variable, it has zero probability of assuming any particular value.] If p^* is of moderate size, say .6, the evidence against the model is not strong. If the model were really correct, deviations this large or larger would occur with probability .6 just on the basis of chance. If p^* is very small, the only defense that can be made for the model is that it really is correct but an extremely unlikely event occurred. The smaller the p-value, the less plausible is this defense and the stronger is the evidence against the model.

One measure of discrepancy that we will use is **Pearson's chi-square statistic:**

$$X^2 = \sum_{\text{cells}} \frac{(O_i - E_i)^2}{E_i}$$

It can be shown that the sampling distribution of X^2 in the case that the model is correct is approximately the chi-square distribution with the number of degrees of freedom (df) given by

$$\text{df} = \textit{number of cells} - \textit{number of independent parameters fitted} - 1 \quad .$$

For the example discussed above, there are 14 degrees of freedom (there are 16 cells and we estimated one parameter, $\lambda = 8.392$), and from the chi-square distribution the p-value may be calculated numerically as .83 $[P(\chi^2_{14} > 8.99) = .83]$. Thus, there is no substantial evidence against the Poisson distribution's holding—if the model were correct, we would see discrepancies of this size or larger 83% of the time.

The chi-square distribution is only an approximation to the real sampling distribution of X^2, and it improves as the number of counts in each cell increases. For this reason, we grouped counts in the first and last cells in the example. The usual rule of thumb is that for the chi-square approximation to be adequate each cell should have at least five expected counts.

We will not at this point try to rationalize the form of the chi-square statistic. It is interesting to note, however, that the division of the squared discrepancy by the expected count tends to equalize the contributions of cells with small expected counts and those with large expected counts. If it were not for this division, the statistic would tend to be dominated by those cells with large expected and observed counts.

This section has informally introduced concepts that will be further elaborated on in this and the next chapter. In this chapter, we will see how a parameter of a probability distribution, such as λ in the example given above, can be estimated from data. In the next chapter, we will find out how goodness of fit can be tested.

8.3   Parameter Estimation

We approach the problem of fitting a probability law to data in the following way: The data to be fit are modeled as a collection of realized values of random variables generated by the probability law under consideration. In particular, we will usually model the n observations as realizations of random variables $X_1, \ldots, X_n$, which are independent and have the same probability distribution. We will refer to the X_i as independent and identically distributed (i.i.d.) or as a random sample from the probability law. Here we are using the term *random sample* in a slightly different way than in chapter 7, where we dealt with random samples from finite populations. We can instead think of $X_1, \ldots, X_n$ as a random sample from an infinite population described by the probability distribution in question. This model of independent and identically distributed observations is used in order to address quantitatively questions of uncertainty in the parameter values and questions of goodness of fit. Such questions arise quite naturally since the nature of data and its collection is such that if observations were taken again, they would not be exactly the same. This sampling model also makes it possible to compare different methods of parameter estimation, for example, in terms of their mean squared errors.

Although the applications are quite different, the mathematical structures of the problems considered in this chapter and the preceding one have many similarities. In chapter 7, we were concerned with estimating population parameters, such as the mean and the total, and the process of random sampling created random variables whose probability distributions depended on the population parameters [for example, $E(X_i) = \mu$]. We considered certain aspects of estimates of those parameters (for example, $\bar{X}$ as an estimate of μ), such as sampling distributions, mean squared errors, and confidence intervals. In this chapter, we consider a model in which data are generated as random variables from a probability distribution; this distribution has a more hypothetical status than the probability distribution in chapter 7, which was induced by deliberate random sampling. Here again, however, we are concerned with estimates of parameters, their sampling distributions, and related questions.

We address the following problem: Given observations of random variables $X_1, \ldots, X_n$, which are independent and identically distributed from a probability law with a cumulative distribution function (cdf) given by $F(x|\theta)$, we want to estimate θ. [We use the notation $F(x|\theta)$ to indicate that the distribution depends on the parameter θ.] The parameter θ may be one-dimensional or a vector. In order to fit the law in question, the parameter must be estimated from the data. If the distribution is discrete, we typically work with the probability mass function, or frequency function; if the distribution is continuous, we work with the density function. We will denote each of these by $f(x|\theta)$. The following are some examples.

EXAMPLE A.   (Poisson Distribution) The Poisson probability mass function is

$$f(x|\lambda) = \frac{\lambda^x e^{-\lambda}}{x!}, \quad x = 0, 1, 2, \ldots$$

The parameter to be estimated is λ. Section 8.2 contained an example of the use of the Poisson distribution as a model.   □

EXAMPLE B.   (Normal Distribution) The normal, or Gaussian, distribution involves two parameters, μ and σ, where μ is the mean of the distribution and σ^2 is the variance:

$$f(x|\mu, \sigma) = \frac{1}{\sigma\sqrt{2\pi}} e^{-1/2((x-\mu)^2/\sigma^2)}, \quad -\infty < x < \infty$$

The normal distribution plays a prominent role in statistics; many simple statistical procedures are optimal if the random variables of interest are normally distributed. These procedures may be far from optimal when the variables are not normally distributed, however.

The use of the normal distribution as a model is usually justified using some version of the central limit theorem, which says that the sum of a large number of independent random variables is approximately normally distributed. For example, Bevan, Kullberg, and Rice (1979) studied random fluctuations of current across a muscle cell membrane. The cell membrane contained a large number of channels, which opened and closed at random and were assumed to operate independently. The net current resulted from ions flowing through open channels and was therefore the sum of a large number of roughly independent currents. As the channels opened and closed, the net current fluctuated randomly. Figure 8-2 includes a smoothed histogram of values obtained from 49,152 observations of the net current and an approximating Gaussian curve. The fit of the Gaussian distribution is quite good, although the smoothed histogram seems to show a slight skewness. In this application, information about the characteristics of the individual channels, such as conductance, was extracted from the estimated parameters μ and σ^2. □

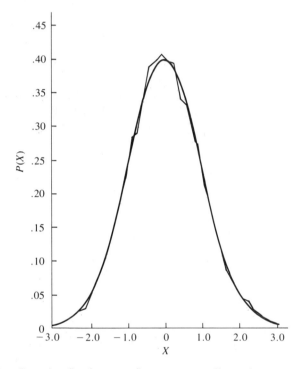

Figure 8-2. Gaussian fit of current flow across a cell membrane to a frequency polygon.

EXAMPLE C. (Gamma Distribution) The gamma distribution depends on two parameters, α and λ:

$$f(x|\alpha, \lambda) = \frac{1}{\Gamma(\alpha)}\lambda^{\alpha}x^{\alpha-1}e^{-\lambda x}, \quad 0 \le x \le \infty$$

The family of gamma distributions provides a flexible set of densities for non-negative random variables.

Figure 8-3 shows how the gamma distribution fits to the amounts of rainfall from different storms (Le Cam and Neyman, 1967). Gamma distributions were fit to rainfall amounts from storms that were seeded and unseeded in an experiment to determine the effects, if any, of seeding. Differences in the distributions between the seeded and unseeded conditions should be reflected in differences in the parameters α and λ.   □

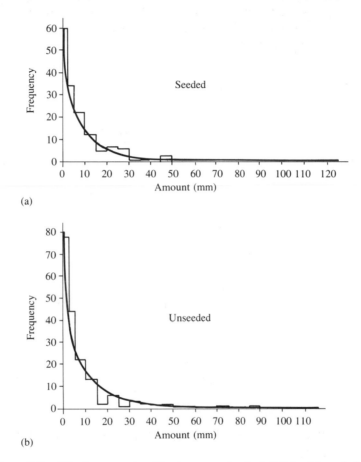

(a)

(b)

Figure 8-3. Fit of gamma densities to amounts of rainfall for (a) seeded and (b) unseeded storms.

It is useful to have general methods of parameter estimation, for otherwise we would have to treat each estimation problem as a unique case. We will discuss two common methods of estimation—the method of moments and the method of maximum likelihood.

Since the data are random, the estimate of a parameter is random as well and has a sampling distribution, which in general is unknown. So that we may assess the precision of an estimate and form approximate confidence intervals for the true parameter value, we will develop some approximate distribution theory for the maximum likelihood estimate. This development is similar to that in chapter 7, where the exact form of the sampling distribution of $\bar{X}$ depended on the actual population values, but we were able to make useful approximations to it by utilizing the central limit theorem.

The method of moments and the method of maximum likelihood do not in general produce identical estimates, and for any given problem it may be possible to propose a large number of *ad hoc* estimates. In that case, we would like to use the "best" estimate, say the one with smallest mean squared error. We will not delve deeply into notions of optimality, but we will argue that the maximum likelihood estimate is "asymptotically efficient," which gives some rationale for preferring the maximum likelihood estimate over its competitors.

8.4   The Method of Moments

The kth moment of a probability law is defined as

$$\mu_k = E(X^k)$$

where X is a random variable following that probability law (of course, this is defined only if the expectation exists). Assume that we have a sample of random variables, $X_1, X_2, \ldots, X_n$, from the probability distribution. To apply the method of moments, we express the parameters in terms of the moments of lowest possible order; we then compute the **sample moments**,

$$\hat{\mu}_k = \frac{1}{n} \sum_{i=1}^{n} X_i^k$$

and estimate the parameters from the sample moments by using the relationships between the two.

Suppose, for example, that we wish to estimate two parameters, θ_1 and θ_2. If θ_1 and θ_2 can be expressed in terms of the first two moments as

$$\theta_1 = f_1(\mu_1, \mu_2)$$

$$\theta_2 = f_2(\mu_1, \mu_2)$$

then the method of moments estimates are

$$\hat{\theta}_1 = f_1(\hat{\mu}_1, \hat{\mu}_2)$$

$$\hat{\theta}_2 = f_2(\hat{\mu}_1, \hat{\mu}_2)$$

To illustrate this procedure, we consider some examples.

EXAMPLE A.   (Poisson Distribution) The first moment for the Poisson distribution is the parameter $\lambda = E(X)$. The first sample moment is

$$\bar{X} = \frac{1}{n} \sum_{i=1}^{n} X_i$$

which is the method of moments estimate of λ.

As a concrete example, let us consider a study done at the National Bureau of Standards (Steel et al., 1980). Asbestos fibers on filters were counted as part of a project to develop measurement standards for asbestos concentration. Asbestos dissolved in water was spread on a filter, and punches of 3-mm diameter were taken from the filter and mounted on a transmission electron microscope. An operator counted the number of fibers in each of 23 grid squares, yielding the following counts:

31	29	19	18	31	28
34	27	34	30	16	18
26	27	27	18	24	22
28	24	21	17	24	

The Poisson distribution would be a plausible model for describing the variability from grid square to grid square in this situation and could be used to characterize the inherent variability in future measurements. The method of moments estimate of λ is simply the arithmetic mean of the numbers listed above, or $\hat{\lambda} = 24.9$.

In chapter 9, we will address the question of whether the Poisson distribution really fits these data. Clearly, we could calculate the average of any batch of numbers, whether or not they were well fit by the Poisson distribution.   □

EXAMPLE B.   (Normal Distribution) The first and second moments for the normal distribution are

$$\mu_1 = E(X) = \mu$$

$$\mu_2 = E(X^2) = \mu^2 + \sigma^2$$

Therefore,

$$\mu = \mu_1$$

$$\sigma^2 = \mu_2 - \mu_1^2$$

The corresponding estimates of μ and σ^2 from the sample moments are

$$\hat{\mu} = \bar{X}$$

$$\hat{\sigma}^2 = \frac{1}{n} \sum_{i=1}^{n} X_i^2 - \bar{X}^2 = \frac{1}{n} \sum_{i=1}^{n} (X_i - \bar{X})^2$$ □

EXAMPLE C. (Gamma Distribution) The first two moments of the gamma distribution are

$$\mu_1 = \frac{\alpha}{\lambda}$$

$$\mu_2 = \frac{\alpha(\alpha + 1)}{\lambda^2}$$

(see Example B in Section 4.5). To apply the method of moments, we must express α and λ in terms of μ_1 and μ_2. From the second equation above:

$$\mu_2 = \mu_1^2 + \frac{\mu_1}{\lambda}$$

or

$$\lambda = \frac{\mu_1}{\mu_2 - \mu_1^2}$$

Also, from the equation for the first moment given above,

$$\alpha = \lambda\mu_1 = \frac{\mu_1^2}{\mu_2 - \mu_1^2}$$

The method of moments estimates are, since $\hat{\sigma}^2 = \hat{\mu}_2 - \hat{\mu}_1^2$,

$$\hat{\lambda} = \frac{\overline{X}}{\hat{\sigma}^2}$$

and

$$\hat{\alpha} = \frac{\overline{X}^2}{\hat{\sigma}^2}$$

As a concrete example, let us consider the fit of the amounts of precipitation during 227 storms in Illinois from 1960 to 1964 to a gamma distribution (Le Cam and Neyman, 1967). The data, listed in Problem 23 at the end of chapter 10, were gathered and analyzed in an attempt to characterize the natural variability in precipitation from storm to storm. A histogram shows that the distribution is quite skewed, so a gamma distribution is a natural candidate for a model. For these data, $\overline{X} = .224$ and $\hat{\sigma}^2 = .1338$, and therefore $\hat{\alpha} = .375$ and $\hat{\lambda} = 1.674$.

The histogram with the fitted density is shown in Figure 8-4. Note that, in order to make visual comparison easy, the density was normalized to have a total

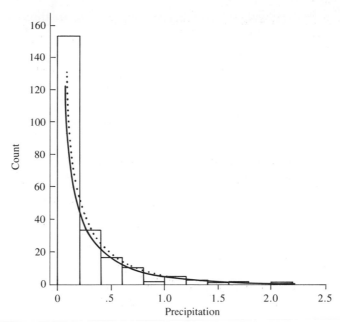

Figure 8-4. Gamma densities fit by the methods of moments and by the method of maximum likelihood to amounts of precipitation; the solid line shows the method of moments estimate and the dotted line the maximum likelihood estimate.

area equal to the total area under the histogram, which is the number of observations times the bin width of the histogram, or $227 \times .2 = 45.4$. Alternatively, the histogram could have been normalized to have a total area of 1. Qualitatively, the fit in Figure 8-4 looks reasonable; we will examine it in more detail in Example C in Section 9.9.   □

EXAMPLE D.   (An Angular Distribution) The angle θ at which electrons are emitted in muon decay has a distribution with the density

$$f(x|\alpha) = \frac{1 + \alpha x}{2}, \quad -1 \le x \le 1 \text{ and } -1 \le \alpha < 1$$

where $x = \cos \theta$. The parameter α is related to polarization. Physical considerations dictate that $|\alpha| \le \frac{1}{3}$, but we note that $f(x|\alpha)$ is a probability density for $|\alpha| \le 1$. The method of moments may be applied to estimate α from a sample of experimental measurements, $X_1, \ldots, X_n$. The mean of the density is

$$\mu = \int_{-1}^{1} x \frac{1 + \alpha x}{2} dx = \frac{\alpha}{3}$$

Thus, the method of moments estimate of α is $\hat{\alpha} = 3\bar{X}$.   □

Reviewing these examples, we see that there are three basic steps to constructing a method of moments estimate.

Step 1: Calculate low-order moments, finding expressions for the moments in terms of the parameters.

Step 2: Invert the expressions found in Step 1, finding new expressions for the parameters in terms of the moments.

Step 3: Insert the sample moments into the expressions obtained in Step 2, thus obtaining estimates of the parameters.

Under reasonable conditions, method of moments estimates have the desirable property of consistency. An estimate, $\hat{\theta}$, is said to be a **consistent** estimate of a parameter, θ, if $\hat{\theta}$ approaches θ as the sample size approaches infinity. The following states this more precisely.

DEFINITION. Let $\hat{\theta}_n$ be an estimate of a parameter θ based on a sample of size n. Then $\hat{\theta}_n$ is said to be consistent in probability if $\hat{\theta}_n$ converges in probability to θ as n approaches infinity; that is, for any $\varepsilon > 0$,

$$P(|\hat{\theta}_n - \theta| > \varepsilon) \to 0 \quad \text{as } n \to \infty$$

The weak law of large numbers implies that the sample moments converge in probability to the population moments. If the functions relating the estimates to the sample moments are continuous, the estimates will converge to the parameters as the sample moments converge to the population moments.

In the preceding examples, the equations for the estimates are simple in form. This is often the case for method of moments estimates, although some problems give rise to systems of nonlinear equations.

In this section, the method of moments was described and illustrated using several examples. In the next section, another method, the method of maximum likelihood, will be described and illustrated with the same examples. We will see that the two methods frequently, but not always, produce the same estimates.

8.5 The Method of Maximum Likelihood

As well as being a useful tool for parameter estimation in our current context, the method of maximum likelihood can be applied to a great variety of other statistical problems, such as curve fitting, for example. This general utility is one of the major reasons for the importance of likelihood methods in statistics. We will later see that maximum likelihood estimates have nice theoretical properties as well.

Suppose that random variables $X_1, \ldots, X_n$ have a joint density or frequency function $f(x_1, x_2, \ldots, x_n | \theta)$. Given observed values $X_i = x_i$, where $i = 1, \ldots, n$, the likelihood of θ as a function of $x_1, x_2, \ldots, x_n$ is defined as

$$\text{lik}(\theta) = f(x_1, x_2, \ldots, x_n | \theta)$$

Note that we consider the joint density as a function of θ rather than as a function of the x_i. If the distribution is discrete, so that f is a frequency function, the likelihood function gives the probability of observing the given data as a function of the parameter θ. The **maximum likelihood estimate (mle)** of θ is that value of θ that maximizes the likelihood—that is, makes the observed data "most probable" or "most likely."

If the X_i are assumed to be i.i.d., their joint density is the product of the marginal densities, and the likelihood is

$$\text{lik}(\theta) = \prod_{i=1}^{n} f(X_i | \theta)$$

Rather than maximizing the likelihood itself, it is usually easier to maximize its natural logarithm (which is equivalent since the logarithm is a monotonic function). For an i.i.d. sample, the **log likelihood** is

$$l(\theta) = \sum_{i=1}^{n} \log[f(X_i | \theta)]$$

Let us find the maximum likelihood estimates for the examples first considered in Section 8.4.

EXAMPLE A.   (Poisson Distribution) If X follows a Poisson distribution with parameter λ, then

$$P(X = x) = \frac{\lambda^x e^{-\lambda}}{x!}$$

If $X_1, \ldots, X_n$ are i.i.d. and Poisson, their joint frequency function is the product of the marginal frequency functions. The log likelihood is thus

$$l(\lambda) = \sum_{i=1}^{n} (X_i \log \lambda - \lambda - \log X_i!)$$

$$= \log \lambda \sum_{i=1}^{n} X_i - n\lambda - \sum_{i=1}^{n} \log X_i!$$

Figure 8-5 is a graph of $l(\lambda)$ for the asbestos counts of Example A in Section 8.4. Setting the first derivative of the log likelihood equal to zero, we find

$$l'(\lambda) = \frac{1}{\lambda} \sum_{i=1}^{n} X_i - n = 0$$

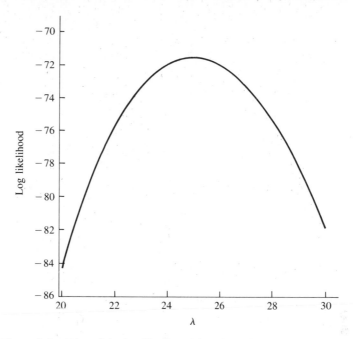

Figure 8-5.   Plot of the log likelihood function of λ for asbestos data.

The mle is then

$$\hat{\lambda} = \bar{X}$$

We can check that this is indeed a maximum [in fact, $l(\lambda)$ is a concave function of λ; see Figure 8-5]. The maximum likelihood estimate agrees with the method of moments estimate for this case.   □

EXAMPLE B.   (Normal Distribution) The log likelihood of an i.i.d. sample is

$$l(\mu, \sigma) = -n \log \sigma - \frac{n}{2} \log 2\pi - \frac{1}{2\sigma^2} \sum_{i=1}^{n} (X_i - \mu)^2$$

The partials with respect to μ and σ are

$$\frac{\partial l}{\partial \mu} = \frac{1}{\sigma^2} \sum_{i=1}^{n} (X_i - \mu)$$

$$\frac{\partial l}{\partial \sigma} = -\frac{n}{\sigma} + \sigma^{-3} \sum_{i=1}^{n} (X_i - \mu)^2$$

Setting the first partial equal to zero and solving for the mle, we obtain

$$\hat{\mu} = \bar{X}$$

Setting the second partial equal to zero and substituting the mle for μ, we find that the mle for σ is

$$\hat{\sigma} = \sqrt{\frac{1}{n} \sum_{i=1}^{n} (X_i - X)^2}$$

Again, these estimates agree with those obtained by the method of moments.   □

EXAMPLE C. (Gamma Distribution) Since the density function of a gamma distribution is

$$f(x|\alpha, \lambda) = \frac{1}{\Gamma(\alpha)} \lambda^{\alpha} x^{\alpha-1} e^{-\lambda x}, \quad 0 \leq x \leq \infty$$

the log likelihood of an i.i.d. sample, $X_1, \ldots, X_n$, is

$$l(\alpha, \lambda) = \sum_{i=1}^{n} [\alpha \log \lambda + (\alpha - 1) \log X_i - \lambda X_i - \log \Gamma(\alpha)]$$

$$= n\alpha \log \lambda + (\alpha - 1) \sum_{i=1}^{n} \log X_i - \lambda \sum_{i=1}^{n} X_i - n \log \Gamma(\alpha)$$

The partial derivatives are

$$\frac{\partial l}{\partial \alpha} = n \log \lambda + \sum_{i=1}^{n} \log X_i - n \frac{\Gamma'(\alpha)}{\Gamma(\alpha)}$$

$$\frac{\partial l}{\partial \lambda} = \frac{n\alpha}{\lambda} - \sum_{i=1}^{n} X_i$$

Setting the second partial equal to zero, we find

$$\hat{\lambda} = \frac{n\hat{\alpha}}{\sum_{i=1}^{n} X_i} = \frac{\hat{\alpha}}{\bar{X}}$$

But when this solution is substituted into the equation for the first partial, we obtain a nonlinear equation for the mle of α:

$$n \log \hat{\alpha} - n \log \bar{X} + \sum_{i=1}^{n} \log X_i - n \frac{\Gamma'(\hat{\alpha})}{\Gamma(\hat{\alpha})} = 0$$

This equation cannot be solved in closed form; an iterative method for finding the roots has to be employed. To start the iterative procedure, we could use the initial value obtained by the method of moments.

For this example, the two methods do not give the same estimates. The mle's are computed from the precipitation data of Example C in Section 8.4 by an iterative procedure (a combination of the secant method and the method of bisection) using the method of moments estimates as starting values. The resulting estimates are $\hat{\alpha} = .471$ and $\hat{\lambda} = 1.97$. In Example C in Section 8.4, the method of moments estimates were found to be $\hat{\alpha} = .375$ and $\hat{\lambda} = 1.674$. Figure 8-4 shows fitted densities from both types of estimates of α and λ. There is clearly little practical difference, especially if we keep in mind that the gamma distribution is only a possible model and should not be taken as being literally true. □

EXAMPLE D. (Muon Decay) From the form of the density given in Example D in Section 8.4, the log likelihood is

$$l(\alpha) = \sum_{i=1}^{n} \log(1 + \alpha X_i) - n \log 2$$

Setting the derivative equal to zero, we see that the mle of α satisfies the following nonlinear equation:

$$\sum_{i=1}^{n} \frac{X_i}{1 + \hat{\alpha} X_i} = 0$$

Again, we would have to use an iterative technique to solve for $\hat{\alpha}$. The method of moments estimate could be used as a starting value. □

In Examples C and D, in order to find the maximum likelihood estimate, we would have to solve a nonlinear equation. In general, in some problems involving several parameters, systems of nonlinear equations must be solved to find the mle's. We will not discuss numerical methods here; a good discussion is found in chapter 6 of Dahlquist and Bjorck (1974).

8.5.1 Maximum Likelihood Estimates of Multinomial Cell Probabilities

The method of maximum likelihood may also be applied to problems involving multinomial cell probabilities. Suppose that $X_1, \ldots, X_m$, the counts in cells $1, \ldots, m$, follow a multinomial distribution with a total count of n and cell probabilities $p_1, \ldots, p_m$. The joint frequency function of $X_1, \ldots, X_m$ is

$$f(x_1, \ldots, x_m | p_1, \ldots, p_m) = \frac{n!}{\prod_{i=1}^{m} x_i!} \prod_{i=1}^{m} p_i^{x_i},$$

Note that the marginal distribution of each X_i is binomial (n, p_i), and that since the X_i are not independent (they are constrained to sum to n), their joint frequency function is not the product of the marginal frequency functons, as it was in the examples considered in the preceding section. We can, however, still use the method of maximum likelihood since we can write an expression for the joint distribution. We assume n is given, and we wish to estimate $p_1, \ldots, p_m$ with the constraint that the p_i sum to 1. From the joint frequency function given above, the log likelihood is

$$l(p_1, \ldots, p_m) = \log n! - \sum_{i=1}^{m} \log x_i! + \sum_{i=1}^{m} x_i \log p_i$$

To maximize this likelihood subject to the constraint, we introduce a Lagrange multiplier and maximize

$$L(p_1, \ldots, p_m, \lambda) = \log n! - \sum_{i=1}^{m} \log x_i! + \sum_{i=1}^{m} x_i \log p_i + \lambda \left(\sum_{i=1}^{m} p_i - 1 \right)$$

Setting the partial derivatives equal to zero, we have the following system of equations:

$$\hat{p}_j = -\frac{x_j}{\lambda}, \quad j = 1, \ldots, m$$

Taking a sum on both sides of this equation, we have

$$1 = \frac{-n}{\lambda}$$

or

$$\lambda = -n$$

Therefore,

$$\hat{p}_j = \frac{x_j}{n}$$

which is an obvious set of estimates.

In some situations, such as frequently occur in the study of genetics, the multinomial cell probabilities are functions of other unknown parameters θ; that is, $p_i = p_i(\theta)$. In such cases, the log likelihood of θ is

$$l(\theta) = \log n! - \sum_{i=1}^{m} \log x_i! + \sum_{i=1}^{m} x_i \log p_i(\theta)$$

EXAMPLE A. (Hardy–Weinberg Equilibrium) If gene frequencies are in equilibrium, the genotypes AA, Aa, and aa occur in a population with frequencies $(1 - \theta)^2$, $2\theta(1 - \theta)$, and θ^2, according to the Hardy–Weinberg Law. In a sample from the Chinese population of Hong Kong in 1937, blood types occurred with the following frequencies, where M and N are erythrocyte antigens:

		Blood Type		
	M	MN	N	Total
Frequency	342	500	187	1029

There are several possible estimates of θ. For example, if we equate θ^2 with $187/1029$, we obtain $.4263$ as an estimate of θ. Intuitively, however, it seems that this procedure ignores some of the information in the other cells. If we let X_1, X_2, and X_3 denote the counts in the three cells and let $n = 1029$, the log likelihood of θ is (you should check this):

$$l(\theta) = \log n! - \sum_{i=1}^{3} \log X_i! + X_1 \log(1 - \theta)^2 + X_2 \log 2\theta(1 - \theta) + X_3 \log \theta^2$$

$$= \log n! - \sum_{i=1}^{3} \log X_i! + (2X_1 + X_2)\log(1 - \theta)$$

$$+ (2X_3 + X_2)\log \theta + X_2 \log 2$$

[We have not explicitly incorporated the constraint that the cell probabilities sum to 1 since the functional form of $p_i(\theta)$ is such that $\sum_{i=1}^{3} p_i(\theta) = 1$.] Setting the derivative equal to zero, we have

$$-\frac{2X_1 + X_2}{1 - \theta} + \frac{2X_3 + X_2}{\theta} = 0$$

Solving this, we obtain the mle:

$$\hat{\theta} = \frac{2X_3 + X_2}{2X_1 + 2X_2 + 2X_3}$$

$$= \frac{2X_3 + X_2}{2n}$$

$$= \frac{2 \times 187 + 500}{2 \times 1029} = .4247$$

How precise is this estimate? Do we have faith in the accuracy of the first, second, third, or fourth decimal place? Do the data in the table actually fit the Hardy–Weinberg Law? We will address these questions in later sections of this chapter and in chapter 9. $\square$

8.5.2   Large Sample Theory for Maximum Likelihood Estimates

Maximum likelihood estimates are functions of the sample values and are therefore random variables. Although we would like to know the sampling distribution in order to ascertain the precision of the estimate and to form confidence intervals for the population parameters, we cannot usually obtain the sampling distribution of the maximum likelihood estimates explicitly. We have already encountered a similar problem in chapter 7, where we wished to know the sampling distribution of the sample mean. In that situation, we used the central limit theorem to approximate the sampling distribution of $\bar{X}$; in this section, we develop some large sample theory to approximate the sampling distributions of maximum likelihood estimates by normal distributions.

The rigorous development of this large sample theory is quite technical; we will simply state some results and give very rough, heuristic arguments for the case of an i.i.d. sample and a one-dimensional parameter. [The arguments for Theorems A and B below may be skipped without loss of continuity. Rigorous proofs may be found in Cramer (1946).]

For an i.i.d. sample of size n, the log likelihood is

$$l(\theta) = \sum_{i=1}^{n} \log f(x_i|\theta)$$

We denote the true value of θ by θ_0. It can be shown that under reasonable conditions $\hat{\theta}$ is a consistent estimate of θ_0; that is, $\hat{\theta}$ converges to θ_0 in probability as n approaches infinity.

THEOREM A.   Under appropriate smoothness conditions on f, the mle is consistent.

Proof.   The following is merely a sketch of the proof. Consider maximizing

$$\frac{1}{n}l(\theta) = \frac{1}{n}\sum_{i=1}^{n} \log f(X_i|\theta)$$

As n tends to infinity, the law of large numbers implies that

$$\frac{1}{n}l(\theta) \to E \log f(X|\theta)$$

$$= \int \log f(x|\theta)f(x|\theta_0)\,dx$$

It is thus plausible that for large n, the θ that maximize $l(\theta)$ should be close to the θ that maximize $E \log f(X|\theta)$. (An involved argument is necessary to establish this.) To maximize $E \log f(X|\theta)$, we consider its derivative:

$$\frac{\partial}{\partial \theta} \int \log f(x|\theta) f(x|\theta_0)\, dx = \int \frac{\frac{\partial}{\partial \theta} f(x|\theta)}{f(x|\theta)} f(x|\theta_0)\, dx$$

If $\theta = \theta_0$, this equation becomes

$$\int \frac{\partial}{\partial \theta_0} f(x|\theta)\, dx = \frac{\partial}{\partial \theta_0} \int f(x|\theta)\, dx = \frac{\partial}{\partial \theta_0}(1) = 0$$

which shows that θ_0 is a stationary point and hopefully a maximum. Note that we have interchanged differentiation and integration and that the assumption of smoothness on f must be strong enough to justify this. □

We will now derive a useful intermediate result.

LEMMA A.   Define $I(\theta)$ by

$$I(\theta) = E\left[\frac{\partial}{\partial \theta} \log f(X|\theta)\right]^2$$

Under appropriate smoothness conditions on f, $I(\theta)$ may also be expressed as

$$I(\theta) = -E\left[\frac{\partial^2}{\partial \theta^2} \log f(X|\theta)\right]$$

Proof.   First, we observe that since $\int f(x|\theta)\, dx = 1$,

$$\frac{\partial}{\partial \theta} \int f(x|\theta)\, dx = 0$$

Combining this with the identity

$$\frac{\partial}{\partial \theta} f(x|\theta) = \left[\frac{\partial}{\partial \theta} \log f(x|\theta)\right] f(x|\theta)$$

we have

$$0 = \frac{\partial}{\partial \theta} \int f(x|\theta)\, dx = \int \left[\frac{\partial}{\partial \theta} \log f(x|\theta)\right] f(x|\theta)\, dx$$

where we have interchanged differentiation and integration (some assumptions must be made in order to do this). Taking second derivatives of the expressions just above, we have

$$0 = \frac{\partial}{\partial\theta}\int\left[\frac{\partial}{\partial\theta}\log f(x|\theta)\right]f(x|\theta)\,dx$$

$$= \int\left[\frac{\partial^2}{\partial\theta^2}\log f(x|\theta)\right]f(x|\theta)\,dx + \int\left[\frac{\partial}{\partial\theta}\log f(x|\theta)\right]^2 f(x|\theta)\,dx$$

From this, the desired result follows.   □

The large sample distribution of a maximum likelihood estimate is approximately normal with mean θ_0 and variance $1/nI(\theta_0)$. Since this is merely a limiting result, which holds as the sample size tends to infinity, we say that the mle is **asymptotically unbiased** and refer to the variance of the limiting normal distribution as the **asymptotic variance of the mle**.

THEOREM B.   Under smoothness conditions on f, the probability distribution of $\sqrt{nI(\theta_0)}(\hat\theta - \theta_0)$ tends to a standard normal distribution.

Proof.   The following is merely a sketch of the proof; the details of the argument are beyond the scope of this book. From a Taylor Series expansion,

$$0 = l'(\hat\theta) \approx l'(\theta_0) + (\hat\theta - \theta_0)l''(\theta_0)$$

$$(\hat\theta - \theta_0) \approx \frac{-l'(\theta_0)}{l''(\theta_0)}$$

$$n^{1/2}(\hat\theta - \theta_0) \approx \frac{-n^{-1/2}l'(\theta_0)}{n^{-1}l''(\theta_0)}$$

First, we consider the numerator of this last expression. Its expectation is

$$E[n^{-1/2}l'(\theta_0)] = n^{-1/2}\sum_{i=1}^{n} E\left[\frac{\partial}{\partial\theta_0}\log f(X|\theta_0)\right]$$

$$= 0$$

as in Theorem A. Its variance is

$$\text{Var}[n^{-1/2}l'(\theta_0)] = \frac{1}{n}\sum_{i=1}^{n} E\left[\frac{\partial}{\partial\theta_0}\log f(X|\theta_0)\right]^2$$

$$= I(\theta_0)$$

Next, we consider the denominator:

$$\frac{1}{n}l''(\theta_0) = \frac{1}{n}\sum_{i=1}^{n}\frac{\partial^2}{\partial\theta_0^2}\log f(x|\theta_0)$$

By the law of large numbers, the latter expression converges to

$$E\left[\frac{\partial^2}{\partial\theta_0^2}\log f(X|\theta_0)\right] = -I(\theta_0)$$

from Lemma A.

We thus have

$$n^{1/2}(\hat{\theta} - \theta_0) \approx \frac{n^{-1/2}l'(\theta_0)}{I(\theta_0)}$$

Therefore,

$$E[n^{1/2}(\hat{\theta} - \theta_0)] \approx 0$$

Furthermore,

$$\mathrm{Var}[n^{1/2}(\hat{\theta} - \theta_0)] \approx \frac{I(\theta_0)}{I^2(\theta_0)}$$

$$= \frac{1}{I(\theta_0)}$$

and thus

$$\mathrm{Var}(\hat{\theta} - \theta_0) \approx \frac{1}{nI(\theta_0)}$$

The central limit theorem may be applied to $l'(\theta_0)$, which is a sum of i.i.d. random variables:

$$l'(\theta_0) = \sum_{i=1}^{n}\frac{\partial}{\partial\theta_0}\log f(X|\theta) \qquad \square$$

Another interpretation of the result of Theorem B is as follows. For an i.i.d. sample, the maximum likelihood estimate is the maximizer of the log likelihood function,

$$l(\theta) = \sum_{i=1}^{n}\log f(X_i|\theta)$$

The asymptotic variance is

$$\frac{1}{nI(\theta_0)} = -\frac{1}{El''(\theta_0)}$$

which may be interpreted as an average radius of curvature of $l(\theta)$ at θ_0. When this radius of curvature is small, the estimate is relatively well resolved and the asymptotic variance is small.

A corresponding result can be proved for the multidimensional case. The vector of maximum likelihood estimates is asymptotically normally distributed. The mean of the asymptotic distribution is the vector of true parameters, and the elements of the vector of estimates have variances and covariances given by the corresponding elements of the matrix $(1/n)I^{-1}(\theta_0)$, where $I(\theta)$ is a matrix with the ij component

$$E\left[\frac{\partial}{\partial \theta_i} \log f(X|\theta) \frac{\partial}{\partial \theta_j} \log f(X|\theta)\right] = -E\left[\frac{\partial^2}{\partial \theta_i\, \partial \theta_j} \log f(X|\theta)\right]$$

The following sections will apply these results in several examples.

8.5.3   Confidence Intervals for Maximum Likelihood Estimates

In chapter 7, confidence intervals for the population mean μ were introduced. Recall that the confidence interval for μ was a random interval that contained μ with the some specified probability. In the current context, we are interested in estimating the parameter θ of a probability distribution. We will develop confidence intervals for θ based on $\hat{\theta}$; these intervals serve essentially the same function as they did in chapter 7 in that they express in a fairly direct way the degree of uncertainty in the estimate $\hat{\theta}$.

In some cases, the exact sampling distribution of a maximum likelihood estimate can be obtained, allowing the derivation of exact confidence intervals; however, this is typically not possible. For moderate to large sample sizes, confidence intervals based on the normal approximation to the sampling distribution developed in Section 8.5.2 may be used. In that section, it was shown that the asymptotic variance of a maximum likelihood estimate depends on $I(\theta_0)$. As it was given there, this result is difficult to use, since θ_0 is not known. The obvious thing to try is to substitute $\hat{\theta}$ for θ_0 and use $I(\hat{\theta})$. In fact, it can be shown that the limiting distribution of $\sqrt{nI(\hat{\theta})}(\hat{\theta} - \theta_0)$ is also the standard normal distribution, so this procedure can be justified.

From essentially the same argument that was used to form approximate confidence intervals for the population mean (see Section 7.3.3), it follows that an approximate $100(1 - \alpha)\%$ confidence interval for θ_0 is

$$\hat{\theta} \pm z(\alpha/2)\frac{1}{\sqrt{nI(\hat{\theta})}}$$

Denoting the estimated standard deviation, or standard error, of $\hat{\theta}$ by $s_{\hat{\theta}}$, we can write this confidence interval as

$$\hat{\theta} \pm z(\alpha/2)s_{\hat{\theta}}$$

This form parallels that of the $100(1 - \alpha)\%$ confidence interval for μ developed in chapter 7, which was $\bar{X} \pm z(\alpha/2)s_{\bar{X}}$. In both cases, the confidence interval is the parameter estimate plus or minus a multiple of its standard error. Many, but not all, confidence intervals are of this form.

Let us consider some examples of exact and approximate confidence intervals.

EXAMPLE A.   (Poisson Distribution) The mle of λ from a sample of size n from a Poisson distribution is

$$\hat{\lambda} = \bar{X}$$

Since the sum of independent Poisson random variables follows a Poisson distribution, the parameter of which is the sum of the parameters of the individual summands, $n\hat{\lambda} = \sum_{i=1}^{n} X_i$ follows a Poisson distribution with mean $n\lambda$. Also, the sampling distribution of $\hat{\lambda}$ is known, although it depends on the true value of λ, which is unknown. Exact confidence intervals for λ may be obtained by using this fact, and special tables are available (Pearson and Hartley, 1966).

For large samples, confidence intervals may be derived as follows. First, we need to calculate $I(\lambda)$. There are two ways to do this. We may use the definition

$$I(\lambda) = E\left[\frac{\partial}{\partial \lambda} \log f(x|\lambda)\right]^2$$

We know that

$$\log f(x|\lambda) = x \log \lambda - \lambda - \log x!$$

and thus

$$I(\lambda) = E\left(\frac{X}{\lambda} - 1\right)^2$$

Rather than evaluate this quantity, we may use the alternative expression for $I(\lambda)$ given by Lemma A of Section 8.5.2:

$$I(\lambda) = -E\left[\frac{\partial^2}{\partial \lambda^2} \log f(X|\lambda)\right]$$

Since

$$\frac{\partial^2}{\partial \lambda^2} \log f(X|\lambda) = -\frac{X}{\lambda^2}$$

$I(\lambda)$ is simply

$$\frac{E(X)}{\lambda^2} = \frac{1}{\lambda}$$

Thus, an approximate $100(1 - \alpha)\%$ confidence interval for λ is

$$\bar{X} \pm z(\alpha/2)\sqrt{\frac{\bar{X}}{n}}$$

Note that the asymptotic variance is in fact the exact variance in this case. The confidence interval, however, is only approximate, since the sampling distribution of $\bar{X}$ is only approximately normal.

As a concrete example, let us return to the study that involved counting asbestos fibers on filters, discussed earlier. In Example A in Section 8.4, we found $\hat{\lambda} = 24.9$. The estimated standard error of $\hat{\lambda}$ is thus ($n = 23$)

$$s_{\hat{\lambda}} = \sqrt{\frac{\hat{\lambda}}{n}} = 1.04$$

An approximate 95% confidence interval for λ is

$$\hat{\lambda} \pm 1.96s_{\hat{\lambda}}$$

or (22.9, 26.9). This interval gives a good indication of the uncertainty inherent in the determination of the average asbestos level using the model that the counts in the grid squares are independent Poisson random variables.   □

EXAMPLE B.   (Normal Distribution) The mle's of μ and σ^2 are

$$\hat{\mu} = \bar{X}$$

$$\hat{\sigma}^2 = \frac{1}{n} \sum_{i=1}^{n} (X_i - \bar{X})^2$$

Exact theory may be used; from Section 6.3,

$$\frac{\sqrt{n}(\bar{X} - \mu)}{s} \sim t_{n-1}$$

where t_{n-1} denotes the t distribution with $n - 1$ degrees of freedom and

$$s^2 = \frac{1}{n - 1} \sum_{i=1}^{n} (X_i - \bar{X})^2$$

Since the t distribution is symmetric, an exact confidence interval for μ is

$$\bar{X} \pm t_{n-1}(\alpha/2)\frac{s}{\sqrt{n}}$$

Here $t_{n-1}(\alpha/2)$ denotes that point beyond which the t distribution with $n-1$ degrees of freedom has probability $\alpha/2$ [the notation is completely analogous to the notation $z(\alpha/2)$ introduced earlier].

An exact confidence interval for σ^2 may be obtained using the fact (see Section 6.3) that

$$\frac{n\hat{\sigma}^2}{\sigma^2} \sim \chi^2_{n-1}$$

Defining $\chi^2_p(\alpha)$ as that point beyond which the chi-square distribution with p degrees of freedom has the area α beneath its tail, we have

$$1 - \alpha = P\left[\chi^2_{n-1}(1 - \alpha/2) \le \frac{n\hat{\sigma}^2}{\sigma^2} \le \chi^2_{n-1}(\alpha/2)\right]$$

$$= P\left[\frac{n\hat{\sigma}^2}{\chi^2_{n-1}(\alpha/2)} \le \sigma^2 \le \frac{n\hat{\sigma}^2}{\chi^2_{n-1}(1 - \alpha/2)}\right]$$

Therefore, a confidence interval for σ^2 is

$$\left(\frac{n\hat{\sigma}^2}{\chi^2_{n-1}(\alpha/2)}, \frac{n\hat{\sigma}^2}{\chi^2_{n-1}(1 - \alpha/2)}\right)$$

Note that the chi-square distribution is not symmetric, and thus the interval is not of the form $\hat{\sigma}^2 \pm c$.   □

It should be kept in mind that the confidence intervals in Examples A and B were derived under the assumption that the observations were samples from the Poisson and normal distributions, respectively. Thus, for example, if the data were not really from a normal distribution but from some other distribution, the confidence interval given for the variance might well not be a confidence interval at the stated level for the variance of the real underlying distribution. In fact, it is known that confidence intervals for variances based on the assumption of a normal distribution may be quite inaccurate where the distributions are actually nonnormal. Confidence intervals for means, however, are much less sensitive to deviations from normality.

In a similar manner, confidence intervals can be obtained for parameters estimated from multinomial counts. In such a case, we do not have an i.i.d. sample of size n, so the analysis is a little different. Recall from Section 8.5.1 that the $\hat{\theta}$ is the maximizer of the log likelihood function,

$$l(\theta) = \log n! - \sum_{i=1}^{m} \log x_i! + \sum_{i=1}^{m} x_i \log p_i(\theta)$$

When $X_1, \ldots, X_n$, the cell counts, are not i.i.d. random variables, the variance is not of the form $1/nI(\theta)$. It can still be shown, however, that

$$\text{Var}(\hat{\theta}) \approx \frac{1}{E[l'(\theta_0)^2]} = \frac{-1}{E[l''(\theta_0)]}$$

and that the maximum likelihood estimate is approximately normally distributed, just as in the i.i.d. case.

EXAMPLE C. (Hardy–Weinberg Equilibrium) Let us return to the example of Hardy–Weinberg equilibrium discussed in Example A in Section 8.5.1. There we found $\hat{\theta} = .4247$. Now,

$$l'(\theta) = -\frac{2X_1 + X_2}{1 - \theta} + \frac{2X_3 + X_2}{\theta}$$

In order to calculate $E[l'(\theta)^2]$, we would have to deal with the variances and covariances of the X_i. This does not look too inviting; it turns out to be easier to calculate $E[l''(\theta)]$.

$$l''(\theta) = -\frac{2X_1 + X_2}{(1 - \theta)^2} - \frac{2X_3 + X_2}{\theta^2}$$

Since the X_i are binomially distributed, we have

$$E(X_1) = n(1 - \theta)^2$$

$$E(X_2) = 2n\theta(1 - \theta)$$

$$E(X_3) = n\theta^2$$

We find, after some algebra, that

$$E[l''(\theta)] = -\frac{2n}{\theta(1 - \theta)}$$

Since θ is unknown, we substitute $\hat{\theta}$ in its place and obtain the estimated standard error of $\hat{\theta}$:

$$s_{\hat{\theta}} = \frac{1}{\sqrt{I(\hat{\theta})}}$$

$$= \sqrt{\frac{\hat{\theta}(1 - \hat{\theta})}{2n}} = .011$$

An approximate 95% confidence interval for θ is $\hat{\theta} \pm 1.96s_{\hat{\theta}}$, or $(.403, .447)$. □

8.6   Efficiency and the Cramer–Rao Lower Bound

In most statistical estimation problems, there are a variety of possible estimates of the relevant parameters. For example, in chapter 7, we considered both the sample mean and a ratio estimate of μ_y. In order to compare the different estimates, a measure such as mean squared error can be used. Other measures are certainly possibilities and might be more desirable in some circumstances; for example, we might compare the probabilities of large deviations from the true parameter values. Mean squared error is the most commonly used measure, however, partly because of its analytic tractability. When the estimates under consideration are unbiased, the mean squared error reduces to the variance.

Given two estimates, $\hat{\theta}$ and $\tilde{\theta}$, of a parameter θ, the **efficiency** of $\hat{\theta}$ relative to $\tilde{\theta}$ is defined to be

$$\text{eff}(\hat{\theta}, \tilde{\theta}) = \frac{\text{Var}(\tilde{\theta})}{\text{Var}(\hat{\theta})}$$

Thus, if the efficiency is smaller than 1, $\hat{\theta}$ has a larger variance than $\tilde{\theta}$ has. This comparison is most meaningful when both $\hat{\theta}$ and $\tilde{\theta}$ are unbiased or when both have the same bias. Frequently, the variances of $\hat{\theta}$ and $\tilde{\theta}$ are of the form

$$\text{Var}(\hat{\theta}) = \frac{c_1}{n}$$

$$\text{Var}(\tilde{\theta}) = \frac{c_2}{n}$$

where n is the sample size. If this is the case, the efficiency can be interpreted as the ratio of sample sizes necessary to obtain the same variance for both $\hat{\theta}$ and $\tilde{\theta}$. (In chapter 7, we compared the efficiencies of estimates of a population mean from a simple random sample, a stratified random sample with proportional allocation, and a stratified random sample with optimal allocation.)

EXAMPLE A.   (Muon Decay) Two estimates have been derived for α in the problem of muon decay. The method of moments estimate is

$$\tilde{\alpha} = 3\bar{X}$$

The maximum likelihood estimate is the solution of the nonlinear equation

$$\sum_{i=1}^{n} \frac{X_i}{1 + \hat{\alpha}X_i} = 0$$

We need to find the variances of these two estimates.

Since the variance of a sample mean is σ^2/n, we compute σ^2:

$$\sigma^2 = E(X^2) - [E(X)]^2$$

$$= \int_{-1}^{1} x^2 \frac{1 + \alpha x}{2} \, dx - \frac{\alpha^2}{9}$$

$$= \frac{1}{3} - \frac{\alpha^2}{9}$$

Thus, the variance of the method of moments estimate is

$$\mathrm{Var}(\tilde{\alpha}) = 9 \, \mathrm{Var}(\bar{X}) = \frac{3 - \alpha^2}{n}$$

The exact variance of the mle, $\hat{\theta}$, cannot be computed in closed form, so we approximate it by the asymptotic variance,

$$\mathrm{Var}(\hat{\alpha}) \approx \frac{1}{nI(\alpha)}$$

and then compare this asymptotic variance to the variance of $\tilde{\alpha}$. The ratio of the former to the latter is called the **asymptotic relative efficiency**. By definition,

$$I(\alpha) = E\left[\frac{\partial}{\partial \alpha} \log f(x|\alpha)\right]^2$$

$$= \int_{-1}^{1} \frac{x^2}{(1 + \alpha x)^2}\left(\frac{1 + \alpha x}{2}\right) dx$$

$$= \frac{\log\left(\dfrac{1 + \alpha}{1 - \alpha}\right) - 2\alpha}{2\alpha^3}, \quad -1 < \alpha < 1, \, \alpha \neq 0$$

$$= \frac{1}{3}, \quad \alpha = 0$$

The asymptotic relative efficiency is thus (for $\alpha \neq 0$)

$$\frac{\mathrm{Var}(\hat{\alpha})}{\mathrm{Var}(\tilde{\alpha})} = \frac{2\alpha^3}{3 - \alpha^2}\left[\frac{1}{\log\left(\dfrac{1 + \alpha}{1 - \alpha}\right) - 2\alpha}\right].$$

The following table gives this efficiency for various values of α between 0 and 1; symmetry would yield the values between -1 and 0.

α	Efficiency
0.0	1.0
.1	.996
.2	.984
.3	.963
.4	.931
.5	.888
.6	.828
.7	.746
.8	.630
.9	.447
.95	.301

As α tends to 1, the efficiency tends to 0. Thus, the mle is not much better than the method of moments estimate for α close to 0 but does increasingly better as α tends to 1.

It must be kept in mind that we have used the asymptotic variance of the mle, so we have actually calculated an asymptotic relative efficiency. To gain more precise information for a given sample size, a simulation of the sampling distribution of the mle could be conducted. This might be especially interesting for $\alpha = 1$, a case for which the formula for the asymptotic variance given above does not appear to make much sense. With a simulation study, the behavior of the bias as n and α vary could be analyzed (we showed that the mle is asymptotically unbiased, but there may be bias for a finite sample size), and the actual distribution could be compared to the approximating normal. ☐

In searching for an optimal estimate, we might ask whether there is a lower bound for the MSE of *any* estimate. If such a lower bound existed, it would function as a benchmark against which estimates could be compared. If an estimate achieved this lower bound, we would know that it could not be improved upon. In the case in which the estimate is unbiased, the Cramer–Rao Inequality provides such a lower bound. We now state and prove the Cramer–Rao Inequality.

THEOREM A. (Cramer–Rao Inequality) Let $X_1, \ldots, X_n$ be i.i.d. with density function $f(x|\theta)$. Let $T = t(X_1, \ldots, X_n)$ be an unbiased estimate of θ. Then, under smoothness assumptions on $f(x|\theta)$,

$$\text{Var}(T) \geq \frac{1}{nI(\theta)}$$

Proof. Let

$$Z = \sum_{i=1}^{n} \frac{\partial}{\partial \theta} \log f(X_i|\theta)$$

$$= \sum_{i=1}^{n} \frac{\frac{\partial}{\partial \theta} f(X_i|\theta)}{f(X_i|\theta)}$$

In Section 8.5.2, we showed that $E(Z) = 0$. Since the correlation coefficient of Z and T is less than or equal to 1 in absolute value

$$\text{Cov}^2(Z, T) \leq \text{Var}(Z)\, \text{Var}(T)$$

It was also shown in Section 8.5.2 that

$$\text{Var}\left[\frac{\partial}{\partial \theta} \log f(x|\theta)\right] = I(\theta)$$

Therefore,

$$\text{Var}(Z) = nI(\theta)$$

The proof will be completed by showing that $\text{Cov}(Z, T) = 1$. Since Z has mean 0,

$$\text{Cov}(Z, T) = E(ZT)$$

$$= \int \cdots \int t(x_1, \ldots, x_n) \left[\sum_{i=1}^{n} \frac{\frac{\partial}{\partial \theta} f(x_i|\theta)}{f(x_i|\theta)}\right] \prod_{j=1}^{n} f(x_j|\theta)\, dx_j$$

Noting that

$$\sum_{i=1}^{n} \frac{\frac{\partial}{\partial \theta} f(x_i|\theta)}{f(x_i|\theta)} \prod_{j=1}^{n} f(x_j|\theta) = \frac{\partial}{\partial \theta} \prod_{i=1}^{n} f(x_i|\theta)$$

we rewrite the expression for the covariance of Z and T as

$$\text{Cov}(Z, T) = \int \cdots \int t(x_1, \ldots, x_n) \frac{\partial}{\partial \theta} \prod_{i=1}^{n} f(x_i|\theta)\, dx_i$$

$$= \frac{\partial}{\partial \theta} \int \cdots \int t(x_1, \ldots, x_n) \prod_{i=1}^{n} f(x_i|\theta)\, dx_i$$

$$= \frac{\partial}{\partial \theta} E(T) = \frac{\partial}{\partial \theta}(\theta) = 1$$

which proves the inequality. [Note the interchange of differentiation and integration that must be justified by the smoothness assumptions on $f(x|\theta)$.] ☐

Theorem A gives a lower bound on the variance of *any* unbiased estimate. An unbiased estimate whose variance achieves this lower bound is said to be **efficient**. Since the asymptotic variance of a maximum likelihood estimate is equal to the lower bound, maximum likelihood estimates are said to be **asymptotically efficient**. For a finite sample size, however, a maximum likelihood estimate may not be efficient, and maximum likelihood estimates are not the only asymptotically efficient estimates.

EXAMPLE B. (Poisson Distribution) In Example A in Section 8.5.3, we found that for the Poisson distribution

$$I(\lambda) = \frac{1}{\lambda}$$

Therefore, by Theorem A, for any unbiased estimate T of λ, based on a sample of independent Poisson random variables, $X_1, \ldots, X_n$,

$$\text{Var}(T) \geq \frac{\lambda}{n}$$

The mle of λ was found to be $\bar{X} = S/n$, where $S = X_1 + \cdots + X_n$. Since S follows a Poisson distribution with parameter $n\lambda$, $\text{Var}(S) = n\lambda$ and $\text{Var}(\bar{X}) = \lambda/n$. Therefore, $\bar{X}$ attains the Cramer–Rao lower bound, and we know that no unbiased estimator of λ can have a smaller variance. In this sense, $\bar{X}$ is optimal for the Poisson distribution. But note that the theorem does not preclude the possibility that there is a biased estimator of λ that has a smaller mean squared error than $\bar{X}$ does. $\qquad\square$

8.6.1 An Example: The Negative Binomial Distribution

The Poisson distribution is often the first model considered for random counts; it has the property that the mean of the distribution is equal to the variance. When it is found that the variance of the counts is substantially larger than the mean, the negative binomial distribution is sometimes instead considered as a model. We consider a generalization of the negative binomial distribution introduced in Section 2.1.3, which is a discrete distribution on the nonnegative integers with a frequency function depending on the parameters m and k:

$$f(x|m, k) = \left(1 + \frac{m}{k}\right)^{-k} \frac{\Gamma(k + x)}{x!\Gamma(k)} \left(\frac{m}{m + k}\right)^x$$

The mean and variance of the negative binomial distribution are

$$\mu = m$$

$$\sigma^2 = m + \frac{m^2}{k}$$

It is apparent that this distribution is over-dispersed relative to the Poisson. We will not derive the mean and variance (they are most easily obtained by using moment-generating functions).

The negative binomial distribution can be used as a model in several cases:

- If k is an integer, the distribution of the number of successes up to the kth failure in a sequence of independent Bernoulli trials with probability of success $p = m/(m + k)$ is negative binomial.
- Suppose that Λ is a random variable following a gamma distribution and that for λ, a given value of Λ, X follows a Poisson distribution with mean λ. It can be shown that the unconditional distribution of X is negative binomial. Thus, for situations in which the rate varies randomly over time or space, the negative binomial distribution might tentatively be considered as a model.
- The negative binomial distribution also arises with a particular type of clustering. Suppose that counts of colonies, or clusters, follow a Poisson distribution and that each colony has a random number of individuals. If the probability distribution of the number of individuals per colony is of a particular form (the logarithmic series distribution), it can be shown that the distribution of counts of individuals is negative binomial. The negative binomial distribution might be a plausible model for the distribution of insect counts if the insects hatch from depositions, or clumps, of larvae.
- The negative binomial distribution can be applied to model population size as a function of a certain birth/death process, the assumption being that the birth rate and death rate per individual are constant and that there is a constant rate of immigration.

Anscombe (1950) discusses estimation of the parameters m and k and compares the efficiencies of several methods of estimation. The simplest method is the method of moments; from the relations of m and k to μ and σ^2 given above, the method of moments estimates of m and k are

$$\hat{m} = \bar{X}$$

$$\hat{k} = \frac{\bar{X}^2}{\hat{\sigma}^2 - \bar{X}}$$

Another relatively simple method of estimation of m and k is based on the number of zeros. The probability of the count being zero is

$$p_0 = \left(1 + \frac{m}{k}\right)^{-k}$$

If m is estimated by the sample mean and there are n_0 zeros out of a sample size of n, then k is estimated by $\hat{k}$, where $\hat{k}$ satisfies

$$\frac{n_0}{n} = \left(1 + \frac{\bar{X}}{\hat{k}}\right)^{-\hat{k}}$$

Although the solution cannot be obtained in closed form, it is not difficult to find by iteration.

Figure 8-6, from Anscombe (1950), shows the asymptotic efficiencies of the two methods of estimation of the negative binomial parameters. In the figure, the method of moments is method 1 and the method based on the number of zeros is method 2. Method 2 is quite efficient when the mean is small—that is, when there are a large number of zeros. Method 1 becomes more efficient as k increases.

The maximum likelihood estimate is asymptotically efficient but is somewhat more difficult to compute. The equations will not be written out here. Bliss and Fisher (1953) discuss computational methods and give several examples. The maximum likelihood estimate of m is the sample mean, but that of k is the solution of a nonlinear equation.

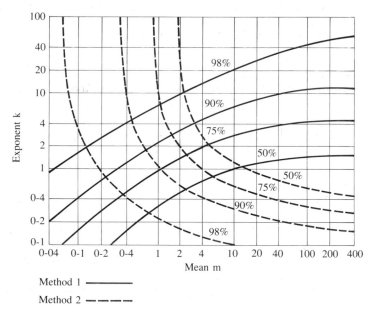

Figure 8-6. Asymptotic efficiencies of estimates of negative binomial parameters.

EXAMPLE A.   (Insect Counts) Let us consider an example from Bliss and Fisher (1953). From each of 6 apple trees in an orchard that had been sprayed, 25 leaves were selected. On each of the leaves, the number of adult female red mites was counted. Intuitively, we might conclude that this situation was too heterogeneous for a Poisson model to fit; the rates of infestation might be different on different trees and at different locations on the same tree. The following table shows the observed counts and the expected counts from fitting Poisson and negative binomial distributions. The mle's for k and m were $\hat{k} = 1.025$ and $\hat{m} = 1.146$.

Number per Leaf	Observed Count	Poisson Distribution	Negative Binomial Distribution
0	70	47.7	69.5
1	38	54.6	37.6
2	17	31.3	20.1
3	10	12.0	10.7
4	9	3.4	5.7
5	3	.75	3.0
6	2	.15	1.6
7	1	.03	.85
8+	0	.00	.95

Casual inspection of this table makes it clear that the Poisson does not fit; there are many more small and large counts observed than are expected for a Poisson distribution. After pooling the last three cells, Pearson's chi-square test gives an extremely small p-value (chi-square value of 82.4 with 5 df). A chi-square test for the goodness of fit of the negative binomial distribution, performed after pooling the last two cells, yields a chi-square statistic of 2.48 with 6 df and a corresponding p-value of .48, so the data are quite consistent with this model.   □

As it is in calculating expected counts for a Poisson distribution, it is useful in fitting the negative binomial distribution to use a recursive relation:

$$p_0 = \left(1 + \frac{m}{k}\right)^{-k}$$

$$p_n = \frac{k + n - 1}{n}\left(\frac{m}{k + m}\right)p_{n-1}$$

8.7 Sufficiency

This section introduces the concept of sufficiency and some of its theoretical implications. Suppose that $X_1, \ldots, X_n$ is a sample from a probability distribution with the density or frequency fucntion $f(x|\theta)$. The concept of sufficiency arises as an attempt to answer the following question: Is there a statistic, a function $T(X_1, \ldots, X_n)$, which contains all the information in the sample about θ? If so, a reduction of the original data to this statistic without loss of information is possible. For example, consider a sequence of independent Bernoulli trials with unknown probability of success, θ. We may have the intuitive feeling that the total number of successes contains all the information about θ that there is in the sample, that the order in which the successes occurred, for example, does not give any additional information. The following definition formalizes this idea.

DEFINITION. A statistic, $T(X_1, \ldots, X_n)$, is used to be *sufficient* for θ if the conditional distribution of $X_1, \ldots, X_n$, given $T = t$, does not depend on θ for any value of t.

In other words, given the value of T, which is called a **sufficient statistic**, we can gain no more knowledge about θ from knowing more about the probability distribution of $X_1, \ldots, X_n$. (Formally, we could envision keeping only T and throwing away all the X_i without any loss of information. Informally, and more realistically, this would make no sense at all. The values of the X_i might indicate that the model did not fit or that something was fishy about the data. What would you think, for example, if you saw 50 ones followed by 50 zeros in a sequence of supposedly independent Bernoulli trials?)

EXAMPLE A.   Let $X_1, \ldots, X_n$ be a sequence of independent Bernoulli random variables with $P(X_i = 1) = \theta$. We will verify that $T = \sum_{i=1}^{n} X_i$ is sufficient for θ.

$$P(X_1 = x_1, \ldots, X_n = x_n | T = t) = \frac{P(X_1 = x_1, \ldots, X_n = x_n, T = t)}{P(T = t)}$$

$$= \frac{\theta^t (1 - \theta)^{n-t}}{\binom{n}{t} \theta^t (1 - \theta)^{n-t}}$$

$$= \frac{1}{\binom{n}{t}}$$

if all the x_i equal 0 or 1 and $t = \sum_{i=1}^{n} x_i$. The conditional distribution thus does not involve θ at all. Given the total number of ones, the probability that they occur on any particular set of t trials is the same.   □

8.7.1   A Factorization Theorem

The above definition of sufficiency is hard to work with, since it does not indicate how to go about finding a sufficient statistic, and given a candidate statistic, T, it would typically be very hard to conclude whether it was sufficient because of the difficulty in evaluating the conditional distribution. The following factorization theorem provides a convenient means of identifying sufficient statistics.

THEOREM A.   A necessary and sufficient condition for $T(X_1, \ldots X_n)$ to be sufficient for a parameter θ is that the joint probability function (density function or frequency function) factors in the form

$$f(x_1, \ldots, x_n | \theta) = g[T(x_1, \ldots, x_n), \theta] h(x_1, \ldots, x_n)$$

Proof.   We give a proof for the discrete case. (The proof for the general case is more subtle and requires regularity conditions, but the basic ideas are the same.) First, suppose that the density function factors as given in the theorem. To simplify notation, we will let $\mathbf{X}$ denote $(X_1, \ldots, X_n)$ and $\mathbf{x}$ denote $(x_1, \ldots, x_n)$. We have

$$P(T = t) = \sum_{T(\mathbf{x})=t} P(\mathbf{X} = \mathbf{x})$$

$$= g(t, \theta) \sum_{T(\mathbf{x})=t} h(\mathbf{x})$$

Here the notation indicates that the sum is over all $\mathbf{x}$ such that $T(\mathbf{x}) = t$. We then have

$$P(\mathbf{X} = \mathbf{x}|T = t) = \frac{P(\mathbf{X} = \mathbf{x}, T = t)}{P(T = t)}$$

$$= \frac{h(\mathbf{x})}{\sum_{T(\mathbf{x})=t} h(\mathbf{x})}$$

This conditional distribution does not depend on θ, as was to be shown.

To show that the conclusion holds in the other direction, suppose that the conditional distribution of $\mathbf{X}$ given T is independent of θ. Let

$$g(t, \theta) = P(T = t|\theta)$$

$$h(\mathbf{x}) = P(\mathbf{X} = \mathbf{x}|T = t)$$

We then have

$$P(\mathbf{X} = \mathbf{x}|\theta) = P(T = t|\theta)P(\mathbf{X} = \mathbf{x}|T = t)$$

$$= g(t, \theta)h(\mathbf{x})$$

as was to be shown.   □

We can demonstrate the utility of Theorem A by applying it to some examples. More examples are included in the problems at the end of this chapter.

EXAMPLE A.   Consider a sequence of independent Bernoulli random variables, $X_1, \ldots, X_n$, where

$$P(X_i = x) = \theta^x(1 - \theta)^{1-x}, \quad x = 0 \text{ or } x = 1$$

Then

$$f(\mathbf{x}|\theta) = \prod_{i=1}^{n} \theta^{x_i}(1 - \theta)^{1-x_i}$$

$$= \theta^{\sum_{i=1}^{n} x_i}(1 - \theta)^{n - \sum_{i=1}^{n} x_i}$$

$$= \left(\frac{\theta}{1 - \theta}\right)^{\sum_{i=1}^{n} x_i}(1 - \theta)^n$$

This function is of the form $g(\sum_{i=1}^{n} x_i, \theta)h(\mathbf{x})$, where $h(\mathbf{x}) = 1$ and

$$g(t, \theta) = \left(\frac{\theta}{1 - \theta}\right)^t (1 - \theta)^n \qquad \square$$

EXAMPLE B. Consider a random sample from a normal distribution that has an unknown mean and variance. We have

$$f(\mathbf{x}|\mu, \sigma) = \prod_{i=1}^{n} \frac{1}{\sigma\sqrt{2\pi}} \exp\left[\frac{-1}{2\sigma^2}(x_i - \mu)^2\right]$$

$$= \frac{1}{\sigma^n(2\pi)^{n/2}} \exp\left[\frac{-1}{2\sigma^2} \sum_{i=1}^{n} (x_i - \mu)^2\right]$$

$$= \frac{1}{\sigma^n(2\pi)^{n/2}} \exp\left[\frac{-1}{2\sigma^2}\left(\sum_{i=1}^{n} x_i^2 - 2\mu \sum_{i=1}^{n} x_i + \mu^2\right)\right]$$

This expression is just a function of $\sum_{i=1}^{n} x_i$ and $\sum_{i=1}^{n} x_i^2$, which are therefore sufficient statistics. $\qquad \square$

A study of the properties of probability distributions that have sufficient statistics of the same dimension as the parameter space regardless of sample size led to the development of what is called the **exponential family** of probability distributions. Many common distributions, including the normal, the binomial, the Poisson, and the gamma, are members of this family. One-parameter members of the exponential family have density or frequency functions of the form

$$f(x|\theta) = \exp[c(\theta)T(x) + d(\theta) + S(x)], \quad x \in A$$

$$= 0, \quad x \notin A$$

where the set A does not depend on θ. Suppose that $X_1, \ldots, X_n$ is a sample from a member of the exponential family; the joint probability function is

$$f(\mathbf{x}|\theta) = \prod_{i=1}^{n} \exp[c(\theta)T(x_i) + d(\theta) + S(x_i)]$$

$$= \exp\left[c(\theta) \sum_{i=1}^{n} T(x_i) + nd(\theta)\right] \exp\left[\sum_{i=1}^{n} S(x_i)\right]$$

From this result, it is apparent by the factorization theorem that $\sum_{i=1}^{n} T(x_i)$ is a sufficient statistic.

EXAMPLE C. The frequency function of the Bernoulli distribution is

$$P(X = x) = \theta^x(1 - \theta)^{1-x}, \quad x = 0 \text{ or } x = 1$$

$$= \exp\left[x\log\left(\frac{\theta}{1 - \theta}\right) + \log(1 - \theta)\right]$$

This is a member of the exponential family with $T(x) = x$, and we have already seen that $\sum_{i=1}^{n} x_i$ is a sufficient statistic for a sample from the Bernoulli distribution. $\square$

A k-parameter member of the exponential family has a density or frequency function of the form

$$f(x|\theta) = \exp\left[\sum_{i=1}^{k} c_i(\theta) T_i(x) + d(\theta) + S(x)\right], \quad x \in A$$

$$= 0, \quad x \notin A$$

where the set A does not depend on θ.

The normal distribution is of this form. A great deal of theoretical work has centered around the exponential family; further discussion of this family can be found in Bickel and Doksum (1977).

We conclude this section with the following corollary of Theorem A.

COROLLARY A.   If T is sufficient for θ, the maximum likelihood estimate is a function of T.

Proof.   From Theorem A, the likelihood is $g(T, \theta)h(\mathbf{x})$, which depends on θ only through T. To maximize this quantity, we need only maximize $g(T, \theta)$. $\square$

Corollary A and the Rao–Blackwell Theorem of the next section may be interpreted as giving some theoretical support to the use of maximum likelihood estimates.

8.7.2   The Rao–Blackwell Theorem

In the preceding section, we argued for the importance of sufficient statistics on essentially qualitative grounds. The Rao–Blackwell Theorem gives a quantitative rationale for basing an estimator of a parameter θ on a sufficient statistic if one exists.

THEOREM A.   (Rao–Blackwell Theorem) Let $\hat{\theta}$ be an estimator of θ with $E(\hat{\theta}^2) < \infty$ for all θ. Suppose that T is sufficient for θ, and let $\tilde{\theta} = E(\hat{\theta}|T)$. Then, for all θ,

$$E(\tilde{\theta} - \theta)^2 \le E(\hat{\theta} - \theta)^2$$

The inequality is strict unless $\hat{\theta} = \tilde{\theta}$.

Proof.   We first note that, from the property of iterated conditional expectation (Theorem A of Section 4.4.1),

$$E(\tilde{\theta}) = E[E(\hat{\theta}|T)] = E(\hat{\theta})$$

Therefore, to compare the mean squared error of the two estimators, we need only compare their variances. From Theorem B of Section 4.4.1, we have

$$\text{Var}(\hat{\theta}) = \text{Var}[E(\hat{\theta}|T)] + E[\text{Var}(\hat{\theta}|T)]$$

or

$$\text{Var}(\hat{\theta}) = \text{Var}(\tilde{\theta}) + E[\text{Var}(\hat{\theta}|T)]$$

Thus, $\text{Var}(\hat{\theta}) < \text{Var}(\tilde{\theta})$ unless $\text{Var}(\hat{\theta}|T) = 0$, which is the case only if $\hat{\theta}$ is a function of T, which would imply $\hat{\theta} = \tilde{\theta}$. $\square$

Since $E(\hat{\theta}|T)$ is a function of the sufficient statistic T, the Rao–Blackwell Theorem gives a strong rationale for basing estimators on sufficient statistics if they exist. If an estimator is not a function of a sufficient statistic, it can be improved.

Suppose that there are two estimates, $\hat{\theta}_1$ and $\hat{\theta}_2$, having the same expectation. Assuming that a sufficient statistic T exists, we may construct two other estimates, $\tilde{\theta}_1$ and $\tilde{\theta}_2$, by conditioning on T. The theory we have developed so far gives no clues as to which one of these two is better. If the probability distribution of T has the property of completeness, $\tilde{\theta}_1$ and $\tilde{\theta}_2$ are identical, by a theorem of Lehmann and Sheffe. We will not define completeness or pursue this topic further; Lehmann (1983) and Bickel and Doksum (1977) discuss this concept.

8.8   Concluding Remarks

Certain key ideas first introduced in the context of survey sampling in chapter 7 have recurred in this chapter. We have viewed an estimate as a random variable having a probability distribution called its sampling distribution. In chapter 7, the estimate was of a parameter, such as the mean, of a finite population; in this chapter, the estimate was of a parameter of a probability distribution. In both cases, characteristics of the sampling distribution, such as the bias and the variance and the large sample approximate form, have been of interest. In both chapters, we have studied confidence intervals for the true value of the unknown parameter. The method of propagation of error, or linearization, has been a useful tool in both chapters. These key ideas will be important in other contexts in later chapters as well.

Important concepts and techniques in estimation theory have been introduced in this chapter. We have discussed two general methods of estimation—the method of moments and the method of maximum likelihood. The latter especially has great general utility in statistics. We have developed and applied some approximate distribution theory for maximum likelihood estimates. Other theoretical developments included the concept of efficiency, the Cramer–Rao lower bound, and the concept of sufficiency and some of its consequences.

Chapter 15 contains further material on estimation theory, from a rather more abstract point of view.

8.9 Problems

1. The following table gives the observed counts in 1-second intervals for Berkson's data (Section 8.2). Use Pearson's chi-square test to assess the goodness of fit to the Poisson distribution.

n	Observed
0	5267
1	4436
2	1800
3	534
4	111
5+	21

2. The Poisson distribution has been used by traffic engineers as a model for light traffic, based on the rationale that if the rate is approximately constant and the traffic is light (so the individual cars move independently of each other), the distribution of counts of cars in a given time interval or space area should be nearly Poisson (Gerlough and Schuhl, 1955). The following table shows the number of right turns during 300 3-minute intervals at a specific intersection. Fit a Poisson distribution and test goodness of fit using Pearson's chi-square statistic. Group the last four cells together for the chi-square test. Comment on the fit. It is useful to know that the 300 intervals were distributed over various hours of the day and various days of the week.

n	Frequency
0	14
1	30
2	36
3	68
4	43
5	43
6	30
7	14
8	10
9	6
10	4
11	1
12	1
13+	0

3. One of the earliest applications of the Poisson distribution was made by Student (1907) in studying errors made in counting yeast cells or blood corpuscles with a haemacytometer. In this study, yeast cells were killed and mixed with water and gelatin; the mixture was then spread on a glass and allowed to cool. Four different concentrations were used. Counts were made on 400 squares, and the data are summarized in the following table:

Number of Cells	Concentration 1	Concentration 2	Concentration 3	Concentration 4
0	213	103	75	10
1	128	143	103	20
2	37	98	121	43
3	18	42	54	53
4	3	8	30	86
5	1	4	13	70
6	0	2	2	54
7	0	0	1	37
8	0	0	0	18
9	0	0	1	10
10	0	0	0	5
11	0	0	0	2
12	0	0	0	2

(a) Estimate the parameter λ for each of the four sets of data.

(b) Find an approximate 95% confidence interval for each estimate.

(c) Test the goodness of fit of the Poisson distribution to these data using Pearson's chi-square test.

4. Suppose that $X \sim \text{bin}(n, p)$.

(a) Show that the mle of p is $\hat{p} = X/n$.

(b) Show that mle of part (a) attains the Cramer–Rao lower bound.

5. Suppose that X follows a geometric distribution,

$$P(X = k) = (1 - p)p^{k-1}$$

and assume a sample of size n.

(a) Find the method of moments estimate of p.

(b) Find the mle of p.

(c) Find the asymptotic variance of the mle.

6. In an ecological study of the feeding behavior of birds, the number of hops between flights was counted for several birds. For the following data, fit a geometric distribution, find an approximate 95% confidence interval for p, and test goodness of fit:

Number of Hops	Frequency
1	48
2	31
3	20
4	9
5	6
6	5
7	4
8	2
9	1
10	1
11	2
12	1

7. The Pareto distribution has been used in economics as a model for a density function with a slowly decaying tail:

$$f(x|x_0, \theta) = \theta x_0^\theta x^{-\theta-1}, \quad x \geq x_0$$

Assume that $x_0 > 0$ is given.
(a) Find the method of moments estimate of θ.
(b) Find the mle of θ.
(c) Find the asymptotic variance of the mle.

8. Consider the following method of estimating λ for a Poisson distribution. Observe that

$$p_0 = P(X = 0) = e^{-\lambda}$$

Letting Y denote the number of zeros from a sample of size n, λ might be estimated by

$$\tilde{\lambda} = -\log\left(\frac{Y}{n}\right)$$

Use the method of propagation of error to obtain approximate expressions for the variance and the bias of this estimate. Compare the variance of this estimate to the variance of the mle, computing relative efficiencies for various values of λ. Note that $Y \sim \text{bin}(n, p_0)$.

9. For the example on muon decay in Section 8.3, suppose that instead of recording $x = \cos\theta$, only whether the electron goes backward ($x < 0$) or forward ($x > 0$) is recorded.
(a) How could α be estimated from n independent observations of this type? (*Hint:* Use the binomial distribution.)
(b) What is the variance of this estimate and its efficiency relative to the method of moments estimate and the mle for $\alpha = 0, .1, .2, .3, .4, .5, .6, .7, .8, .9$?

10. Let $X_1, \ldots, X_n$ be a sample from a Rayleigh distribution with parameter θ:

$$f(x|\theta) = \frac{x}{\theta^2} e^{-x^2/2\theta^2}, \quad x \geq 0$$

(a) Find the method of moments estimate of θ.
(b) Find the mle of θ.
(c) Find the asymptotic variance of the mle.

11. The double exponential distribution is

$$f(x|\theta) = \frac{1}{2} e^{-|x-\theta|}, \quad -\infty < x < \infty$$

For a sample of size $n = 2m + 1$, show that the mle of θ is the median of the sample (the observation such that half of the rest of the observations are smaller and half are larger). [*Hint:* The function $g(x) = |x|$ is not differentiable. Draw a picture for a small value of n to try to understand what is going on.]

12. Let $X_1, \ldots, X_n$ be i.i.d. random variables with the density function

$$f(x|\theta) = (\theta + 1)x^\theta, \quad 0 \le x \le 1$$

   (a) Find the method of moments estimate of θ.
   (b) Find the mle of θ.
   (c) Find the asymptotic variance of the mle.

13. Let $X_1, \ldots, X_n$ be uniform on $[0, \theta]$.
   (a) Find the method of moments estimate of θ and its mean and variance.
   (b) Find the mle of θ.
   (c) Find the probability density of the mle, and calculate its mean and variance. Compare the variance, the bias, and the mean squared error to those of the method of moments estimate.
   (d) Find a modification of the mle that renders it unbiased.

14. Suppose that a sample of size 15 from a normal distribution gives $\bar{X} = 10$ and $s^2 = 25$. Find 90% confidence intervals for μ and σ^2.

15. For two factors—starchy or sugary and green base leaf or white base leaf—the following counts for the progeny of self-fertilized heterozygotes were observed (Fisher, 1958):

Type	Count
Starchy-green	1997
Starchy-white	906
Sugary-green	904
Sugary-White	32

   According to genetic theory, the cell probabilities are $.25(2 + \theta)$, $.25(1 - \theta)$, $.25(1 - \theta)$, and $.25\theta$, where $\theta(0 < \theta < 1)$ is a parameter related to the linkage of the factors.
   (a) Find the mle of θ and its asymptotic variance.
   (b) Form an approximate 95% confidence interval for θ.

16. Referring to Problem 15, consider two other estimates of θ. (1) The expected number of counts in the first cell is $n(2 + \theta)/4$; if this expected number is equated to the count X_1, the following estimate is obtained:

$$\tilde{\theta}_1 = \frac{4X_1}{n} - 2$$

(2) The same procedure done for the last cell yields

$$\tilde{\theta}_2 = \frac{4X_4}{n}$$

Compute these estimates. Using that X_1 and X_4 are binomial random variables, show that these estimates are unbiased, and obtain expressions for their variances. Evaluate the estimated standard errors and compare them to the estimated standard error of the mle.

17. This problem is concerned with the estimation of the variance of a normal distribution with unknown mean from a sample $X_1, \ldots, X_n$ of i.i.d. normal random variables. In answering the following questions, use the fact that (from Theorem B of Section 6.3)

$$\frac{(n-1)s^2}{\sigma^2} \sim \chi^2_{n-1}$$

and that the mean and variance of a chi-square random variable with r df are r and $2r$, respectively.

(a) Which of the following estimates is unbiased?

$$s^2 = \frac{1}{n-1} \sum_{i=1}^{n} (X_i - \bar{X})^2$$

$$\hat{\sigma}^2 = \frac{1}{n} \sum_{i=1}^{n} (X_i - \bar{X})^2$$

(b) Which of the estimates given in part (a) has the smaller MSE?

(c) For what value of ρ does $\rho \sum_{i=1}^{n} (X_i - \bar{X})^2$ have the minimal MSE?

18. If gene frequencies are in equilibrium, the genotypes AA, Aa, and aa occur with probabilities $(1-\theta)^2, 2\theta(1-\theta)$, and θ^2, respectively. Plato et al. (1964) published the following data on haptoglobin type in a sample of 190 people:

Haptoglobin Type		
Hp1-1	Hp1-2	Hp2-2
10	68	112

Find the mle of θ and its asymptotic variance. Find an approximate 99% confidence interval for θ.

19. Suppose that in the population of twins, males (M) and females (F) are equally likely to occur and that the probability that twins are identical is α. If twins are not identical, their genders are independent.

(a) Show that

$$P(MM) = P(FF) = \frac{1+\alpha}{4}$$

$$P(MF) = \frac{1-\alpha}{2}$$

(b) Suppose that n twins are sampled. It is found that n_1 are MM, n_2 are FF, and n_3 are MF, but it is not known which twins are identical. Find the mle of α and its variance.

20. Let $X_1, \ldots, X_n$ be a sample from an exponential distribution with the density function

$$f(x|\tau) = \frac{1}{\tau}e^{-x/\tau}, \quad 0 \le x < \infty$$

(a) Find the mle of τ.

(b) Show that the mle is unbiased, and find its exact variance. (*Hint:* The sum of the X_i follows a gamma distribution.)

(c) Find the Cramer–Rao lower bound, and compare it to the mle of part (b).

(d) Find the form of an approximate confidence interval for τ.

(e) Find the form of an exact confidence interval for τ.

21. Evans (1953) considered fitting the negative binomial distribution and other distributions to a number of data sets that arose in ecological studies. Two of these sets will be used in this problem. The first data set gives frequency counts of *Glaux maritima* made in 500 contiguous 20-cm^2 quadrants. For the second data set, a plot of potato plants 48 rows wide and 96 feet long was examined. The area was split into 2304 sampling units consisting of 2-foot lengths of row and in each unit the number of potato beetles was counted. Fit Poisson and negative binomial distributions, and comment on the goodness of fit. For these data, the method of moments should be fairly efficient.

Count	Glaux maritima	Potato Beetles
0	1	190
1	15	264
2	27	304
3	42	260
4	77	294
5	77	219
6	89	183
7	57	150
8	48	104
9	24	90
10	14	60
11	16	46
12	9	29
13	3	36
14	1	19
15		12
16		11
17		6
18		10
19		2
20		4
21		1
22		3
23		4
24		1
25		1
26		0
27		0
28		1

22. Let $X_1, \ldots, X_n$ be a random sample from a Poisson distribution with mean λ, and let $T = \sum_{i=1}^{n} X_i$. Show that the distribution of $X_1, \ldots, X_n$ given T is independent of λ, and conclude that T is sufficient for λ. Show that X_1 is not sufficient.

23. Use the factorization theorem (Theorem A in Section 8.7.1) to conclude that $T = \sum_{i=1}^{n} X_i$ is a sufficient statistic when the X_i are a random sample from a geometric distribution.

24. Use the factorization theorem to find a sufficient statistic for the exponential distribution.

25. Let $X_1, \ldots, X_n$ be a random sample from a distribution with the density function

$$f(x|\theta) = \frac{1}{2}\left[\frac{\theta}{(1 + x)^{\theta+1}}\right], \quad 0 < \theta < \infty \text{ and } -\infty < x < \infty$$

Find a sufficient statistic for θ.

26. Show that $\prod_{i=1}^{n} X_i$ and $\sum_{i=1}^{n} X_i$ are sufficient statistics for the gamma distribution.

27. Find a sufficient statistic for the Rayleigh density,

$$f(x|\theta) = \frac{x}{\theta^2} e^{-x^2/2\theta^2}, \quad x \geq 0$$

28. Show that the binomial distribution belongs to the exponential family.

29. Show that the gamma distribution belongs to the exponential family.

30. Let T_1 and T_2 be sufficient statistics for θ, and suppose that $T_2 = g(T_1)$ for some function g. Let U be an unbiased estimate of θ, and let

$$V_1 = E(U|T_1)$$

$$V_2 = E(V_1|T_2)$$

Show that $\text{Var}(V_2) \leq \text{Var}(V_1)$.

9

Testing Hypotheses and Assessing Goodness of Fit

9.1 Introduction

The first part of this chapter will develop the concepts and theory of statistical hypothesis testing. This theory and other less formal techniques will then be applied to the problem of assessing goodness of fit. In later chapters, we will encounter many other examples of hypothesis testing.

Statistical hypothesis testing is a formal means of distinguishing between probability distributions on the basis of random variables generated from one of the distributions. For example, consider the following problem: Given a sample $X_1, \ldots, X_n$ from a normal distribution having a given variance and a mean equal to either μ_1 or μ_2, decide whether the mean is μ_1 or μ_2. For the most part, we will follow a paradigm developed by Neyman and Pearson for the treatment of such problems, although we will point out certain practical deficiencies of this paradigm and give a point of view advocated by R. A. Fisher. A decision-theoretic model, developed by Wald, will be covered in chapter 15.

In the Neyman–Pearson approach, the probability distributions are grouped into two aggregates, one of which is called the **null hypothesis** and is denoted by

H_0 and the other of which is called the **alternative hypothesis** and is denoted by H_A. (There are several common variations on this notation; the alternative hypothesis, for example, is sometimes denoted as H_1 or even K.) There is an asymmetry between the null and alternative hypotheses, which we will explore below; for the moment, let us merely regard these terms as labels. For the problem given in the preceding paragraph, H_0 might state that the distribution was $N(\mu_1, \sigma^2)$ and H_A might state that the distribution was $N(\mu_2, \sigma^2)$. Since σ^2 is known, each of these hypotheses completely specifies the probability distribution —such hypotheses are called **simple hypotheses**.

Let us consider some additional examples.

EXAMPLE A. (Goodness of Fit Test) Let $X_1, \ldots, X_n$ be a sample from a discrete probability distribution. The null hypothesis might state that the distribution is Poisson with an unknown mean, and the alternative hypothesis might state that the distribution is not Poisson. Neither of these hypotheses completely specify the probability distribution (under H_0, λ is not specified)—they are examples of **composite hypotheses**. If H_0 were changed so as to specify λ precisely, it would then be a simple hypothesis. $\square$

EXAMPLE B. Consider a hypothetical experiment for a study of ESP in which a subject is asked to guess, without looking at them, the suits of 52 cards chosen randomly with replacement from a deck. Letting T denote the total number of successes, the null hypothesis states that T is a binomial random variable with probability .25 of success and 52 trials. The null hypothesis essentially states that the subject is merely guessing and has no extrasensory ability. The alternative hypothesis might be that T is binomial with $p > .25$ and 52 trials. Here H_0 is simple and H_A is composite. $\square$

Hypotheses often specify, or partially specify, the value of a parameter of a probability distribution—in Example B, the alternative hypothesis specified that $p > .25$. We could also have considered the alternative hypothesis that $p \neq .25$. These alternatives are, respectively, examples of **one-sided** and **two-sided alternative hypotheses**. As a further example, consider a normal testing problem with known variance for which the null hypothesis is $H_0: \mu = 0$, and the alternative hypothesis is $H_A: \mu > 0$. Here the alternative is one-sided. The alternative $H_A: \mu \neq 0$ is two-sided.

9.2 The Neyman–Pearson Paradigm

According to the Neyman–Pearson Paradigm, a decision as to whether or not to reject H_0 in favor of H_A is made on the basis of $T(\mathbf{X})$, where $\mathbf{X}$ denotes the sample values and $T(\mathbf{X})$ is a statistic. The sets of values of T for which H_0 is accepted and rejected are called, respectively, the **acceptance region** and the

rejection region of the test. Two types of errors may be incurred in applying this paradigm:

1. H_0 may be rejected when it is true. Such an error is called a **type I error**, and its probability is denoted by α. If H_0 is simple, α is called the **significance level** of the test. If H_0 is composite, the probability of a type I error will generally depend on which particular member of H_0 is true; in this case, the significance level is defined to be the maximum (or more generally the least upper bound) of these probabilities.

2. H_0 may be accepted when it is false. Such an error is called a **type II error**, and its probability is denoted by β. If H_A is composite, β depends on which particular member of H_A holds.

The probability that H_0 is rejected when it is false is called the **power** of the test. Clearly, the power equals $1 - \beta$.

An "ideal" test would have $\alpha = \beta = 0$, but this can be achieved only in trivial cases. Furthermore, in practice, it is always the case that, for fixed sample size, in order to decrease α, β must be increased, and vice versa. The Neyman–Pearson approach resolves this conflict by imposing an asymmetry between the null and alternative hypotheses. The significance level is fixed in advance, usually at a rather small number, and then an attempt is made to construct a test yielding a small value for β.

We can illustrate these ideas with the following two examples.

EXAMPLE A.   Let us consider the problem of testing the value of the parameter p for a binomial random variable with $n = 10$ trials. We wish to test

$$H_0: p = .5$$

versus

$$H_A: p > .5$$

(Note that in structure this is the ESP problem of Example B of Section 9.1 but with different n and p.) We will use the number of successes, X, as a test statistic; the rejection region will consist of large values of X, those values that are relatively unlikely under H_0 and more likely under H_A. To determine the precise rejection region for a given value of α, we can use this table of cumulative binomial probabilities $[P(X \le n)]$:

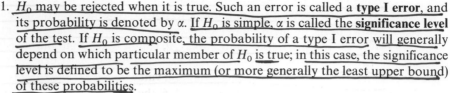

p	0	1	2	3	4	5	6	7	8	9	10
.7	.0000	.0001	.0016	.0106	.0474	.1503	.3504	.6172	.8507	.9718	1.0000
.6	.0001	.0017	.0123	.0548	.1662	.3669	.6177	.8327	.9536	.9940	1.0000
.5	.0010	.0107	.0547	.1719	.3770	.6230	.8281	.9453	.9893	.9990	1.0000

Suppose that the rejection region consists of the points $\{8, 9, 10\}$. The significance level of the test, α, is the probability of rejecting H_0 when it is true; from the last row of the table ($p = .5$), we see that

$$\alpha = P(X > 7) = 1 - P(X \le 7) = .0547$$

If the rejection region consists of $\{7, 8, 9, 10\}$, the significance level of the test is $\alpha = .172$.

The Neyman–Pearson approach sets a value for α first; suppose that we choose to set $\alpha = .0547$. If the true value of p is $.6$, the power of the test is the probability that X is greater than or equal to 8; that is, the power is $.1673$. If the true value is $.7$, the power is $.3828$. The power is thus a function of p, and it is not difficult to see that the power tends to 1 as p approaches 1 and that the power tends to α as p approaches $.5$. $\square$

EXAMPLE B. Let us again consider the problem of testing goodness of fit to a Poisson distribution. The null hypothesis is that the data are a sample from a Poisson distribution, and the alternative hypothesis is that they are from some other unprescribed discrete distribution. We have already seen a similar example in Section 8.2, which was concerned with fitting a Poisson distribution to counts of alpha particle emissions. After grouping the data into bins, we can use Pearson's chi-square statistic to test H_0. If H_0 is true, X^2 has an approximate chi-square distribution with m degrees of freedom, say. A test with significance level α can be constructed in the following way: Let $\chi_m^2(\alpha)$ be the α point of the chi-square distribution with m degrees of freedom (that is, the point such that the area under the density function to its right is equal to α). By construction, the test that rejects if $X^2 > \chi_m^2(\alpha)$ has significance level α. The power of the test is more difficult to evaluate than in Example A. If the real distribution is not exactly Poisson but is very close, the power will be small; if the distribution is quite different from a Poisson distribution, the power will be larger. $\square$

These two examples have certain important features in common. In both cases, the test is based on the value of a test statistic that has the property that its distribution is known (at least approximately) if the null hypothesis is true. (The probability distribution of a test statistic under the null hypothesis is called its **null distribution**.) Knowing this distribution makes it possible to define a rejection region that has probability α under the null hypothesis. Also, in both cases, the rejection region is of the form $\{T > t\}$, where T is the test statistic. In such a case, the number t is called the **critical value** of the test; the critical value separates the rejection region and the acceptance region.

Reconsidering Example B, we can see the relationship of the p-value, introduced in chapter 8 in conjunction with Pearson's chi-square statistic, to the significance level, α. Recall that if the observed value of the test statistic X^2 is c, then the p-value is defined as $p^* = P(X^2 > c)$. Therefore, if $c > \chi^2(\alpha)$, then $p^* < \alpha$; whereas if $c < \chi^2(\alpha)$, then $p^* > \alpha$. The test thus rejects if and only if $p^* < \alpha$. Putting the same thing another way, *the p-value is the smallest value of*

α *for which the null hypothesis will be rejected.* Also, in Example A, if we observe $X = n$, the *p*-value is $P(X \geq n)$, where the probability is calculated for the value of *p* specified by the null hypothesis ($p = .5$ in the example). If we observed 9 successes, the *p*-value would be .0107; if we observed 8 successes, the *p*-value would be .0547. In many applications, it makes more sense to report a *p*-value than merely to report whether or not the null hypothesis was rejected. (We will return to this point in the context of some examples later in this chapter.)

As should be clear by now, there is an asymmetry in the Neyman–Pearson Paradigm between the null and alternative hypotheses. It is natural to wonder at this point how one goes about determining which is the null and which the alternative. This is not a mathematical decision, and the choice typically depends on reasons of custom and convenience. These matters will gradually become clearer in this and ensuing chapters. We make only the following remarks here.

• In Example B, we chose as the null hypothesis the hypothesis that the distribution was Poisson and as the alternative hypothesis the hypothesis that the distribution was not Poisson. In this case, the null hypothesis is simpler than the alternative, which in a sense contains more distributions than does the null. It is conventional to choose the simpler of two hypotheses as the null.

• The consequences of incorrectly rejecting one hypothesis may be graver than those of incorrectly rejecting the other. In such a case, the former should be chosen as the null hypothesis, since the probability of falsely rejecting it could be controlled by choosing α. Examples of this kind arise in screening new drugs; frequently, it must be documented rather conclusively that a new drug is superior before it is accepted for general use.

• In scientific investigations, the null hypothesis is often a simple explanation that must be discredited in order to demonstrate the presence of some physical phenomenon or effect. The hypothetical ESP experiment referred to earlier falls in this category; the null hypothesis states that the subject is merely guessing, that there is no ESP. This must be conclusively disproved in order to convince a skeptic that there is any ESP effect. We will see other examples of this type, beginning in chapter 11.

The Neyman–Pearson approach is predicated on setting α in advance, but the theory itself gives no guidance as to how to make this choice. It is almost always the case that the choice of α is essentially arbitrary but is heavily influenced by custom. Small values of α, such as .1, .05, and .01, are commonly used.

9.3   Optimal Tests: The Neyman–Pearson Lemma

There are typically many tests with significance level α possible for a null hypothesis versus an alternative hypothesis. From among them, we would like to be able select the "best" one. A best test would be the one that had the correct significance level α and was as or more powerful than *any* other test with significance level α. The Neyman–Pearson Lemma shows that for testing a simple hypothesis versus a simple hypothesis such a test exists and is based on the ratio of the likelihoods under the two hypotheses. Suppose that H_0 specifies that the density or frequency function of the data is $f_0(\mathbf{x})$ and that H_A specifies that it is $f_A(\mathbf{x})$. The test rejects for small values of the **likelihood ratio**, $f_0(\mathbf{x})/f_A(\mathbf{x})$. Let us now state the Neyman–Pearson Lemma.

LEMMA A. (Neyman–Pearson Lemma)   Let d^* be a test that accepts if

$$\frac{f_0(\mathbf{x})}{f_A(\mathbf{x})} > c$$

and let α^* be the significance level of d^*. Let d be another test, which has the significance level $\alpha \le \alpha^*$. Then the power of d is less than or equal to the power of d^*.

A proof of the Neyman–Pearson Lemma will be given in chapter 15. We illustrate the likelihood ratio test with an example.

EXAMPLE A.   Let $X_1, \ldots, X_n$ be a random sample from a normal distribution having known variance σ^2. Consider two simple hypotheses:

$$H_0: \mu = \mu_0$$

$$H_A: \mu = \mu_A$$

where μ_0 and μ_A are given constants. Let the significance level α be prescribed. The Neyman–Pearson Lemma states that among all tests with significance level α, the test that rejects for small values of the likelihood ratio is most powerful. We thus calculate the likelihood ratio, which is

$$\frac{f_0(\mathbf{X})}{f_A(\mathbf{X})} = \frac{\exp\left[\dfrac{-1}{2\sigma^2}\sum_{i=1}^{n}(X_i - \mu_0)^2\right]}{\exp\left[\dfrac{-1}{2\sigma^2}\sum_{i=1}^{n}(X_i - \mu_A)^2\right]}$$

since the multipliers of the exponentials cancel. Small values of this statistic correspond to small values of $\sum_{i=1}^{n}(X_i - \mu_A)^2 - \sum_{i=1}^{n}(X_i - \mu_0)^2$. Expanding the squares, we see that the latter expression reduces to

$$2n\bar{X}(\mu_0 - \mu_A) + n\mu_A^2 + n\mu_0^2$$

Now, if $\mu_0 - \mu_A > 0$, the likelihood ratio is small if $\bar{X}$ is small; if $\mu_0 - \mu_A < 0$, the likelihood ratio is small if $\bar{X}$ is large. To be concrete, let us assume the latter case. We then know that the likelihood ratio is a function of $\bar{X}$ and is small when

$\bar{X}$ is large. The Neyman–Pearson Lemma tells us that the most powerful test rejects for $\bar{X} > c$ for some c, and we choose c so as to give the test the desired level α. That is, c is chosen so that $P(\bar{X} > c) = \alpha$ if H_0 is true. Under H_0 in this example, $\bar{X}$ follows a normal distribution with mean μ_0 and variance σ^2/n, so c can be chosen from tables of the standard normal distribution. $\square$

Example A is typical of the way the Neyman–Pearson Lemma is used. We write down the likelihood ratio and observe that small values of it correspond in a one-to-one manner with large values of a statistic, $T(\mathbf{X})$. Knowing the null distribution of T makes it possible to choose a critical value that produces a desired level α.

In real situations, we are seldom presented with the problem of testing two simple hypotheses. Typically, one or both hypotheses are composite and the Neyman–Pearson Lemma does not apply. In some situations, the theory can be extended to include composite one-sided hypotheses. Suppose that H_A is composite. A test that is most powerful for *every* simple alternative in H_A is said to be **uniformly most powerful**. Uniformly most powerful tests exist for some common one-sided alternatives.

EXAMPLE B. Referring to Example A, let us now consider testing $H_0: \mu = \mu_0$ versus $H_A: \mu > \mu_0$. In Example A, we saw that for a particular simple alternative, $\mu = \mu_A > \mu_0$, the most powerful test rejects for $\bar{X} > c$, where c depends on μ_0, n, and σ^2 but *not* on μ_A. Since this test is most powerful and is the same for every simple alternative in H_A, it is uniformly most powerful. $\square$

It can also be argued that the test of Example A is uniformly most powerful for testing $H_0: \mu \le \mu_0$ versus $H_A: \mu > \mu_0$. But it is not uniformly most powerful for testing $H_0: \mu = \mu_0$ versus $H_A: \mu \ne \mu_0$. This follows from further examination of the example, which shows that the most powerful test of H_0 against a simple alternative, $H_A: \mu = \mu_A$, rejects for large values of $\bar{X}$ if $\mu_A > \mu_0$ but rejects for small values of $\bar{X}$ if $\mu_A < \mu_0$ and is thus not the same test for every simple alternative.

For further discussion of the optimality properties of hypothesis tests, see Bickel and Doksum (1977).

9.4 The Duality of Confidence Intervals and Hypothesis Tests

There is a duality between confidence intervals (more generally, confidence sets) and hypothesis tests. In this section, we will show that a confidence set can be obtained by "inverting" a hypothesis test, and vice versa. Before presenting the general structure, we consider an example.

EXAMPLE A. Let $X_1, \ldots, X_n$ be a random sample from a normal distribution having unknown mean μ and known variance σ^2. We consider testing the following hypotheses:

$$H_0: \mu = \mu_0$$

$$H_A: \mu \neq \mu_0$$

Consider a test at a specific level α that rejects for $|\bar{X} - \mu_0| > c$, where c is determined so that $P(|\bar{X} - \mu_0| > c) = \alpha$ if H_0 is true: $c = \sigma_{\bar{X}} z(\alpha/2)$. The test thus accepts when

$$|\bar{X} - \mu_0| < \sigma_{\bar{X}} z(\alpha/2)$$

or

$$-\sigma_{\bar{X}} z(\alpha/2) < \bar{X} - \mu_0 < \sigma_{\bar{X}} z(\alpha/2)$$

or

$$\bar{X} - \sigma_{\bar{X}} z(\alpha/2) < \mu_0 < \bar{X} + \sigma_{\bar{X}} z(\alpha/2)$$

A $100(1 - \alpha)\%$ confidence interval for μ_0 is

$$[\bar{X} - \sigma_{\bar{X}} z(\alpha/2), \bar{X} + \sigma_{\bar{X}} z(\alpha/2)]$$

Comparing the acceptance region of the test to the confidence interval, we see that μ_0 lies in the confidence interval if and only if the hypothesis test accepts. In other words, *the confidence interval consists precisely of all those values of μ_0 for which the null hypothesis $H_0: \mu = \mu_0$ is accepted.*   ◻

We now demonstrate that this duality holds more generally. Let θ be a parameter of a family of probability distributions, and denote the set of all possible values of θ by Θ. Denote the random variables constituting the data by **X**.

THEOREM A.   Suppose that for every value θ_0 in Θ there is a test at level α of the hypothesis $H: \theta = \theta_0$. Denote the acceptance region of the test by $A(\theta_0)$. Then the set

$$C(\mathbf{X}) = \{\theta: \mathbf{X} \in A(\theta)\}$$

is a $100(1 - \alpha)\%$ confidence region for θ.

Proof.   Since A is the acceptance region of a test at level α,

$$P[\mathbf{X} \in A(\theta_0)|\theta = \theta_0] = 1 - \alpha$$

Now,

$$P[\theta_0 \in C(\mathbf{X})|\theta = \theta_0] = P[\mathbf{X} \in A(\theta_0)|\theta = \theta_0]$$
$$= 1 - \alpha$$

by the definition of $C(\mathbf{X})$. □

It is helpful to state Theorem A in words: A $100(1 - \alpha)\%$ confidence region for θ consists of all those values of θ_0 for which the hypothesis that θ equals θ_0 will not be rejected at level α.

THEOREM B.   Suppose that $C(\mathbf{X})$ is a $100(1 - \alpha)\%$ confidence region for θ; that is, for every θ_0,

$$P[\theta_0 \in C(\mathbf{X})|\theta = \theta_0] = 1 - \alpha$$

Then an acceptance region for a test at level α of the hypothesis $H: \theta = \theta_0$ is

$$A(\theta_0) = \{\mathbf{X}|\theta_0 \in C(\mathbf{X})\}.$$

Proof.   The test has level α since

$$P(\mathbf{X} \in A(\theta_0)|\theta = \theta_0) = P(\theta_0 \in C(\mathbf{X})|\theta = \theta_0) = 1 - \alpha. \qquad \square$$

In words, Theorem B says that the hypothesis that θ equals θ_0 is accepted if θ_0 lies in the confidence region.

This duality can be quite useful. In some situations, it is possible to form confidence intervals for parameters of probability distributions and then use those intervals to test hypotheses about the values of those parameters. In other situations, it may be relatively easy to test hypotheses and then determine the acceptance regions for the test to form confidence intervals that might have been quite difficult to derive in a more direct manner. We will see examples of both types of situations in later chapters.

9.5   Generalized Likelihood Ratio Tests

The likelihood ratio test is optimal for testing a simple versus a simple hypothesis. In this section, we will develop a generalization of this test for use in situations in which the hypotheses are not simple. Such tests are not generally optimal, but they are typically nonoptimal in situations for which no optimal test exists, and they usually perform reasonably well. Generalized likelihood ratio tests have wide utility; they play the same role in testing as maximum likelihood estimates do in estimation.

It is frequently the case that the hypotheses under consideration specify, or partially specify, the values of parameters of the probability distribution that has generated the data. Specifically, suppose that the observations $\mathbf{X} = (X_1, \ldots, X_n)$ have a joint density or frequency function $f(\mathbf{x}|\theta)$. Then H_0 may specify that $\theta \in \omega_0$, where ω_0 is a subset of the set of all possible values of θ, and H_A may specify that $\theta \in \omega_1$, where ω_1 is disjoint from ω_0. Let $\Omega = \omega_0 \cup \omega_1$. Based on the data, a plausible measure of the relative tenability of the hypotheses is the ratio of their likelihoods. If the hypotheses are composite, each likelihood is evaluated at that value of θ that maximizes it, yielding the generalized likelihood ratio

$$\Lambda^* = \frac{\max\limits_{\theta \in \omega_0} [\mathrm{lik}(\theta)]}{\max\limits_{\theta \in \omega_1} [\mathrm{lik}(\theta)]}$$

Small values of Λ^* tend to discredit H_0.

It is preferable for certain technical reasons to use the test statistic

$$\Lambda = \frac{\max\limits_{\theta \in \omega_0} [\mathrm{lik}(\theta)]}{\max\limits_{\theta \in \Omega} [\mathrm{lik}(\theta)]}$$

rather than Λ^*. Note that $\Lambda = \min(\Lambda^*, 1)$ so small values of Λ^* correspond to small values of Λ. The rejection region for a likelihood ratio test consists of small values of Λ, for example, all $\Lambda \leq \lambda_0$.

We now illustrate the construction of a likelihood ratio test with a simple example.

EXAMPLE A.   (Testing a Normal Mean) Let $X_1, \ldots, X_n$ be i.i.d. and normally distributed with mean μ and variance σ^2, where σ is known. We wish to test $H: \mu = \mu_0$ against $H_A: \mu \neq \mu_0$, where μ_0 is a prescribed number. The role of θ is played by μ, and $\omega_0 = \{\mu_0\}$, $\omega_1 = \{\mu | \mu \neq \mu_0\}$, and $\Omega = \{-\infty < \mu < \infty\}$.

Since ω_0 consists of only one point, the numerator of the likelihood ratio statistic is

$$\frac{1}{(\sigma\sqrt{2\pi})^n} e^{-(1/2\sigma^2)\sum_{i=1}^{n}(X_i - \mu_0)^2}$$

For the denominator, we have to maximize the likelihood for $\mu \in \Omega$, which is achieved when μ is the mle $\bar{X}$. The denominator is

$$\frac{1}{(\sigma\sqrt{2\pi})^n} e^{-(1/2\sigma^2)\sum_{i=1}^{n}(X_i - \bar{X})^2}$$

The likelihood ratio statistic is therefore

$$\Lambda = \exp\left(-\frac{1}{2\sigma^2}\left[\sum_{i=1}^{n}(X_i - \mu_0)^2 - \sum_{i=1}^{n}(X_i - \bar{X})^2\right]\right)$$

Rejecting for small values of Λ is equivalent to rejecting for large values of

$$-2\log\Lambda = \frac{1}{\sigma^2}\left(\sum_{i=1}^{n}(X_i - \mu_0)^2 - \sum_{i=1}^{n}(X_i - \bar{X})^2\right)$$

Using the identity

$$\sum_{i=1}^{n}(X_i - \mu_0)^2 = \sum_{i=1}^{n}(X_i - \bar{X})^2 + n(\bar{X} - \mu_0)^2$$

we see that the likelihood ratio test rejects for large values of $n(\bar{X} - \mu_0)^2/\sigma^2$. Since $\bar{X} \sim N(\mu_0, \sigma^2/n)$ under H_0, the distribution of $-2\log\Lambda$ under H_0 is chi-square with 1 degree of freedom (recall that the distribution of the square of a standard normal random variable is chi-square with 1 degree of freedom). Knowing the null distribution of the test statistic makes possible the construction of a rejection region for any significance level α: The test rejects when

$$\frac{n}{\sigma^2}(\bar{X} - \mu_0)^2 > \chi_1^2(\alpha)$$

Again using the fact that a chi-square random variable with 1 degree of freedom is the square of a standard normal random variable, we can rewrite this relation to show that the rejection region for the test is

$$|\bar{X} - \mu_0| \geq \frac{\sigma}{\sqrt{n}}z(\alpha/2) \qquad \square$$

In order for the likelihood ratio test to have the significance level α, λ_0 must be chosen so that $P(\Lambda \leq \lambda_0) = \alpha$ if H_0 is true. If the sampling distribution of Λ under H_0 is known, we can choose λ_0. Generally, the sampling distribution is not of a simple form, but in many situations the following theorem can be used.

THEOREM A.   Under smoothness conditions on the probability density or frequency functions involved, the null distribution of $-2\log\Lambda$ tends to a chi-square distribution with degrees of freedom equal to $\dim\Omega - \dim\omega_0$ as the sample size tends to infinity.

Proof.   The proof, which is beyond the scope of this book, is based on a second-order Taylor Series expansion.

In the statement of Theorem A, $\dim \Omega$ and $\dim \omega_0$ are the numbers of free parameters under Ω and ω_0, respectively. In Example A, the null hypothesis completely specifies μ and σ^2; there are no free parameters under ω_0, so $\dim \omega_0 = 0$. Under Ω, σ is fixed but μ is free, so $\dim \Omega = 1$. For this example, the null distribution of $-2 \log \Lambda$ is exactly χ_1^2.

9.6   Likelihood Ratio Tests for the Multinomial Distribution

In a multinomial goodness-of-fit test, the null hypothesis, H_0, specifies the cell probabilities or places some restriction on them, such as that they depend on an unknown parameter. Using the terminology above, H_0 specifies that $p = p(\theta) \in \omega_0$, where p is the vector of cell probabilities and θ is a parameter. The alternative hypothesis, H_A, specifies that H_0 is not true. Therefore, Ω allows the probabilities to be free, with the constraint that they sum to 1.

The numerator of the likelihood ratio is

$$\max_{\theta \in \omega_0} \left(\frac{n!}{x_1! \cdots x_m!} \right) p_1(\theta)^{x_1} \cdots p_m(\theta)^{x_m}$$

where the x_i are the observed counts in the m cells. By the definition of the maximum likelihood estimate, this likelihood is maximized when $\hat{\theta}$ is the maximum likelihood estimate of θ. The corresponding probabilities will be denoted by $p_i(\hat{\theta})$.

Since the probabilities are unrestricted under Ω, the denominator is maximized by the unrestricted maximum likelihood estimates, or

$$\hat{p}_i = \frac{x_i}{n}$$

The likelihood ratio is therefore

$$\Lambda = \frac{\dfrac{n!}{x_1! \cdots x_m!} p_1(\hat{\theta})^{x_1} \cdots p_m(\hat{\theta})^{x_m}}{\dfrac{n!}{x_1! \cdots x_m!} \hat{p}_1^{x_1} \cdots \hat{p}_m^{x_m}}$$

$$= \prod_{i=1}^{m} \left(\frac{p_i(\hat{\theta})}{\hat{p}_i} \right)^{x_i}$$

Also, since $x_i = n\hat{p}_i$,

$$-2 \log \Lambda = -2n \sum_{i=1}^{m} \hat{p}_i \log \left(\frac{p_i(\hat{\theta})}{\hat{p}_i} \right)$$

$$= 2 \sum_{i=1}^{m} O_i \log \left(\frac{O_i}{E_i} \right)$$

where $O_i = n\hat{p}_i$ and $E_i = np_i(\hat{\theta})$ denote the observed and expected counts, respectively.

Under Ω, the cell probabilities are allowed to be free, with the constraint that they sum to 1, so $\dim \Omega = m - 1$. If, under H_0, the probabilities $p_i(\hat{\theta})$ depend on a k-dimensional parameter θ that has been estimated from the data, $\dim \omega_0 = k$. The large sample distribution of Ω is thus a chi-square distribution with $m - k - 1$ degrees of freedom (the number of cells minus the number of estimated parameters minus 1).

In Section 8.2, Pearson's chi-square statistic was introduced as a test for goodness of fit

$$X^2 = \sum_{i=1}^{m} \frac{[x_i - np_i(\hat{\theta})]^2}{np_i(\hat{\theta})}$$

Pearson's statistic and the likelihood ratio are asymptotically equivalent under H_0. To indicate heuristically why this is so, we will go through a Taylor Series argument. To begin,

$$-2 \log \Lambda = 2n \sum_{i=1}^{m} \hat{p}_i \log\left(\frac{\hat{p}_i}{p_i(\hat{\theta})}\right)$$

If H_0 is true and n is large, $\hat{p}_i \approx p_i(\hat{\theta})$. The Taylor Series expansion of the function

$$f(x) = x \log\left(\frac{x}{x_0}\right)$$

about x_0 is

$$f(x) = (x - x_0) + \frac{1}{2}(x - x_0)^2 \frac{1}{x_0} + \cdots$$

Thus,

$$-2 \log \Lambda \approx 2n \sum_{i=1}^{m} [\hat{p}_i - p_i(\hat{\theta})] + n \sum_{i=1}^{m} \frac{[\hat{p}_i - p_i(\hat{\theta})]^2}{p_i(\hat{\theta})}$$

The first term on the right-hand side is equal to 0 since the probabilities sum to 1, and the second term on the right-hand side may be expressed as

$$\sum_{i=1}^{m} \frac{[x_i - np_i(\hat{\theta})]^2}{np_i(\hat{\theta})}$$

since x_i, the observed count, equals $n\hat{p}_i$ for $i = 1, \ldots, m$.

We have argued for the approximate equivalence of two test statistics. Pearson's test has been more commonly used than the likelihood ratio test since it is somewhat easier to calculate without the use of a computer.

Let us consider some examples.

EXAMPLE A.   (Hardy–Weinberg Equilibrium) Hardy–Weinberg equilibrium was first introduced in Example A in Section 8.5.1. There we found the mle of θ, and in Example C in Section 8.5.3 we found an approximate 95% confidence interval for θ. Here we will investigate whether the equilibrium model fits the data. Using the estimate $\hat{\theta} = .4247$, we find the expected frequencies

	Blood Type		
	M	MN	N
Observed	342	500	187
Expected	340.6	502.8	185.6

We can analyze these data using the formalism of the Neyman–Pearson approach in order to illustrate those concepts. The null hypothesis will be that the multinomial distribution is as specified by the Hardy–Weinberg equilibrium frequencies, with unknown parameter θ. The alternative hypothesis will be that the multinomial distribution does not have probabilities of that specified form. With this approach, we first choose a value for α, the significance level for the test (recall that the significance level is the probability of falsely rejecting the hypothesis that the multinomial cell probabilities are as specified by genetic theory). In this application, there is no compelling reason to choose any particular value of α, so we will follow convention and let $\alpha = .05$. This means that our decision rule will falsely reject H_0 in only 5% of the cases.

We will use Pearson's chi-square test, and therefore X^2 as our test statistic. The null distribution of X^2 is approximately chi-square with 1 degree of freedom (there are two independent cells and one parameter has been estimated from the data). Since, from Table 3 in Appendix B, the point defining the upper 5% of the chi-square distribution with 1 degree of freedom is 3.84, the test rejects if $X^2 > 3.84$. We next calculate X^2:

$$X^2 = \sum \frac{(O - E)^2}{E} = .091$$

Thus, the null hypothesis is not rejected.

There is a certain unnecessary rigidity in this procedure, since it is not clear that such a decision (to reject or not) has to be made at all. There is also a certain arbitrariness: There was no strong reason to let $\alpha = .05$, but that choice essentially determined our decision. If we had let $\alpha = .01$, the decision would have been the same since $\chi^2(.01) > \chi^2(.05)$, but what if we had let $\alpha = .10$, or .20? It is here that the concept of the p-value becomes useful. Recall that the p-value is the smallest significance level at which the null hypothesis would be rejected.

From a table of the chi-square distribution (or from a table of the normal distribution since a chi-square distribution with 1 degree of freedom is the square of a standard normal random variable), $.09 = \chi_1^2(.76)$, so the p-value is $.76$. Another interpretation of this p-value is that if the model were correct, deviations this large or larger would occur 76% of the time. Thus, the data give us no reason to doubt the model.

In comparison, the likelihood ratio test statistic is

$$-2 \log \Lambda = 2n \sum_{i=1}^{3} O_i \log\left(\frac{O_i}{E_i}\right) = .032$$

The corresponding p-value is $.86$. The two tests lead to the same conclusion. □

EXAMPLE B. (Bacterial Clumps) In testing milk for bacterial contamination, .01 mL of milk is spread over an area of 1 cm^2 on a slide. The slide is mounted on a microscope, and counts of bacterial clumps within grid squares are made. The Poisson model appears quite reasonable for the distribution of the clumps at first glance: The clumps are presumably mixed uniformly throughout the milk, and there is no reason to suspect that the clumps bunch together. However, on closer examination, two possible problems are noted. First, bacteria held by surface tension on the lower surface of the drop may adhere to the glass slide on contact, producing increased concentrations in that area of the film. Second, the film is not of uniform thickness, being thicker in the center and thinner at the edges, giving rise to nonuniform concentrations of bacteria. The following table, taken from Bliss and Fisher (1953), summarizes the counts of clumps on 400 grid squares.

Number per Square	0	1	2	3	4	5	6	7	8	9	10	19
Frequency	56	104	80	62	42	27	9	9	5	3	2	1

To fit a Poisson distribution to these data, we compute the mle, $\hat{\lambda}$, which is the mean of the 400 counts:

$$\hat{\lambda} = \frac{0 \times 56 + 1 \times 104 + 2 \times 80 + \cdots + 19 \times 1}{400}$$

$$= 2.40$$

The following table shows the observed and expected counts and the components of chi-square test statistic. (The last several cells were grouped together so that the minimum expected count would be nearly 5.)

Observed	56	104	80	62	42	27	9	20
Expected	36.4	87.2	104.4	83.6	50.0	24.0	9.6	4.8
Component of X^2	10.55	3.24	5.70	5.58	1.28	.38	.04	48.13

The chi-square statistic is $X^2 = 74.90$. With 6 degrees of freedom (there are eight cells and one parameter has been estimated from the data), the null hypothesis is conclusively rejected [$\chi_6^2(.001) = 22.5$, so the p-value is less than .001]. When a goodness-of-fit test rejects, it is instructive to find out why; where does the model fail to fit? This can be seen by looking at the cells that make large contributions to X^2 and the signs of the observed minus the expected counts for those cells. We see here that the greatest contributions to X^2 come from the first and last cells of the table—there are too many small counts and too many large counts relative to what is expected for a Poisson distribution.   □

EXAMPLE C.   (Fisher's Reexamination of Mendel's Data) In one of his famous experiments, Mendel crossed 556 smooth, yellow male peas with wrinkled, green female peas. According to now established genetic theory, the relative frequencies of the progeny should be as given in the following table.

Type	Frequency
Smooth-yellow	$\frac{9}{16}$
Smooth-green	$\frac{3}{16}$
Wrinkled-yellow	$\frac{3}{16}$
Wrinkled-green	$\frac{1}{16}$

Mendel recorded the counts given in this table (the expected counts have also been listed so they can be used later):

Type	Observed Count	Expected Count
Smooth-yellow	315	312.75
Smooth-green	108	104.25
Wrinkled-yellow	102	104.25
Wrinkled-green	31	34.75

Calculating the likelihood ratio test statistic, we obtain

$$-2 \log \Lambda = 2 \sum_{i=1}^{4} O_i \log \left(\frac{O_i}{E_i} \right) = .618$$

Comparing this value with the chi-square distribution with 3 degrees of freedom (three independent parameters are estimated under Ω and none under ω_0), we have a p-value of slightly less than .9. Pearson's chi-square statistic is .604, which is quite close to the value from the likelihood ratio test. We interpret the p-value as meaning that, even if the model were correct, discrepancies this large or larger would be expected to occur on the basis of chance about 90% of the time. There is thus no reason to reject the hypothesis that the counts come from a multinomial distribution with the prescribed probabilities. We would only tend to doubt this hypothesis for small p-values.

The p-value can also be interpreted to mean that on the basis of chance we would expect agreement this close or closer about only 10% of the time. There is some validity to the suggestion that the data agree with the model too well; if the *p*-value had been .999, for example, we would definitely be suspicious.

Fisher pooled the results of all of Mendel's experiments in the following way. Suppose that two independent experiments give chi-square statistics with *p* and *r* degrees of freedom, respectively. Then, under the null hypothesis that the models were correct, the sum of those two test statistics would follow a chi-square distribution with $p + r$ degrees of freedom. Fisher added the chi-square statistics for all the independent experiments and compared the result with the chi-square distribution with degrees of freedom equal to the sum of all the degrees of freedom. The resulting *p*-value was .99996. Such close agreement would only occur 4 times out of 100,000 on the basis of chance!

What happened? Did Mendel deliberately or unconsciously fudge the data? Did he have an overzealous lab technician who was hoping for a recommendation to medical school? Was there divine intervention? Perhaps the best explanation is that Mendel continued experimenting until the results looked good and then stopped. The statistical analysis here assumes that the sample size is fixed before the data are collected. □

Mendel is not the only scientist whose data are too good to be true. Cyril Burt was an English psychologist whose work had a great impact on the debate concerning the genetic basis for intelligence. His many papers and extensive data argue for such a basis. In 1946, Burt became the first psychologist to be knighted; however, during the 1970s, his work came under increasing attack, and he was accused of actually fabricating data. One of his most famous studies was of the intelligence and occupational status of 40,000 fathers and sons. Dorfman (1978) studied the goodness of fit of these intelligence scores to a normal distribution, using Pearson's chi-square test. The *p*-values for fathers and sons were greater than $1 - 10^{-7}$ and $1 - 10^{-6}$, respectively! Dorfman concluded that "it may well be that Burt's frequency distributions are the most normally distributed in the history of anthropometric measurement."

9.7 The Poisson Dispersion Test

The likelihood ratio test and Pearson's chi-square test are carried out with respect to the general alternative hypothesis that the cell probabilites are completely free. If one has a specific alternative hypothesis in mind, better power can usually be obtained by testing against that alternative rather than against a more general alternative. Such a test is developed in this section for the hypothesis that a distribution is Poisson. The test is quite useful, and its construction affords another illustration of a generalized likelihood ratio test.

The two key assumptions underlying the Poisson distribution are that the rate is constant and that the counts in one interval of time or space are independent of the counts in disjoint intervals. These assumptions are often not met. For example, suppose that insects are counted on leaves of plants. The leaves are of different sizes and occur at various locations on different plants; the rate of infestation may well not be constant over the different locations. Furthermore, if the insects hatched from eggs that were deposited in groups, there might be clustering of the insects and the independence assumption might fail. If counts occurring over time are being recorded, the underlying rate of the phenomenon being studied might not be constant. Motor vehicle counts for traffic studies, for example, typically vary cyclically over time.

Given counts $x_1, \ldots, x_n$, we consider testing the null hypothesis that the counts are Poisson with the common parameter λ versus the alternative hypothesis that they are Poisson but have different rates, $\lambda_1, \ldots, \lambda_n$. Under ω_0, the maximum likelihood estimate of λ is $\hat{\lambda} = \bar{X}$. Under Ω, the maximum likelihood estimates of the λ_i are $x_1, \ldots, x_n$; we denote these estimates by $\tilde{\lambda}_i$. The likelihood ratio is thus

$$\Lambda = \frac{\displaystyle\prod_{i=1}^{n} \hat{\lambda}^{x_i} e^{-\hat{\lambda}} / x_i!}{\displaystyle\prod_{i=1}^{n} \tilde{\lambda}_i^{x_i} e^{-\tilde{\lambda}_i} / x_i!}$$

$$= \prod_{i=1}^{n} \left(\frac{\bar{x}}{x_i}\right)^{x_i} e^{x_i - \bar{x}}$$

The likelihood ratio test statistic is

$$-2 \log \Lambda = -2 \sum_{i=1}^{n} \left[x_i \log\left(\frac{\bar{x}}{x_i}\right) + (x_i - \bar{x}) \right]$$

$$= 2 \sum_{i=1}^{n} x_i \log\left(\frac{x_i}{\bar{x}}\right)$$

A nearly equivalent form of this statistic is produced using the Taylor Series argument given in Section 9.6:

$$-2 \log \Lambda \approx \frac{1}{\bar{x}} \sum_{i=1}^{n} (x_i - \bar{x})^2$$

Under Ω, there are n independent parameters, $\lambda_1, \ldots, \lambda_n$, so dim $\Omega = n$. Under ω_0, there is only one parameter, λ, so dim $\omega_0 = 1$, and the degrees of freedom are $n - 1$.

The last equation given above for the test statistic may be interpreted as being the ratio of n times the estimated variance to the estimated mean. For the Poisson distribution, the variance equals the mean; for the types of alternatives discussed above, the variance is typically greater than the mean. For this reason the test is often called the **Poisson dispersion test**. It is sensitive to—that is, has high power against—alternatives that are overdispersed relative to the Poisson, such as the negative binomial distribution. The ratio $\hat{\sigma}^2/\bar{x}$ is sometimes used as a measure of clustering. This test is often used when there is not enough data to be accumulated into several cells so that Pearson's chi-square test can be used.

EXAMPLE A. (Asbestos Fibers) In Example A in Section 8.4, we considered whether counts of asbestos fibers on grid squares could be modeled as a Poisson distribution. Applying the Poisson dispersion test, we find that

$$\frac{1}{\bar{x}} \sum (x_i - \bar{x})^2 = 26.56$$

or, if the likelihood ratio statistic is used,

$$2 \sum x_i \log\left(\frac{x_i}{\bar{x}}\right) = 27.11$$

Since there are 23 observations, there are 22 degrees of freedom, and $\chi_{22}^2(.250) = 26.04$, so the p-value is about .25. The evidence against the null hypothesis is not persuasive; however, the sample size is small and the test may have low power.

□

EXAMPLE B. (Bacterial Clumps) In Example B in Section 9.6, we applied Pearson's chi-square test to test whether counts of bacteria clumps in milk were fit by a Poisson distribution. There we found $\bar{x} = 2.40$. The sample variance is

$$\hat{\sigma}^2 = \frac{0^2 \times 56 + 1^2 \times 104 + \cdots + 19^2 \times 1}{400} - \bar{x}^2$$

$$= 3.36$$

The ratio of the variance to the mean is 1.4 rather than 1; the test statistic is

$$T = \frac{n\hat{\sigma}^2}{\bar{x}}$$

$$= \frac{400 \times 3.36}{2.40} = 560$$

Under the null hypothesis, the statistic approximately follows a chi-square distribution with 399 degrees of freedom. Since a chi-square random variable with m degrees of freedom is the sum of the squares of m independent $N(0, 1)$ random variables, the central limit theorem implies that for large values of m the chi-square distribution with m degrees of freedom is approximately normal with mean equal to the number of degrees of freedom and variance equal to twice the number of degrees of freedom. The p-value can thus be found by standardizing the statistic and using tables of the standard normal distribution:

$$P(T \geq 560) = P\left(\frac{T - 399}{\sqrt{2 \times 399}} \geq \frac{560 - 399}{\sqrt{2 \times 399}}\right)$$

$$\approx 1 - \Phi(5.7) \approx 0$$

Thus, there is almost no doubt that the Poisson distribution fails to fit the data.

$\square$

9.8 Hanging Rootograms

In this and next section, we develop additional informal techniques for assessing goodness of fit. The first of these is the hanging rootogram. **Hanging rootograms** are a graphical display of the differences between observed and fitted values in histograms. To illustrate the construction and interpretation of hanging rootograms, we will use a set of data from the field of clinical chemistry (Martin, Gudzinowicz, and Fanger, 1975). The following table gives the empirical distribution of 152 serum potassium levels. In clinical chemistry, distributions such as this are often tabulated to establish a range of "normal" values against which the level of the chemical found in a patient can be compared to determine whether it is abnormal. The tabulated distributions are often fit to parametric forms such as the normal distribution.

Serum potassium levels

Interval Midpoint	Frequency
3.2	2
3.3	1
3.4	3
3.5	2
3.6	7
3.7	8
3.8	8
3.9	14
4.0	14
4.1	18
4.2	16
4.3	15
4.4	10
4.5	8
4.6	8
4.7	6
4.8	4
4.9	1
5.0	1
5.1	1
5.2	4
5.3	1

Figure 9-1(a) is a histogram of the frequencies. The plot looks roughly bell-shaped, but the normal distribution is not the only bell-shaped distribution. In order to evaluate their distribution more exactly, we must compare the observed frequencies to frequencies fit by the normal distribution. This can be done in the following way. Suppose that the parameters μ and σ of the normal distribution are estimated from the data by $\bar{x}$ and $\hat{\sigma}$. If the jth interval has the left boundary x_{j-1} and the right boundary x_j, then according to the model, the probability that an observation falls in that interval is

$$\hat{p}_j = \Phi\left(\frac{x_j - \bar{x}}{\hat{\sigma}}\right) - \Phi\left(\frac{x_{j-1} - \bar{x}}{\hat{\sigma}}\right)$$

If the sample is of size n, the predicted, or fitted, count in the jth interval is

$$\hat{n}_j = n\hat{p}_j$$

which may be compared to the observed counts, n_j.

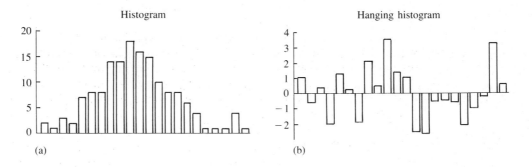

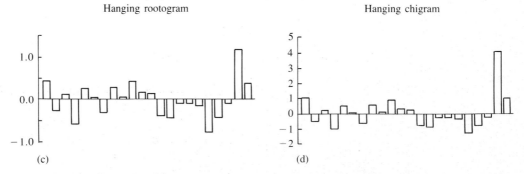

Figure 9-1. (a) Histogram, (b) hanging histogram, (c) hanging rootogram, and (d) hanging chi-gram for normal fit to serum potassium data.

Figure 9-1(b) is a "hanging histogram" of the differences: observed count (n_j) minus expected count ($\hat{n}_j$). These differences are difficult to interpret since the variability is not constant from cell to cell. If we neglect the variability in the estimated expected counts, we have

$$\text{Var}(n_j - \hat{n}_j) = \text{Var}(n_j)$$
$$= np_j(1 - p_j)$$
$$= np_j - np_j^2$$

In this case, the p_j are small, so

$$\text{Var}(n_j - \hat{n}_j) \approx np_j$$

Thus, cells with large values of p_j (equivalent to large values of n_j if the model is at all close) have more variable differences, $n_j - \hat{n}_j$. In a hanging histogram, we expect larger fluctuations in the center than in the tails. This unequal variability makes it difficult to assess and compare the fluctuations, since a large deviation may indicate real misfit of the model or may be merely caused by large random variability.

   To put the differences between observed and expected values on a scale on which they all have equal variability, a **variance-stabilizing transformation** may be used. (Such transformations will be used in later chapters as well.) Suppose that a random variable X has mean μ and variance $\sigma^2(\mu)$, which depends on μ. If $Y = f(X)$, the method of propagation of error shows that

$$\text{Var}(Y) \approx \sigma^2(\mu)[f'(\mu)]^2$$

Thus if f is chosen so that $\sigma^2(\mu)[f'(\mu)]^2$ is constant, the variance of Y will not depend on μ. The transformation f that accomplishes this is called a variance-stabilizing transformation.

   Let us apply this idea to the case we have been considering:

$$E(n_j) = np_j = \mu$$
$$\text{Var}(n_j) \approx np_j = \sigma^2(\mu)$$

That is, $\sigma^2(\mu) = \mu$, so f will be a variance-stabilizing transformation if $\mu[f'(\mu)]^2$ is constant. The function $f(x) = \sqrt{x}$ does the job, and

$$E(\sqrt{n_j}) \approx \sqrt{np_j}$$
$$\text{Var}(\sqrt{n_j}) \approx \tfrac{1}{4}$$

if the model is correct.

Figure 9-1(c) shows a hanging rootogram, a display showing

$$\sqrt{n_j} - \sqrt{\hat{n}_j}$$

The advantage of the hanging rootogram is that the deviations from cell to cell have approximately the same statistical variability. To assess the deviations, we may use the rough rule of thumb that a deviation of more than 2 or 3 standard deviations (more than 1.0 or 1.5 in this case) is "large." The most striking feature of the hanging rootogram in Figure 9-1(c) is the large deviation in the right tail. Generally, deviations in the center have been down-weighted and those in the tails emphasized by the transformation. Also, it is noteworthy that although the deviations other than the one in the right tail are not especially large, they have a certain systematic character: Note the run of positive deviations followed by the run of negative deviations. This may indicate some asymmetry in the distribution.

A possible alternative to the rootogram is what can be called a "hanging chi-gram," a plot of the components of Pearson's chi-square statistic:

$$\frac{n_j - \hat{n}_j}{\sqrt{\hat{n}_j}}$$

Since $\text{Var}(n_j) \approx \hat{n}_j$,

$$\text{Var}\left(\frac{n_j - \hat{n}_j}{\sqrt{\hat{n}_j}}\right) \approx 1$$

so this technique also stabilizes variance. Figure 9-1(d) is a hanging chi-gram for the case we have been considering; it is quite similar in overall character to the hanging rootogram, but the deviation in the right tail is emphasized even more.

9.9 Probability Plots

Probability plots are an extremely useful graphical tool for qualitatively assessing the fit of data to a theoretical distribution. Consider a sample of size n from a uniform distribution on $[0, 1]$. Denote the *ordered* sample values by $X_{(1)} < X_{(2)} \cdots < X_{(n)}$. These values are called **order statistics**. It can be shown (see Problem 15 at the end of chapter 4) that

$$E(X_{(j)}) = \frac{j}{n + 1}$$

This suggests plotting the ordered observations $X_{(1)}, \ldots, X_{(n)}$ against the points

$1/(n + 1), \ldots, n/(n + 1)$. If the underlying distribution is uniform, the plot should look roughly linear. Figure 9-2 is such a plot for a sample of size 100 from a uniform distribution.

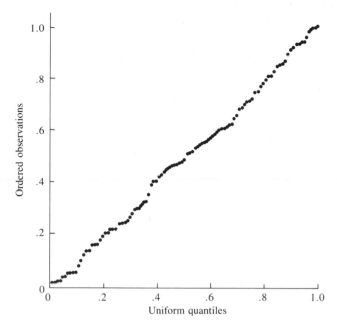

Figure 9-2. Uniform-uniform probability plot.

Now suppose that a sample $Y_1, \ldots, Y_{100}$ is generated in which each Y is half the sum of two independent uniform random variables. The distribution of Y is no longer uniform but triangular:

$$f(y) = \begin{cases} 4y, & 0 \le y \le \tfrac{1}{2} \\ 4 - 4y, & \tfrac{1}{2} \le y \le 1 \end{cases}$$

The ordered observations $Y_{(1)}, \ldots, Y_{(n)}$ are plotted against the points $1/(n + 1)$, $\ldots, n/(n + 1)$. The graph in Figure 9-3 shows a clear deviation from linearity and enables us to describe qualitatively the deviation of the distribution of the Y's from the uniform distribution. Note that in the left tail of the plotted distribution (near 0) the observations are larger than expected for a uniform distribution and in the right tail (near 1) they are smaller, indicating that the tails of the distribution of the Y's decrease more quickly (are "lighter") than the tails of the uniform distribution.

The technique can be extended to other continuous probability laws by means of Proposition C of Section 2.3, which states that if X is a continuous random variable with a strictly increasing cumulative distribution function, F_X, and if

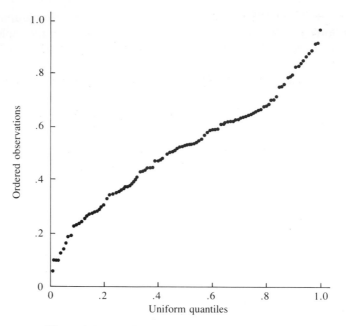

Figure 9-3. Uniform-triangular probability plot.

$Y = F_X(X)$, then Y has a uniform distribution on $[0, 1]$. The transformation $Y = F_X(X)$ is known as the **probability integral transform**.

The following procedure is suggested by the proposition just referred to. Suppose that it is hypothesized that X follows a certain distribution, F. Given a sample $X_1, \ldots, X_n$, we plot

$$F(X_{(k)}) \text{ vs. } \frac{k}{n+1}$$

or equivalently

$$X_{(k)} \text{ vs. } F^{-1}\left(\frac{k}{n+1}\right)$$

In some cases, F is of the form

$$F(x) = G\left(\frac{x - \mu}{\sigma}\right)$$

where μ and σ are called location and scale parameters, respectively. The normal distribution is of this form. We could plot

$$\frac{X_{(k)} - \mu}{\sigma} \text{ vs. } G^{-1}\left(\frac{k}{n+1}\right)$$

or if we plotted

$$X_{(k)} \text{ vs. } G^{-1}\left(\frac{k}{n+1}\right)$$

the result would be approximately a straight line if the model were correct:

$$X_{(k)} \approx \sigma G^{-1}\left(\frac{k}{n+1}\right) + \mu$$

Slight modifications of this procedure are sometimes used. For example, rather than $G^{-1}[k/(n+1)]$, $E(X_{(k)})$, the expected value of the kth smallest observation, can be used. But it can be argued that

$$E(X_{(k)}) \approx F^{-1}\left(\frac{k}{n+1}\right)$$

$$= \sigma G^{-1}\left(\frac{k}{n+1}\right) + \mu$$

so this modification yields very similar results to the original procedure.

The procedure can be viewed from another perspective. Recall from Section 2.2 that $F^{-1}[k/(n+1)]$ is the $k/(n+1)$ quantile of the distribution F; that is, it is the point such that the probability that a random variable with distribution function F is less than it is $k/(n+1)$. We are thus plotting the ordered observations (which may be viewed as the observed or empirical quantiles) versus the quantiles of the theoretical distribution.

EXAMPLE A.   We can illustrate the procedure just described using a set of 100 observations, which are Michelson's determinations of the velocity of light made from June 5, 1879 to July 2, 1879; 299,000 has been subtracted from the determinations to give the values listed (data from Stigler, 1977):

850	960	880	890	890	740
940	880	810	840	900	960
880	810	780	1070	940	860
820	810	930	880	720	800
760	850	800	720	770	810
950	850	620	760	790	980
880	860	740	810	980	900
970	750	820	880	840	950
760	850	1000	830	880	910
870	980	790	910	920	870
930	810	850	890	810	650
880	870	860	740	760	880
840	880	810	810	830	840
720	940	1000	800	850	840
950	1000	790	840	850	800
960	760	840	850	810	960
800	840	780	870		

Figure 9-4 shows the normal probability plot. The plot looks straight, showing that the normal distribution gives a reasonable fit.

A word of caution is in order here: Probability plots are by nature monotone-increasing and they all tend to look fairly straight. Some experience is necessary in gauging "straightness." Simulations, which are easily done, are very helpful in sharpening one's skill. Some find it useful to hold the plot so that they are looking down the plotted line as if it were a roadway; this often makes curvature much more apparent.   □

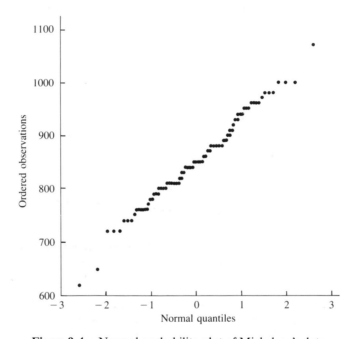

Figure 9-4. Normal probability plot of Michelson's data.

EXAMPLE B.   In order to be able to interpret probability plots, it is useful to see how they are shaped for samples from nonnormal distributions. Figure 9-5 is a normal probability plot of 500 pseudorandom variables from a double exponential distribution:

$$f(x) = \tfrac{1}{2}e^{-|x|}, \quad -\infty < x < \infty$$

This density is symmetric about zero, but its tails die off at the rate $\exp(-|x|)$. This rate is slower than that for the tails of the normal distribution, which decay at the rate $\exp(-x^2)$. Note how the plot in Figure 9-5 bends down at the left and up at the right, indicating that the observations in the left tail were more negative than expected for a normal distribution and the observations in the right tail were more positive. In other words, the extreme observations were larger in magnitude than extreme observations from a normal distribution would be. This effect results because the tails of the double exponential are "heavier" than those of a normal distribution.

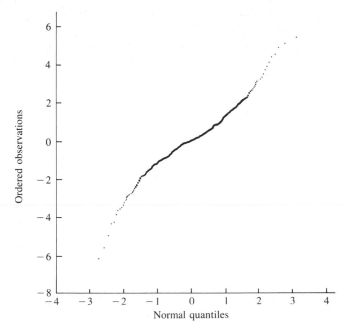

Figure 9-5. Normal probability plot of 500 pseudorandom variables from a double exponential distribution.

Figure 9-6 is a normal probability plot of 500 pseudorandom numbers from a gamma distribution with the shape parameter $\alpha = 5$ and the scale parameter $\lambda = 1$. As can be seen in Figure 2-11, the gamma density with $\alpha = 5$ is nonsymmetric, or skewed, and this is reflected by the bowlike appearance of the probability plot. $\square$

EXAMPLE C. As an example for a nonnormal distribution, Figure 9-7 is a gamma probability plot of the precipitation amounts of Example C in Section 8.5. Qualitatively, the fit looks reasonable, since there is no gross systematic deviation from a straight line. $\square$

Probability plots can also be constructed for grouped data, such as the data on serum potassium levels in Section 9.8. Since the ordered observations are not all available in such a case, the procedure must be modified. Suppose that the grouping gives the points $x_1, x_2, \ldots, x_{m+1}$ for the histogram's bin boundaries and that in the interval $[x_i, x_{i+1})$ there are n_i counts, where $i = 1, \ldots, m$. We denote the cumulative frequencies by $N_j = \sum_{i=1}^{j} n_i$. Then $N_1 < N_2 < \cdots < N_m$ and $N_m = n$, which is the total sample size. We thus plot

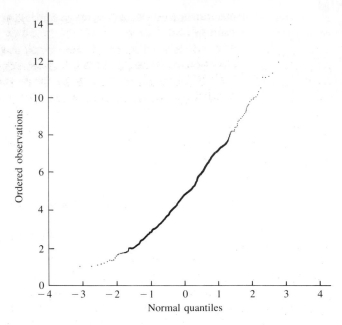

Figure 9-6.   Normal probability plot of 500 pseudorandom variables from a gamma distribution with the shape parameter $\alpha = 5$.

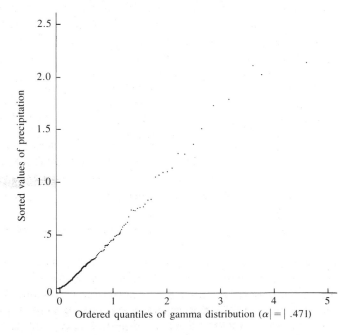

Figure 9-7.   Gamma probability plot of rainfall distribution.

$$x_{j+1} \text{ vs. } G^{-1}\left(\frac{N_j}{n+1}\right), \quad j = 1, \dots, m$$

EXAMPLE D. Figure 9-8 shows a probability plot for the serum potassium data of Section 9.8. The cumulative frequencies are found by summing the frequencies in each bin. The deviations in the right tail are immediately apparent. ☐

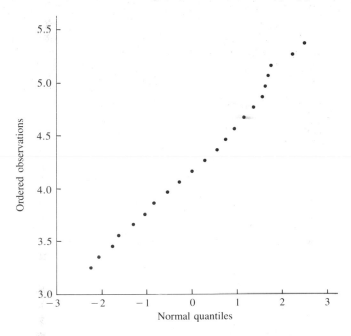

Figure 9-8. Normal probability plot of serum potassium data.

9.10 Tests for Normality

A wide variety of tests are available for testing goodness of fit to the normal distribution. We discuss some of them in this section; more discussion may be found in the works referred to.

If the data are grouped into bins, with several counts in each bin, Pearson's chi-square test for goodness of fit may be applied. But if the parameters are estimated from ungrouped data and the expected counts in each bin are calculated using the estimated parameters, the limiting distribution of the test statistic is no longer chi-square. In order for the limiting distribution to be chi-square, the parameters must be estimated from the grouped data. This was pointed out by Chernoff and Lehmann (1954) and is further discussed by Dahiya and Gurland (1972). Generally speaking, it seems rather artificial and wasteful of information to group continuous data.

Departures from normality often take the form of asymmetry, or skewness. Suppose that we wish to test the null hypothesis that $X_1, \ldots, X_n$ are independent normally distributed random variables with the same mean and variance. A goodness-of-fit test can be based on the **coefficient of skewness** for the sample,

$$b_1 = \frac{\frac{1}{n} \sum_{i=1}^{n} (X_i - \bar{X})^3}{s^3}$$

The test rejects for large values of $|b_1|$.

Symmetric distributions can depart from normality by being heavy-tailed or light-tailed or too peaked or too flat in the center. These forms of departures may be detected by the **coefficient of kurtosis** for the sample,

$$b_2 = \frac{\frac{1}{n} \sum_{i=1}^{n} (X_i - \bar{X})^4}{s^4}$$

If either of these measures is to be used as a test statistic, its sampling distributions when the distribution generating the data is normal must be determined. The hypothesis test rejects normality when the observed value of the statistic is in the tails of the sampling distribution. These sampling distributions are difficult to evaluate in closed form but have been approximated by simulations.

A goodness-of-fit test may also be based on the linearity of the probability plot, as measured by the correlation coefficient, r, of the x and y components of the points of the probability plot. The test rejects for small values of r. The sampling distribution of r under normality has been approximated by simulations and is tabled in Filliben (1975). Ryan and Joiner (unpublished) give a short table for the null sampling distribution of r from normal probability plots with critical values for the correlation coefficient corresponding to significance levels .1, .05, and .01:

n	.1	.05	.01
4	.8951	.8734	.8318
5	.9033	.8804	.8320
10	.9347	.9180	.8804
15	.9506	.9383	.9110
20	.9600	.9503	.9290
25	.9662	.9582	.9408
30	.9707	.9639	.9490
40	.9767	.9715	.9597
50	.9807	.9764	.9664
60	.9836	.9799	.9710
75	.9865	.9835	.9752

They also report the results of some simulations of the power of r as the test statistic for certain alternative distributions. For example, the power against a uniform distribution with a significance level of .1 is .13 for $n = 10$ and .20 for $n = 20$. This is somewhat discouraging—the test only rejects 13% of the time and 20% of the time for the given sample sizes if the real underlying distribution is uniform. The moral is that it may be quite difficult to detect departure from normality in small samples. On a more positive note, the power of r against an exponential distribution is 53% for $n = 10$ and 89% for $n = 20$.

Pearson, D'Agostino, and Bowman (1977) report the results of quite extensive simulations of the power for several alternative distributions. Further references are contained in this paper.

For Michelson's data (see Example A in Section 9.9), the correlation coefficient is .995. From the tables in Filliben (1975), this falls between the 50th and 75th percentiles of the null sampling distribution, giving no reason to reject the hypothesis of normality. It may not be very realistic, however, to model the 100 observations of the velocity of light as a sample of 100 independent random variables from some probability distribution and to use this model to test goodness of fit. We have little information about how these data were collected and processed. For example, since the observations were made sequentially, it is quite possible that the measurement process drifted in time or that successive errors were correlated. It is also possible that Michelson discarded some obviously bad data.

9.11 Concluding Remarks

Two very important concepts, estimation and hypothesis testing, have been introduced in this chapter and the last. They have been introduced here in the context of fitting probability distributions but will recur throughout the rest of this book in various other contexts. Generally, observations are taken from a probability law that depends on a parameter, θ. Estimation theory is concerned with estimating θ from the data; the theory of hypothesis testing is concerned with testing hypotheses about the value of θ. Methods based on likelihood, maximum likelihood estimation and likelihood ratio tests, have also been introduced. These methods are much more generally useful than has been demonstrated by the specific purposes to which they have been put in these chapters.

The fundamental concepts and techniques of hypothesis testing have been introduced in this chapter. We have seen how to test a null hypothesis by choosing a test statistic and a rejection region such that, under the null hypothesis, the probability that the test statistic falls in the rejection region is α, the significance level of the test. The choice of this region is determined by knowing, at least approximately, the null distribution of the test statistic. The test statistic is frequently, but not always, a likelihood ratio; when the exact distribution of the likelihood ratio cannot be found, we can use the chi-square distribution as a large-sample approximation. We have also explored the relation of the p-value of the test statistic to the significance level. In some situations, the p-value is a less rigid summary of the evidence than is a decision whether to reject the null hypothesis.

With the increasing availability of flexible computer programs and inexpensive terminals and output devices, graphical methods are being used more and more in statistics. The last part of this chapter introduced two graphical techniques: hanging rootograms and probability plots. Other graphical techniques will be introduced in chapter 10. Such informal techniques are often of more practical use than are more formal techniques, such as hypothesis testing according to the Neyman–Pearson paradigm. Literally testing for goodness of fit is often rather artificial—a parametric distribution is usually entertained only as a model for the distribution of data values, and it is clear that the data do not really come from that distribution. If enough data were available, the goodness-of-fit test would certainly reject. Rather than test a hypothesis that no one literally believes could hold, it is usually more useful to ascertain qualitatively where the model fits and where and how it fails to fit.

Some concepts introduced in chapters 7 and 8 have been elaborated on in this chapter. In chapter 7, we introduced confidence intervals for the parameters of finite populations; in chapter 8, we considered confidence intervals for parameters of probability distributions. In this chapter, we have introduced hypothesis testing and developed a relation between hypothesis tests and confidence intervals. The method of propagation of error, used in chapter 7 as a tool for analyzing the statistical behavior of ratio estimates, has been used in this chapter in connection with variance-stabilizing transformations.

9.12   Problems

1. A coin is thrown independently 10 times to test the hypothesis that the probability of heads is $\frac{1}{2}$ versus the alternative that the probability is not $\frac{1}{2}$. The test rejects if either 0 or 10 heads are observed.
   (a) What is the significance level of the test?
   (b) If in fact the probability of heads is .1, what is the power of the test?

2. Which of the following hypotheses are simple, and which are composite?
   (a) X follows a uniform distribution on $[0, 1]$.
   (b) A die is unbiased.
   (c) X follows a normal distribution with mean 0 and variance $\sigma^2 > 10$.
   (d) X follows a normal distribution with mean $\mu = 0$.

3. Suppose that $X \sim \text{bin}(100, p)$. Consider the test that rejects $H_0: p = .5$ in favor of $H_A: p \neq .5$ for $|X - 50| > 10$. Use the normal approximation to the binomial distribution to answer the following:
   (a) What is α?
   (b) Graph the power as a function of p.

4. Let X have one of the following distributions:

X	H_1	H_2
x_1	.2	.1
x_2	.3	.4
x_3	.3	.1
x_4	.2	.4

(a) Compare the likelihood ratio, Λ, for each possible value X and order the x_i according to Λ.

(b) What is the likelihood ratio test of H_1 versus H_2 at level $\alpha = .2$? What is the test at level $\alpha = .5$?

5. Let $X_1, \ldots, X_n$ be a sample from a Poisson distribution. Find the likelihood ratio for testing $H_0: \lambda = \lambda_0$ versus $H_A: \lambda = \lambda_1$, where $\lambda_1 > \lambda_0$. Use the fact that the sum of independent Poisson random variables follows a Poisson distribution to explain how to determine a rejection region for a test at level α.

6. Show that the test of Problem 5 is uniformly most powerful for testing $H_0: \lambda = \lambda_0$ versus $H_A: \lambda > \lambda_0$.

7. Let $X_1, \ldots, X_{25}$ be a sample from a normal distribution having a variance of 100. Find the rejection region for a test at level $\alpha = .10$ of $H_0: \mu = 0$ versus $H_A: \mu = 1.5$. What is the power of the test? Repeat for $\alpha = .01$.

8. Suppose that $X_1, \ldots, X_n$ form a random sample from a density function, $f(x|\theta)$, for which T is a sufficient statistic for θ. Show that the likelihood ratio test of $H_0: \theta = \theta_0$ versus $H_A: \theta = \theta_1$ is a function of T. Explain how, if the distribution of T is known under H_0, the rejection region of the test may be chosen so that the test has the level α.

9. Suppose that $X_1, \ldots, X_{25}$ form a random sample from a normal distribution having a variance of 100. Graph the power of the likelihood ratio test of $H_0: \mu = 0$ versus $H_A: \mu \neq 0$ as a function of μ, at significance levels .10 and .05. Do the same for a sample size of 100. Compare the graphs.

10. Let $X_1, \ldots, X_n$ be a random sample from an exponential distribution with the density function $f(x|\theta) = \theta \exp[-\theta x]$. Derive a likelihood ratio test of $H_0: \theta = \theta_0$ versus $H_A: \theta \neq \theta_0$, and show that the rejection region is of the form $\{\bar{X} \exp[-\theta_0 \bar{X}] \leq c\}$. Is this region symmetric about θ_0?

11. Derive a likelihood ratio test for $H_0: \sigma^2 = \sigma_0^2$ versus $H_A: \sigma^2 \neq \sigma_0^2$ for a sample from a normal distribution with an unknown mean.

12. Test the goodness of fit of the data to the genetic model given in Problem 18 of chapter 8.

13. Test the goodness of fit of the data to the genetic model given in Problem 21 of chapter 8.

14. The National Center for Health Statistics (1970) gives the following data on distribution of suicides in the United States by month in 1970. Is there any evidence that the suicide rate varies seasonally, or are the data consistent with the hypothesis that the rate is constant? (*Hint:* Under the latter hypothesis, model the number of suicides in each month as a multinomial random variable with the appropriate probabilities and conduct a goodness-of-fit test. Look at the signs of the deviations, $O_i - E_i$, and see if there is a pattern.)

Month	Number of Suicides	Days/Month
Jan.	1867	31
Feb.	1789	28
Mar.	1944	31
Apr.	2094	30
May	2097	31
June	1981	30
July	1887	31
Aug.	2024	31
Sept.	1928	30
Oct.	2032	31
Nov.	1978	30
Dec.	1859	31

15. The following table gives the number of deaths due to accidental falls for each month during 1970. Is there any evidence for a departure from uniformity in the rate over time? That is, is there a seasonal pattern to this death rate? If so, describe its pattern and speculate as to causes.

Month	Number of Deaths
Jan.	1668
Feb.	1407
Mar.	1370
Apr.	1309
May	1341
June	1338
July	1406
Aug.	1446
Sept.	1332
Oct.	1363
Nov.	1410
Dec.	1526

16. Consider testing goodness of fit for a multinomial distribution with two cells. Denote the number of observations in each cell by X_1 and X_2 and let the hypothesized probabilities be p_1 and p_2. Pearson's chi-square statistic is equal to

$$\sum_{i=1}^{2} \frac{(X_i - np_i)^2}{np_i}$$

Show that this may be expressed as

$$\frac{(X_1 - np_1)^2}{np_1(1 - p_1)}$$

Since X_1 is binomially distributed, the following holds approximately under the null hypothesis:

$$\frac{X_1 - np_1}{\sqrt{np_1(1 - p_1)}} \sim N(0, 1)$$

Thus, the square of the quantity on the left-hand side is approximately distributed as a chi-square random variable with 1 degree of freedom.

17. Let $X_i \sim \text{bin}(n_i, p_i)$, for $i = 1, \ldots, m$. Derive a likelihood ratio test for the hypothesis

$$H_0: p_1 = p_2 = \cdots = p_m$$

against the alternative hypothesis that the p_i are not all equal. What is the large-sample distribution of the test statistic?

18. Nylon bars were tested for brittleness (Bennett and Franklin, 1954). Each of 280 bars was molded under similar conditions and was tested in five places. Assuming that each bar has uniform composition, the number of breaks on a given bar should be binomially distributed with five trials and an unknown probability p of failure. If the bars are all of the same uniform strength, p should be the same for all of them; if they are of different strengths, p should vary from bar to bar. Thus, the null hypothesis is that the p's are all equal. The following table summarizes the outcome of the experiment:

Breaks/Bar	Frequency
0	157
1	69
2	35
3	17
4	1
5	1

(a) Under the given assumption, the data in the table consist of 280 observations of independent binomial random variables. Find the mle of p.

(b) Pooling the last three cells, test the agreement of the observed frequency distribution with the binomial distribution using Pearson's chi-square test.

(c) Apply the test procedure derived in Problem 17.

19. (a) In 1965, a newspaper carried a story about a high school student who reported getting 9207 heads and 8743 tails in 17,950 coin tosses. Is this a significant discrepancy from the null hypothesis $H_0: p = \frac{1}{2}$?

(b) Jack Youden, a statistician at the National Bureau of Standards, contacted the student and asked him exactly how he had performed the experiment (Youden, 1974). To save time, the student had tossed groups of five coins at a time, and a younger brother had recorded the results, shown in the following table:

Number of Heads	Frequency
0	100
1	524
2	1080
3	1126
4	655
5	105

Are the data consistent with the hypothesis that all the coins were fair ($p = \frac{1}{2}$)?

(c) Are the data consistent with the hypothesis that all five coins had the same probability of heads but that this probability was not necessarily $\frac{1}{2}$? (*Hint:* Use the binomial distribution.)

20. Derive and carry out a likelihood ratio test of the hypothesis $H_0 : \theta = 0$ versus $H_1 : \theta \neq 0$ for Problem 18 of chapter 8.

21. In a classic genetics study, Geissler (1889) studied hospital records in Saxony and compiled data on the gender ratio. The following table shows the number of male children in 6115 families with 12 children. If the genders of successive children are independent and the probabilties remain constant over time, the number of males born to a particular family of 12 children should be a binomial random variable with 12 trials and an unknown probability p of success. If the probability of a male child is the same for each family, the table represents the occurrence of 6115 binomial random variables. Test whether the data agree with this model. Why might the model fail?

Number	Frequency
0	7
1	45
2	181
3	478
4	829
5	1112
6	1343
7	1033
8	670
9	286
10	104
11	24
12	3

22. Show that the transformation $Y = \sin^{-1} \sqrt{\hat{p}}$ is variance-stabilizing if $\hat{p} = X/n$, where $X \sim \text{bin}(n, p)$.

23. Let X follow a Poisson distribution with mean λ. Show that the transformation $Y = \sqrt{X}$ is variance-stabilizing.

24. Suppose that $E(X) = \mu$ and $\text{Var}(X) = c\mu^2$, where c is a constant. Find a variance-stabilizing transformation.

25. An English naturalist collected data on the lengths of cuckoo eggs, measuring to the nearest .5 mm. Examine the normality of this distribution by (a)

constructing a histogram and superposing a normal density, (b) plotting on normal probability paper, and (c) constructing a hanging rootogram.

Length	Frequency
18.5	0
19	1
19.5	3
20.0	33
20.5	39
21.0	156
21.5	152
22.0	392
22.5	288
23.0	286
23.5	100
24.0	86
24.5	21
25.0	12
25.5	2
26.0	0
26.5	1

26. Burr (1974) gives the following data on the percentage of manganese in iron made in a blast furnace. For 24 days, a single analysis was made on each of five casts. Examine the normality of this distribution by making a normal probability plot and a hanging rootogram. (As a prelude to topics that will be taken up in later chapters, you might also informally examine whether the percentage of manganese is roughly constant from one day to the next or whether there are significant trends over time.)

Day 1	Day 2	Day 3	Day 4	Day 5	Day 6	Day 7	Day 8	Day 9	Day 10	Day 11	Day 12
1.40	1.40	1.80	1.54	1.52	1.62	1.58	1.62	1.60	1.38	1.34	1.50
1.28	1.34	1.44	1.50	1.46	1.58	1.64	1.46	1.44	1.34	1.28	1.46
1.36	1.54	1.46	1.48	1.42	1.62	1.62	1.38	1.46	1.36	1.08	1.28
1.38	1.44	1.50	1.52	1.58	1.76	1.72	1.42	1.38	1.58	1.08	1.18
1.44	1.46	1.38	1.58	1.70	1.68	1.60	1.38	1.34	1.38	1.36	1.28

Day 13	Day 14	Day 15	Day 16	Day 17	Day 18	Day 19	Day 20	Day 21	Day 22	Day 23	Day 24
1.26	1.52	1.50	1.42	1.32	1.16	1.24	1.30	1.30	1.48	1.32	1.44
1.50	1.50	1.42	1.32	1.40	1.34	1.22	1.48	1.52	1.46	1.22	1.28
1.52	1.46	1.38	1.48	1.40	1.40	1.20	1.28	1.76	1.48	1.72	1.10
1.38	1.34	1.36	1.36	1.26	1.16	1.30	1.18	1.16	1.42	1.18	1.06
1.50	1.40	1.38	1.38	1.26	1.54	1.36	1.28	1.28	1.36	1.36	1.10

27. Examine the probability plot in Figure 9-4 and explain why there are several sets of horizontal bands of points.

28. The following table gives values of two abundance ratios for different isotopes of potassium from several samples of minerals (H. Ku, private communication). Examine whether each of the ratios appears normally distributed by first making histograms and superposing normal densities and then making probability plots.

$^{39}K/^{41}K$	$^{41}K/^{40}K$	$^{39}K/^{41}K$	$^{41}K/^{40}K$	$^{39}K/^{41}K$	$^{41}K/^{40}K$
13.8645	576.369	13.8689	578.277	13.8724	576.017
13.8695	578.012	13.8593	574.708	13.8665	574.881
13.8659	575.597	13.8742	573.630	13.8566	578.508
13.8622	575.244	13.8703	576.069	13.8555	576.796
13.8696	575.567	13.8472	575.637	13.8534	580.394
13.8604	576.836	13.8555	575.971	13.8685	576.772
13.8672	576.236	13.8439	576.403	13.8694	576.501
13.8598	575.291	13.8646	576.179	13.8599	574.950
13.8641	576.478	13.8702	575.129	13.8605	577.614
13.8673	576.992	13.8606	577.084	13.8619	574.506
13.8597	578.335	13.8622	576.749	13.9641	576.317
13.8604	576.767	13.8588	576.669	13.8597	575.665
13.8591	576.571	13.8547	575.869	13.8617	575.815
13.8472	576.617	13.8597	577.793	13.861	576.109
13.863	575.885	13.8663	577.770	13.8615	576.144
13.8566	576.651	13.8597	577.697	13.8469	576.820
13.8503	575.974	13.8604	576.299	13.8582	576.672
13.8553	577.255	13.8634	575.903	13.8645	576.169
13.8642	574.664	13.8658	574.773	13.8713	575.390
13.8613	576.405	13.8547	577.391	13.8593	575.108
13.8706	574.306	13.8519	577.057	13.8522	576.663
13.8601	577.095	13.863	577.286	13.8489	578.358
13.866	576.957	13.8581	575.510	13.8609	575.371
13.8655	576.434	13.8644	576.509	13.857	575.851
13.8612	575.211	13.8665	574.300	13.8566	575.644
13.8598	576.630	13.8648	575.846	13.864	574.462

29. Hoaglin (1980) suggested a "Poissoness plot"—a simple visual method for assessing goodness of fit. The expected frequencies for a sample of size n from a Poisson distribution are

$$E_k = nP(X = k) = ne^{-\lambda}\frac{\lambda^k}{k!}$$

or

$$\log E_k = \log n - \lambda + k \log \lambda - \log k!$$

Thus, a plot of $\log(O_k) + \log k!$ versus k should yield nearly a straight line with a slope of $\log \lambda$ and an intercept of $\log n - \lambda$. Construct such plots for the data of Problems 1, 2, and 3 of chapter 8. Comment on how straight they are.

30. A random variable X is said to follow a lognormal distribution if $Y = \log(X)$ follows a normal distribution. The lognormal is sometimes used as a model for heavy-tailed skewed distributions.
    (a) Calculate the density function of the lognormal distribution.
    (b) Examine whether the lognormal roughly fits the following data (Robson, 1929), which are the dorsal lengths in millimeters of taxonomically distinct octopods.

110	15	60	54	19	115	73
190	57	43	44	18	37	43
55	19	23	82	175	50	80
65	63	36	16	10	17	52
43	70	22	95	20	41	17
15	12	11	29	29	61	22
40	17	26	30	16	116	28
32	33	29	27	16	55	8
11	49	82	85	20	67	27
44	16	6	35	17	26	32
76	150	21	5	6	51	75
23	29	64	22	47	9	10
28	18	84	52	130	50	45
12	21	73				

31. (a) Generate samples of size 25, 50, and 100 from a normal distribution. Construct probability plots and hanging rootograms. Do this several times to get an idea of how probability plots behave when the underlying distribution is really normal.
    (b) Repeat part (a) for a chi-square distribution with 10 df.
    (c) Repeat part (a) for $Y = Z/U$, where $Z \sim N(0, 1)$ and $U \sim U[0, 1]$ and Z and U are independent.
    (d) Repeat part (a) for a uniform distribution.
    (e) Repeat part (a) for an exponential distribution.
    (f) Can you distinguish between the normal distribution of part (a) and the subsequent nonnormal distributions?

32. Suppose that a sample is taken from a symmetric distribution whose tails decrease more slowly than those of the normal distribution. What would be the qualitative shape of a normal probability plot of this sample?

33. The Cauchy distribution has the probability density function

$$f(x) = \frac{1}{\pi}\left(\frac{1}{1 + x^2}\right), \quad -\infty < x < \infty$$

What would be the qualitative shape of a normal probability plot of a sample from this distribution?

34. Show how probability plots for the exponential distribution, $F(x) = 1 - e^{-\lambda x}$, may be constructed on semilog paper. Berkson (1966) recorded times between events and fit them to an exponential distribution. (The times

between events in a Poisson process are exponentially distributed.) The following table comes from Berkson's paper. Make an exponential probability plot, and evaluate its "straightness."

Time Interval (sec)	Observed Frequency
0–60	115
60–120	104
120–181	99
181–243	106
243–306	113
306–369	104
369–432	101
432–497	106
497–562	104
562–628	96
628–698	512
968–1130	524
1130–1714	468
1714–2125	531
2125–2567	461
2567–3044	526
3044–3562	506
3562–4130	509
4130–4758	520
4758–5460	540
5460–6255	542
6255–7174	499
7174–8260	494
8260–9590	500
9590–11,304	550
11,304–13,719	465
13,719–14,347	104
14,347–15,049	97
15,049–15,845	101
15,845–16,763	104
16,763–17,849	92
17,849–19,179	102
19,179–20,893	103
20,893–23,309	110
23,309–27,439	112
27,439+	100

35. Construct a hanging rootogram from the data of Problem 34 in order to compare the observed distribution to an exponential distribution.
36. The exponential distribution is widely used in studies of reliability as a model for lifetimes, largely because of its mathematical simplicity. Barlow, Toland, and Freeman (1984) analyzed data on the strength of Kevlar 49/epoxy, a material used in the space shuttle. The times to failure (in hours) of 101 strands tested at a stress level of 90% are given below.

.01	.01	.02	.02	.02
.03	.03	.04	.05	.06
.07	.07	.08	.09	.09
.10	.10	.11	.11	.12
.13	.18	.19	.20	.23
.80	.80	.83	.85	.90
.92	.95	.99	1.00	1.01
1.02	1.03	1.05	1.10	1.10
1.11	1.15	1.18	1.20	1.29
1.31	1.33	1.34	1.40	1.43
1.45	1.50	1.51	1.52	1.53
1.54	1.54	1.55	1.58	1.60
1.63	1.64	1.80	1.80	1.81
2.02	2.05	2.14	2.17	2.33
3.03	3.03	3.24	4.20	4.69
7.89				

(a) Construct a probability plot of the data against the quantiles of an exponential distribution to assess qualitatively whether the exponential is a reasonable model. Can you explain the peculiar appearance of the plot?

(b) Compare the data to the exponential distribution by means of a hanging rootogram.

10

Summarizing Data

10.1 Introduction

This chapter deals with methods of describing and summarizing data that are in the form of one or more samples, or batches. These procedures, many of which generate graphical products, are useful in revealing the structure of data that are initially in the form of numbers printed in columns on a page or recorded on a tape or disk as a computer file. The statistical properties of the methods will also be discussed, since such considerations are relevant if it is reasonable to model the data stochastically. In the absence of a stochastic model, the methods are useful for purely descriptive purposes.

We will first discuss methods that are sample analogues of the cumulative distribution function of a random variable. These methods are useful in displaying the distribution of sample values. Next, we will discuss the histogram and related graphical tools that play the role for data that the probability density or frequency function plays for a random variable, giving a different view of the distribution of data values than that provided by the cumulative distribution function. We then discuss simpler numerical summaries of the data, numbers that indicate a typical or central value of the data and a quantification of the spread. Such statistics provide a more condensed summary than do the cumulative distribution function and the histogram. We will pay particular attention to the effect of extreme data points on these measures. Finally, we will introduce boxplots, graphical summaries that combine in a simple form information about the central values, spread, and shape of a distribution. Boxplots facilitate comparisons of several data sets.

10.2 Methods Based on the Cumulative Distribution Function

10.2.1 The Empirical Cumulative Distribution Function

Suppose that $x_1, \ldots, x_n$ is a batch of numbers (we will use the word *sample* in the case that the x_i are independently and identically distributed with some distribution function; the word *batch* will imply no such commitment to a stochastic model). The **empirical cumulative distribution function (ecdf)** is defined as

$$F_n(x) = \frac{1}{n}(\# x_i \leq x)$$

$F_n(x)$ gives the proportion of the data less than or equal to x. Note that it is a step function with a jump of height $1/n$ at each point x_i. (With this definition, F_n is right-continuous; in the Soviet Union and Eastern Europe, the ecdf is usually defined to be left-continuous.) The ecdf is to a sample what the cumulative distribution function is to a random variable.

EXAMPLE A. As an example of the use of the ecdf, let us consider data taken from a study by White, Riethof, and Kushnir (1960) of the chemical properties of beeswax. The aim of the study was to investigate chemical methods for detecting the presence of synthetic waxes that had been added to beeswax. For example, the addition of microcrystalline wax raises the melting point of beeswax. If all pure beeswax had the same melting point, its determination would be a reasonable way to detect dilutions. The melting point and other chemical properties of beeswax, however, vary from one beehive to another. The authors obtained samples of pure beeswax from 59 sources, measured several chemical properties, and examined the variability of the measurements. The 59 melting points (in °C) are listed below.

63.78	63.45	63.58	63.08	63.40	64.42	63.27	63.10
63.34	63.50	63.83	63.63	63.27	63.30	63.83	63.50
63.36	63.86	63.34	63.92	63.88	63.36	63.36	63.51
63.51	63.84	64.27	63.50	63.56	63.39	63.78	63.92
63.92	63.56	63.43	64.21	64.24	64.12	63.92	63.53
63.50	63.30	63.86	63.93	63.43	64.40	63.61	63.03
63.68	63.13	63.41	63.60	63.13	63.69	63.05	62.85
63.31	63.66	63.60					

As a summary of these measurements, the ecdf is plotted in Figure 10-1. This graph conveniently summarizes the natural variability in melting points. For example, we can see from the graph that about 90% of the samples had melting points less than 64.2 °C and that about 12% had melting points less than 63.2 °C.

White, Riethof, and Kushnir showed that the addition of 5% microcrystalline wax raised the melting point of beeswax by .85 °C and the addition of 10% raised the melting point by 2.22 °C. From Figure 10-1, we can see that an addition of 5% microcrystalline wax might well be difficult to detect, especially if it was made to beeswax that had a low melting point, but that an addition of 10% would be detectable. In further calculations, the investigators modeled the distribution of melting points as Gaussian. How reasonable does this model appear to be? □

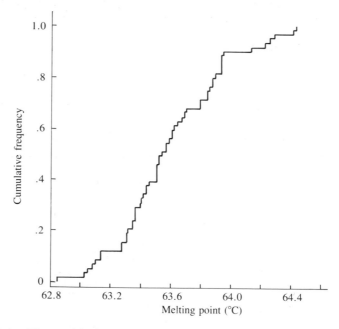

Figure 10-1. The empirical cumulative distribution function of the melting points of beeswax.

Let us briefly consider some of the elementary statistical properties of the ecdf in the case in which $X_1, \ldots, X_n$ is a random sample from a continuous distribution function, F. For purposes of analysis, it is convenient to express F_n in the following way:

$$F_n(x) = \frac{1}{n} \sum_{i=1}^{n} I_{(-\infty, x]}(X_i)$$

where

$$I_{(-\infty, x]}(X_i) = \begin{cases} 1, & \text{if } X_i \le x \\ 0, & \text{if } X_i > x \end{cases}$$

The random variables $I_{(-\infty, x]}(X_i)$ are independent Bernoulli random variables:

$$I_{(-\infty, x]}(X_i) = \begin{cases} 1, & \text{with probability } F(x) \\ 0, & \text{with probability } 1 - F(x) \end{cases}$$

Thus, $nF_n(x)$ is a binomial random variable and

$$E[F_n(x)] = F(x)$$

$$\text{Var}[F_n(x)] = \frac{1}{n} F(x)[1 - F(x)]$$

As an estimate of $F(x)$, $F_n(x)$ is unbiased and has a maximum variance at that value of x such that $F(x) = .5$, that is, at the median. As x becomes very large or very small, the variance tends to zero.

In the preceding paragraph, we considered $F_n(x)$ for fixed x; the results can be applied to form a confidence interval for $F(x)$ for any given value of x. Much deeper analysis focuses on the stochastic behavior of F_n as a random function; that is, all values of x are considered simultaneously. It turns out, somewhat surprisingly, that the distribution of

$$\max_{-\infty < x < \infty} |F_n(x) - F(x)|$$

does not depend on F if F is continuous. This result makes possible the construction of a simultaneous confidence band about F_n, which can be used to test goodness of fit. (For further details, refer to Section 9.6 of Bickel and Doksum, 1977.) It is important to realize the difference between the simultaneous confidence band and the individual confidence intervals that may be constructed using the binomial distribution. Each such individual confidence interval covers F at one point with a certain probability, say, $1 - \alpha$, but the probability that all such intervals cover F simultaneously is not necessarily $1 - \alpha$. We will encounter other phenomena of this type in later chapters.

10.2.2 The Survival Function

The **survival function** is equivalent to the cumulative distribution function and is defined as

$$S(t) = P(T > t) = 1 - F(t)$$

where T is a random variable with cdf F. In applications where the data consist of times until failure or death and are thus nonnegative, it is often customary to work with the survival function rather than the cumulative distribution function, although the two give equivalent information. Data of this type occur in medical and reliability studies. In these cases, $S(t)$ is simply the probability that the lifetime will be longer than t. We will be concerned with the sample analogue of S,

$$S_n(t) = 1 - F_n(t)$$

which gives the proportion of the data greater than t.

EXAMPLE A.   As an example, let us consider the use of the survival function with a study of the lifetimes of guinea pigs infected with varying doses of tubercle bacilli (Bjerkdahl, 1960). In one study, five groups of 72 animals each were inoculated with the bacilli at increasing dosages, and a control group of 107 animals was used. We denote the inoculated groups by I, II, III, IV, and V, in order of increasing dose. The animals were observed over a two-year period, and their times of death were recorded. The data are given below. Note that not all the animals in the lower-dosage regimens died.

Control Lifetimes

18	36	50	52	86	87	89	91
102	105	114	114	115	118	119	120
149	160	165	166	167	167	173	178
189	209	212	216	273	278	279	292
341	355	367	380	382	421	421	432
446	455	463	474	506	515	546	559
576	590	603	607	608	621	634	634
637	638	641	650	663	665	688	725
735							

Dose I Lifetimes

76	93	97	107	108	113	114	119
136	137	138	139	152	154	154	160
164	164	166	168	178	179	181	181
183	185	194	198	212	213	216	220
225	225	244	253	256	259	265	268
268	270	283	289	291	311	315	326
326	361	373	373	376	397	398	406
452	466	592	598				

Dose II Lifetimes

72	72	78	83	85	99	99	110
113	113	114	114	118	119	123	124
131	133	135	137	140	142	144	145
154	156	157	162	162	164	165	167
171	176	177	181	182	187	192	196
211	214	216	216	218	228	238	242
248	256	257	262	264	267	267	270
286	303	309	324	326	334	335	358
409	473	550					

Dose III Lifetimes

10	33	44	56	59	72	74	77
92	93	96	100	100	102	105	107
107	108	108	108	109	112	113	115
116	120	121	122	122	124	130	134
136	139	144	146	153	159	160	163
163	168	171	172	176	183	195	196
197	202	213	215	216	222	230	231
240	245	251	253	254	254	278	293
327	342	347	361	402	432	458	555

Dose IV Lifetimes

43	45	53	56	56	57	58	66
67	73	74	79	80	80	81	81
81	82	83	83	84	88	89	91
91	92	92	97	99	99	100	100
101	102	102	102	103	104	107	108
109	113	114	118	121	123	126	128
137	138	139	144	145	147	156	162
174	178	179	184	191	198	211	214
243	249	329	380	403	511	522	598

Dose V Lifetimes

12	15	22	24	24	32	32	33
34	38	38	43	44	48	52	53
54	54	55	56	57	58	58	59
60	60	60	60	61	62	63	65
65	67	68	70	70	72	73	75
76	76	81	83	84	85	87	91
95	96	98	99	109	110	121	127
129	131	143	146	146	175	175	211
233	258	258	263	297	341	341	376

A plot (Figure 10-2) of the empirical survival functions provides a convenient summary of the data. The proportions surviving beyond given times are plotted; it is not necessary to know the actual lifetimes of the animals that survived beyond the termination of the study. The graph is a much more effective presentation of the data than the tabular listings.

One of Bjerkdahl's primary interests was comparing the effect increased exposure had on guinea pigs that had different levels of resistivity. Comparing groups III and V, for example, we see that the difference in lifetimes of the weakest guinea pigs (say the 10% weakest) from the two groups was about 50 days, whereas the difference in lifetimes for stronger animals increases to about 100 days.   □

Survival plots may also be used for informal examinations of the **hazard function**, which may be interpreted as the instantaneous death rate for individuals who have survived up to a given time. If an individual is alive at time *t*, the

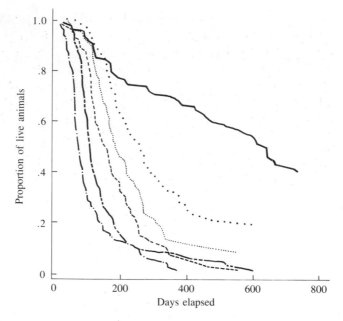

Figure 10-2. Survival functions for guinea pig lifetimes. For purposes of visual clarity, the points have been joined by lines: The solid line corresponds to the control group, the dotted line to group I, the short-dash line to group II, the long-dash line to group III, the dot-and-long-dash line to group IV, and the short-and-long-dash line to group V.

probability that that individual will die in the time interval $(t, t + \delta)$ is, assuming that the density function f is continuous at t,

$$P(t \leq T \leq t + \delta | T \geq t) = \frac{P(t \leq T \leq t + \delta)}{P(T \geq t)}$$

$$= \frac{F(t + \delta) - F(t)}{1 - F(t)}$$

$$\approx \frac{\delta f(t)}{1 - F(t)}$$

The hazard function is defined as

$$h(t) = \frac{f(t)}{1 - F(t)}$$

and may be thought of as the instantaneous rate of mortality for an individual alive at time t. It may also be expressed as

$$h(t) = -\frac{d}{dt}\log[1 - F(t)] = -\frac{d}{dt}\log S(t)$$

which reveals that it is the negative of the slope of the log of the survival function.

EXAMPLE B.   For the data of Example A, Figure 10-3 is a plot of the log of the empirical survival functions. To avoid the singularity of the logarithm, we have modified the survival function slightly, plotting the ordered survival times $T_{(i)}$ versus $\log[1 - i/(n + 1)]$. From the slopes of these curves, we see that the hazard functions are initially fairly small. As the dosage level increases, the instantaneous mortality rates both increase more quickly and reach higher levels. The increased mortality rate sets in at an earlier age for the high-dosage group and seems greater (the slope is greater). (To see this, hold the figure at an angle so that you are "looking down" the curves.)   □

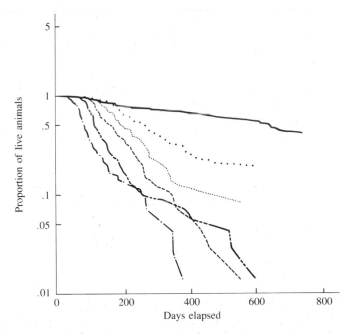

Figure 10-3.   Log survival functions for guinea pig lifetimes. For purposes of visual clarity, the points have been joined by lines: The solid line corresponds to the control group, the dotted line to group I, the short-dash line to group II, the long-dash line to group III, the dot-and-long-dash line to group IV, and the short-and-long-dash line to group V.

When interpreting plots such as that presented in Figure 10-3, we will find it useful to keep in mind the variability of the empirical log survival function. Using the method of propagation of error, we have

$$\text{Var}\{\log[1 - F_n(t)]\} \approx \frac{\text{Var}[1 - F_n(t)]}{[1 - F(t)]^2}$$

$$= \frac{1}{n}\left(\frac{F(t)[1 - F(t)]}{[1 - F(t)]^2}\right)$$

$$= \frac{1}{n}\left(\frac{F(t)}{1 - F(t)}\right)$$

From this expression, we see that for large values of t, the empirical log survival function is extremely unreliable, since $1 - F(t)$ is then very small. Thus, in practice, the last few data points are disregarded (note the large fluctuations of the log survival functions in Figure 10-3 for large times).

10.2.3 Quantile-Quantile Plots

Quantile-quantile (Q-Q) plots are useful for comparing distribution functions. If X is a continuous random variable with a strictly increasing distribution function, F, the pth quantile of the distribution was defined in Section 2.2 to be that value of x such that

$$F(x) = p$$

or

$$x_p = F^{-1}(p)$$

In a Q-Q plot, the quantiles of one distribution are plotted against those of another. Suppose, for purposes of discussion, that one cdf (F) is a model for observations of a control group and another (G) is a model for observations of a group that has received some treatment. Let the observations of the control group be denoted by x with cdf F, and let the observations of the treatment group be denoted by y with cdf G. The simplest effect that the treatment could have would be to increase the expected response of every member of the treatment group by the same amount, say h units. That is, both the weakest and the strongest individuals would have their responses changed by h. In this case, a simple relationship would hold (see Figure 10-4):

$$G(y) = F(y - h)$$

The quantiles would be simply related as

$$y_p = x_p + h$$

and a Q-Q plot would be a straight line with slope 1 and intercept h.

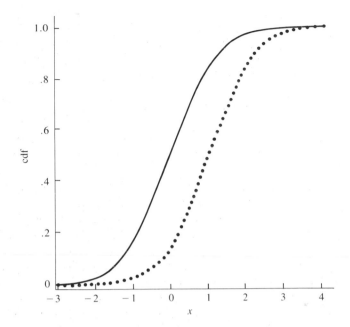

Figure 10-4.   An additive treatment effect. The solid line is $F(y)$, and the dotted line is $G(y) = F(y - h)$.

Another possible effect of a treatment would be multiplicative: The response (such as lifetime or strength) is multiplied by a constant, c. For this model, the distribution functions would be related as follows:

$$G(y) = F\left(\frac{y}{c}\right)$$

(See Figure 10-5.) The relation of the quantiles would be

$$y_p = cx_p$$

and the Q-Q plot would be a straight line with slope c and intercept 0.

A simple summary of a treatment effect for the additive model would be of the form "the treatment increases lifetime by two months." For the multiplicative model, one might say something like "the treatment increases lifetime by 25%."

The effect of a treatment can, of course, be much more complicated than either of these two simple models. For example, a treatment could benefit weaker individuals and be to the detriment of stronger individuals. An educational program that places very heavy emphasis on elementary or basic skills might be expected to have this sort of effect relative to a regular program.

Given a batch of numbers, or a sample from a probability distribution, quantiles are constructed from the order statistics. Given n observations and the

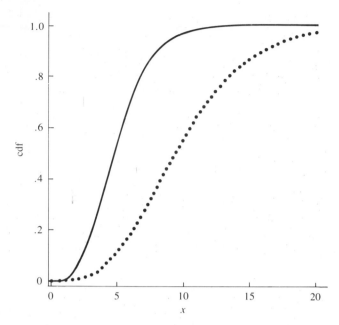

Figure 10-5. A multiplicative treatment effect. The solid line is $F(y)$, and the dotted line is $G(y) = F(y/c)$.

order statistics $X_{(1)}, \ldots, X_{(n)}$, the $k/(n + 1)$ quantile of data is assigned to $X_{(k)}$. (This convention is not unique; sometimes, for example, the quantile assigned to $X_{(k)}$ is defined as $(k - .5)/n$. For descriptive purposes, it makes little difference which definition we accept. In constructing probability plots in chapter 9, we plotted sample quantiles defined as just described versus the quantiles of a theoretical distribution, such as the normal, and used these plots to informally assess goodness of fit.

To compare two batches of n numbers with order statistics $X_{(1)}, \ldots, X_{(n)}$ and $Y_{(1)}, \ldots, Y_{(n)}$, a Q-Q plot is simply constructed by plotting the points $(X_{(i)}, Y_{(i)})$. If the batches are of unequal size, an interpolation process can be used. A procedure for interpolating intermediate quantiles is described in the end-of-chapter problems.

EXAMPLE A. Cleveland et al. (1974) used Q-Q plots in a study of air pollution. They plotted the quantiles of distributions of the values of various variables on Sunday against the quantiles for weekdays (Figure 10-6). The Q-Q plot of the ozone maxima shows that the very highest quantiles occur on weekdays but that all the other quantiles are larger on Sundays. For carbon monoxide, nitrogen oxide, and aerosols, the differences in the quantiles increase with increasing concentration. The very high and very low quantiles of solar radiation are about the same

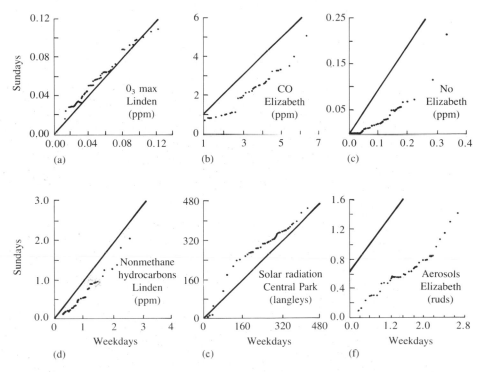

Figure 10-6. Q-Q plots of air pollution variables: (a) ozone maxima (ppm), (b) carbon monoxide concentration (ppm), (c) nitrogen oxide concentration (ppm), (d) nonmethane hydrocarbons (ppm), (e) solar radiation (langleys), (f) aerosols (ruds).

on Sundays and weekdays (presumably corresponding to very clear days and days with heavy cloud cover), but for intermediate quantiles, the Sunday quantiles are larger.   □

EXAMPLE B.   Figure 10-7 is a Q-Q plot for groups III and V of Bjerkdahl (see Example A in Section 10.2.2). It shows that the difference in the quantiles increases for the larger quantiles; this is consistent with the observations we made above. From his analysis of the data, Bjerkdahl concluded that the increases were proportionally the same for animals with little, average, or great resistance—that is, that the treatment effect is multiplicative in the sense defined earlier. If this were the case, the Q-Q plot would be a straight line. For times up to about 200 days, the animals in group III live approximately twice as long as those in group V, but beyond 200 days the difference is roughly constant.   □

Further discussion and examples of Q-Q plots can be found in Wilk and Gnanadesikan (1968).

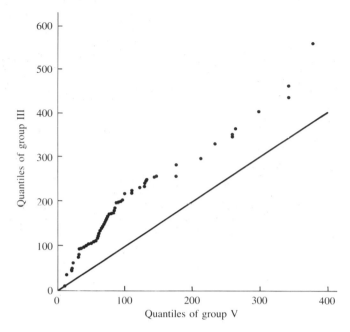

Figure 10-7. Q-Q plot of groups III and IV from Bjerkdahl (1960). For reference, the line $y = x$ has been added.

10.3 Histograms, Density Curves, and Stem-and-Leaf Plots

The histogram, a time-honored method of displaying data, has already been introduced. It displays the shape of the distribution of data values in the same sense that a density function displays probabilities. The range of the data is divided into intervals, or bins, and the number or proportion of the observations falling in each bin is plotted. If the bins are not of equal size, the resulting histogram can be misleading. A procedure that is often recommended is to plot the proportion of observations falling in the bin divided by the bin width; if this procedure is used, the area under the histogram is 1.

Figure 10-8 shows three histograms of the melting points of beeswax from Example A in Section 10.2.1 with increasingly larger bin width. If the bin width is too small, the histogram is too ragged; if it is too wide, the shape is over-smoothed and obscured. The choice of bin width is usually made subjectively in an attempt to strike a balance between a histogram that is too ragged and one that oversmooths. Rudemo (1982) discusses automatic methods for choosing the bin width.

Histograms are frequently used to display data for which there is no assumption of any stochastic model—for example, populations of U.S. cities. If the data are modeled as a random sample from some continuous distribution, the histo-

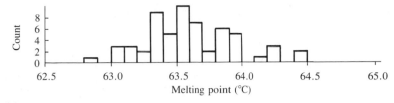

(a)

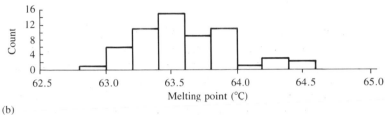

(b)

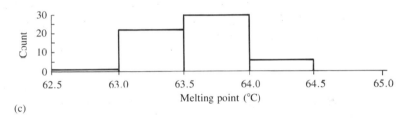

(c)

Figure 10-8. Histograms of melting points of beeswax: (a) bin width = .1, (b) bin width = .2, (c) bin width = .5.

gram may be viewed as an estimate of the probability density. Regarded in this light, it suffers from not being smooth.

A smooth probability density estimate can be constructed in the following way. Let $w(x)$ be a nonnegative, symmetric weight function, centered at zero and integrating to 1. For example, $w(x)$ can be the standard normal density. The function

$$w_h(x) = \frac{1}{h} w\left(\frac{x}{h}\right)$$

is a rescaled version of w. As h approaches zero, w_h becomes more concentrated and peaked about zero. As h approaches infinity, w_h becomes more spread out and flatter. If $w(x)$ is the standard normal density, then $w_h(x)$ is the normal density with standard deviation h. If $X_1, \ldots, X_n$ is a sample from a probability density function, f, an estimate of f is

$$f_h(x) = \frac{1}{n} \sum_{i=1}^{n} w_h(x - X_i)$$

This estimate, called a **kernel probability density estimate**, consists of the super-position of "hills" centered over the observations. In the case where $w(x)$ is the standard normal density, $w_h(x - X_i)$ is the normal density with mean X_i and standard deviation h.

The parameter h, the **bandwidth** of the estimating function, controls its smooth-ness and corresponds to the bin width of the histogram. If h is too small, the estimate is too rough; if it is too large, the shape of f is smeared out too much. Figure 10-9 shows estimates of the probability density of the melting points of beeswax (from Example A in Section 10.2.1) for various values of h. Making a reasonable choice of the bandwidth is important, just as is choosing the bin width

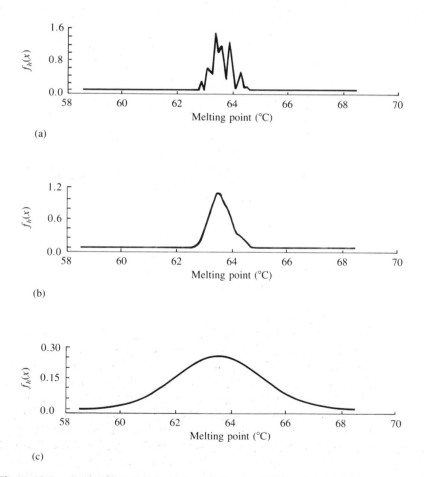

Figure 10-9. Probability density estimates from melting point data. The kernel w is the standard normal density with standard deviation (a) .025, (b) .125, and (c) 1.25.

for a histogram. From Figure 10-9, we see that too small a bandwidth yields a ragged curve and too large a bandwidth obscures the shape and spreads the probability mass out too much. The bandwidth is usually chosen subjectively, but automatic methods are receiving increasing attention (Rudemo, 1982). Although a histogram can be drawn with pencil and paper, the calculations involved in a probability density estimate are only practical with a computer.

One disadvantage of a histogram or a probability density estimate is that information is lost; neither tool allows the reconstruction of the original data. Furthermore, a histogram does not allow one to calculate a statistic such as a median; one can only tell from a histogram in which bin the median lies and not its actual value.

Stem-and-leaf plots (Tukey, 1977) convey information about shape while retaining the numerical information. It is easiest to define this type of plot by an example, a stem-and-leaf plot of the beeswax melting-point data (the decimal point is one place to the left of the colon):

		STEM LEAF
1	1	628 : 5
1	0	629 :
4	3	630 : 358
7	3	631 : 033
9	2	632 : 77
18	9	633 : 001446669
23	5	634 : 01335
	10	635 : 0000113668
26	7	636 : 0013689
19	2	637 : 88
17	6	638 : 334668
11	5	639 : 22223
6	0	640 :
6	1	641 : 2
5	3	642 : 147
2	0	643 :
2	2	644 : 02

The first three digits of the melting points have been selected to form the stem and are listed in the third column. The leaves on each stem are the fourth digit of all numbers with that stem. For example, the first stem is 628, and its leaf indicates the presence of the number 62.85 in the data. The third stem is 630, and its leaves indicate the presence of the numbers 63.03, 63.05, and 63.08. This stem-and-leaf plot was constructed by a computer, but they are very easy to make by hand. The second column of numbers gives the number of leaves on each stem. The first column of numbers facilitates finding order statistics, such as quartiles and the median; starting at the top of the plot and continuing down to the stem containing the median, the cumulative numbers of observations out to the smallest observation are listed. The numbering process is then extended symmetrically from the stem containing the median to the largest observation of the data.

Straightforward stem-and-leaf plots do not work well for data that range over several orders of magnitude. In such a situation, it is better to make a stem-and-leaf plot of the logarithms of the data.

10.4   Measures of Location

Sections 10.2 and 10.3 were concerned with sample analogues of the cumulative distribution and density functions and with related curves, which convey visual information about the shape of the distribution of the data. Here and in Section 10.5, we discuss simple numerical summaries of data that are useful when there is not enough data to justify constructing a histogram or an ecdf or when a more concise summary is desired.

A measure of location is a measure of the center of a batch of numbers. If the numbers result from different measurements of the same quantity, a measure of location is often used in hopes that it is a reliable estimate of that quantity. In other situations, a measure of location is used as a simple summary of the numbers—for example, "the average grade on the exam was 72." In this section, we will discuss several common measures of location and their relative advantages and disadvantages.

10.4.1   The Arithmetic Mean

The most commonly used measure of location is the arithmetic mean,

$$\bar{x} = \frac{1}{n} \sum_{i=1}^{n} x_i$$

To illustrate the concept of the mean, we will consider a set of 26 measurements of the heat of sublimation of platinum from an experiment done by Hampson and Walker (1961). The data are listed below:

Heats of Sublimation of Platinum (kcal/mole)

136.2	136.6	135.8	135.4	134.7	135.0	134.1	143.3
147.8	148.8	134.8	135.2	134.9	146.5	141.2	135.4
134.8	135.8	135.0	133.7	134.2	134.9	134.8	134.5
134.3	135.2						

The 26 measurements are all attempts to measure the "true" heat of sublimation, and we see that there is variability among them. Intuitively, it may seem that a measure of location or center for this batch of numbers would give a more accurate estimate of the heat of sublimation than any one of the numbers alone.

A common statistical model for the variability of a measurement process is the following:

$$X_i = \mu + \beta + \varepsilon_i$$

(See Section 4.2.1.) Here, X_i is the value of the ith measurement, μ is the true value of the heat of sublimation, β represents bias in the measurement procedure, and ε_i is the random error. The ε_i are usually assumed to be independent and identically distributed random variables with mean 0 and variance σ^2. The efficacy of measures of location is often judged by comparing their performances (mean squared error, for example) with this model. Note that with this model the data alone tell us nothing about β, the bias in the measurement procedure, which in some cases may be as or more important than the random variability.

The observations are listed across rows in the order in which the experiments were done. When observations are acquired sequentially, it is often informative to plot them in order, as in Figure 10-10. From this plot, we see that the first few observations were somewhat high. The most striking aspect of the plot is the presence of five extreme observations that occurred in groups of three and two. Such observations, which are quite far from the bulk of the data, are called **outliers**. Outliers occur all too frequently, even in carefully conducted studies. The outliers in this case might have been caused by improperly calibrated equipment, for example. Outliers can also be caused by recording and transcrip-

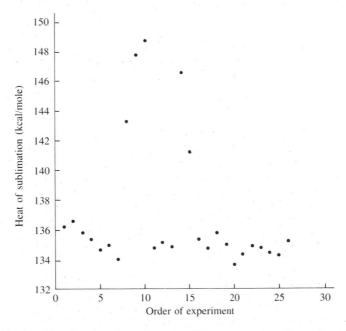

Figure 10-10.   Plot showing time sequence of measurements of heat of sublimation of platinum.

tion errors or by equipment malfunctions. It is important to detect outliers, since they may have an undue influence on subsequent calculations. Graphical presentation is an effective means of detection. Careful reexamination of the data and the circumstances under which they were obtained can sometimes uncover the causes behind the outliers. Although outliers are often unexplainable aberrations, an examination of them and their causes can sometimes deepen an investigator's understanding of the phenomenon under study.

Figure 10-10 also makes us doubt that the model for measurement error given above is appropriate for this set of data. The fact that the outliers occur in groups of two and three, rather than being randomly scattered, makes the independence model somewhat implausible.

A stem-and-leaf plot provides another summary of this data (the decimal point is at the colon):

$$
\begin{array}{rrl}
1 & 1 & 133:7 \\
4 & 3 & 134:134 \\
11 & 7 & 134:5788899 \\
& 6 & 135:002244 \\
9 & 2 & 135:88 \\
7 & 1 & 136:3 \\
6 & 1 & 136:6
\end{array}
$$

High: 141.2   143.3   146.5   147.8   148.8

On this stem-and-leaf plot, the outlying observations have been isolated and flagged as high.

In their analysis, Hampson and Walker set aside the seven largest observations and the smallest observation and found the average of the remaining observations to be 134.9. Calculated from all the observations, the arithmetic mean is 137.05. Note from the stem-and-leaf plot that this number is larger than the bulk of the data and is clearly not a good descriptive measure of the "center" of this batch of numbers. We would not be satisfied with it as an estimate of the true heat of sublimation.

If the data are modeled as a sample from a probability law, as with the measurement error model described above, an approximate $100(1 - \alpha)\%$ confidence interval for the population mean can be obtained from the central limit theorem as in chapter 7. The interval is of the form

$$\bar{x} \pm z(\alpha/2)s_{\bar{x}}$$

Applying this formula to the platinum data, with $\alpha = .05$, we obtain the interval 137.05 ± 1.71, or $(135.3, 138.8)$. Note where this interval falls on the stem-and-leaf plot!

Although the example presented here may be somewhat extreme, it illustrates the sensitivity of the sample mean to outlying observations. In fact, by changing a single number, the arithmetic mean of a batch of numbers can be made arbitrarily large or small. Thus, if used blindly, without careful attention to the data, the arithmetic mean can produce misleading results. When data are automatically acquired, stored as files on disks or tapes, and not visually examined, this danger increases. For this reason, there has been increasing attention to and advocacy of measures of location that are **robust**, or insensitive to outliers.

10.4.2   The Median

If the sample size is an odd number, the median is defined to be the middle value; if the example size is even, the median is the average of the two middle values. Clearly, moving the extreme observations does not affect the sample median at all, so it is quite robust. The median of the platinum data is 135.2, which, as can be seen from the stem-and-leaf plot, is more reasonable than the mean as a measure of the center.

When the data are a sample from a continuous probability law, the sample median can be viewed as an estimate of the population median, η, for which a simple confidence interval can be formed. We will now demonstrate that this interval is of the form

$$(X_{(k)}, X_{(n-k+1)})$$

The coverage probability of this interval is

$$P(X_{(k)} \le \eta \le X_{(n-k+1)}) = 1 - P(\eta < X_{(k)} \text{ or } \eta > X_{(n-k+1)})$$
$$= 1 - P(\eta < X_{(k)}) - P(\eta > X_{(n-k+1)})$$

since the events are mutually exclusive. To evaluate these terms, we first note that

$$P(\eta > X_{(n-k+1)}) = \sum_{j=0}^{k-1} P(j \text{ observations are greater than } \eta)$$

$$P(\eta < X_{(k)}) = \sum_{j=0}^{k-1} P(j \text{ observations are less than } \eta)$$

Since, by definition, the median satisfies

$$P(X_i > \eta) = P(X_i < \eta) = \tfrac{1}{2}$$

and since the n observations $X_1, \ldots, X_n$ are independent and identically distributed, the distribution of the number of observations greater than the median is binomial with n trials and probability $\frac{1}{2}$ of success on each trial. Thus,

$$P(\text{exactly } j \text{ observations are greater than } \eta) = \frac{1}{2^n} \binom{n}{j}$$

and

$$P(\eta > X_{(n-k+1)}) = \frac{1}{2^n} \sum_{j=0}^{k-1} \binom{n}{j}$$

From symmetry, we then have that the coverage probability of the interval in question is

$$1 - \frac{1}{2^{n-1}} \sum_{j=0}^{k-1} \binom{n}{j}$$

These probabilities can be found from tables of the cumulative binomial distribution since

$$\frac{1}{2^n} \sum_{j=0}^{k-1} \binom{n}{j} = P(Y \le k - 1)$$

where Y is a binomial random variable with n trials and probability of success equal to $\frac{1}{2}$.

EXAMPLE A. As a concrete example, with $n = 26$, we have the following cumulative binomial probabilities:

k	$P(Y \le k)$
5	.0012
6	.0047
7	.0145
8	.0378
9	.0843

If we choose $k = 8$,

$$P(Y \le k - 1) = .0145$$

and since $2 \times .0145 = .029$, the interval $(X_{(8)}, X_{(19)})$ is a 97% confidence interval. Note that this confidence interval is exact, not approximate, and does not depend on the form of the underlying cdf but only on the assumption that the cdf is continuous and that the observations are independent.

For the platinum data, this confidence interval is $(134.8, 135.8)$. Compare this interval to the interval based on the sample mean. (But as we have noted, there is reason to doubt the independence assumption for the platinum data, so these calculations should be viewed as an illustrative numerical exercise.) □

10.4.3 The Trimmed Mean

Another simple and robust measure of location is the **trimmed mean**. The $100\alpha\%$ trimmed mean is easy to calculate: Order the data, discard the lowest $100\alpha\%$ and the highest $100\alpha\%$, and take the arithmetic mean of the remaining data. It is generally recommended that the value chosen for α be from .1 to .2. Formally, we may write the trimmed mean as

$$\bar{x}_\alpha = \frac{x_{([n\alpha]+1)} + \cdots + x_{(n-[n\alpha])}}{n - 2[n\alpha]}$$

where $[n\alpha]$ denotes the greatest integer less than or equal to $n\alpha$. Note that the median can be regarded as a 50% trimmed mean.

The variance of the trimmed mean depends on the variances and covariances of the order statistics. Using large-sample theory for order statistics, it may be shown that if the observations are a sample from a probability law with density f, symmetric about μ, the trimmed mean is asymptotically normally distributed with mean μ and variance σ_α^2/n, where

$$\sigma_\alpha^2 = (1 - 2\alpha)^{-2}\left[\int_{x_\alpha}^{x_{1-\alpha}} t^2 f(t)\,dt + \alpha x_\alpha^2 + \alpha x_{1-\alpha}^2\right]$$

In this equation, x_α and $x_{1-\alpha}$ are the α and $1 - \alpha$ quantiles, and f is taken, without loss of generality, to be symmetric about zero, since the variance of the trimmed mean does not depend on μ.

The variance of the trimmed mean can be estimated from the data, and approximate confidence intervals can be constructed in the usual way from the normal distribution. The variance is estimated by imitating the equation above in the following way:

Step 1: Move the order statistics for the discarded values over to the first nondiscarded values and calculate the sample variance of the resulting n numbers about the trimmed mean.
Step 2: Divide this result by $(1 - 2\alpha)^2 n$.

More discussion of this procedure may be found in Section 9.5 of Bickel and Doksum (1977).

The 20% trimmed mean for the platinum data listed in Section 10.4.1 is formed by discarding the highest and lowest five observations ($.2 \times 26 = 5.2$) and averaging the rest. The result is 135.29; for the same data, the median was 135.2 and the mean was 137.05. Following the steps above, we find that the standard deviation of the trimmed mean is approximately .254. Multiplying this by 1.96, we find an approximate 95% confidence interval of (134.93, 135.93). This estimate and confidence interval are quite comparable to those produced by the median.

10.4.4   M Estimates

The sample mean is the maximum likelihood estimate of μ, the location parameter when the underlying distribution is normal. The sample mean is the minimizer of

$$\sum_{i=1}^{n} \left(\frac{X_i - \mu}{\sigma} \right)^2$$

This is the simplest case of a least squares estimate (we will discuss least squares estimates in more detail in the context of curve fitting). Outliers have a great effect on this estimate, since the deviation of μ from X_i is measured by the square of their difference. In contrast, the median is the minimizer of (see Problem 14 of the end-of-chapter problems)

$$\sum_{i=1}^{n} \left| \frac{X_i - \mu}{\sigma} \right|$$

Here, large deviations are not weighted as heavily, and it is this property that causes the median to be robust.

Huber (1981) proposed a class of estimates, **M estimates**, which are the minimizers of

$$\sum_{i=1}^{n} \Psi \left(\frac{X_i - \mu}{\sigma} \right)$$

where the weight function Ψ is a compromise between the weight functions for the mean and the median. A wide variety of weight functions have been proposed. Huber discusses weight functions that are quadratic near zero and are linear beyond a cutoff point, k. Thus, $k = \infty$ corresponds to the mean and $k = 0$ to the median. A common choice is $k = 1.5$. With this choice, the influence of observations more than $1.5(\sigma)$ away from the center is reduced. In practice, a robust estimate of σ, such as those discussed in Section 10.5, must be used.

The computation of an M estimate is a nonlinear minimization problem and must be done iteratively (using the Newton–Raphson method, for example). If M is a convex function, the minimizer will be unique. Fairly simple computer programs that do this are becoming more common in statistical packages. The M estimate ($k = 1.5$) for the platinum data we have been considering is 135.38, close to the median (135.2) and the trimmed mean (135.29) but quite different from the mean (137.05). Approximate confidence intervals based on M estimates can be constructed, but we will not pursue this topic here.

10.4.5   Comparison of Location Estimates

We have introduced several location estimates (and there are many others). Which one is best? There is no simple answer to this question. It is always important to bear in mind what is being estimated by the location estimates and to what purpose the estimate is being put. If the underlying distribution is symmetric, the trimmed mean, the sample mean, the sample median, and an M estimate all estimate the center of symmetry. If the underlying distribution is not symmetric, however, the four statistics estimate four different population parameters: the population mean, the population median, the population trimmed mean, and a functional of the cdf determined by the weight function Ψ. Moreover, there is no single estimate that is best for all symmetric distributions. Simulations have been done to compare estimates for a variety of distributions. Andrews et al. (1972) report the results of a large number of simulations from symmetric distributions. Their results show that the 10% or 20% trimmed mean is quite an effective estimate: Its variance is never much larger than the variance of the ordinary mean (even in the Gaussian case for which the mean is optimal) and can be quite a lot smaller when the underlying distribution is heavy-tailed relative to the Gaussian. The median, although quite robust, has a substantially larger variance in the Gaussian case than does the trimmed mean. The trimmed mean and the median have a certain appealing simplicity and are easy to explain to someone who has little formal statistical training. M estimates performed quite well in the simulations of the Andrews et al. study, and they do generalize more naturally to other problems such as curve fitting. But they are somewhat more difficult to compute and have less immediate intuitive appeal. For the purpose of simply summarizing data, it is often useful to compute more than one measure of location and compare the results.

10.5   Measures of Dispersion

A measure of dispersion, or scale, gives a numerical indication of the "scattered-ness" of a batch of numbers. Simple summaries of data often consist of a measure of location and a measure of dispersion. The most commonly used measure is the sample standard deviation, s, which is the square root of the sample variance,

$$s^2 = \frac{1}{n-1} \sum_{i=1}^{n} (X_i - \bar{X})^2$$

Using $n - 1$ as the divisor rather than the more obvious divisor n is based on the rationale that s^2 is an unbiased estimate of the population variance if the observations are independent and identically distributed with variance σ^2. (But s is not an unbiased estimate of σ since the square root is a nonlinear function.) If n is of moderate to large size, it makes little difference whether n or $n - 1$ is used.

If the observations are a sample from a normal distribution with variance σ^2,

$$\frac{(n-1)s^2}{\sigma^2} \sim \chi^2_{n-1}$$

This distributional result may be used to construct confidence intervals for σ in the normal case (compare with Example B in Section 8.5.3), but the result is not robust against deviations from normality.

Like the sample mean, the sample standard deviation is sensitive to outlying observations. Two simple robust measures of dispersion are the **interquartile range (IQR),** or the difference between the two sample quartiles, and the **median absolute deviation from the median (MAD).** If the data are $x_1, \ldots, x_n$ with median $\tilde{x}$, the MAD is defined to be the median of the numbers $|x_i - \tilde{x}|$. These two measures of dispersion, the IQR and the MAD, can be converted into estimates of σ for a normal distribution by dividing them by 1.35 and .65, respectively. David (1981) discusses a method for finding a confidence interval for the population interquartile range, using reasoning similar to that used in Section 10.4.2 for developing a confidence interval for the population median.

Let us compare all three measures of dispersion for the platinum data:

$$s = 4.45$$

$$\frac{\text{IQR}}{1.35} = 1.19$$

$$\frac{\text{MAD}}{.65} = .970$$

The two robust estimates are similar. From the stem-and-leaf plot of the platinum values presented earlier, we can see that both the IQR and the MAD give a measure of the spread of the central portion of the data, whereas the standard deviation is heavily influenced by the outliers.

10.6   Boxplots

A **boxplot** is a graphical display invented by Tukey that shows a measure of location (the median), a measure of dispersion (the interquartile range), and the presence of possible outliers and also gives an indication of the symmetry or skewness of the distribution. Figure 10-11 is a boxplot of the platinum data.

We outline the construction of a boxplot:

1. Horizontal lines are drawn at the median and at the upper and lower quartiles and are joined by vertical lines to produce the box.
2. A vertical line is drawn up from the upper quartile to the most extreme data point that is within a distance of 1.5 (IQR) of the upper quartile. A similarly defined vertical line is drawn down from the lower quartile. Short horizontal lines are added to mark the ends of these vertical lines.
3. Each data point beyond the ends of the vertical lines is marked with an asterisk or dot (* or ·).

Boxplots are not uniformly standardized, but the basic structure is as outlined above, perhaps with additional embellishments or small variations. A boxplot

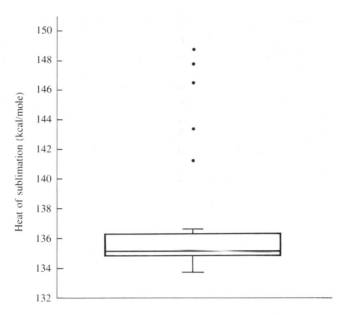

Figure 10-11. Boxplot of the platinum data.

thus gives an indication of the center of the data (the median), the spread of the data (the interquartile range), and the presence of outliers, and indicates the symmetry or asymmetry of the distribution of data values (the location of the median relative to the quartiles). In Figure 10-11, the five outliers of the platinum data are clearly displayed, and we see an indication that the central part of the distribution is somewhat skewed toward high values.

EXAMPLE A.   Figure 10-12 is taken from Chambers et al. (1983). The data plotted are daily maximum concentrations in parts per billion of sulfur dioxide in Bayonne, N.J. from November 1969 to October 1972 grouped by month. There are thus 36 batches, each of size about 30. The investigators concluded:

> The boxplots ... show many properties of the data rather strikingly. There is a general reduction in sulphur dioxide concentration through time due to the gradual conversion to low sulphur fuels in the region. The decline is most dramatic for the highest quantiles. Also, there are higher concentrations during the winter months due to the use of heating oil. In addition, the boxplots show that the distributions are skewed toward high values and that the spread of the distributions ... is larger when the general level of concentration is higher.

The boxplot is clearly a very effective method of presenting and summarizing these data. As they are in this example, boxplots are generally useful for comparing batches of numbers, a purpose to which they will be put in the next two chapters.   □

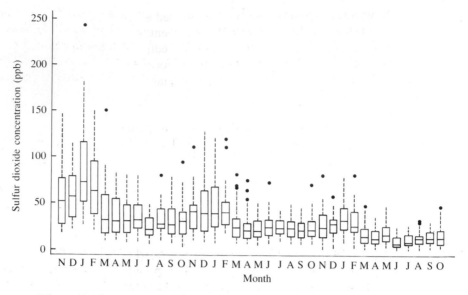

Figure 10-12. Boxplots of daily maximum concentrations of sulfur dioxide.

10.7 Concluding Remarks

This chapter has introduced several tools for summarizing data, some of which are graphical in nature. Under the assumption of a stochastic model for the data, some aspects of the sampling distributions of these summaries have been discussed. Summaries are very important in practice; an intelligent summary of data is often sufficient to fulfill the purposes for which the data were gathered, and more formal techniques such as confidence intervals or hypothesis tests sometimes add little to an investigator's understanding. Effective summaries can also point to "bad" data or to unexpected aspects of data that might have gone unnoticed if the data had been blindly crunched by a computer.

10.8 Problems

1. Let $X_1, \ldots, X_n$ be a sample from a distribution function, F, and let F_n denote the ecdf. Show that

$$\text{Cov}[F_n(u), F_n(v)] = \frac{1}{n}[F(m) - F(u)F(v)]$$

where $m = \min(u, v)$. Conclude that $F_n(u)$ and $F_n(v)$ are positively correlated: If $F_n(u)$ overshoots $F(u)$, then $F_n(v)$ will tend to overshoot $F(v)$.

2. Various chemical tests were conducted on beeswax by White, Riethof, and Kushnir (1960). In particular, the percentage of hydrocarbons in each sample of wax was determined. (a) Plot the ecdf, a histogram, and a normal probability plot of the percentages of hydrocarbons given in the following table. Find the .90, .75, .50, .25, and .10 quantiles. Does the distribution appear Gaussian?

14.27	14.80	12.28	17.09	15.10	12.92	15.56	15.38
15.15	13.98	14.90	15.91	14.52	15.63	13.83	13.66
13.98	14.47	14.65	14.73	15.18	14.49	14.56	15.03
15.40	14.68	13.33	14.41	14.19	15.21	14.75	14.41
14.04	13.68	15.31	14.32	13.64	14.77	14.30	14.62
14.10	15.47	13.73	13.65	15.02	14.01	14.92	15.47
13.75	14.87	15.28	14.43	13.96	14.57	15.49	15.13
14.23	14.44	14.57					

(b) The average percentage of hydrocarbons in microcrystalline wax (a synthetic commercial wax) is 85%. Suppose that beeswax was diluted with 1% microcrystalline wax. Could this be detected? What about a 3% or a 5% dilution? (Such questions were one of the main concerns of the beeswax study.)

3. Use the method of propagation of error to derive an approximation to the bias of the log survival function. Where is this bias large, and what is its sign?

4. Let $X_1, \ldots, X_n$ be a sample from F and denote the order statistics by $X_{(1)}$, $X_{(2)}, \ldots, X_{(n)}$. We will assume that F is continuous, with density function f. From Theorem A in Section 3.7, the density function of $X_{(k)}$ is

$$f_k(x) = n \binom{n-1}{k-1} [F(x)]^{k-1} [1 - F(x)]^{n-k} f(x)$$

(a) Find the mean and variance of $X_{(k)}$ from a uniform distribution. You will need to use the fact that the density of $X_{(k)}$ integrates to 1. Show that

$$\text{Mean} = \frac{k}{n+1}$$

$$\text{Variance} = \frac{1}{n+2} \left(\frac{k}{n+1} \right) \left(1 - \frac{k}{n+1} \right)$$

(b) Find the approximate mean and variance of $Y_{(k)}$, the kth order statistic of a sample of size n from F. To do this, let

$$X_i = F(Y_i)$$

or

$$Y_i = F^{-1}(X_i)$$

The X_i are a sample from a $U[0, 1]$ distribution (why?). Use the propagation of error formula,

$$Y_{(k)} = F^{-1}(X_{(k)})$$

$$\approx F^{-1}\left(\frac{k}{n+1}\right) + \left(X_{(k)} - \frac{k}{n+1}\right)\frac{d}{dx}F^{-1}(x)\bigg|_{k/(n+1)}$$

and argue that

$$EY_{(k)} \approx F^{-1}\left(\frac{k}{n+1}\right)$$

$$\text{Var}(Y_{(k)}) \approx \frac{k}{n+1}\left(1 - \frac{k}{n+1}\right)\frac{1}{(f\{F^{-1}[k/(n+1)]\})^2}\left(\frac{1}{n+2}\right)$$

(c) Use the results of parts (a) and (b) to show that the variance of the pth sample quantile is approximately

$$\frac{1}{nf^2(x_p)}p(1-p)$$

where x_p is the pth quantile.

(d) Use the result of part (c) to find the approximate variance of the median of a sample of size n from a $N(\mu, \sigma^2)$ distribution. Compare to the variance of the mean.

5. Calculate the hazard function for

$$F(t) = 1 - e^{-\alpha t^\beta}, \quad t \geq 0$$

6. Let f denote the density function and h the hazard function of a nonnegative random variable. Show that

$$f(t) = h(t)e^{-\int_0^t h(s)ds}$$

that is, that the hazard function uniquely determines the density.

7. For what probability law is the hazard function constant?

8. A certain chemotherapy treatment for cancer tends to lengthen the lifetimes of very seriously ill patients and decrease the lifetimes of the least ill patients. Suppose that an experiment is done that compares this treatment to a placebo. Draw a sketch showing the qualitative behavior of a Q-Q plot.

9. Consider the two cdfs:

$$F(x) = x, \quad 0 \leq x \leq 1$$

$$G(x) = x^2, \quad 0 \leq x \leq 1$$

Sketch a Q-Q plot of F versus G.

10. Sketch what you would expect the qualitative shape of the hazard function of human mortality to look like.

11. Make Q-Q plots for other pairs of treatment groups from Bjerkdahl's data (see Example A in Section 10.2.2). Does the model of a multiplicative effect appear reasonable?

12. By examining the survival function of group V of Bjerkedahl's data (see Example A in Section 10.2.2), make a rough sketch of the qualitative shape of a histogram. Then make a histogram, and compare it to your guess.

13. Hampson and Walker (1960) also made measurements of the heats of sublimation of rhodium and iridium. Analyze the two sets of data given below. Make histograms, stem-and-leaf plots, and boxplots. Plot the observations in the order of the experiment; the observations are listed across rows in the order in which they occurred in the experiment. Find the means, medians, and 10% and 20% trimmed means and compare. Find confidence intervals for these location estimates, and compare them. Does the statistical model of independent and identically distributed measurement errors given in Section 10.4.1 appear reasonable?

Iridium

136.6	145.2	151.5	162.7	159.1	159.8	160.8	173.9	160.1
160.4	161.1	160.6	160.2	159.5	160.3	159.2	159.3	159.6
160.0	160.2	160.1	160.0	159.7	159.5	159.5	159.6	159.5

Rhodium

126.4	135.7	132.9	131.5	131.1	131.1	131.9	132.7
133.3	132.5	133.0	133.0	132.4	131.6	132.6	132.2
131.3	131.2	132.1	131.1	131.4	131.2	131.1	131.1
134.2	133.8	133.3	133.5	133.4	133.5	133.0	132.8
132.6	133.3	133.5	133.5	132.3	132.7	132.9	134.1

14. Show that the median is an M estimate if $\Psi(x) = |x|$. For what symmetric density function is this the mle of the mean?

15. What proportion of the observations from a normal sample would you expect to be marked by an asterisk on a boxplot?

16. Explain why the IQR and the MAD are divided by 1.35 and .65, respectively, to estimate σ for a normal sample.

17. Find the mean, median, and 10% and 20% trimmed means for the data of Problem 2 and compare them. Find the standard deviation, s, and the IQR and the MAD as estimates of σ, and compare them.

18. The Cauchy distribution has the density function

$$f(x) = \frac{1}{\pi}\left(\frac{1}{1 + x^2}\right), \quad -\infty < x < \infty$$

which is symmetric about zero. This distribution has very heavy tails, which cause the arithmetic mean to be a very poor estimate of location. Simulate the distribution of the arithmetic mean and of the median from a sample of size 25 from the Cauchy distribution by drawing 100 samples of size 25 and compare. From Example B in Section 3.6.1, if Z_1 and Z_2 are independent and $N(0, 1)$, then their quotient follows a Cauchy distribution. (This gives a simple way of generating Cauchy random variables.)

19. Simiu and Filliben (1975), in a statistical analysis of extreme winds, present the data in Table 10.1. Construct boxplots to examine and compare the forms of the distributions across cities and across years.

20. Olson, Simpson, and Eden (1975) discuss the analysis of data obtained from a cloud seeding experiment. A cloud was deemed "seedable" if it satisfied certain criteria; for each seedable cloud a decision was made at random whether or not to actually seed. The nonseeded clouds are referred to as control clouds. The following table presents the rainfall from 26 seeded and 26 control clouds. Make Q-Q plots for rainfall vs. rainfall and log rainfall vs. log rainfall. What do these plots suggest about the effect, if any, of seeding?

Seeded Clouds

129.6	31.4	2745.6	489.1	430.0	302.8	119.0	4.1
92.4	17.5	200.7	274.7	274.7	7.7	1656.0	978.0
198.6	703.4	1697.8	334.1	118.3	255.0	115.3	242.5
32.7	40.6						

Control Clouds

26.1	26.3	87.0	95.0	372.4	0.01	17.3	24.4
11.5	321.2	68.5	81.5	47.3	28.6	830.1	345.5
1202.6	36.6	4.9	4.9	41.1	29.0	163.0	244.3
147.8	21.7						

21. Construct a nonparametric confidence interval for a quantile x_p by using the same reasoning as in the derivation of a confidence interval for a median.

22. In the examples of Q-Q plots in the text, we only discussed the case in which quantiles of equal size batches are compared. From two batches of size n the $k/(n + 1)$ quantiles are estimated as $X_{(k)}$ and $Y_{(k)}$, so one merely has to plot $X_{(k)}$ vs. $Y_{(k)}$. Write down a linear interpolation formula for the pth quantile where $k/(n + 1) \leq p \leq (k + 1)/(n + 1)$. Now suppose that the batch sizes are not the same, being m and n, $m < n$ say. A Q-Q plot may be constructed by fixing the quantiles $k/(m + 1)$ of the smaller data set and interpolating these quantiles for the larger data set.

Interpolate to find the upper and lower quartiles of the following batch of numbers: 1, 2, 3, 4, 5, 6.

23. In a study of the natural variability of rainfall, the rainfall of summer storms was measured by a network of rain gauges in southern Illinois for the years 1960–1964 (Changnon and Huff, in LeCam and Neyman, 1967). The tables on page 344 give the average amount of rainfall (in inches) from each storm, by year.

Table 10.1   Strongest maximum wind speeds (mph)

	1912	13	14	15	16	17	18	19	20	21	22	23	24	25	26	27	28	29	30	31	32	33	34	35	36	37	38	39	40	41	42	43	44	45	46	47	48
Cairo	35	38	33	35	40	38	45	60	42	47	38	38	47	43	41	47	37	45	37	30	35	40	35	38	40	51	34	39	43	34	42	37	34	40	43	43	45
Alpena	38	43	41	39	41	38	43	47	37	46	43	40	44	38	43	41	38	45	42	42	47	41	37	43	38	43	38	42	47	44	38	47	42	45	50	42	
Tatoush Island	68	51	65	61	68	54	57	70	71	84	59	69	62	60	61	68	68	56	66	62	65	63	71	59	59	68	68	61	78	71	74	64	62	66	59	61	66
Williston	38	50	40	35	38	41	38	50	38	38	44	41	50	35	34	35	42	36	35	37	40	46	49	40	37	46	37	39	33	44	34	34	35	42	46	41	44
Richmond	46	48	41	43	37	47	47	36	34	32	37	36	37	38	41	37	42	38	45	37	38	46	39	34	38	34	38	33	32	35	44	46	40	34	32	41	38
Burlington	40	47	43	40	41	50	44	47	49	49	44	53	40	46	47	43	50	43	43	40	42	41	42	43	43	44	47	38	43	38	41	41	40	42	42	43	46
Eastport	53	41	54	49	60	54	52	56	48	39	57	46	46	51	38	50	45	50	48	52	42	46	51	48	46	46	45	47	46	49	46	42	51	55	44	52	48
Canton	51	53	50	44	46	44	49	47	46	39	54	43	49	43	62	39	47	48	36	38	39	39	38	34	44	35	38	39	35	40	38	42	32	34	39	34	35
Yuma	32	32	29	32	32	37	32	34	30	33	33	34	29	35	30	30	34	38	34	35	30	34	35	35	32	29	34	34	29	29	31	29	30	33	37	34	41
Duluth	54	49	46	50	45	51	45	44	49	54	54	53	50	45	54	54	57	52	47	39	49	51	59	51	49	53	54	50	52	68	55	50	54	49	61	49	49
Valentine	38	44	44	43	39	36	39	44	49	43	41	46	41	42	41	44	36	45	36	37	42	56	35	39	34	43	41	41	35	37	38	42	38	36	46	39	38
Charleston	52	49	46	43	50	37	38	41	32	35	43	35	38	43	40	43	40	55	36	35	40	51	53	47	40	34	43	47	66	33	44	33	60	57	43	45	34
Eureka	35	35	46	46	35	35	40	35	32	38	39	37	34	38	36	30	34	31	34	38	34	32	34	38	30	34	32	33	35	34	37	41	37	35	38	34	35
Oklahoma City	56	41	44	57	48	43	57	46	43	37	40	38	46	35	36	38	34	38	34	38	29	32	32	34	38	30	34	32	33	31	33	26	34	28	31	37	33
Baker	27	30	29	28	28	30	38	30	28	28	30	27	25	30	40	37	29	26	29	27	28	35	26	32	29	28	30	38	28	27	34	28	25	30	28	30	33
Sheridan	44	38	41	43	40	38	32	32	34	46	37	37	37	31	32	33	33	34	33	31	32	37	33	32	33	32	28	33	32	33	38	53	49	51	52	59	56
Block Island	60	54	65	63	56	59	54	54	56	56	59	47	60	56	59	54	52	54	51	59	60	57	62	58	47	82	56	50	52	57	70	82	63	56	67	57	
Winnemucca	36	32	36	32	38	41	37	36	43	38	48	38	38	31	32	37	34	32	37	34	34	32	48	30	35	36	32	35	36	35	58	52	45	33	42	39	40
North Head	69	65	70	63	73	65	68	65	57	95	60	68	64	70	73	65	66	63	66	66	79	70	87	69	73	72	67	70	84	67	65	77	65	64	67	66	69
Key West	32	40	39	44	41	40	40	84	40	38	41	40	51	41	49	44	37	61	34	46	40	43	32	46	38	43	33	33	42	40	35	40	56	40	35	54	78
Corpus Christi	47	41	41	40	90	38	50	95	41	51	43	39	44	38	37	39	37	36	43	35	47	49	47	39	49	42	50	38	43	37	61	54	45	56	43	41	35

1960

.02	.001	.001	.12	.08	.42	1.72	.05
.01	.01	.003	.001	.003	.27	.001	.06
.05	2.13	.04	1.10	.02	.001	.14	.08
.21	.07	.32	.24	.29	.001	.29	1.13
.003	.01	.19	.002	.01	.04	.002	.07
.45	.01	.18	.67	.003	.01	.04	.002

1961

.49	.02	.02	.34	.14	.37	.33	.33
.35	.01	.50	.76	1.06	.002	.06	.16
.27	.25	.29	.02	.05	.46	.07	.41
.02	.08	.21	.01	.44	.02	.05	.11
1.50	.003	.18	.01	.002	.24	.01	.75
.01	.14	.13	.01	.01	.27	.45	1.78

1962

.25	.24	.004	.21	.17	.83	.15	.03
.03	.50	.04	.09	.04	.06	.06	.12
.003	.003	.40	.02	.51	.003	.02	.02
.02	.01	.001	.14	.001	.10	.01	1.09
.01	.002	.001	.84	.03	.35	.07	.001
.002	.002	.20	.06	.14	.01	.02	.02
.002	.001	.55	.13	.19	2.10	.09	.35

1963

.79	.32	1.35	.17	.02	.002	.01	.25
.23	.17	.01	.02	.001	.01	.02	.11
.21	1.26	.01	.73	.10	.09	.007	.36
.77	.21	1.27	.07	.08	.16	.26	.01
.23	.08	.02	.01	.29			

1964

.01	.01	.07	.40	.002	.003	.01	.09
.16	.04	.27	.73	.41	.03	.12	.03
1.04	.06	.09	.73	.04	.16	.59	.003
.002	.02	.004	.01	.001	.06	.62	.01
.52	.11	.003	.60	.002	.05		

(a) Is the form of the distribution of rainfall per storm skewed or symmetric?

(b) What is the average rainfall per storm? What is the median rainfall per storm? Explain why these measures differ, using the results of part (a).

(c) You may have read statements like "10% of the storms account for 90% of the rain." Construct a graph that shows such a relationship for these data.

(d) Compare the years using boxplots.

(e) Which years were wet and which were dry? Are the wet years wet because there were more storms, because individual storms produced more rain, or for both of these reasons?

24. Barlow, Toland, and Freeman (1984) studied the lifetimes of Kevlar 49/epoxy strands subjected to sustained stress. (The space shuttle uses Kevlar/epoxy spherical vessels in an environment of sustained pressure.) The following data (reported in Andrews et al., 1985) give the lifetimes (in hours) of strands tested at various stress levels. What do these data indicate about the nature of the distribution of lifetimes and the effect of increasing stress?

Times to Failure at 90% Stress Level

.01	.01	.02	.02	.02
.03	.03	.04	.05	.06
.07	.07	.08	.09	.09
.10	.10	.11	.11	.12
.13	.18	.19	.20	.23
.80	.80	.83	.85	.90
.92	.95	.99	1.00	1.01
1.02	1.03	1.05	1.10	1.10
1.11	1.15	1.18	1.20	1.29
1.31	1.33	1.34	1.40	1.43
1.45	1.50	1.51	1.52	1.53
1.54	1.54	1.55	1.58	1.60
1.63	1.64	1.80	1.80	1.81
2.02	2.05	2.14	2.17	2.33
3.03	3.03	3.24	4.20	4.69
7.89				

Times to Failure at 80% Stress Level

.18	3.1	4.2	6.0	7.5
8.2	8.5	10.3	10.6	24.2
29.6	31.7	41.9	44.1	49.5
50.1	59.7	61.7	64.4	69.7
70.0	77.8	80.5	82.3	83.5
84.2	87.1	87.3	93.2	103.4
104.6	105.5	108.8	112.6	116.8
118.0	122.3	123.5	124.4	125.4
129.5	130.4	131.6	132.8	133.8
137.0	140.2	140.9	148.5	149.2
152.2	152.8	157.7	160.0	163.6
166.9	170.5	174.9	177.7	179.2
183.6	183.8	194.3	195.1	195.3
202.6	220.0	221.3	227.2	251.0
266.5	267.9	269.2	270.4	272.5
285.9	292.6	295.1	301.1	304.3
316.8	329.8	334.1	346.2	351.2
353.3	369.3	372.3	381.3	393.5
451.3	461.5	574.2	656.3	663.0
669.8	739.7	759.6	894.7	974.9

Times to Failure at 70% Stress Level

1051	1337	1389	1921	1942
2322	3629	4006	4012	4063
4921	5445	5620	5817	5905
5956	6068	6121	6473	7501
7886	8108	8546	8666	8831
9106	9711	9806	10205	10396
10861	11026	11214	11362	11604
11608	11745	11762	11895	12044
13520	13670	14110	14496	15395
16179	17092	17568	17568	

11

Comparing Two Samples

11.1 Introduction

This chapter is concerned with methods for comparing samples from distributions that may be different and especially with methods for making inferences about how the distributions differ. In many applications, the samples are drawn under different conditions, and inferences must be made about possible effects of these differences. We will be primarily concerned with treatment effects that tend to increase or decrease the average level of response.

For example, in the end-of-chapter problems, we will consider some experiments performed to determine to what degree, if any, cloud seeding increases precipitation. In cloud-seeding experiments, some storms are selected for seeding, other storms are left unseeded, and the amount of precipitation from each storm is measured. This amount varies widely from storm to storm, and in the face of this natural variability, it is difficult to tell whether seeding has a systematic effect. The average precipitation from the seeded storms might be slightly higher than that from the unseeded storms, but a skeptic might not be convinced that the difference was due to anything but chance. We will develop statistical methods

347

to deal with this type of problem based on a stochastic model that treats the amounts of precipitation as random variables. We will also see how a process of randomization allows us to make inferences about treatment effects even in the case where the observations are not modeled as samples from populations or probability laws.

This chapter will be concerned with analyzing measurements that are continuous in nature (such as temperature); chapter 13 will take up the analysis of qualitative data. This chapter will conclude with some general discussion of the design and interpretation of experimental studies.

11.2   Comparing Two Independent Samples

In many experiments, the two samples may be regarded as being independent of each other. In a medical study, for example, a sample of subjects may be assigned to a particular treatment, and another independent sample may be assigned to a control (or placebo) treatment. In later sections, we will discuss methods that are appropriate when there is some pairing, or dependence, between the samples, such as might occur if each person receiving the treatment were paired with an individual of similar weight in the control group.

Many experiments are such that if they were repeated, the measurements would not be exactly the same. To deal with this problem a statistical model is often employed: The observations from the control group are modeled as independent random variables with a common distribution, F, and the observations from the treatment group are modeled as being independent of each other and of the controls and having their own common distribution function, G. Analyzing the data thus entails making inferences about the comparison of F and G. In many experiments, the primary effect of the treatment is to change the overall level of the responses, so that analysis focuses on the difference of the means of F and G. When only a small amount of data is available, it may not be practical to do much more than this.

11.2.1   Methods Based on the Normal Distribution

In this section, we will assume that a sample, $X_1, \ldots, X_n$, is drawn from a normal distribution that has mean μ_X and variance σ^2, and that an independent sample, $Y_1, \ldots, Y_m$, is drawn from another normal distribution that has mean μ_Y and the same variance, σ^2. If we think of the X's as having received a treatment and the Y's as being the control group, the effect of the treatment is characterized by the difference $\mu_X - \mu_Y$. A natural estimate of $\mu_X - \mu_Y$ is $\bar{X} - \bar{Y}$; in fact, this is the maximum likelihood estimate. Since $\bar{X} - \bar{Y}$ may be expressed as a linear combination of independent normally distributed random variables, it is normally distributed:

$$\bar{X} - \bar{Y} \sim N\left[\mu_X - \mu_Y, \sigma^2\left(\frac{1}{n} + \frac{1}{m}\right)\right]$$

If σ^2 were known, a confidence interval for $\mu_X - \mu_Y$ could be based on

$$Z = \frac{(\bar{X} - \bar{Y}) - (\mu_X - \mu_Y)}{\sigma\sqrt{\frac{1}{n} + \frac{1}{m}}}$$

which follows a standard normal distribution. The confidence interval would be of the form

$$(\bar{X} - \bar{Y}) \pm z(\alpha/2)\sigma\sqrt{\frac{1}{n} + \frac{1}{m}}$$

This confidence interval is of the same form as those introduced in chapters 7 and 8—a statistic ($\bar{X} - \bar{Y}$ in this case) plus or minus a multiple of its standard deviation.

Generally, σ^2 will not be known and must be estimated from the data by calculating the **pooled sample variance**,

$$s_p^2 = \frac{(n-1)s_X^2 + (m-1)s_Y^2}{m + n - 2}$$

Note the s_p^2 is a weighted average of the sample variances of the X's and Y's, with the weights proportional to the degrees of freedom. This weighting is appropriate since if one sample is much larger than the other, the estimate of σ^2 from that sample is more reliable and should receive greater weight. The following theorem gives the distribution of a statistic that will be used for forming confidence intervals and performing hypothesis tests.

THEOREM A.   Suppose that $X_1, \ldots, X_n$ are independent and normally distributed random variables with mean μ_X and variance σ^2 and that $Y_1, \ldots, Y_m$ are independent and normally distributed random variables with mean μ_Y and variance σ^2 and that the Y_i are independent of the X_i. The statistic

$$t = \frac{(\bar{X} - \bar{Y}) - (\mu_X - \mu_Y)}{s_p\sqrt{\frac{1}{n} + \frac{1}{m}}}$$

follows a t distribution with $m + n - 2$ degrees of freedom.

Proof.   According to the definition of the t distribution in Section 6.2, we have to show that the statistic is the quotient of a standard normal random variable and the square root of an independent chi-square random variable with $n + m - 2$ degrees of freedom. First, we note from Theorem B in Section 6.3 that $(n - 1)s_X^2/\sigma^2$ and $(m - 1)s_Y^2/\sigma^2$ are distributed as chi-square random variables with $n - 1$ and $m - 1$ degrees of freedom, respectively, and are independent since the X_i and Y_i are. Their sum is thus chi-square with $m + n - 2$ df. Now, we express the statistic as the ratio U/V, where

$$U = \frac{(\bar{X} - \bar{Y}) - (\mu_X - \mu_Y)}{\sigma \sqrt{\dfrac{1}{n} + \dfrac{1}{m}}}$$

$$V = \sqrt{\left[\frac{(n - 1)s_X^2}{\sigma^2} + \frac{(m - 1)s_Y^2}{\sigma^2}\right] \frac{1}{m + n - 2}}$$

U follows a standard normal distribution and from the argument above V has the distribution of the square root of a chi-square random variable divided by its degrees of freedom. The independence of U and V follows from Corollary A in Section 6.3.   □

It is convenient and suggestive to use the following notation:

$$s_{\bar{X} - \bar{Y}} = s_p \sqrt{\frac{1}{n} + \frac{1}{m}}$$

A confidence interval for $\mu_X - \mu_Y$ follows as a corollary to Theorem A.

COROLLARY A.   Under the assumptions of Theorem A, a $100(1 - \alpha)\%$ confidence interval for $\mu_X - \mu_Y$ is

$$(\bar{X} - \bar{Y}) \pm t_{m+n-2}(\alpha/2)s_{\bar{X} - \bar{Y}}$$

EXAMPLE A.   Two methods, A and B, were used in a determination of the latent heat of fusion of ice (Natrella, 1963). The investigators wished to find out whether the methods differed. The following table gives the change in total heat from ice at $-.72°$ C to water at $0°$ C in calories per gram of mass:

Method A	Method B
79.98	80.02
80.04	79.94
80.02	79.98
80.04	79.97
80.03	79.97
80.03	80.03
80.04	79.95
79.97	79.97
80.05	
80.03	
80.02	
80.00	
80.02	

It is fairly obvious from the table and from boxplots (Figure 11-1) that there is a difference between the two methods (we will test this more formally later). If we assume the conditions of Theorem A, we can form a 95% confidence interval to estimate the magnitude of the average difference between the two methods. From the table, we calculate

$$\bar{X}_A = 80.02 \qquad s_A = .024$$

$$\bar{X}_B = 79.98 \qquad s_B = .031$$

$$s_p^2 = \frac{12 \times s_A^2 + 7 \times s_B^2}{19} = .0007178$$

$$s_p = .027$$

From Corollary A, a 95% confidence interval for $\mu_A - \mu_B$ is

$$(\bar{X}_A - \bar{X}_B) \pm t_{19}(.025)s_p\sqrt{\tfrac{1}{12} + \tfrac{1}{8}}$$

or $(.02, .07)$.   □

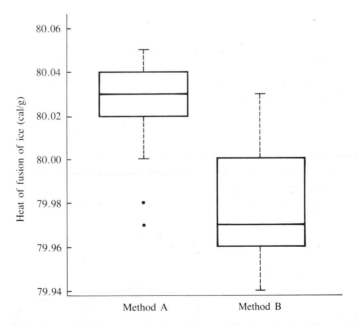

Figure 11-1.   Boxplots of measurements of heat of fusion obtained by methods A and B.

We will now discuss hypothesis testing for the two-sample problem. Although the hypotheses under consideration are different from those of chapter 9, the general conceptual framework is the same (you should review that framework at this time). In the current case, the null hypothesis to be tested is

$$H_0: \mu_X = \mu_Y$$

This asserts that there is no difference between the distributions of the X's and Y's. If one group is a treatment group and the other a control, for example, this hypothesis asserts that there is no treatment effect. In order to conclude that there is a treatment effect, the null hypothesis must be rejected.

There are three common alternative hypotheses for the two-sample case:

$$H_1: \mu_X \neq \mu_Y$$
$$H_2: \mu_X > \mu_Y$$
$$H_3: \mu_X < \mu_Y$$

The first of these is a **two-sided alternative**, and the other two are **one-sided alternatives**. The first hypothesis is appropriate if deviations could in principle go in either direction, and one of the latter two is appropriate if it is believed that any deviation must be in one direction or the other. In practice, such *a priori* information is not usually available, and it is more prudent to conduct two-sided tests.

The test statistic that will be used to make a decision whether or not to reject the null hypothesis is

$$t = \frac{\bar{X} - \bar{Y}}{s_{\bar{X} - \bar{Y}}}$$

This statistic plays the same role in the comparison of two samples as is played by the chi-square statistic in testing goodness of fit. Just as we rejected for large values of the chi-square statistic, we will reject in this case for extreme values of t. The distribution of t under H_0, its null distribution, is, from Theorem A, the t distribution with $m + n - 2$ degrees of freedom. Knowing this null distribution allows us to determine a rejection region for a test at level α, just as knowing that the null distribution of the chi-square statistic was chi-square with the appropriate degrees of freedom allowed the determination of a rejection region for testing goodness of fit. The rejection regions for the three alternatives listed above are

$$\text{For } H_1, |t| > t_{n+m-2}(\alpha/2)$$
$$\text{For } H_2, t > t_{n+m-2}(\alpha)$$
$$\text{For } H_3, t < -t_{n+m-2}(\alpha)$$

Note how the rejection regions are tailored to the particular alternatives and how knowing the null distribution of t allows us to determine the rejection region for any value of α.

EXAMPLE B.   Let us continue Example A. To test $H_0: \mu_A = \mu_B$ versus a two-sided alternative, we form and calculate the following test statistic:

$$t = \frac{\bar{X}_A - \bar{X}_B}{s_p \sqrt{\dfrac{1}{n} + \dfrac{1}{m}}}$$

$$= 3.33$$

From the tables of the t distribution with 19 df, a two-sided test rejects at level $\alpha = .01$. There is little doubt that there is a difference between the two methods.

$\square$

In chapter 9, we developed a general duality between hypothesis tests and confidence intervals. In the case of the testing and confidence interval methods considered in this section, the t test rejects if and only if the confidence interval does not include zero (see Problem 2 at the end of this chapter).

We will now demonstrate that the test of H_0 versus H_1 is equivalent to a likelihood ratio test. (The rather long argument is sketched here and should be read with paper and pencil in hand.) Ω is the set of all possible parameter values:

$$\Omega = \{-\infty < \mu_X < \infty, -\infty < \mu_Y < \infty, 0 < \sigma < \infty\}$$

The unknown parameters are $\theta = (\mu_X, \mu_Y, \sigma)$. Under H_0, $\theta \in \omega_0$, where $\omega_0 = \{\mu_X = \mu_Y, 0 < \sigma < \infty\}$. The likelihood of the two samples $X_1, \ldots, X_n$ and $Y_1, \ldots, Y_m$ is

$$\text{lik}(\mu_X, \mu_Y, \sigma^2) = \prod_{i=1}^{n} \frac{1}{\sqrt{2\pi\sigma^2}} e^{-(1/2)[(X_i - \mu_X)^2/\sigma^2]} \prod_{j=1}^{m} \frac{1}{\sqrt{2\pi\sigma^2}} e^{-(1/2)[(Y_i - \mu_Y)^2/\sigma^2]}$$

and the log likelihood is

$$l(\mu_X, \mu_Y, \sigma^2) = -\frac{(m + n)}{2} \log 2\pi - \frac{(m + n)}{2} \log \sigma^2$$

$$- \frac{1}{2\sigma^2} \left[\sum_{i=1}^{n} (X_i - \mu_X)^2 + \sum_{j=1}^{m} (Y_j - \mu_Y)^2 \right]$$

We must maximize the likelihood under ω_0 and under Ω and then calculate the ratio of the two maximized likelihoods, or the difference of their logarithms.

Under ω_0, we have a sample of size $m + n$ from a normal distribution with unknown mean μ_0 and unknown variance σ_0^2. The maximum likelihood estimates of μ_0 and σ_0^2 are thus

$$\hat{\mu}_0 = \frac{1}{m+n}\left(\sum_{i=1}^{n} X_i + \sum_{j=1}^{m} Y_j\right)$$

$$\hat{\sigma}_0^2 = \frac{1}{m+n}\left[\sum_{i=1}^{n} (X_i - \hat{\mu}_0)^2 + \sum_{j=1}^{m} (Y_j - \hat{\mu}_0)^2\right]$$

The corresponding value of the maximized log likelihood is, after some cancellation,

$$l(\hat{\mu}_0, \hat{\sigma}_0^2) = -\frac{m+n}{2}\log 2\pi - \frac{m+n}{2}\log \hat{\sigma}_0^2 - \frac{m+n}{2}$$

To find the maximum likelihood estimates $\hat{\mu}_X$, $\hat{\mu}_Y$, and $\hat{\sigma}_1^2$ under Ω, we first differentiate the log likelihood and obtain the equations

$$\sum_{i=1}^{n} (X_i - \hat{\mu}_X) = 0$$

$$\sum_{j=1}^{m} (Y_j - \hat{\mu}_Y) = 0$$

$$-\frac{m+n}{2\hat{\sigma}_1^2} + \frac{1}{2\hat{\sigma}_1^4}\left[\sum_{i=1}^{n} (X_i - \hat{\mu}_X)^2 + \sum_{j=1}^{m} (Y_j - \hat{\mu}_Y)^2\right] = 0$$

The maximum likelihood estimates are therefore

$$\hat{\mu}_X = \bar{X}$$

$$\hat{\mu}_Y = \bar{Y}$$

$$\hat{\sigma}_1^2 = \frac{1}{m+n}\left[\sum_{i=1}^{n} (X_i - \hat{\mu}_X)^2 + \sum_{j=1}^{m} (Y_j - \hat{\mu}_Y)^2\right]$$

When these are substituted into the log likelihood, we obtain

$$l(\hat{\mu}_X, \hat{\mu}_Y, \hat{\sigma}_1^2) = -\frac{m+n}{2}\log 2\pi - \frac{m+n}{2}\log \hat{\sigma}_1^2 - \frac{m+n}{2}$$

The log of the likelihood ratio is thus

$$\frac{m+n}{2}\log\left(\frac{\hat{\sigma}_1^2}{\hat{\sigma}_0^2}\right)$$

and the likelihood ratio test rejects for large values of

$$\frac{\hat{\sigma}_0^2}{\hat{\sigma}_1^2} = \frac{\sum_{i=1}^{n}(X_i - \hat{\mu}_0)^2 + \sum_{j=1}^{m}(Y_j - \hat{\mu}_0)^2}{\sum_{i=1}^{n}(X_i - \bar{X})^2 + \sum_{j=1}^{m}(Y_j - \bar{Y})^2}$$

We now find an alternative expression for the numerator of this ratio, by using the identities

$$\sum_{i=1}^{n}(X_i - \hat{\mu}_0)^2 = \sum_{i=1}^{n}(X_i - \bar{X})^2 + n(\bar{X} - \hat{\mu}_0)^2$$

$$\sum_{j=1}^{m}(Y_j - \hat{\mu}_0)^2 = \sum_{j=1}^{m}(Y_j - \bar{Y})^2 + m(\bar{Y} - \hat{\mu}_0)^2$$

We obtain

$$\hat{\mu}_0 = \frac{1}{m+n}(n\bar{X} + m\bar{Y})$$

$$= \frac{n}{m+n}\bar{X} + \frac{m}{m+n}\bar{Y}$$

Therefore,

$$\bar{X} - \hat{\mu}_0 = \frac{m(\bar{X} - \bar{Y})}{m+n}$$

$$\bar{Y} - \hat{\mu}_0 = \frac{n(\bar{Y} - \bar{X})}{m+n}$$

The alternative expression for the numerator of the ratio is thus

$$\sum_{i=1}^{n}(X_i - \bar{X})^2 + \sum_{j=1}^{m}(Y_j - \bar{Y})^2 + \frac{mn}{m+n}(\bar{X} - \bar{Y})^2$$

and the test rejects for large values of

$$1 + \frac{mn}{m+n}\left(\frac{(\bar{X} - \bar{Y})^2}{\sum_{i=1}^{n}(X_i - \bar{X})^2 + \sum_{j=1}^{m}(Y_j - \bar{Y})^2}\right)$$

or, equivalently, for large values of

$$\frac{|\bar{X} - \bar{Y}|}{\sqrt{\sum_{i=1}^{n} (X_i - \bar{X})^2 + \sum_{j=1}^{m} (Y_j - \bar{Y})^2}}$$

which is the t statistic apart from constants that do not depend on the data. Thus, the likelihood-ratio test is equivalent to the t test, as claimed.

We have used the assumption that the two populations have the same variance. If the two variances are not assumed to be equal, a natural estimate of $Var(\bar{X} - \bar{Y})$ is

$$\frac{s_X^2}{n} + \frac{s_Y^2}{m}$$

If this estimate is used in the denominator of the t statistic, the distribution of that statistic is no longer the t distribution. But it has been shown that its distribution can be closely approximated by the t distribution with degrees of freedom calculated in the following way and then rounded to the nearest integer:

$$df = \frac{[(s_X^2/n) + (s_Y^2/m)]^2}{\dfrac{(s_X^2/n)^2}{n} + \dfrac{(s_Y^2/m)^2}{m}} - 2$$

EXAMPLE C.   Let us rework Example B, but without the assumption that the variances are equal. Using the formula above, we find the degrees of freedom to be 12, rather than 19. The t statistic is 3.25, and from tables of the t distribution, the test still rejects at level $\alpha = .01$.   □

If the underlying distributions are not normal and the sample sizes are large, the use of the t distribution or the normal distribution is justified by the central limit theorem, and the probability levels of confidence intervals and hypothesis tests are approximately valid. In such a case, however, there is little difference between the t and normal distributions. If the sample sizes are small, however, and the distributions are not normal, conclusions based on the assumption of normality may not be valid. Unfortunately, if the sample sizes are small, the assumption of normality cannot be tested effectively unless the deviation is quite gross, as we saw in chapter 9.

11.2.1.1   An Example—A Study of Iron Retention

An experiment was performed to determine whether two forms of iron (Fe^{2+} and Fe^{3+}) are retained differently. (If one form of iron were retained especially well, it would be the better dietary supplement.) The investigators divided 108 mice randomly into 6 groups of 18 each; 3 groups were given Fe^{2+} in three different concentrations, 10.2, 1.2, and .3 millimolar, and 3 groups were given Fe^{3+} at the same three concentrations. The mice were given the iron orally; the iron was radioactively labeled so that a counter could be used to measure the initial amount given. At a later time, another count was taken for each mouse, and the percentage of iron retained was calculated. The data for the two forms of iron are listed in the table below. We will look at the data for the concentration 1.2 millimolar. (In chapter 12, we will discuss methods for analyzing all the groups simultaneously.)

Fe^{3+}			Fe^{2+}		
10.2	*1.2*	*.3*	*10.2*	*1.2*	*.3*
.71	2.20	2.25	2.20	4.04	2.71
1.66	2.93	3.93	2.69	4.16	5.43
2.01	3.08	5.08	3.54	4.42	6.38
2.16	3.49	5.82	3.75	4.93	6.38
2.42	4.11	5.84	3.83	5.49	8.32
2.42	4.95	6.89	4.08	5.77	9.04
2.56	5.16	8.50	4.27	5.86	9.56
2.60	5.54	8.56	4.53	6.28	10.01
3.31	5.68	9.44	5.32	6.97	10.08
3.64	6.25	10.52	6.18	7.06	10.62
3.74	7.25	13.46	6.22	7.78	13.80
3.74	7.90	13.57	6.33	9.23	15.99
4.39	8.85	14.76	6.97	9.34	17.90
4.50	11.96	16.41	6.97	9.91	18.25
5.07	15.54	16.96	7.52	13.46	19.32
5.26	15.89	17.56	8.36	18.4	19.87
8.15	18.3	22.82	11.65	23.89	21.60
8.24	18.59	29.13	12.45	26.39	22.25

As a summary of the data, boxplots (Figure 11-2) show that the data are quite

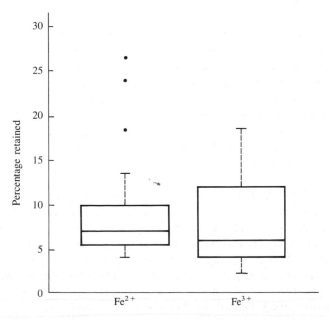

Figure 11-2. Boxplots of the percentages of iron retained for the two forms.

skewed to the right. This is not uncommon with percentages or other variables that are bounded below by zero. Three observations from the Fe^{2+} group are flagged as possible outliers. The median of the Fe^{2+} group is slightly larger than the median of the Fe^{3+} group, but the two distributions overlap substantially.

Another view of these data is provided by normal probability plots (Figure 11-3). These plots also indicate the skewness of the distributions. We should obviously doubt the validity of using normal distribution theory (for example, the t test) for this problem even though the combined sample size is fairly large (36).

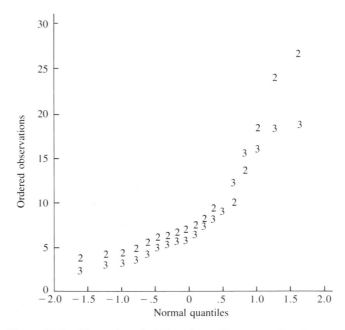

Figure 11-3.   Normal probability plots of iron retention data.

The mean and standard deviation of the Fe^{2+} group are 9.63 and 6.69; for the Fe^{3+} group, the mean is 8.20 and the standard deviation is 5.45. To test the hypothesis that the two means are equal, we can use a t test without assuming that the population standard deviations are equal. The approximate degrees of freedom, calculated as described at the end of Section 11.2.1, are 32. The t statistic is .702, which corresponds to a p-value of .49 for a two-sided test; if the two populations had the same mean, values of the t statistic this large or larger would occur 49% of the time. There is thus insufficient evidence to reject the null hypothesis. A 95% confidence interval for the difference of the two population means is $(-2.7, 5.6)$. But the t test assumes that the underlying populations are normally distributed, and we have seen there is reason to doubt this assumption. It is sometimes advocated that skewed data be transformed to a more symmetric shape before normal theory is applied. Transformations such as taking the log or the square root can be effective in symmetrizing skewed distributions because they spread out small values and compress large ones. Figures 11-4 and 11-5 show boxplots and normal probability plots for the natural logs of the iron retention data we have been considering. The transformation was fairly successful in symmetrizing these distributions, and the probability plots are more linear than those in Figure 11-3, although some curvature is still evident.

The following model is natural for the log transformation:

$$X_i = \mu_X(1 + \varepsilon_i), \quad i = 1, \ldots, n$$

$$Y_j = \mu_Y(1 + \delta_j), \quad j = 1, \ldots, m$$

$$\log X_i = \log \mu_X + \log(1 + \varepsilon_i)$$

$$\log Y_i = \log \mu_Y + \log(1 + \delta_j)$$

Here the ε_i and δ_j are independent random variables with mean zero. This model implies that if the variances of the errors are σ^2, then

$$\left.\begin{array}{l} E(X_i) = \mu_X \\ E(Y_j) = \mu_Y \\ \sigma_X = \mu_X\sigma \\ \sigma_Y = \mu_Y\sigma \end{array}\right\}$$

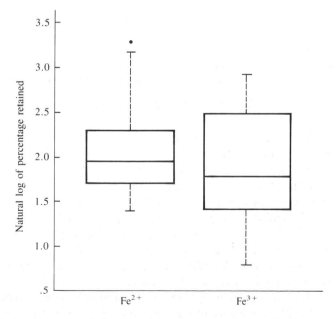

Figure 11-4. Boxplots of natural logs of percentages of iron retained.

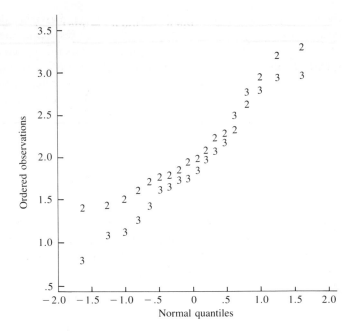

Figure 11-5.   Normal probability plots of natural logs of iron retention data.

or that

$$\frac{\sigma_X}{\mu_X} = \frac{\sigma_Y}{\mu_Y}$$

If the ε_i and δ_i have the same distribution, $\mathrm{Var}(\log X) = \mathrm{Var}(\log Y)$. The ratio of the standard deviation of a distribution to the mean is called the **coefficient of variation (CV)**; it expresses the standard deviation as a fraction of the mean. Coefficients of variation are sometimes expressed as percentages. For the iron retention data we have been considering, the CV's are .69 and .67 for the Fe^{2+} and Fe^{3+} groups; these values are quite close. These data are quite "noisy"—the standard deviation is nearly 70% of the mean for both groups.

For the transformed iron retention data, the means and standard deviations are given in the following table:

	Fe^{2+}	Fe^{3+}
Mean	2.08	1.90
Standard Deviation	.659	.574

For the transformed data, the t statistic is .917, which gives a p-value of .37. Again, there is no reason to reject the null hypothesis. A 95% confidence interval is $(-.61, .23)$. Using the model above, this is a confidence interval for

$$\log \mu_X - \log \mu_Y = \log\left(\frac{\mu_X}{\mu_Y}\right)$$

The interval is

$$-.61 \leq \log\left(\frac{\mu_X}{\mu_Y}\right) \leq .23$$

or

$$.54 \leq \frac{\mu_X}{\mu_Y} \leq 1.26$$

Other transformations, such as raising all values to some power, are sometimes used. Attitudes toward the use of transformations vary: Some view them as a very useful tool in statistics and data analysis, and others regard them as questionable manipulation of the data.

11.2.2 Power

Calculations of power are an important part of planning experiments in order to determine how large sample sizes should be. The power of a test is the probability of rejecting the null hypothesis when it is false. The power of the two-sample t test depends on four factors:

1. The real difference, $\Delta = |\mu_X - \mu_Y|$. The larger this difference, the greater the power.
2. The significance level α at which the test is done. The larger the significance level, the more powerful the test.
3. The population standard deviation σ, which is the amplitude of the "noise" that hides the "signal." The smaller the standard deviation, the larger the power.
4. The sample sizes n and m. The larger the sample sizes, the greater the power.

Before continuing, you should try to understand intuitively why these statements are true. We will express them quantitatively below.

The necessary sample sizes can be determined from the significance level of the test, the standard deviation, and the desired power against an alternative hypothesis,

$$H_1: \mu_X - \mu_Y = \Delta$$

In order to use the t distribution, special charts, or tables, of the noncentral t distribution are required. If the sample sizes are reasonably large, the normal distribution may be used instead of the t, and the calculations are relatively straightforward, as we will now demonstrate.

Suppose that σ, α, and Δ are given and that the samples are both of size n. The test at level α of H_0: $\mu_X = \mu_Y$ against the alternative H_1: $\mu_X \neq \mu_Y$ is based on the test statistic

$$Z = \frac{\bar{X} - \bar{Y}}{\sigma\sqrt{2/n}}$$

The rejection region for this test is $|Z| > z(\alpha/2)$, or

$$|\bar{X} - \bar{Y}| > z(\alpha/2)\sigma\sqrt{\frac{2}{n}}$$

The power of the test if $\mu_X - \mu_Y = \Delta$ is the probability that the test statistic falls in the rejection region, or

$$P\left[|\bar{X} - \bar{Y}| > z(\alpha/2)\sigma\sqrt{\frac{2}{n}} \right]$$

$$= P\left[\bar{X} - \bar{Y} > z(\alpha/2)\sigma\sqrt{\frac{2}{n}} \right] + P\left[\bar{X} - \bar{Y} < -z(\alpha/2)\sigma\sqrt{\frac{2}{n}} \right]$$

since the two events are mutually exclusive. Both probabilities on the right-hand side are calculated by standardizing. For the first one, we have

$$P\left[\bar{X} - \bar{Y} > z(\alpha/2)\sigma\sqrt{\frac{2}{n}} \right] = P\left[\frac{(\bar{X} - \bar{Y}) - \Delta}{\sigma\sqrt{2/n}} > \frac{z(\alpha/2)\sigma\sqrt{2/n} - \Delta}{\sigma\sqrt{2/n}} \right]$$

$$= 1 - \Phi\left[z(\alpha/2) - \frac{\Delta}{\sigma}\sqrt{\frac{n}{2}} \right]$$

where Φ is the standard normal cdf. Similarly, the second probability is

$$\Phi\left[-z(\alpha/2) - \frac{\Delta}{\sigma}\sqrt{\frac{n}{2}} \right]$$

Thus, the probability that the test statistic falls in the rejection region is equal to

$$1 - \Phi\left[z(\alpha/2) - \frac{\Delta}{\sigma}\sqrt{\frac{n}{2}} \right] + \Phi\left[-z(\alpha/2) - \frac{\Delta}{\sigma}\sqrt{\frac{n}{2}} \right]$$

Typically, as Δ moves away from zero, one of these terms will be negligible with respect to the other. For example, if Δ is greater than zero, the first term will be dominant. For fixed n, this expression can be evaluated as a function of Δ; or for fixed Δ, it can be evaluated as a function of n.

EXAMPLE A.   As an example, let us consider a situation similar to an idealized form of the iron retention experiment. Assume that we have samples of size 18 from two normal distributions whose standard deviations are both 5, and we calculate the power for various values of Δ when the null hypothesis is tested at a significance level of .05. The results of the calculations are displayed in Figure 11-6. We see from the plot that if the mean difference in retention is only 1%, the probability of rejecting the null hypothesis is quite small, only 9%. A mean difference of 5% in retention rate gives a more satisfactory power of 85%.

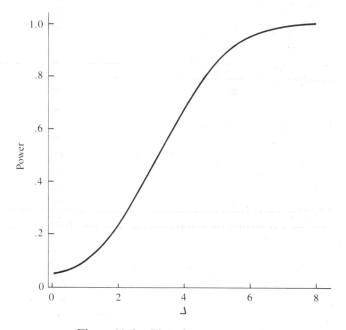

Figure 11-6.   Plot of power versus Δ.

Suppose that we wanted to be able to detect a difference of $\Delta = 1$ with probability .9. What sample size would be necessary? Using only the dominant term in the expression for the power, the sample size should be such that

$$\Phi\left(1.96 - \frac{\Delta}{\sigma}\sqrt{\frac{n}{2}}\right) = .1$$

From the tables for the normal distribution, $.1 = \Phi(-1.28)$, so

$$1.96 - \sigma\sqrt{\frac{2}{n}} = -1.28$$

Solving for n, we find that the necessary sample size would be 524! This is clearly

unfeasible; if in fact the experimenters wished to detect such a difference, some modification of the experimental technique to reduce σ would be necessary.   □

11.2.3   A Nonparametric Method—The Mann–Whitney Test

Nonparametric methods do not assume that the data follow any particular distributional form. Many of them are based on replacement of the data by ranks. With this replacement, the results are invariant under any monotonic transformation; in comparison, we saw that the p-value of a t test may change if the log of the measurements is analyzed rather than the measurements on the original scale. Replacing the data by ranks also has the effect of moderating the influence of outliers.

For purposes of discussion, we will develop the **Mann–Whitney test** (also sometimes called the Wilcoxon rank sum test) in a specific context. Suppose that we have $m + n$ experimental units to assign to a treatment group and a control group. The assignment is made at random: n units are randomly chosen and assigned to the control, and the remaining m units are assigned to the treatment. We are interested in testing the null hypothesis that the treatment has no effect. If the null hypothesis is true, then any difference in the outcomes under the two conditions is due to the randomization.

A test statistic is calculated in the following way. First, we group all $m + n$ observations together and rank them in order of increasing size (we will assume for simplicity that there are no ties, although the argument holds even in the presence of ties). We next calculate the sum of the ranks of those observations that came from the control group. If this sum is too small or too large, we will reject the null hypothesis.

In order to find the critical values that define the rejection region, we need to calculate the distribution of the test statistic if the null hypothesis is true. If the null hypothesis is true, any assignment of ranks to the $m + n$ observations is equally likely. There are $(m + n)!$ such assignments, and for each one the test statistic (the sum of the ranks of the control group) can be calculated. This gives us a list of $(m + n)!$ values (not all distinct), each of which occurs with probability $1/(m + n)!$, and we therefore have the null distribution of the test statistic. The rejection region is located in the tails of this distribution.

It is important to note that we have not made any assumption that the observations from the control and treatment groups are samples from a probability distribution. Probability has only entered in as a result of the random assignment of experimental units to treatment and control groups (this is similar to the way that probability enters into survey sampling). We should also note that, although we chose the sum of control ranks as the test statistic, any other test statistic could have been used and its null distribution computed in the same fashion. The rank sum is easy to compute and is sensitive to a treatment effect that tends to make responses larger or smaller. Also, its null distribution has only to be computed once and tabled; if we worked with the actual numerical values, the null distribution would depend on those particular values.

Tables of the null distribution of the rank sum are widely available and vary in format. Note that since the sum of the two rank sums is the sum of the integers from 1 to $m + n$, which is $[(m + n)(m + n + 1)/2]$, knowing one rank sum tells us the other. Some tables are given in terms of the rank sum of the smaller of the two groups, and some are in terms of the smaller of the two rank sums (the advantage of the latter scheme is that only one tail of the distribution has to be tabled). Table 8 of Appendix B makes use of additional symmetries. Let n_1 be the smaller sample size and let R be the sum of the ranks from that sample. Let $R' = n_1(m + n + 1) - R$ and $R^* = \min(R, R')$. The table gives critical values for R^*.

When it is more appropriate to model the control values, $X_1, \ldots, X_n$, as a sample from some probability distribution F and the experimental values, $Y_1, \ldots, Y_m$, as a sample from some distribution G, the Mann–Whitney test is a test of the null hypothesis $H_0: F = G$. The reasoning is exactly the same: Under H_0, any assignment of ranks to the pooled $m + n$ observations is equally likely, etc.

We have assumed here that there are no ties among the observations. If there are only a small number of ties, tied observations are assigned average ranks (the average of the ranks for which they are tied); the significance levels are not greatly affected.

EXAMPLE A.   Let us illustrate the Mann–Whitney test by referring to the data on latent heats of fusion of ice considered earlier (Example A in Section 11.2.1). The sample sizes are fairly small (13 and 8), so in the absence of any prior knowledge concerning the adequacy of the assumption of a normal distribution, it would seem safer to use a nonparametric method. The following table exhibits the ranks given to the measurements for each method (refer to Example A in Section 11.2.1 for the original data):

Method A	Method B
7.5	11.5
19.0	1.0
11.5	7.5
19.0	4.5
15.5	4.5
15.5	15.5
19.0	2.0
4.5	4.5
21.0	
15.5	
11.5	
9.0	
11.5	

Table 8 of Appendix B is used as follows. The sum of the ranks of the smaller sample is $R = 51$.

$$R' - 8(8 + 13 + 1) - R$$

$$= 125$$

Thus, $R^* = 51$. From the table, 53 is the critical value for a two-tailed test with $\alpha = .01$, and 60 is the critical value for $\alpha = .05$. The Mann–Whitney test thus rejects at the .01 significance level.   □

The Mann–Whitney test can also be derived starting from a slightly different point of view. Suppose that the X's are a sample from F and the Y's a sample from G, and consider estimating, as a measure of the effect of the treatment,

$$\pi = P(X < Y)$$

where X and Y are independently distributed with distribution functions F and G, respectively. The value π is the probability that an observation from the distribution F is smaller than an independent observation from the distribution G. An estimate of π can be obtained by comparing all n values of X to all m values of Y and calculating the proportion of the comparisons for which X was less than Y:

$$\hat{\pi} = \frac{1}{mn} \sum_{i=1}^{n} \sum_{j=1}^{m} Z_{ij}$$

where

$$Z_{ij} = \begin{cases} 1, & \text{if } X_i < Y_j \\ 0, & \text{otherwise} \end{cases}$$

To see the relationship of $\hat{\pi}$ to the rank sum introduced earlier, we will find it convenient to work with

$$V_{ij} = \begin{cases} 1, & \text{if } X_{(i)} < Y_{(j)} \\ 0, & \text{otherwise} \end{cases}$$

Clearly,

$$\sum_{i=1}^{n} \sum_{j=1}^{m} Z_{ij} = \sum_{i=1}^{n} \sum_{j=1}^{m} V_{ij}$$

since the V_{ij} are just a reordering of the Z_{ij}. Also,

$$\sum_{i=1}^{n} \sum_{j=1}^{m} V_{ij} = (\text{number of } X\text{'s that are less than } Y_{(1)})$$

$$+ (\text{number of } X\text{'s that are less than } Y_{(2)})$$

$$+ \cdots + (\text{number of } X\text{'s that are less than } Y_{(m)})$$

If the rank of $Y_{(k)}$ in the combined sample is denoted by R_{yk}, then the number of X's less than $Y_{(1)}$ is $R_{y1} - 1$, the number of X's less than $Y_{(2)}$ is $R_{y2} - 2$, etc. Therefore,

$$\sum_{i=1}^{n} \sum_{j=1}^{m} V_{ij} = (R_{y1} - 1) + (R_{y2} - 2) + \cdots + (R_{ym} - m)$$

$$= \sum_{i=1}^{m} R_{yi} - \sum_{i=1}^{m} i$$

$$= \sum_{i=1}^{m} R_{yi} - \frac{m(m + 1)}{2}$$

Thus, $\hat{\pi}$ may be expressed in terms of the rank sum of the Y's (or in terms of the rank sum of the X's, since the two rank sums add up to a constant).

This representation makes calculations of moments fairly straightforward. We will calculate the mean and variance of

$$U_Y = \sum_{i=1}^{n} \sum_{j=1}^{m} Z_{ij}$$

under the null hypothesis $H_0: F = G$.

THEOREM A.   Under the null hypothesis $H_0: F = G$,

$$E(U_Y) = \frac{mn}{2}$$

$$\text{Var}(U_Y) = \frac{mn(m + n + 1)}{12}$$

Proof.   First, we note that Z_{ij} is a Bernoulli random variable, so

$$E(Z_{ij}) = P(Z_{ij} = 1) = P(X_i < Y_j) = \tfrac{1}{2}$$

and thus

$$E(U_Y) = \sum_{i=1}^{n} \sum_{j=1}^{m} E(Z_{ij}) = \frac{mn}{2}$$

The variance is harder to calculate:

$$\text{Var}(U_Y) = \text{Cov}\left(\sum_{i=1}^{n} \sum_{j=1}^{m} Z_{ij}, \sum_{k=1}^{n} \sum_{l=1}^{m} Z_{kl} \right)$$

$$= \sum_{i=1}^{n} \sum_{j=1}^{m} \sum_{k=1}^{n} \sum_{l=1}^{m} \text{Cov}(Z_{ij}, Z_{kl})$$

This covariance can be expressed as

$$\mathrm{Cov}(Z_{ij}, Z_{kl}) = E(Z_{ij}Z_{kl}) - E(Z_{ij})E(Z_{kl})$$

$$= E(Z_{ij}Z_{kl}) - \tfrac{1}{4}$$

Since

$$Z_{ij}Z_{kl} = \begin{cases} 1, & \text{if } X_i < Y_j \text{ and } X_k < Y_l \\ 0, & \text{otherwise} \end{cases}$$

we have

$$E(Z_{ij}Z_{kl}) = P(X_i < Y_j \text{ and } X_k < Y_l)$$

Now, we claim that

$$P(X_i < Y_j \text{ and } X_k < Y_l) = \begin{cases} \tfrac{1}{2}, & i = k, j = l \\ \tfrac{1}{4}, & i \neq k, j \neq l \\ \tfrac{1}{3}, & i = k, j \neq l \text{ or } i \neq k, j = l \end{cases}$$

The first two cases are straightforward. The third case follows since of three i.i.d. random variables, any one is equally likely to be the smallest. We thus have

$$\mathrm{Cov}(Z_{ij}, Z_{kl}) = \begin{cases} \tfrac{1}{4}, & i = k, j = l \\ 0, & i \neq k, j \neq l \\ \tfrac{1}{12}, & i = k, j \neq l \text{ or } i \neq k, j = l \end{cases}$$

The case for which $i = k$ and $j = l$ occurs mn times; the case for which $i \neq k$ and $j = l$ occurs $n(n-1)m$ times; the case for which $i = k$ and $j \neq l$ occurs $m(m-1)n$ times. We thus have

$$\mathrm{Var}(U_Y) = \frac{mn}{4} + \frac{n(n-1)m + m(m-1)n}{12}$$

$$= \frac{mn(m+n+1)}{12} \qquad \square$$

For m and n both greater than 10, the null distribution of U is quite well approximated by a normal distribution,

$$\frac{U_Y - E(U_Y)}{\sqrt{\mathrm{Var}(U_Y)}} \sim N(0, 1)$$

(Note that this does not follow immediately from the ordinary central limit theorem; although U_Y is a sum of random variables, they are not independent.) Similarly, the distribution of the rank sum of the X's or Y's may be approximated by a normal distribution, since these rank sums differ from U_Y only by constants. In particular, we have the following corollary.

COROLLARY A.   Let T_Y denote the sum of the m ranks of the Y's. Then, under H_0: $F = G$,

$$E(T_Y) = \frac{m(m + n + 1)}{2}$$

$$\mathrm{Var}(T_Y) = \frac{mn(m + n + 1)}{12}$$

EXAMPLE B.   Referring to Example A, let us use a normal approximation to the distribution of the rank sum from method B. For $n = 8$ and $m = 13$, we have from Corollary A that under the null hypothesis,

$$E(T) = \frac{8(8 + 13 + 1)}{2} = 88$$

$$\sigma_T = \sqrt{\frac{8 \times 13(8 + 13 + 1)}{12}} = 13.8$$

T is the sum of the ranks from method B, or 51, and the normalized test statistic is

$$\frac{T - E(T)}{\sigma_T} = -2.75$$

From the tables of the normal distribution, this corresponds to a p-value of .006 for a two-sided test, so the null hypothesis is rejected at level $\alpha = .01$, just as it was when we used the exact distribution. For this set of data, we have seen that the t test with the assumption of equal variances, the t test without that assumption, the exact Mann–Whitney test, and the approximate Mann–Whitney test all reject at level $\alpha = .01$.   □

The Mann–Whitney test can be inverted to form confidence intervals. Let us consider a "shift" model: $G(x) = F(x - \Delta)$. This model says that the effect of the treatment (the Y's) is to add a constant Δ to what the response would have been with no treatment (the X's). (This is a very simple model, and we have already seen cases for which it is not appropriate.) We now derive a confidence interval for Δ. To test H_0: $F = G$, we used the statistic U_Y equal to the number of the $X_i - Y_j$ that are less than zero. To test the hypothesis that the shift parameter is Δ, we can similarly use

$$U_Y(\Delta) = \#[X_i - (Y_j - \Delta) < 0] = \#(Y_j - X_i > \Delta)$$

It can be argued that the acceptance region for the test is of the form

$$k < U_Y(\Delta) < mn - k$$

where k is chosen so that the test has level α. Thus, from the results of the Section 9.4, a confidence region for α is of the form

$$C = \{\Delta | k < U_Y(\Delta) < mn - k\}$$

that is, it is the set of values of Δ for which the null hypothesis would not be rejected.

We can find an explicit form for this confidence region. Let $D_{(1)}, \ldots, D_{(nm)}$ denote the ordered mn differences $Y_j - X_i$. We will show that

$$C = [D_{(k)}, D_{(mn-k+1)})]$$

First, if $\Delta = D_{(k)}$, then $nm - k$ of the differences $Y_j - X_i$ are greater than Δ; if Δ is less than $D_{(k)}$, then at least $nm - k + 1$ of those differences are greater than Δ. Thus, $D_{(k)}$ is the leftmost point of the interval. Second, if $\Delta = D_{(mn-k)}$, then only k of the differences are greater than Δ. If Δ is between $D_{(mn-k)}$ and $D_{(mn-k+1)}$ then there are still only k differences greater than Δ; if Δ is greater than or equal to $D_{(mn-k+1)}$, then at most $k - 1$ of the differences are greater than Δ. The confidence interval is thus

$$[D_{(k)}, D_{(mn-k+1)})$$

Since the distributions are assumed to be continuous, the closed interval has the same probability of coverage as the open interval. Computation of such intervals is best done with a computer.

EXAMPLE C. We return to the data on iron retention (Section 11.2.1.1). The earlier analysis using the t test rested on the assumption that the populations were normally distributed, which, in fact, seemed rather dubious. The Mann–Whitney test does not make this assumption. The sum of the ranks of the Fe^{2+} group is used as a test statistic (we could have as easily used the U statistic). The rank sum is 362. Using the normal approximation to the null distribution of the rank sum, we get a p-value of .36. Again, there is insufficient evidence to reject the null hypothesis that there is differential retention. The 95% confidence interval for the shift between the two distributions is $(-1.6, 3.7)$, which overlaps zero substantially. Note that this interval is shorter than the interval based on the t distribution; the latter was inflated by the contributions of the large observations to the sample variance. ☐

11.3   Comparing Paired Samples

In Section 11.2, we considered the problem of analyzing two independent samples. In many experiments, the samples are paired. In a medical experiment, for example, subjects might be matched by age or weight or severity of condition, and then one member of each pair randomly assigned to the treatment group and the other to the control group. In a biological experiment, the paired subjects might be littermates. In some applications, the pair consists of a "before" and an "after" measurement on the same object. Since pairing causes the samples to be dependent, the analysis of Section 11.2 does not apply.

Pairing can be an effective experimental technique. Let us denote the pairs as (X_i, Y_i), where $i = 1, \ldots, n$, and assume the X's and Y's have means μ_X and μ_Y and variances σ_X^2 and σ_Y^2. We will assume that different pairs are independently distributed and that $\text{Cov}(X_i, Y_i) = \sigma_{XY}$. We will work with the differences $D_i = X_i - Y_i$, which are independent with

$$E(D_i) = \mu_X - \mu_Y$$

$$\text{Var}(D_i) = \sigma_X^2 + \sigma_Y^2 - 2\sigma_{XY}$$

A natural estimate of $\mu_X - \mu_Y$ is $\bar{D}$, the average difference. From the properties of D_i, it follows that

$$E(\bar{D}) = \mu_X - \mu_Y$$

$$\text{Var}(\bar{D}) = \frac{1}{n}(\sigma_X^2 + \sigma_Y^2 - 2\sigma_{XY})$$

Suppose that an experiment had been done by taking a sample of n X's and an independent sample of n Y's. Then $\mu_X - \mu_Y$ would be estimated by $\bar{X} - \bar{Y}$ and

$$E(\bar{X} - \bar{Y}) = \mu_X - \mu_Y$$

$$\text{Var}(\bar{X} - \bar{Y}) = \frac{1}{n}(\sigma_X^2 + \sigma_Y^2)$$

Comparing the variances of the two estimates, we see that the variance of $\bar{D}$ is smaller if the covariance term is positive—that is, if the X's and Y's are positively correlated. In this circumstance, pairing is the more effective experimental design. In the simple case in which $\sigma_X = \sigma_Y = \sigma$, the two variances may be more simply expressed as

$$\text{Var}(\bar{D}) = \frac{2\sigma^2(1 - \rho)}{n}$$

$$\text{Var}(\bar{X} - \bar{Y}) = \frac{2\sigma^2}{n}$$

and the relative efficiency is

$$\frac{\text{Var}(\bar{D})}{\text{Var}(\bar{X} - \bar{Y})} = 1 - \rho$$

If the correlation coefficient is .5, for example, a paired design with n pairs of subjects yields the same precision as an unpaired design with $2n$ subjects per treatment. This additional precision results in shorter confidence intervals and more powerful tests if the degrees of freedom for estimating σ^2 are sufficiently large.

11.3.1 Methods Based on the Normal Distribution

In this section, we assume that the differences are a sample from a normal distribution with

$$E(D_i) = \mu_X - \mu_Y = \mu_D$$

$$\text{Var}(D_i) = \sigma_D^2$$

Generally, σ_D will be unknown, and inferences will be based on

$$t = \frac{\bar{D} - \mu_D}{s_{\bar{D}}}$$

which follows a t distribution with $n - 1$ degrees of freedom. Following familiar reasoning, a $100(1 - \alpha)\%$ confidence interval for μ_D is

$$\bar{D} \pm t_{n-1}(\alpha/2)s_{\bar{D}}$$

A two-sided test of the null hypothesis H_0: $\mu_D = 0$ (the natural null hypothesis for testing no treatment effect) at level α has the rejection region

$$|\bar{D}| > t_{n-1}(\alpha/2)s_{\bar{D}}$$

If the sample size n is large, the approximate validity of the confidence interval and hypothesis test follows from the central limit theorem. If the sample size is small and the true distribution of the differences is far from normal, the stated probability levels may be considerably in error.

EXAMPLE A. To study the effect of cigarette smoking on platelet aggregation, Levine (1973) drew blood samples from 11 individuals before and after they smoked a cigarette and measured the extent to which the blood platelets aggregated. Platelets are involved in the formation of blood clots, and it is known that smokers suffer more often from disorders involving blood clots than do non-smokers. The data are shown in the following table, which gives the maximum percentage of all the platelets that aggregated after being exposed to a stimulus.

Before	After	Difference
25	27	2
25	29	4
27	37	10
44	56	12
30	46	16
67	82	15
53	57	4
53	80	27
52	61	9
60	59	−1
28	43	15

From the column of differences, $\bar{D} = 10.27$ and $s_{\bar{D}} = 2.40$. The t statistic is 4.28, and the p-value for a two-sided test is .0016. There is little doubt that smoking increased platelet aggregation.

The experiment was actually more complex than we have indicated. Some subjects also smoked cigarettes made of lettuce leaves and "smoked" unlit cigarettes. (You should reflect on why these additional experiments were done.)

Figure 11-7 is a plot of the after values versus the before values. They are

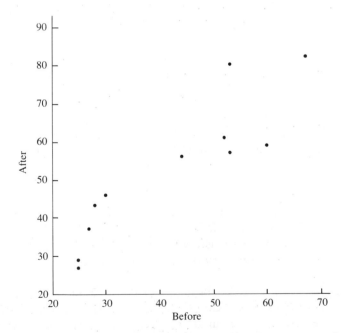

Figure 11-7. Plot of platelet aggregation after smoking versus aggregation before smoking.

correlated, with a correlation coefficient of .90. Pairing was a natural and effective experimental design in this case.   □

11.3.2   A Nonparametric Method—The Signed Rank Test

A nonparametric test based on ranks can be constructed for paired samples. The test statistic is calculated in the following way:

1. Rank the absolute values of the differences. (For now, assume there are no ties.) Let R_i be the rank of $|D_i|$, for $i = 1, \ldots, n$.
2. Restore the signs of the D_i to the ranks, obtaining signed ranks.
3. Calculate W_+, the sum of those ranks that have positive signs.

The idea behind the **signed rank test** (sometimes called the Wilcoxon signed rank test) is intuitively simple. If there is no difference between the two paired conditions, we expect about half the D_i to be positive and half negative, and W_+ will not be too small or too large. If one condition tends to produce larger values than the other, W_+ will tend to be more extreme. We therefore can use W_+ as a test statistic and reject for extreme values.

Before continuing, we need to specify more precisely the null hypothesis we are testing with the signed rank test: H_0 states that the distribution of the D_i is symmetric about zero. This will be true if the members of pairs of experimental units are assigned randomly to treatment and control conditions, and the treatment has no effect at all.

As usual, in order to define a rejection region for a test at level α, we need to know the sampling distribution of W_+ if the null hypothesis is true. The rejection region will be located in the tails of this null distribution in such a way that the test has level α. The null distribution may be calculated in the following way. If H_0 is true, it makes no difference which member of the pair corresponds to treatment and which to control. The difference $X_i - Y_i = D_i$ has the same distribution as the difference $Y_i - X_i = -D_i$, so the distribution of D_i is symmetric about zero. The kth largest value of D is thus equally likely to be positive or negative, and any particular assignment of signs to the integers $1, \ldots, n$ (the ranks) is equally likely. There are 2^n such assignments, and for each we can calculate W_+. We obtain a list of 2^n values (not all distinct) of W_+, each of which occurs with probability $1/2^n$. The probability of each distinct value of W_+ may thus be calculated, giving the desired null distribution.

The preceding argument has assumed that the D_i are a sample from some continuous probability distribution. If we do not wish to regard the X_i and Y_i as random variables and if the assignments to treatment and control have been made at random, the hypothesis that there is no treatment effect may be tested in exactly the same manner, except that inferences are based on the distribution induced by the randomization, as was done for the Mann–Whitney test.

Tables of the null distribution of W_+ for various sample sizes are available.

EXAMPLE A.   The signed rank test can be applied to the data on platelet aggregation considered previously (Example A in Section 11.3.1). In this case, it is easier to work with W_- rather than W_+ since W_- is clearly 1. From Table 9 of Appendix B, the two-sided test is significant at $\alpha = .01$. □

If the sample size is greater than 20, there is a normal approximation to the null distribution. To find this, we calculate the mean and variance of W_+.

THEOREM A. Under the null hypothesis that the D_i are independent and symmetrically distributed about zero,

$$E(W_+) = \frac{n(n + 1)}{4}$$

$$\text{Var}(W_+) = \frac{n(n + 1)(2n + 1)}{24}$$

Proof. To facilitate the calculation, we represent W_+ in the following way:

$$W_+ = \sum_{k=1}^{n} kI_k$$

where

$$I_k = \begin{cases} 1, & \text{if the } k\text{th largest } |D_i| \text{ has } D_i > 0 \\ 0, & \text{otherwise} \end{cases}$$

Under H_0, the I_k are independent Bernoulli random variables with $p = \frac{1}{2}$, so

$$E(I_k) = \tfrac{1}{2}$$

$$\text{Var}(I_k) = \tfrac{1}{4}$$

We thus have

$$E(W_+) = \tfrac{1}{2} \sum_{k=1}^{n} k = \frac{n(n + 1)}{4}$$

$$\text{Var}(W_+) = \tfrac{1}{4} \sum_{k=1}^{n} k^2 = \frac{n(n + 1)(2n + 1)}{24}$$

as was to be shown. □

If some of the differences are equal to zero, the most common technique is to discard those observations. If there are ties, each $|D_i|$ is assigned the average value of the ranks for which it is tied. If there are not too many ties, the significance level of the test is not greatly affected. If there are a large number of ties, modifications must be made. For further information on these matters, see Hollander and Wolfe (1973) or Lehmann (1975).

11.3.3 An Example—Measuring Mercury Levels in Fish

Kacprzak and Chvojka (1976) compared two methods of measuring mercury levels in fish. A new method, which they called "selective reduction," was compared to an established method, referred to as "the permanganate method." One advantage of selective reduction is that it allows simultaneous measurement of both inorganic mercury and methyl mercury. The mercury in each of 25 juvenile black marlin was measured by both techniques. The 25 measurements for each method (in ppm of mercury) and the differences are given in the table below.

Fish	Selective Reduction	Permanganate	Difference	Signed Rank
1	.32	.39	.07	+15.5
2	.40	.47	.07	+15.5
3	.11	.11	.00	
4	.47	.43	−.04	−11
5	.32	.42	.10	+19
6	.35	.30	−.05	−13.5
7	.32	.43	.11	+20
8	.63	.98	.35	+23
9	.50	.86	.36	+24
10	.60	.79	.19	+22
11	.38	.33	−.05	−13.5
12	.46	.45	−.01	−2.5
13	.20	.22	.02	+6.5
14	.31	.30	−.01	−2.5
15	.62	.60	−.02	−6.5
16	.52	.53	.01	+2.5
17	.77	.85	.08	+17.5
18	.23	.21	−.02	−6.5
19	.30	.33	.03	+9.0
20	.70	.57	−.13	−21
21	.41	.43	.02	+6.5
22	.53	.49	−.04	−11
23	.19	.20	.01	+2.5
24	.31	.35	.04	+11
25	.48	.40	−.08	−17.5

In analyzing such data, it is often informative to check whether the differences depend in some way on the level or size of the quantity being measured. The differences versus the permanganate values are plotted in Figure 11-8. This plot is quite interesting. It appears that the differences are small for low permanganate values and larger for higher permanganate values. It is striking that the differences are all positive and large for the highest four values. The investigators do not comment on these phenomena. It is not uncommon for the size of errors to increase as the value being measured increases; the percent error may remain

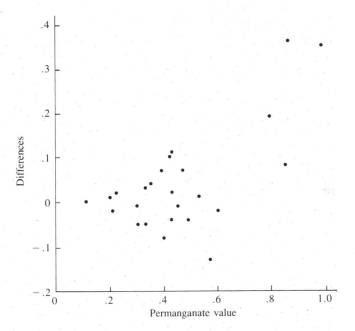

Figure 11-8. Plot of differences versus permanganate values.

nearly constant but the actual error does not. For this reason, data of this nature are often analyzed on a log scale.

Since the observations are paired (two measurements on each fish), we will use the paired t test for a parametric test. The sample size is large enough that the test should be robust against nonnormality. The mean difference is .04, and the standard deviation of the differences is .116. The t statistic is 1.745; with 24 degrees of freedom, this corresponds to a p-value of .094 for a two-sided test. Although this p-value is fairly small, the evidence against H_0: $\mu_D = 0$ is not overwhelming. The test does not reject at the significance level .05.

The signed ranks are shown in the last column of the table above. Note that the single zero difference was set aside, and also note how the tied ranks were handled. The test statistic W_+ is 194.5. Under H_0, its mean and variance are

$$E(W_+) = \frac{24 \times 25}{4} = 150$$

$$\mathrm{Var}(W_+) = \frac{24 \times 25 \times 49}{24} = 1225$$

Since n is greater than 20, we use the normalized test statistic, or

$$Z = \frac{W_+ - E(W_+)}{\sqrt{\mathrm{Var}(W_+)}} = 1.27$$

The p-value for a two-sided test from the normal approximation is .20, which is not strong evidence against the null hypothesis. It is possible to correct for the presence of ties, but in this case the correction only amounts to changing the standard deviation of W_+ from 35 to 34.95.

Neither the parametric nor the nonparametric test gives conclusive evidence that there is any systematic difference between the two methods of measurement. The informal graphical analysis does suggest, however, that there may be a difference for high concentrations of mercury.

11.4   Experimental Design

This section covers some basic principles of the interpretation and design of experimental studies and illustrates them with case studies.

11.4.1   Mammary Artery Ligation

A person with coronary artery disease suffers from chest pain during exercise because the constricted arteries cannot deliver enough oxygen to the heart. The treatment of ligating the mammary arteries enjoyed a brief vogue; the basic idea was that ligating these arteries forced more blood to flow into the heart. This procedure had the advantage of being quite simple surgically, and it was widely publicized in an article in *Reader's Digest* (Ratcliffe, 1957). Two years later, the results of a more careful study (Cobb et al., 1959) were published. In this study, a control group and an experimental group were established in the following way. When a prospective patient entered surgery, the surgeon made the necessary preliminary incisions prior to tying off the mammary artery. At that point, the surgeon opened a sealed envelope that contained instructions as to whether to complete the operation by tying off the artery. Neither the patient nor his attending physician knew whether the operation had actually been carried out. The study showed essentially no difference after the operation between the control group (no ligation) and the experimental group (ligation), although there was some suggestion that the control group had done better.

The Ratcliffe and Cobb studies differ in that in the earlier one there was no control group and thus no benchmark by which to gauge improvement. The reported improvement of the patients in this earlier study could have been due to the placebo effect, which we discuss next. The design of the later study protected against possible unconscious biases by randomly assigning the control and experimental groups and by concealing from the patients and their physicians the actual nature of the treatment. Such a design is called a double-blind, randomized controlled experiment.

11.4.2   The Placebo Effect

The **placebo effect** refers to the effect produced by any treatment, including dummy pills (placebos), when the subject believes that he or she has been given an effective treatment. The possibility of a placebo effect makes the use of a blind design necessary in many experimental investigations.

The placebo effect may not be due entirely to psychological factors, as was shown in an interesting experiment by Levine, Gordon, and Fields (1978). A group of subjects had teeth extracted. During the extraction, they were given nitrous oxide and local anesthesia. In the recovery room, they rated the amount of pain they were experiencing on a numerical scale. Two hours after surgery, the subjects were given a placebo and were again asked to rate their pain. An hour later, some of the subjects were given a placebo and some were given naloxene, a morphine antagonist. It is known that there are specific receptors to morphine in the brain and that the body can also release endorphins that bind to these sites. Naloxene blocks the morphine receptors. In the study, it was found that when those subjects who responded positively to the placebo received naloxene, they experienced an increase in pain that made their pain levels comparable to those of the patients who did not respond to the placebo. The implication is that those who responded to the placebo had produced endorphins, the actions of which were subsequently blocked by the naloxene.

The use of controls does not in itself ensure a valid experimental design; the allocation of subjects to treatment and control groups should be done by randomization. Wilson (1952) relates a story of a test of a pill to prevent seasickness. Prior to the voyage, the use of controls was carefully explained to the captain. On returning, he reported that the pills had been a marvelous success—most of the controls had been sick and the treatment group had fared well. But when further questioned, he revealed that he had given the pills to his seamen and had used the passengers as controls.

11.4.3 The Lanarkshire Milk Experiment

The importance of the randomized assignment of individuals (or other experimental units) to treatment and control groups is illustrated by a famous study known as the Lanarkshire milk experiment. In the spring of 1930, an experiment was carried out in Lanarkshire, England to determine the effect of providing free milk to schoolchildren. In each participating school, some children (treatment group) were given free milk and others (controls) were not. The assignment of children to control or treatment was initially done at random; however, teachers were allowed to use their judgment in switching children between treatment and control to obtain a better balance of undernourished and well-nourished individuals in the groups.

A paper by Gosset (1931), who published under the name Student (as in Student's t test), is a very interesting critique of the experiment. An examination of the data revealed that at the start of the experiment the controls were heavier and taller. Student conjectured that the teachers, perhaps unconsciously, had adjusted the initial randomization in a manner that placed more of the undernourished children in the treatment group. A further complication was caused by weighing the children with their clothes on. The experimental data were weight gains measured in late spring relative to early spring or late winter. The more well-to-do children probably tended to be better nourished and may have had heavier winter clothing than the poor children. Thus, the well-to-do children's weight gains were vitiated as a result of differences in clothing, which may have influenced comparisons between the treatment and control groups.

11.4.4   The Portocaval Shunt

Cirrhosis of the liver, to which alcoholics are prone, is a condition in which resistance to blood flow causes blood pressure in the liver to build up to dangerously high levels. Vessels may rupture, which may cause death. Surgeons have attempted to relieve this condition by connecting the portal artery, which feeds the liver, to the vena cava, one of the main veins returning to the heart, thus reducing blood flow through the liver. This procedure, called the portocaval shunt, had been used for more than 20 years when Grace, Muench, and Chalmers (1966) published an examination of 51 studies of the method. They examined the design of each study (presence or absence of a control group and presence or absence of randomization) and the investigators' conclusions (categorized as markedly enthusiastic, moderately enthusiastic, or not enthusiastic). The results are summarized in the following table, which speaks for itself:

| | *Enthusiasm* | | |
Design	*Marked*	*Moderate*	*No*
No Controls	24	7	1
Nonrandomized Controls	10	3	2
Randomized Controls	0	1	3

The differences between the experiments that used controls and those that did not is not entirely surprising since the placebo effect was probably operating. The importance of randomized assignment to treatment and control groups is illustrated by comparing the conclusions for the randomized and nonrandomized controlled experiments. Randomization can help to ensure against subtle unconscious biases that may creep into an experiment. For example, a physician might tend to recommend surgery for patients who are somewhat more robust than the average. Articulate patients might be more likely to have an influence on the decision as to which group they are assigned to.

11.4.5   FD&C Red No. 40

This discussion follows Lagakos and Mosteller (1981). During the middle and late 1970s, experiments were conducted to determine possible carcinogenic effects of a widely used food coloring, FD&C Red No. 40. One of the experiments involved 500 male and 500 female mice. Both genders were divided into five groups: two control groups, a low-dose group, a medium-dose group, and a high-dose group. The mice were bred in the following way: Males and females were paired and before and during mating were given their prescribed dose of Red No. 40. The regime was continued during gestation and weaning of the young. From litters that had at least three pups of each sex, three of each sex were selected randomly and continued on their parents' dosage throughout their lives. After 109–111 weeks, all the mice still living were killed. The presence or absence of reticuloendothelial tumors was of particular interest. Although there were significant differences between some of the treatment groups, the results were rather confusing. For example, there was a significant difference between the incidence rates for the two male control groups, and among the males the medium-dose group had the lowest incidence.

Several experts were asked to examine the results of this and other experiments. Among them were Lagakos and Mosteller, who requested information on how the cages that housed the mice were arranged. There were three racks of cages, each containing five rows of seven cages in the front and five rows of seven cages in the back. Five mice were housed in each cage. The mice were assigned to the cages in a systematic way: The first male control group was in the top of the front of rack 1; the first female control group was in the bottom of the front of rack 1; and so on, ending with the high-dose females in the bottom of the back of rack 3 (Figure 11-9). Lagakos and Mosteller showed that there were effects due to cage position that could not be explained by gender or by dosage group. A random assignment of cage positions would have eliminated this confounding. Lagakos and Mosteller also suggested some experimental designs to systematically control for cage position.

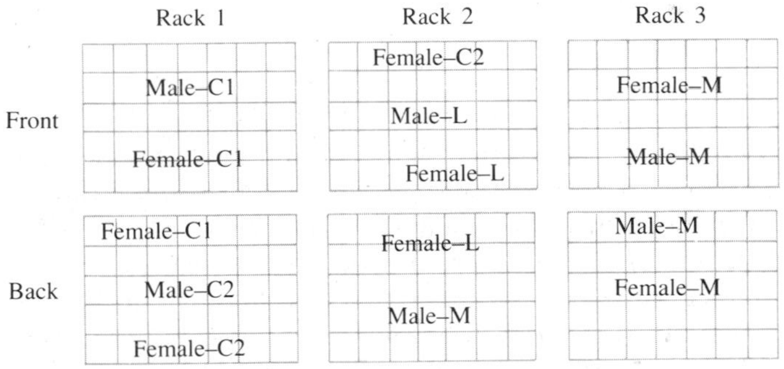

Figure 11-9. Location of mice cages in racks.

It was also possible that a litter effect might be complicating the analysis, since littermates received the same treatment and littermates of the same sex were housed in the same or contiguous cages. In the presence of a litter effect, mice from the same litter might show less variability than that present among mice from different litters. This reduces the effective sample size—in the extreme case in which littermates react identically, the effective sample size is the number of litters, not the total number of mice. One way around this problem would have been to use only one mouse from each litter.

The presence of a possible selection bias is another problem. Since mice were only included in the experiment if they came from a litter with at least three males and three females, offspring of possibly less healthy parents were excluded. This could be a serious problem since exposure to Red No. 40 might affect the parents' health and the birth process. If, for example, among the high-dose mice, only the most hardy produced large enough litters, their offspring might be hardier than the controls' offspring.

11.4.6   Further Remarks on Randomization

As well as guarding against possible biases on the part of the experimenter, the process of randomization tends to balance any factors that may be influential but are not explicitly controlled in the experiment. Time is often such a factor; background variables such as temperature, equipment calibration, line voltage, and chemical composition can change slowly with time. In experiments that are run over some period of time, therefore, it is important to randomize the assignments to treatment and control over time. Time is not the only factor that should be randomized, however. In agricultural experiments, the positions of test plots in a field are often randomly assigned. In biological experiments with test animals, the locations of the animals' cages may have an effect, as illustrated in the preceding section.

Generally, if it is anticipated that a variable will have a significant effect, that variable should be included as one of the controlled factors in the experimental design. The matched-pairs design of this chapter can be used to control for a single factor. To control for more than one factor, factorial designs, which are briefly introduced in the next chapter, may be used.

11.4.7   Observational Studies, Confounding, and Bias in Graduate Admissions

It is not always possible to conduct controlled experiments or use randomization. In evaluating some medical therapies, for example, a randomized, controlled experiment would be unethical if one therapy was strongly believed to be superior. For many problems of psychological interest (effects of parental modes of discipline, for example), it is impossible to conduct controlled experiments. In such situations, recourse is often made to observational studies. Hospital records may be examined to compare the outcomes of different therapies, or psychological records of children raised in different ways may be analyzed. Although such studies may be valuable, the results are seldom unequivocal. Since there is no randomization, it is always possible that the groups under comparison differ in respects other than their "treatments."

As an example, let us consider a study of gender bias in admissions to graduate school at the University of California at Berkeley (Bickel and O'Connell, 1975). In the fall of 1973, 8442 men applied for admission to graduate studies at Berkeley, and 44% were admitted; 4321 women applied, and 35% were admitted. If the men and women were similar in every respect other than sex, this would be strong evidence of sex bias. This was not a controlled, randomized experiment, however; sex was not randomly assigned to the applicants. As will be seen, the male and female applicants differed in other respects, which influenced admission.

The table below shows admission rates for the six most popular majors on the Berkeley campus.

Major	Men		Women	
	Number of Applicants	Percentage Admitted	Number of Applicants	Percentage Admitted
A	825	62	108	82
B	560	63	25	68
C	325	37	593	34
D	417	33	375	35
E	191	28	393	34
F	373	6	341	7

If the percentages admitted are compared, women do not seem to be unfavorably treated. But when the combined admission rates for all six majors are calculated, it is found that 44% of the men and only 30% of the women were admitted, which seems paradoxical. The resolution of the paradox lies in the observation that the women tended to apply to majors that had low admission rates (C through F) and the men to majors that had relatively high admission rates (A and B). This factor was not controlled for, since the study was observational in nature; it was also "confounded" with the factor of interest, sex; randomization, had it been possible, would have tended to balance out the confounded factor.

Confounding also plays an important role in studies of the effect of coffee drinking. Several studies have claimed to show a significant association of coffee consumption with coronary disease. Clearly, randomized, controlled trials are not possible here—a randomly selected individual cannot be told that he or she is in the treatment group and must drink 10 cups of coffee a day for the next five years. Also, it is known that heavy coffee drinkers also tend to smoke more than average, so smoking is confounded with coffee drinking. Hennekens et al. (1976) review several studies in this area.

11.4.8   Fishing Expeditions

Another problem that sometimes flaws observational studies, and controlled experiments as well, is that they engage in "fishing expeditions." For example, consider a hypothetical study of the effects of birth control pills. In such a case, it would be impossible to assign women to a treatment or a placebo at random, but a nonrandomized study might be conducted by carefully matching controls to treatments on such factors as age and medical history. The two groups might be followed up on for some time, with many variables being recorded for each subject such as blood pressure, psychological measures, and incidences of various medical problems. After termination of the study, the two groups might be compared on each of these variables, and it might be found, say, that there was a "significant difference" in the incidence of melanoma. The problem with this study is the following. Suppose that 100 independent two sample t tests are conducted at the .05 level and that, in fact, all the null hypotheses are true. We would expect that five of the tests would produce a "significant" result. Although each of the tests has probability .05 of type I error, as a collection they do not simultaneously have $\alpha = .05$. The combined significance level is the probability that at least one of the null hypotheses is rejected:

$$\alpha = P\{\text{at least one } H_0 \text{ rejected}\}$$

$$= 1 - P\{\text{no } H_0 \text{ rejected}\}$$

$$= 1 - .95^{100}$$

$$= .994$$

Thus, with very high probability, at least one "significant" result will be found, even if all the null hypotheses are true.

There are no simple cures for this problem. One possibility is to regard the results of a fishing expedition as merely providing suggestions for further experiments. Alternatively, and in the same spirit, the data could be split randomly into two halves, one half for fishing in and the other half to be locked safely away, unexamined. "Significant" results from the first half could then be tested on the second half. A third alternative is to conduct each individual hypothesis test at a small significance level. To see how this works, suppose that all null hypotheses are true and that each of n null hypotheses is tested at level α. Let R_i denote the event that the ith null hypothesis is rejected, and let α^* denote the overall probability of a type I error. Then

$$\alpha^* = P\{R_1 \text{ or } R_2 \text{ or } \cdots \text{ or } R_n\}$$

$$\leq P\{R_1\} + P\{R_2\} + \cdots + P\{R_n\}$$

$$= n\alpha$$

Thus, if each of the n null hypotheses is tested at level α/n, the overall significance level is less than or equal to α. This is often called the **Bonferroni method**.

11.5   Concluding Remarks

This chapter has been concerned with the problem of comparing two samples. Within this context, the fundamental statistical concepts of estimation and hypothesis testing, which were introduced in earlier chapters, have been extended and utilized. The chapter also showed how informal descriptive and analytic techniques are used in supplementing more formal analysis of data. Chapter 12 will extend the techniques of this chapter to deal with multisample problems.

Chapter 13 is concerned with similar problems that arise in the analysis of qualitative data.

We considered two types of experiments, those with two independent samples and those with matched pairs. For the case of independent samples, we developed the t test, based on an assumption of normality, as well as a modification of the t test that takes into account possibly unequal variances. The Mann–Whitney test, based on ranks, was presented as a nonparametric method, that is, a method that is not based on an assumption of a particular distribution. Similarly, for the matched-pairs design, we developed a parametric t test and a nonparametric test, the signed rank test.

We have discussed methods based on an assumption of normality and rank methods, which do not make this assumption. It turns out, rather surprisingly, that even if the normality assumption holds, the rank methods are quite powerful relative to the t test. Lehmann (1975) shows that the efficiency of the rank tests relative to that of the t test—that is, the ratio of sample sizes required to attain the same power—is typically around .95 if the distributions are normal. Thus, a rank test using a sample of size 100 is as powerful as a t test based on 95 observations. Collecting the extra 5 pieces of data is a small price to pay for a safeguard against nonnormality.

The chapter concluded with a discussion of experimental design, which emphasized the importance of incorporating controls and randomization in investigations. Possible problems associated with observational studies were discussed. Finally, the difficulties encountered in making many comparisons from a single data set were pointed out; such problems of multiplicity will come up again in chapter 12.

11.6   Problems

1. Referring to the data in Section 11.2.2, compare iron retention at concentrations of 10.2 and .3 millimolar using graphical procedures and parametric and nonparametric tests. Write a brief summary of your conclusions.
2. Verify that the two-sample t test at level α of $H_0: \mu_X = \mu_Y$ versus $H_A: \mu_X \neq \mu_Y$ rejects if and only if the confidence interval for $\mu_X - \mu_Y$ does not contain zero.
3. Explain how to modify the t test of Section 11.2.1 to test $H_0: \mu_X = \mu_Y + \Delta$ versus $H_A: \mu_X \neq \mu_Y + \Delta$.
4. An equivalence between hypothesis tests and confidence intervals was demonstrated in chapter 9. In chapter 10, a nonparametric confidence interval for the median, η, was derived. Explain how to use this confidence interval to test the hypothesis $H_0: \eta = \eta_0$. In the case where $\eta_0 = 0$, show that using this approach on a sample of differences from a paired experiment is equivalent to the **sign test**. The sign test counts the number of positive differences and uses the fact that in the case that the null hypothesis is true, the distribution of the number of positive differences is binomial with $(n, .5)$. Apply the sign test to the data from the measurement of mercury levels, listed in Section 11.3.3.

5. Let $X_1, \ldots, X_{25}$ be i.i.d. $N(.3, 1)$. Consider testing the null hypothesis $H_0: \mu = 0$ versus $H_A: \mu > 0$ at significance level $\alpha = .05$. Compare the power of the sign test and the power of the test based on normal theory assuming that σ is known.

6. Suppose that $X_1, \ldots, X_n$ are i.i.d. $N(\mu, \sigma^2)$. To test the null hypothesis $H_0: \mu = \mu_0$, the t test is often used:

$$ t = \frac{\bar{X} - \mu_0}{s_{\bar{X}}} $$

Under H_0, t follows a t distribution with $n - 1$ df. Show that the likelihood ratio test of this H_0 is equivalent to the t test.

7. Suppose that n measurements are to be taken under a treatment condition and another n measurements are to be taken independently under a control condition. It is thought that the standard deviation of a single observation is about 10 under both conditions. How large should n be so that a 95% confidence interval for $\mu_X - \mu_Y$ has a width of 2? Use the normal distribution rather than the t distribution, since n will turn out to be rather large.

8. Referring to Problem 7, how large should n be so that the test of $H_0: \mu_X = \mu_Y$ against the one-sided alternative $H_1: \mu_X > \mu_Y$ has a power of .5 if $\mu_X - \mu_Y = 2$ and $\alpha = .10$?

9. Consider conducting a two-sided test of the null hypothesis $H_0: \mu_X = \mu_Y$ as described in Problem 8. Sketch power curves for (a) $\alpha = .05, n = 20$; (b) $\alpha = .10, n = 20$; (c) $\alpha = .05, n = 40$; (d) $\alpha = .10, n = 40$. Compare the curves.

10. Two independent samples are to be compared to see if there is a difference in their means. If a total of m subjects are available for the experiment, how should this total be allocated between the two samples in order to (a) provide the shortest confidence interval for $\mu_X - \mu_Y$ and (b) make the test of $H_0: \mu_X = \mu_Y$ as powerful as possible? Assume that the observations in the two samples are normally distributed with the same variance.

11. A study was done to compare the performances of engine bearings made of different compounds (McCool, 1979). Ten bearings of each type were tested. The following table gives the times until failure (in units of millions of cycles):

Type I	Type II
3.03	3.19
5.53	4.26
5.60	4.47
9.30	4.53
9.92	4.67
12.51	4.69
12.95	12.78
15.21	6.79
16.04	9.37
16.84	12.75

(a) Use normal theory to test the hypothesis that there is no difference between the two types of bearings.

(b) Test the same hypothesis using a nonparametric method.

(c) Which of the methods—that of part (a) or that of part (b)—do you think is better in this case?

12. An experiment was done to compare two methods of measuring the calcium content of animal feeds. The standard method uses calcium oxalate precipitation followed by titration and is quite time-consuming. A new method using flame photometry is faster. Measurements of the percent calcium content made by each method on 118 routine feed samples (Heckman, 1960) are given in the following table. Analyze the data to see if there is any systematic difference between the two methods. Use both parametric and nonparametric tests and graphical methods.

Oxalate	Flame	Oxalate	Flame	Oxalate	Flame	Oxalate	Flame
4.20	4.15	5.70	5.65	1.16	1.15	4.70	4.65
1.31	1.30	2.73	2.65	1.44	1.48	5.80	5.70
.83	.68	.27	.27	6.62	6.60	7.09	6.80
1.24	1.22	1.20	1.18	2.72	2.60	3.12	3.15
4.70	4.60	2.18	2.16	3.75	3.70	2.34	2.35
4.43	4.18	2.66	2.60	2.54	2.50	4.17	4.08
2.08	2.05	1.94	1.90	2.11	2.12	1.11	1.13
2.45	2.45	3.04	2.95	2.98	3.00	1.87	1.85
4.77	4.65	3.62	3.65	1.73	1.65	4.62	4.35
3.46	3.75	5.35	5.30	2.34	2.35	2.91	2.95
1.55	1.56	.87	.89	1.62	1.77	1.78	1.83
1.80	1.92	1.90	1.90	1.90	1.83	1.75	1.80
1.81	1.77	1.78	1.88	1.23	1.23	1.23	1.18
1.16	1.20	1.26	1.23	1.20	1.20	1.86	1.80
2.38	2.35	5.55	5.30	.68	.70	.80	.74
3.44	3.25	1.41	1.20	1.86	1.80	1.27	1.23
2.21	2.15	2.69	2.65	6.16	5.90	4.05	3.90
1.23	1.18	3.55	3.35	2.42	2.40	1.26	1.20
3.46	3.25	1.91	1.87	1.93	1.87	5.15	4.90
2.12	2.00	2.84	2.85	2.28	2.20	4.72	5.10
2.94	2.85	1.31	1.33	3.39	3.35	2.56	2.55
1.39	1.45	1.39	1.42	1.28	1.30	1.30	1.30
1.24	1.25	1.25	1.20	1.21	1.22	1.28	1.30
1.20	1.20	1.35	1.35	2.13	2.10	1.66	1.67
1.86	1.72	.86	.86	.89	.82	1.02	.95
.88	.89	1.90	1.83	2.08	2.07	1.84	1.78
1.22	1.20	1.26	1.28	1.02	1.00	1.23	1.20
5.16	5.20	1.52	1.50	2.36	2.25	3.04	3.06
4.68	4.60	3.00	2.96	3.03	3.05	1.23	1.22
1.07	1.06	.35	.36				

13. Let $X_1, \ldots, X_n$ be i.i.d. with cdf F, and let $Y_1, \ldots, Y_m$ be i.i.d. with cdf G. The hypothesis to be tested is that $F = G$. Suppose for simplicity that $m + n$ is even so that in the combined sample of X's and Y's, $(m + n)/2$ observations are less than the median and $(m + n)/2$ are greater.

(a) As a test statistic, consider T, the number of X's less than the median of the combined sample. Show that T follows a hypergeometric distribution under the null hypothesis:

$$P(T = t) = \frac{\binom{(m+n)/2}{t}\binom{(m+n)/2}{n-t}}{\binom{m+n}{n}}$$

Explain how to form a rejection region for this test.

(b) Show how to find a confidence interval for the difference between the median of F and the median of G under the shift model, $G(x) = F(x - \Delta)$. (*Hint:* Use the order statistics.)

(c) Apply the results (a) and (b) to the data of Problem 11.

14. Find the exact null distribution of the Mann–Whitney statistic, U_Y, in the case where $m = 3$ and $n = 2$.

15. Referring to Example A in Section 11.2.1, (a) if the smallest observation for method B (79.94) is made arbitrarily small, will the t test still reject? (b) If the largest observation for method B (80.03) is made arbitrarily large, will the t test still reject? (c) Answer the same questions for the Mann–Whitney test.

16. Let $X_1, \ldots, X_n$ be a sample from an $N(0, 1)$ distribution and let $Y_1, \ldots, Y_n$ be an independent sample from an $N(1, 1)$ distribution.
    (a) Determine the expected rank sum of the X's.
    (b) Determine the variance of the rank sum of the X's.

17. Find the exact null distribution of W_+ in the case where $n = 4$.

18. For $n = 10, 20$, and 30, find the .05 and .01 critical values for a two-sided signed rank test from the tables and then by using the normal approximation. Compare the values.

19. To compare two variances in the normal case, let $X_1, \ldots, X_n$ be i.i.d. with $N(\mu_X, \sigma_X^2)$, and let $Y_1, \ldots, Y_m$ be i.i.d. with $N(\mu_Y, \sigma_Y^2)$, where the X's and Y's are independent samples. Argue that under $H_0: \sigma_X = \sigma_Y$,

$$\frac{s_X^2}{s_Y^2} \sim F_{n-1, m-1}$$

    (a) Construct rejection regions for one- and two-sided tests of H_0.
    (b) Construct a confidence interval for the ratio σ_X^2/σ_Y^2.
    (c) Apply the results of parts (a) and (b) to Example A in Section 11.2.1. (*Caution:* This test and confidence interval are not robust against violations of the assumption of normality.)

20. This problem contrasts the power functions of paired and unpaired designs. Graph and compare the power curves for testing $H_0: \mu_X = \mu_Y$ for the following two designs.
    (a) Paired: $\text{Cov}(X_i, Y_i) = 50, \sigma_X = \sigma_Y = 10, i = 1, \ldots, 25$.
    (b) Unpaired: $X_1, \ldots, X_{25}$ and $Y_1, \ldots, Y_{25}$ are independent with variance as in part (a).

21. An experiment was done to measure the effects of ozone, a component of smog. A group of 22 70-day-old rats were kept in an environment containing ozone for seven days, and their weight gains were recorded. Another group of 23 rats of a similar age were kept in an ozone-free environment for a similar

time, and their weight gains were recorded. The data (in grams) are given below. Analyze the data to determine the effect of ozone. Write a summary of your conclusions. (This problem is from Doksum and Sievers, 1976, who provide an interesting analysis.)

Controls			Ozone		
41.0	38.4	24.9	10.1	6.1	20.4
25.9	21.9	18.3	7.3	14.3	15.5
13.1	27.3	28.5	−9.9	6.8	28.2
−16.9	17.4	21.8	17.9	−12.9	14.0
15.4	27.4	19.2	6.6	12.1	15.7
22.4	17.7	26.0	39.9	−15.9	54.6
29.4	21.4	22.7	−14.7	44.1	−9.0
26.0	26.6		−9.0		

22. Lin, Sutton, and Qurashi (1979) compared microbiological and hydroxylamine methods for the analysis of ampicillin dosages. In one series of experiments, pairs of tablets were analyzed by the two methods. The data in the following table give the percentages of claimed amount of ampicillin found by the two methods in several pairs of tablets. Analyze the data to determine if there is a systematic difference between the two methods.

Microbiological Method	Hydroxylamine Method
97.2	97.2
105.8	97.8
99.5	96.2
100.0	101.8
93.8	88.0
79.2	74.0
72.0	75.0
72.0	67.5
69.5	65.8
20.5	21.2
95.2	94.8
90.8	95.8
96.2	98.0
96.2	99.0
91.0	100.2

23. Stanley and Walton (1961) ran a controlled clinical trial to investigate the effect of the drug stelazine on chronic schizophrenics. The trials were conducted on chronic schizophrenics in two closed wards. In each of the wards, the patients were divided into two groups matched for age, length of time in the hospital, and score on a behavior rating sheet. One member of each pair was given stelazine, and the other a placebo. Only the hospital pharmacist knew which member of each pair received the actual drug. The table below gives the behavioral rating scores for the patients at the beginning of the trial and after three months. High scores are good.

Ward A

Stelazine		*Placebo*	
Before	*After*	*Before*	*After*
2.3	3.1	2.4	2.0
2.0	2.1	2.2	2.6
1.9	2.45	2.1	2.0
3.1	3.7	2.9	2.0
2.2	2.54	2.2	2.4
2.3	3.72	2.4	3.18
2.8	4.54	2.7	3.0
1.9	1.61	1.9	2.54
1.1	1.63	1.3	1.72

Ward B

Stelazine		*Placebo*	
Before	*After*	*Before*	*After*
1.9	1.45	1.9	1.91
2.3	2.45	2.4	2.54
2.0	1.81	2.0	1.45
1.6	1.72	1.5	1.45
1.6	1.63	1.5	1.54
2.6	2.45	2.7	1.54
1.7	2.18	1.7	1.54

(a) For each of the wards, test whether stelazine is associated with improvement in the patients' scores.

(b) Test if there is any difference in improvement between the wards. (These data are also presented in Lehmann, 1975, who discusses methods of combining the data from the wards.)

24. Bailey, Cox, and Springer (1978) used high-pressure liquid chromatography to measure the amounts of various intermediates and by-products in food dyes. The following table gives the percentages added and found for two substances in the dye FD&C Yellow No. 5. Is there any evidence that the amounts found differ systematically from the amounts added?

Sulfanilic Acid		*Pyrazolone-T*	
Percentage Added	*Percentage Found*	*Percentage Added*	*Percentage Found*
.048	.060	.035	.031
.096	.091	.087	.084
.20	.16	.19	.16
.19	.16	.19	.17
.096	.091	.16	.15
.18	.19	.032	.040
.080	.070	.060	.076
.24	.23	.13	.11
0	0	.080	.082
.040	.042	0	0
.060	.056		

25. This and the next two problems are based on discussions and data in Le Cam and Neyman (1967), which is devoted to the analysis of weather modification experiments. The examples illustrate some ways in which principles of experimental design have been used in this field. During the summers of 1957 through 1960, a series of randomized cloud-seeding experiments were carried out in the mountains of Arizona. Of each pair of successive days, one day was randomly selected for seeding to be done. The seeding was done during a two-hour to four-hour period starting at midday, and rainfall during the afternoon was measured by a network of 29 guages. The data for the four years are given in the table below (in inches). Observations in this table are listed in chronological order.

(a) Analyze the data for each year and for the years pooled together to see if there appears to be any effect due to seeding. You should use graphical descriptive methods to get a qualitative impression of the results and hypothesis tests to assess the significance of the results.

(b) Why should the day on which seeding is to be done be chosen at random rather than just alternating seeded and unseeded days? Why should the days be paired at all, rather than just deciding randomly which days to seed?

1957		1958		1959		1960	
Seeded	Unseeded	Seeded	Unseeded	Seeded	Unseeded	Seeded	Unseeded
0	.154	.152	.013	.015	0	0	.010
.154	0	0	0	0	0	0	0
.003	.008	0	.445	0	.086	.042	.057
.084	.033	.002	0	.021	.006	0	0
.002	.035	.007	.079	0	.115	0	.093
.157	.007	.013	.006	.004	.090	0	.183
.010	.140	.161	.008	.010	0	.152	0
0	.022	0	.001	0	0	0	0
.002	0	.274	.001	.055	0	0	0
.078	.074	.001	.025	.004	.076	0	0
.101	.002	.122	.046	.053	.090	0	0
.169	.318	.101	.007	0	0	0	0
.139	.096	.012	.019	0	.078	.008	0
.172	0	.002	0	.090	.121	.040	.060
0	0	.066	0	.028	1.027	.003	.102
0	.050	.040	.012	0	.104	.011	.041
				.032	.023		
				.133	.172		
				.083	.002		
				0	0		

26. The National Weather Bureau's ACN Cloud-Seeding Project was carried out in the states of Oregon and Washington. Cloud seeding was accomplished by dispersing dry ice from an aircraft; only clouds that were deemed "ripe" for seeding were candidates for seeding. On each occasion, a decision was made at random whether to seed, the probability of seeding being $\frac{2}{3}$. This

resulted in 22 seeded and 13 control cases. Three types of targets were considered, two of which are dealt with in this problem. Type I targets were large geographical areas downwind from the seeding; type II targets were sections of type I targets located so as to have, theoretically, the greatest sensitivity to cloud seeding. The following table gives the average target rainfalls (in inches) for the seeded and control cases, listed in chronological order. Is there evidence that seeding had an effect on either type of target? In what ways is the design of this experiment different from that of the one in Problem 25?

Control Cases		Seeded Cases	
Type I	Type II	Type I	Type II
.0080	.0000	.1218	.0200
.0046	.0000	.0403	.0163
.0549	.0053	.1166	.1560
.1313	.0920	.2375	.2885
.0587	.0220	.1256	.1483
.1723	.1133	.1400	.1019
.3812	.2880	.2439	.1867
.1720	.0000	.0072	.0233
.1182	.1058	.0707	.1067
.1383	.2050	.1036	.1011
.0106	.0100	.1632	.2407
.2126	.2450	.0788	.0666
.1435	.1529	.0365	.0133
		.2409	.2897
		.0408	.0425
		.2204	.2191
		.1847	.0789
		.3332	.3570
		.0676	.0760
		.1097	.0913
		.0952	.0400
		.2095	.1467

27. During 1963 and 1964, an experiment was carried out in France; its design differed somewhat from those of the previous two problems. A 1500-km target area was selected, and an adjacent area of about the same size was designated as the control area; 33 ground generators were used to produce silver iodide to seed the target area. Precipitation was measured by a network of guages for each suitable "rainy period," which was defined as a sequence of periods of continuous precipitation between dry spells of a specified length. When a forecaster determined that the situation was favorable for seeding, he telephoned an order to a service agent, who then opened a sealed envelope that contained an order to actually seed or not. The envelopes had been prepared in advance, using a table of random numbers. The following table gives precipitation (in inches) in the target and control areas for the seeded and unseeded periods.

(a) Analyze the data, which are listed in chronological order, to see if there is an effect of seeding.

(b) The analysis done by the French investigators used the square root transformation in order to make normal theory more applicable. Do you think that taking the square root was an effective transformation for this purpose?

(c) Reflect on the nature of this design. In particular, what advantage is there to using the control area? Why not just compare seeded and unseeded periods on the target area?

Seeded		Unseeded	
Target	Control	Target	Control
1.6	1.0	1.1	2.2
28.1	27.0	3.5	5.2
7.8	.3	2.6	0.0
4.0	6.0	2.6	2.0
9.6	12.6	9.8	4.9
0.2	0.5	5.6	8.5
18.7	8.7	.1	3.5
16.5	21.5	0.0	1.1
4.6	13.9	17.7	11.0
9.3	6.7	19.4	19.8
3.5	4.5	8.9	5.3
0.1	0.7	10.6	8.9
11.5	8.7	10.2	4.5
0.0	0.0	16.0	13.0
9.3	10.7	9.7	21.1
5.5	4.7	21.4	15.9
70.2	29.1	6.1	19.5
0.7	1.9	24.3	16.3
38.6	34.7	20.9	6.3
11.3	10.2	60.2	47.0
3.3	2.7	15.2	10.8
8.9	2.8	2.7	4.8
11.1	4.3	0.3	0.0
64.3	38.7	12.2	5.7
16.6	11.1	2.2	5.1
7.3	6.5	23.3	30.6
3.2	3.0	9.9	3.7
23.9	13.6		
0.6	0.1		

28. The media often present short reports of the results of experiments. To the critical reader or listener, such reports often raise more questions than they answer. Comment on possible pitfalls in the interpretation of each of the following.

(a) It is reported that patients whose hospital rooms have a window recover faster than those whose rooms do not.

(b) Nonsmoking wives whose husbands smoke have a cancer rate twice that of wives whose husbands do not smoke.

(c) A two-year study in North Carolina found that 75% of all industrial accidents in the state happened to workers who had skipped breakfast.

(d) A school integration program involved busing children from minority schools to majority (primarily white) schools. Participation in the program was voluntary. It was found that the students who were bused scored lower on standardized tests than did their peers who chose not to be bused.

(e) When a group of students were asked to match pictures of newborns with pictures of their mothers, they were correct 36% of the time.

(f) A survey found that those who drank a moderate amount of beer were healthier than those who totally abstained from alcohol.

29. Explain why in Levine's experiment (Example A in Section 11.3.1) subjects also smoked cigarettes made of lettuce leaves and unlit cigarettes.

30. This example is taken from an interesting article by Joiner (1981) and from data in Ryan, Joiner, and Ryan (1976). The National Bureau of Standards supplies standard materials of many varieties to manufacturers and other parties, who use these materials to calibrate their own testing equipment. Great pains are taken to make these reference materials as homogeneous as possible. In an experiment, a long homogeneous steel rod was cut into 4-inch lengths, 20 of which were randomly selected and tested for oxygen content. Two measurements were made on each piece. The 40 measurements were made over a period of five days, with eight measurements per day. In order to avoid possible bias from time-related trends, the sequence of measurements was randomized. The following table gives the measurements. There is an unexpected systematic source of variability in these data. Can you find it by making an appropriate plot? Would this effect have been detectable if the measurements had not been randomized over time?

Day	Piece Number	Oxygen (ppm)
1	17	5.6
	11	5.9
	10	6.8
	15	7.5
	5	4.7
	6	4.0
	19	4.4
2	4	6.6
	7	4.9
	1	5.5
	13	4.9
	3	6.3
	18	4.2
	9	3.3
	14	4.8

Day	Piece Number	Oxygen (ppm)
3	16	6.1
	12	5.3
	8	5.2
	2	4.3
	11	4.0
	19	6.2
	17	5.1
	9	3.3
4	13	5.4
	18	6.1
	2	5.7
	10	6.1
	4	5.7
	20	4.6
	7	5.7
	6	4.4
	12	4.1
5	16	6.3
	15	7.5
	8	6.1
	14	6.4
	3	5.1
	20	5.7
	5	3.8
	1	3.8

12

The Analysis of Variance

12.1 Introduction

Chapter 11 was concerned with the analysis of data arising from experimental designs with two samples. Experiments frequently involve more than two samples; they may compare several treatments, such as different drugs, and perhaps other factors, such as sex, at the same time. This chapter is an introduction to the statistical analysis of such experiments. The methods we will discuss are called "analysis of variance." Contrary to what this phrase seems to say, we will be primarily concerned with the comparison of the means of the data, not their variances. We will consider the two most elementary multisample designs, the one-way and two-way layouts. Methods based on the normal distribution and nonparametric methods will be developed.

12.2 The One-Way Layout

A **one-way layout** is an experimental design in which independent measurements are made under each of several treatments. The techniques we will introduce are

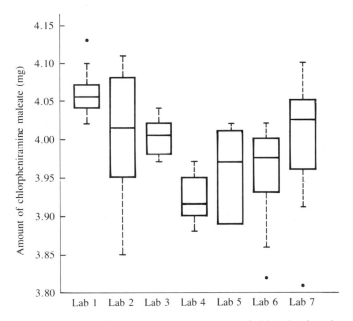

Figure 12-1. Boxplots of determinations of amounts of chlorpheniramine maleate in tablets by seven laboratories.

thus generalizations of the techniques for comparing two independent samples that were covered in chapter 11.

Throughout this section, examples will use data from Kirchhoefer (1979), a study of the use of a semiautomated method for measuring the amount of chlorpheniramine maleate in tablets. For each of four manufacturers, composites were prepared by grinding and mixing together tablets that had nominal dosage levels of 4 milligrams. Seven labs were asked to make 10 determinations on each composite; each determination was made on a portion of composite whose weight was equivalent to that of one tablet. The purpose of the experiment was to study the consistency between labs and the variability of the measurement process. The data for one manufacturer are shown in the following table:

Lab 1	Lab 2	Lab 3	Lab 4	Lab 5	Lab 6	Lab 7
4.13	3.86	4.00	3.88	4.02	4.02	4.00
4.07	3.85	4.02	3.88	3.95	3.86	4.02
4.04	4.08	4.01	3.91	4.02	3.96	4.03
4.07	4.11	4.01	3.95	3.89	3.97	4.04
4.05	4.08	4.04	3.92	3.91	4.00	4.10
4.04	4.01	3.99	3.97	4.01	3.82	3.81
4.02	4.02	4.03	3.92	3.89	3.98	3.91
4.06	4.04	3.97	3.90	3.89	3.99	3.96
4.10	3.97	3.98	3.97	3.99	4.02	4.05
4.04	3.95	3.98	3.90	4.00	3.93	4.06

Figure 12-1, a boxplot of these data, shows some variation in the medians among the seven labs, as well as some variation in the interquartile ranges. It appears from the figure that there may be some systematic differences between the labs and that there is less variability in some labs than in others. We will discuss the following question: Are the differences in the means of the measurements from the various labs significant, or might they be due to chance?

12.2.1 Normal Theory; the F Test

We first discuss the analysis of variance and the F test in the case of I samples, each of the same size, J. The I groups will be referred to generically as "treatments" or "levels." (We will discuss the case of unequal sample sizes later.)

We must define some notation and introduce the basic model. Let

$$Y_{ij} = \text{the } j\text{th observation of the } i\text{th treatment}$$

We will assume that the observations are corrupted by random errors and that the error in one observation is independent of the errors in the other observations. The errors will be assumed to have mean 0 and standard deviation σ. The statistical model is

$$Y_{ij} = \mu + \alpha_i + \varepsilon_{ij}$$

Here μ is the overall mean level, α_i is the differential effect of the ith treatment, and ε_{ij} is the random error in the jth observation under the ith treatment. The α_i are normalized:

$$\sum_{i=1}^{I} \alpha_i = 0$$

Thus, if $\alpha_i = 0$, for $i = 1, \ldots, I$, all treatments have the same expected response, and, in general, $\alpha_i - \alpha_j$ is the difference between the mean values of treatments i and j. We will derive a test for the null hypothesis, which is that all the means are equal.

The analysis of variance is based on the following identity:

$$\sum_{i=1}^{I} \sum_{j=1}^{J} (Y_{ij} - \bar{Y}_{..})^2 = \sum_{i=1}^{I} \sum_{j=1}^{J} (Y_{ij} - \bar{Y}_{i.})^2 + J \sum_{i=1}^{I} (\bar{Y}_{i.} - \bar{Y}_{..})^2$$

where

$$\bar{Y}_{i.} = \frac{1}{J} \sum_{j=1}^{J} Y_{ij}$$

is the average of the observations under the ith treatment and

$$\bar{Y}_{..} = \frac{1}{IJ} \sum_{i=1}^{I} \sum_{j=1}^{J} Y_{ij}$$

is the overall average. The terms appearing in the first identity above are called "sums of squares," and the identity may be symbolically expressed as

$$SS_{TOT} = SS_W + SS_B$$

In words, the total sum of squares equals the sum of squares within groups plus the sum of squares between groups. The terminology reflects that SS_W is a measure of the variation of the data within the treatment groups and that SS_B is a measure of the variation of the treatment means among treatments.

To establish the identity, we express the left-hand side as

$$\sum_{i=1}^{I} \sum_{j=1}^{J} (Y_{ij} - \bar{Y}_{..})^2 = \sum_{i=1}^{I} \sum_{j=1}^{J} [(Y_{ij} - \bar{Y}_{i.}) + (\bar{Y}_{i.} - \bar{Y}_{..})]^2$$

$$= \sum_{i=1}^{I} \sum_{j=1}^{J} (Y_{ij} - \bar{Y}_{i.})^2 + \sum_{i=1}^{I} \sum_{j=1}^{J} (\bar{Y}_{i.} - \bar{Y}_{..})^2$$

$$+ 2 \sum_{i=1}^{I} \sum_{j=1}^{J} (Y_{ij} - \bar{Y}_{i.})(\bar{Y}_{i.} - \bar{Y}_{..})$$

$$= \sum_{i=1}^{I} \sum_{j=1}^{J} (Y_{ij} - \bar{Y}_{i.})^2 + \sum_{i=1}^{I} \sum_{j=1}^{J} (\bar{Y}_{i.} - \bar{Y}_{..})^2$$

$$+ 2 \sum_{i=1}^{I} (\bar{Y}_{i.} - \bar{Y}_{..}) \sum_{j=1}^{J} (Y_{ij} - \bar{Y}_{i.})$$

The last term of the final expression vanishes since the sum of deviations from a mean is zero.

The basic idea underlying analysis of variance is the comparison of the sizes of various sums of squares. We can calculate the expected values of the sums of squares discussed above using the following lemma.

LEMMA A. Let X_i, where $i = 1, \ldots, n$, be independent random variables with $E(X_i) = \mu_i$ and $Var(X_i) = \sigma^2$. Then

$$E(X_i - \bar{X})^2 = (\mu_i - \bar{\mu})^2 + \frac{n-1}{n} \sigma^2$$

where

$$\bar{\mu} = \frac{1}{n} \sum_{i=1}^{n} \mu_i$$

Proof.   We use the fact that $E(U^2) = [E(U)]^2 + \text{Var}(U)$ for any random variable U with finite variance. The first term on the right-hand side of the equation in the lemma follows immediately. For the second term, we have to calculate $\text{Var}(X_i - \bar{X})$:

$$\text{Var}(X_i - \bar{X}) = \text{Var}(X_i) + \text{Var}(\bar{X}) - 2\,\text{Cov}(X_i, \bar{X})$$

and

$$\text{Var}(X_i) = \sigma^2$$

$$\text{Var}(\bar{X}) = \frac{1}{n}\sigma^2$$

$$\text{Cov}(X_i, \bar{X}) = \text{Cov}\left(X_i, \frac{1}{n}\sum_{j=1}^{n} X_j\right) = \frac{1}{n}\sigma^2$$

Putting these results together proves the lemma.   □

Lemma A may be applied to the sums of squares discussed above, yielding the following theorem.

THEOREM A.   Under the assumptions for the model stated at the beginning of this section,

$$E(SS_W) = \sum_{i=1}^{I} \sum_{j=1}^{J} E(Y_{ij} - \bar{Y}_{i.})^2$$

$$= \sum_{i=1}^{I} \sum_{j=1}^{J} \frac{J-1}{J}\sigma^2$$

$$= I(J-1)\sigma^2$$

$$E(SS_B) = J \sum_{i=1}^{I} E(\bar{Y}_i - \bar{Y}_{..})^2$$

$$= J \sum_{i=1}^{I} \left[\alpha_i^2 + \frac{(I-1)\sigma^2}{IJ}\right]$$

$$= J \sum_{i=1}^{I} \alpha_i^2 + (I-1)\sigma^2$$

SS_W may be used to estimate σ^2; the estimate is

$$s_p^2 = \frac{SS_W}{I(J-1)}$$

which is unbiased. The subscript p stands for "pooled." Estimates of σ^2 from the I treatments are pooled together, since SS_W can be written as

$$SS_W = \sum_{i=1}^{I} (J - 1)s_i^2$$

where s_i^2 is the sample variance in the ith group.

If all the α_i are equal to zero, then the expectation of $SS_B/(I - 1)$ is also σ^2. Thus, in this case, $SS_W/I(J - 1)$ and $SS_B/(I - 1)$ should be about equal. If some of the α_i are nonzero, SS_B will be inflated. We next develop a method of comparing the two sums of squares to find a test statistic for testing the null hypothesis that all the α_i are equal. Under the assumption that the errors are normally distributed, the probability distributions of the sums of squares can be calculated.

THEOREM B.   If the errors are independent and normally distributed with means 0 and variances σ^2, then SS_W/σ^2 follows a chi-square distribution with $I(J - 1)$ degrees of freedom. If, additionally, the α_i are all equal to zero, then SS_B/σ^2 follows a chi-square distribution with $I - 1$ degrees of freedom and is independent of SS_W.

Proof.   We first consider SS_W. From Theorem B of Section 6.3,

$$\frac{1}{\sigma^2} \sum_{j=1}^{J} (Y_{ij} - \bar{Y}_{i.})^2$$

follows a chi-square distribution with $J - 1$ degrees of freedom. There are I such sums in SS_W, and they are independent of each other since the observations are independent. The sum of I independent chi-square random variables that each have $J - 1$ degrees of freedom follows a chi-square distribution with $I(J - 1)$ degrees of freedom. Theorem B of Section 6.3 can also be applied to SS_B, noting that $\text{Var}(\bar{Y}_{i.}) = \sigma^2/J$.

We next prove that the two sums of squares are independent of each other. SS_W is a function of the vector $\mathbf{U}$, which has elements $Y_{ij} - \bar{Y}_{i.}$, where $i = 1, \ldots,$ I and $j = 1, \ldots, J$. SS_B is a function of the vector $\mathbf{V}$, whose elements are $\bar{Y}_{i.}$, where $i = 1, \ldots, I$, since $\bar{Y}_{..}$ can be obtained from the $\bar{Y}_{i.}$. Thus, it is sufficient to show that these two vectors are independent of each other. First, if $i \neq i'$, $Y_{ij} - \bar{Y}_{i.}$ and $Y_{i'.}$ are independent since they are functions of different observations. Second, $Y_{ij} - \bar{Y}_{i.}$ and $\bar{Y}_{i.}$ are independent by Theorem A of Section 6.3. This completes the proof of the theorem.   □

The statistic

$$F = \frac{SS_B/(I - 1)}{SS_W/I(J - 1)}$$

is used to test the following null hypothesis:

$$H_0: \alpha_1 = \alpha_2 = \cdots = \alpha_I = 0$$

If the null hypothesis is true, the F statistic should be close to 1; if it is false, the statistic should tend to be greater than 1 because the expectation of the numerator is inflated. The decision rule is thus to reject H_0 for large values of F. As usual, to apply the test we must know the null distribution of the test statistic.

THEOREM C.   Under the assumption that the errors are normally distributed, the null distribution of F is the F distribution with $(I - 1)$ and $I(J - 1)$ degrees of freedom.

Proof.   The theorem follows from Theorem B and from the definition of the F distribution (Section 6.2), since, under H_0, F is the ratio of two independent chi-square random variables divided by their degrees of freedom.   □

Percentage points of the F distribution are widely tabled. It can shown that, under the normality assumption, the F test is equivalent to the likelihood ratio test.

EXAMPLE A.   We can illustrate the use of the F statistic by applying it to the tablet data from Section 12.2. The sums of squares defined above are calculated and presented in a table called the **analysis of variance table**:

Source	df	SS	MS	F
Labs	6	.125	.021	5.66
Error	63	.231	.0037	
Total	69	.356		

In the table, SS_W is the sum of squares due to error, and SS_B is the sum of squares due to labs. MS stands for "mean square" and equals the sum of squares divided by the degrees of freedom. The column headed F gives the F statistic for testing the null hypothesis that there is no systematic difference among the seven labs. The F statistic has 6 df and 63 df, and from tables of the F distribution, the p-value is less than .005. We may thus conclude that the means of the measurements from the various labs are significantly different.

Figure 12-2 is a normal probability plot of the residuals from the analysis of variance model (the residuals are formed by simply subtracting from the measurements of each lab the mean value for that lab). There is some indication of deviation from normality in the lower tail of the distribution, but the data do not appear grossly nonnormal.   □

We now outline the procedure for the case in which the numbers of observations under the various treatments are not necessarily equal. The only difficulties

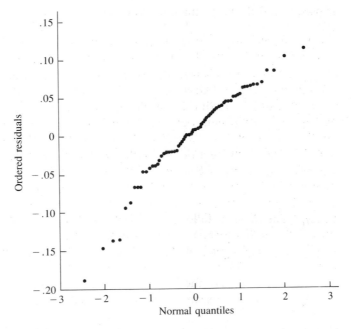

Figure 12-2.   Normal probability plot of residuals from one-way analyses of variance of tablet data.

with this case are algebraic; conceptually, the analysis is the same as for the case of equal sample sizes. Suppose that there are J_i observations under treatment i, for $i = 1, \ldots, I$. The basic identity still holds; that is, we have

$$\sum_{i=1}^{I} \sum_{j=1}^{J_i} (Y_{ij} - \bar{Y}_{..})^2 = \sum_{i=1}^{I} \sum_{j=1}^{J_i} (Y_{ij} - \bar{Y}_{i.})^2 + \sum_{i=1}^{I} J_i(\bar{Y}_i - \bar{Y}_{..})^2$$

By reasoning similar to that used above for the simple case, it can be shown that

$$E(SS_W) = \sigma^2 \sum_{i=1}^{I} (J_i - 1)$$

$$E(SS_B) = (I - 1)\sigma^2 + \sum_{i=1}^{I} J_i\alpha_i^2$$

The degrees of freedom for these sums of squares are $\sum_{i=1}^{I} J_i - I$ and $I - 1$, respectively. It may be argued, as in the proof of Theorem B, that the normalized sums of squares follows chi-square distributions and that the ratio of mean squares follows an F distribution under the null hypothesis of no treatment differences.

To conclude this section, let us review the basic assumptions and comment on their importance. The model is

$$Y_{ij} = \mu + \alpha_i + \varepsilon_{ij}$$

We assume the following:

1. The ε_{ij} are normally distributed. The F test, like the t test, remains approximately valid for moderate to large samples from moderately nonnormal distributions.
2. The error variance, σ^2, is constant. In many applications, the error variances may be different in different groups. For example, Figure 12-1 suggests that some labs may be more precise in their measurements than others. Fortunately, if there are an equal number of observations in each group, the F test is not strongly affected.
3. The ε_{ij} are independent. This assumption is very important, both for normal theory and for the nonparametric analysis we will present later.

12.2.2   The Problem of Multiple Comparisons

The application of the F test in Example A in Section 12.2.1 has an anticlimactic character. We concluded that the means of measurements from different labs are not all equal, but the test gives no information about how they differ, in particular about which pairs are different. In many applications, the null hypothesis is a "strawman" that is not seriously entertained. Real interest may be focused on comparing pairs or groups of treatments and estimating the treatment means and their differences. A naive approach would be to compare all pairs of treatment means using t tests. The difficulty with such a procedure was pointed out in the section on experimental design in chapter 11: Although each individual comparison would have a type I error rate of α, the collection of all comparisons considered simultaneously would not. In this section, we discuss two solutions to this problem—Tukey's method and the Bonferroni method. More discussion can be found in Miller (1981).

12.2.2.1   *Tukey's Method*

If the sample sizes are all equal and the errors are normally distributed with a constant variance, the centered sample means, $\overline{Y}_{i.} - \mu_i$, are independent and normally distributed with means 0 and variances σ^2/J, which may be estimated by s_p^2/J. Tukey's method is based on the probability distribution of the random variable

$$\max_{i_1, i_2} \frac{|(\overline{Y}_{i_1.} - \mu_{i_1}) - (\overline{Y}_{i_2.} - \mu_{i_2})|}{s_p/\sqrt{J}}$$

where the maximum is taken over all pairs i_1 and i_2. This distribution is called the **studentized range distribution** with parameters I (the number of samples being compared) and $I(J - 1)$ (the degrees of freedom in s_p.) The upper 100α percentage point of the distribution is denoted by $q_{I, I(J-1)}(\alpha)$. Now,

$$P\left[|(\bar{Y}_{i_1.} - \mu_{i_1}) - (\bar{Y}_{i_2.} - \mu_{i_2})| \le q_{I, I(J-1)}(\alpha)\frac{s_p}{\sqrt{J}}, \quad \text{for all } i_1 \text{ and } i_2\right]$$

$$= P\left[\max_{i_1, i_2} |(\bar{Y}_{i_1.} - \mu_{i_1}) - (\bar{Y}_{i_2.} - \mu_{i_2})| \le q_{I, I(J-1)}(\alpha)\frac{s_p}{\sqrt{J}}\right]$$

By definition, this latter probability equals $1 - \alpha$. The idea is that all the differences are less than some number if and only if the largest difference is. The above probability statement can be converted directly into a set of confidence intervals that hold simultaneously for all differences $\mu_{i_1} - \mu_{i_2}$ with confidence level $100(1 - \alpha)\%$. The intervals are

$$(\bar{Y}_{i_1.} - \bar{Y}_{i_2.}) \pm q_{I, I(J-1)}(\alpha)\frac{s_p}{\sqrt{J}}$$

By the duality of confidence intervals and hypothesis tests, if the $100(1 - \alpha)\%$ confidence interval for $(\bar{Y}_{i_1.} - \bar{Y}_{i_2.})$ does not include zero—that is, if

$$|\bar{Y}_{i_1.} - \bar{Y}_{i_2.}| > q_{I, I(J-1)}(\alpha)\frac{s_p}{\sqrt{J}}$$

the null hypothesis that there is no difference between μ_{i_1} and μ_{i_2} may be rejected at level α. Also, all such hypothesis tests considered collectively have level α.

EXAMPLE A. We can illustrate Tukey's method by applying it to the tablet data of Section 12.2. We list the labs in decreasing order of the mean of their measurements:

Lab	Mean
1	4.062
3	4.003
7	3.998
2	3.997
5	3.957
6	3.955
4	3.920

We obtain $s_p = .06$ and $q_{7, 63}(.05) = 4.31$. Two of these means are thus significantly different at the .05 level if they differ by more than

$$q_{7, 63}(.05)\frac{s_p}{\sqrt{J}} = .086$$

The mean from lab 1 is thus significantly different from those from labs 4, 5, and 6; no other comparisons are significant at the .05 level.

It is interesting to note that a price is paid here for performing multiple comparisons simultaneously. If separate t tests had been conducted using the pooled sample variance, labs would have been declared significantly different if their means had differed by more than

$$t_{63}(.025)s_p\sqrt{\frac{2}{J}} = .053 \qquad \qquad \square$$

12.2.2.2   *The Bonferroni Method*

The Bonferroni method was briefly introduced in Section 11.4.8. The idea is very simple. If k null hypotheses are to be tested, a desired overall type I error rate of at most α can be guaranteed by testing each null hypothesis at level α/k. Equivalently, if k confidence intervals are each formed to have confidence level $100(1 - \alpha/k)\%$, they all hold simultaneously with confidence level at least $100(1 - \alpha)\%$.

The method is simple and versatile and, although crude, gives surprisingly good results if k is not too large.

EXAMPLE A.   To apply the Bonferroni method to the data on tablets, we note that there are $k = \binom{7}{2} = 21$ pairwise comparisons among the seven labs. A set of simultaneous 95% confidence intervals for the pairwise comparisons is

$$(\bar{Y}_{i_1.} - \bar{Y}_{i_2.}) \pm s_p \frac{t_{63}(.025/21)}{\sqrt{5}}$$

Special tables for such values of the t distribution have been prepared; from Table 7 of Appendix B, we find

$$t_{63}\left(\frac{.025}{21}\right) = 3.16$$

giving confidence intervals

$$(\bar{Y}_{i_1.} - \bar{Y}_{i_2.}) \pm .089$$

Given the crude nature of the Bonferroni method, these are surprisingly close to the intervals produced by Tukey's method. Here, too, we conclude that lab 1 produced significantly higher measurements than those of labs 4, 5, and 6. $\square$

A significant advantage of the Bonferroni method over Tukey's method is that it does not require equal sample sizes in each treatment.

12.2.3   A Nonparametric Method—The Kruskal–Wallis Test

The Kruskal–Wallis test is a generalization of the Mann–Whitney test that is conceptually quite simple. The observations are pooled together and ranked. Let

$$R_{ij} = \text{the rank of } Y_{ij} \text{ in the combined sample}$$

Let

$$\bar{R}_{i.} = \frac{1}{J_i} \sum_{j=1}^{J_i} R_{ij}$$

be the average rank in the ith group. Let

$$\bar{R}_{..} = \frac{1}{N} \sum_{i=1}^{I} \sum_{j=1}^{J_i} R_{ij}$$

$$= \frac{N + 1}{2}$$

where N is the total number of observations. As in the analysis of variance, let

$$SS_B = \sum_{i=1}^{I} J_i (\bar{R}_{i.} - \bar{R}_{..})^2$$

be a measure of the dispersion of the $\bar{R}_{i.}$. SS_B may be used to test the null hypothesis that the probability distributions generating the observations under the various treatments are identical. The larger SS_B is, the stronger is the evidence against the null hypothesis. The exact null distribution of this statistic for various combinations of I and J_i can be enumerated, as for the Mann–Whitney test. Tables are given in Lehman (1975) and in references therein. For $I = 3$ and $J_i \geq 5$ or $I > 3$ and $J_i \geq 4$, a chi-square approximation to a normalized version of SS_B is fairly accurate. Under the null hypothesis that the probability distributions of the I groups are identical, the statistic

$$K = \frac{12}{N(N + 1)} SS_B$$

is approximately distributed as a chi-square random variable with $I - 1$ degrees of freedom. The value of K can be found by running the ranks through an analysis of variance program and multiplying SS_B by $12/[N(N + 1)]$. It can be shown that K can also be expressed as

$$K = \frac{12}{N(N + 1)} \left(\sum_{i=1}^{I} \frac{\bar{R}_{i.}^2}{J_i} \right) - 3(N + 1)$$

which is easier to compute by hand.

EXAMPLE A.   For the data on the tablets, $K = 29.51$. Referring to a table of the chi-square distribution with 6 df, we find that the p-value is less than .001. The nonparametric analysis, too, indicates that there is a systematic difference among the labs.   □

Multiple comparison procedures for nonparametric methods are discussed in detail in Miller (1981). The Bonferroni method requires no special discussion—it can be applied to all comparisons tested by Mann–Whitney tests.

Like the Mann–Whitney test, the Kruskal–Wallis test makes no assumption of normality and thus has a wider range of applicability than does the F test. It is especially useful in small-sample situations. Also, since data are replaced by their ranks, outliers will have less influence on this nonparametric test than on the F test. In some applications, the data consist of ranks—for example, in a wine tasting, judges usually rank the wines—which makes the use of the Kruskal–Wallis test natural.

12.3   The Two-Way Layout

A **two-way layout** is an experimental design involving two factors, each at two or more levels. The levels of one factor might be various drugs, for example, and the levels of the other factor might be genders. If there are I levels of one factor and J of the other, there are $I \times J$ combinations. We will assume that K independent observations are taken for each of these combinations. (The last section of this chapter will outline the advantages of such an experimental design.)

The next section defines the parameters that we might wish to estimate from a two-way layout. Later sections present statistical methods based on normal theory and nonparametric methods.

12.3.1   Additive Parametrization

To develop and illustrate the ideas in this section, we will use a portion of the data contained in a study of electric range energy consumption (Fechter and Porter, 1978). The following table shows the mean number of kilowatt-hours used by three electric ranges in cooking on each of three menu days (means are over several cooks).

Menu Day	Range 1	Range 2	Range 3
1	3.97	4.24	4.44
2	2.39	2.61	2.82
3	2.76	2.75	3.01

We wish to describe the variation in the numbers in this table in terms of the effects of different ranges and different menu days. Denoting the number in the

ith row and jth column by Y_{ij}, we first calculate a grand average

$$\hat{\mu} = \bar{Y}_{..} = \frac{1}{9} \sum_{i=1}^{3} \sum_{j=1}^{3} Y_{ij} = 3.22$$

This gives a measure of typical energy consumption per menu day.

The menu day means, averaged over the ranges, are

$$\bar{Y}_{1.} = 4.22$$

$$\bar{Y}_{2.} = 2.61$$

$$\bar{Y}_{3.} = 2.84$$

We will define the differential effect of a menu day as the difference between the mean for that day and the overall mean; we will denote these differential effects by $\hat{\alpha}_i$, where $i = 1, 2,$ or 3.

$$\hat{\alpha}_1 = \bar{Y}_{1.} - \bar{Y}_{..} = 1.00$$

$$\hat{\alpha}_2 = \bar{Y}_{2.} - \bar{Y}_{..} = -.61$$

$$\hat{\alpha}_3 = \bar{Y}_{3.} - \bar{Y}_{..} = -.38$$

(Note that, except for rounding error, the α_i would sum to zero.) In words, on menu day 1, 1 kwh more than the average is consumed, and so on.

The range means, averaged over the menu days, are

$$\bar{Y}_{.1} = 3.04$$

$$\bar{Y}_{.2} = 3.20$$

$$Y_{.3} = 3.42$$

The differential effects of the ranges are

$$\hat{\beta}_1 = \bar{Y}_{.1} - \bar{Y}_{..} = -.18$$

$$\hat{\beta}_2 = \bar{Y}_{.2} - \bar{Y}_{..} = -.02$$

$$\hat{\beta}_3 = \bar{Y}_{.3} - \bar{Y}_{..} = .20$$

The effects of the ranges are smaller than the effects of the menu days.

The preceding description of the values in the table incorporates an overall average level plus differential effects of ranges and menu days. This is a simple **additive model**. Note that this model may not give an accurate description of some tables; according to the model, for example, the difference between ranges

1 and 2 on menu day 1 is the same as the difference on menu day 2, which need not really be the case. The standard way of checking how well a model fits is to examine the residuals, which are observed values minus fitted values. The fitted values for the electric range data are

$$\hat{Y}_{ij} = \hat{\mu} + \hat{\alpha}_i + \hat{\beta}_j$$

Calculating these fitted values and subtracting them from the values in the original table, we obtain the following table of residuals:

Menu Day	Range 1	Range 2	Range 3
1	−.07	.04	.02
2	−.04	.02	.01
3	.10	−.07	−.03

The residuals are small relative to the main effects, with the possible exception of those for menu day 3. The simple plot in Figure 12-3 shows that relative to ranges 2 and 3, range 1 used more energy on day 3 than it did on days 1 and 2. This fact may be ascertained from the table of residuals but is more readily apparent from the plot. This phenomenon is called an **interaction** between menu

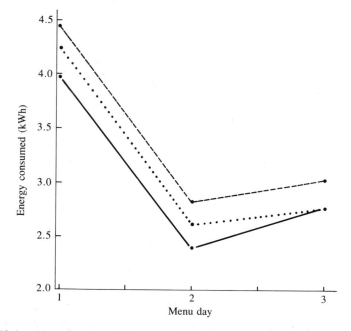

Figure 12-3. Plot of energy consumption versus menu day for three electric ranges. The dashed line corresponds to range 3, the dotted line to range 2, and the solid line to range 1.

days and ranges—it is as if there were something about menu day 3 especially that adversely affected the performance of range 1.

Interactions can be incorporated into the model to make it fit the data exactly. The residual in cell ij is

$$
\begin{aligned}
Y_{ij} - \hat{\mu} - \hat{\alpha}_i - \hat{\beta}_j &= Y_{ij} - Y_{..} - (\bar{Y}_{i.} - \bar{Y}_{..}) - (\bar{Y}_{.j} - \bar{Y}_{..}) \\
&= Y_{ij} - \bar{Y}_{i.} - \bar{Y}_{.j} + \bar{Y}_{..} \\
&= \hat{\delta}_{ij}
\end{aligned}
$$

Note that

$$
\sum_{i=1}^{3} \delta_{ij} = \sum_{j=1}^{3} \delta_{ij} = 0
$$

In the table of residuals above, the row and column sums are not exactly zero because of rounding errors. The model

$$
Y_{ij} = \hat{\mu} + \hat{\alpha}_i + \hat{\beta}_j + \hat{\delta}_{ij}
$$

thus fits the data exactly; it is merely another way of expressing the numbers listed in the table.

An additive model is simple and easy to interpret, especially in the absence of interactions. Transformations of the data are sometimes used to improve the adequacy of an additive model. The logarithmic transformation, for example, converts a multiplicative model into an additive one. Transformations are also used to stabilize the variance (to make the variance independent of the mean) and to make normal theory more applicable. There is no guarantee, of course, that a given transformation will accomplish all these aims.

The discussion in this section has centered around the parametrization and interpretation of the additive model as used in the analysis of variance. We have not taken into account the possibility of random errors and their effects on the inferences about the parameters, but will do so in the next section.

12.3.2 Normal Theory for the Two-Way Layout

In this section, we will assume that there are $K > 1$ observations per cell in a two-way layout. A design with an equal number of observations per cell is called **balanced**. Let Y_{ijk} denote the kth observation in cell ij; the statistical model is

$$
Y_{ijk} = \mu + \alpha_i + \beta_j + \delta_{ij} + \varepsilon_{ijk}
$$

We will assume that the random errors, ε_{ijk}, are independent and normally distributed with mean 0 and common variance σ^2. The parameters satisfy the following constraints:

$$\sum_{i=1}^{I} \alpha_i = 0$$

$$\sum_{j=1}^{J} \beta_j = 0$$

$$\sum_{i=1}^{I} \delta_{ij} = \sum_{j=1}^{J} \delta_{ij} = 0$$

We now find the maximum likelihood estimates of the unknown parameters, which we denote by θ. Since the observations in cell ij are normally distributed with mean $\mu + \alpha_i + \beta_j + \delta_{ij}$ and variance σ^2, and since all the observations are independent, the log likelihood of θ is

$$l(\theta) = \frac{IJK}{2}\log(2\pi\sigma^2) - \frac{\sigma^2}{2}\sum_{i=1}^{I}\sum_{j=1}^{J}\sum_{k=1}^{K}(Y_{ijk} - \mu - \alpha_i - \beta_j - \delta_{ij})^2$$

Maximizing the likelihood subject to the constraints given above yields the following estimates (see Problem 9 at the end of this chapter):

$$\hat{\mu} = \bar{Y}_{...}$$

$$\hat{\alpha}_i = \bar{Y}_{i..} - \bar{Y}_{...}, i = 1, \ldots, I$$

$$\hat{\beta}_j = \bar{Y}_{.j.} - \bar{Y}_{...}, j = 1, \ldots, J$$

$$\hat{\delta}_{ij} = \bar{Y}_{ij.} - \bar{Y}_{i..} - \bar{Y}_{.j.} + \bar{Y}_{...}$$

as is expected from the discussion in Section 12.3.1.

Like one-way analysis of variance, two-way analysis of variance is conducted by comparing various sums of squares. The sums of squares are as follows:

$$SS_A = JK\sum_{i=1}^{I}(\bar{Y}_{i..} - \bar{Y}_{...})^2$$

$$SS_B = IK\sum_{j=1}^{J}(\bar{Y}_{.j.} - \bar{Y}_{...})^2$$

$$SS_{AB} = K\sum_{i=1}^{I}\sum_{j=1}^{J}(\bar{Y}_{ij.} - \bar{Y}_{i..} - \bar{Y}_{.j.} + \bar{Y}_{...})^2$$

$$SS_E = \sum_{i=1}^{I}\sum_{j=1}^{J}\sum_{k=1}^{K}(Y_{ijk} - \bar{Y}_{ij.})^2$$

$$SS_{TOT} = \sum_{i=1}^{I}\sum_{j=1}^{J}\sum_{k=1}^{K}(Y_{ijk} - \bar{Y}_{...})^2$$

The sums of squares satisfy this algebraic identity:

$$SS_{TOT} = SS_A + SS_B + SS_{AB} + SS_E$$

This identity may be proved by writing

$$Y_{ijk} - \bar{Y}_{...} = (Y_{ijk} - \bar{Y}_{ij.}) + (\bar{Y}_{i..} - \bar{Y}_{...}) + (\bar{Y}_{.j.} - \bar{Y}_{...})$$
$$+ (\bar{Y}_{ij.} - \bar{Y}_{i..} - \bar{Y}_{.j.} + \bar{Y}_{...})$$

and then squaring both sides, summing, and verifying that the cross products vanish.

The following theorem gives the expectations of these sums of squares.

THEOREM A.   Under the assumption that the errors are independent with mean 0 and variance σ^2,

$$E(SS_A) = (I - 1)\sigma^2 + JK \sum_{i=1}^{I} \alpha_i^2$$

$$E(SS_B) = (J - 1)\sigma^2 + IK \sum_{j=1}^{J} \beta_j^2$$

$$E(SS_{AB}) = (I - 1)(J - 1)\sigma^2 + K \sum_{i=1}^{I} \sum_{j=1}^{J} \delta_{ij}^2$$

$$E(SS_E) = IJ(K - 1)\sigma^2$$

Proof.   The results for SS_A, SS_B, and SS_E follow from Lemma A of Section 12.2.1. Applying the lemma to SS_{TOT}, we have

$$E(SS_{TOT}) = E \sum_{i=1}^{I} \sum_{j=1}^{J} \sum_{k=1}^{K} (Y_{ijk} - \bar{Y}_{...})^2$$

$$= (IJK - 1)\sigma^2 + \sum_{i=1}^{I} \sum_{j=1}^{J} \sum_{k=1}^{K} (\alpha_i + \beta_j + \delta_{ij})^2$$

$$= (IJK - 1)\sigma^2 + JK \sum_{i=1}^{I} \alpha_i^2 + IK \sum_{j=1}^{J} \beta_j^2 + K \sum_{i=1}^{I} \sum_{j=1}^{J} \delta_{ij}^2$$

In the last step, we used the constraints on the parameters. The desired expression for $E(SS_{AB})$ now follows, since

$$E(SS_{TOT}) = E(SS_A) + E(SS_B) + E(SS_{AB}) + E(SS_E) \qquad \square$$

The distributions of these sums of squares are given by the following theorem.

THEOREM B.   Assume that the errors are independent and normally distributed with means 0 and variances σ^2. Then

(a) SS_E/σ^2 follows a chi-square distribution with $IJ(K-1)$ degrees of freedom.

(b) Under the null hypothesis

$$H_A: \alpha_i = 0, i = 1, \ldots, I$$

SS_A/σ^2 follows a chi-square distribution with $I-1$ degrees of freedom.

(c) Under the null hypothesis

$$H_B: \beta_j = 0, j = 1, \ldots, J$$

SS_B/σ^2 follows a chi-square distribution with $J-1$ degrees of freedom.

(d) Under the null hypothesis

$$H_{AB}: \delta_{ij} = 0, i = 1, \ldots, I, j = 1, \ldots, J$$

SS_{AB}/σ^2 follows a chi-square distribution with $(I-1)(J-1)$ degrees of freedom.

(e) The sums of squares are independently distributed.

Proof.   We will not give a full proof of this theorem. The results for SS_A, SS_B, and SS_E follow from arguments similar to those used in proving Theorem B of Section 12.2.1. The result for SS_{AB} requires some additional argument.   □

F tests of the various null hypotheses are conducted by comparing the appropriate sums of squares to the sum of squares for error, as was done for the simpler case of the one-way layout.

EXAMPLE A.   As an example, we return to the experiment on iron retention discussed in Section 11.2.1.1. In the complete experiment, there were $I = 2$ forms of iron, $J = 3$ dosage levels, and $K = 18$ observations per cell. In Section 11.2.1.1, we discussed a logarithmic transformation of the data to make it more nearly normal and to stabilize the variance. Figure 12-4 shows boxplots of the data on the original scale; boxplots of the log data are given in Figure 12-5. The distribution of the log data is more symmetrical, and the interquartile ranges are less variable. Figure 12-6 is a plot of cell standard deviations versus cell means for the untransformed data; it shows that the error variance changes with the mean. Figure 12-7 is a plot of cell standard deviations versus means for the log data; it shows that the transformation is successful in stabilizing the variance.

Figure 12-8 is a plot of the cell means of the transformed data versus the dosage levels for the two forms of iron. If there is no interaction, the two curves should be parallel except for random variation. This appears to be roughly the case, although there is a hint that the difference in retention of the two forms of iron increases with dosage level. To check this, we will perform a quantitative test for interaction.

In the analysis of variance table below, SS_A is the sum of squares due to the

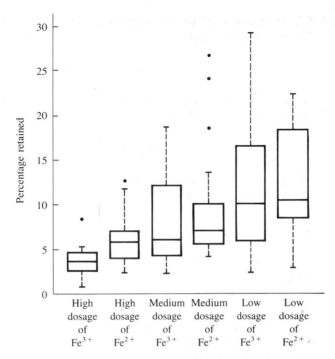

Figure 12-4.   Boxplots of iron retention for two forms of iron at three dosage levels.

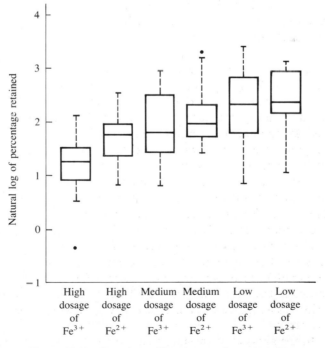

Figure 12-5.   Boxplots of log data on iron retention.

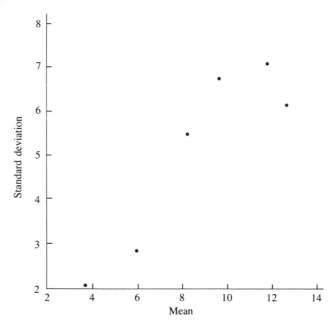

Figure 12-6.   Plot of cell standard deviations versus cell means for iron retention data.

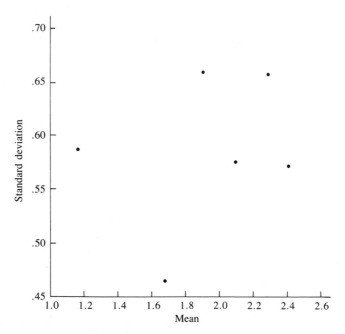

Figure 12-7.   Plot of cell standard deviations versus cell means for log data on iron retention.

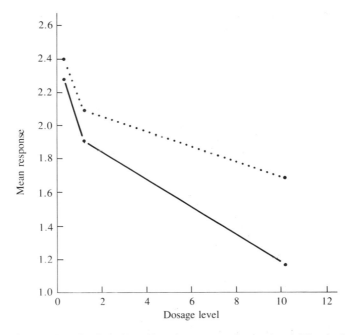

Figure 12-8. Plot of cell means of log data versus dosage level. The dashed line corresponds to Fe^{2+} and the solid line to Fe^{3+}.

form of iron, SS_B is the sum of squares due to dosage, and SS_{AB} is the sum of squares due to interaction. The F statistics were found by dividing the appropriate mean square by the mean square for error.

Analysis of variance table

Source	df	SS	MS	F
Iron form	1	2.074	2.074	5.99
Dosage	2	15.588	7.794	22.53
Interaction	2	.810	.405	1.17
Error	102	35.296	.346	
Total	107	53.768		

To test the effect of the form of iron, we test

$$H_A: \alpha_1 = \alpha_2 = 0$$

using the statistic

$$F = \frac{SS_{IRON}/1}{SS_E/102} = 5.99$$

From tables of the F distribution with 1 and 102 df, the p-value is less than .025.

There is an effect due to the form of iron. An estimate of the difference $\alpha_1 - \alpha_2$ is

$$\bar{Y}_{1..} = \bar{Y}_{2..} = .28$$

and a confidence interval for the difference may be obtained by noting that $\bar{Y}_{1..}$ and $\bar{Y}_{2..}$ are uncorrelated since they are averages over different observations and that

$$\text{Var}(\bar{Y}_{1..}) = \text{Var}(\bar{Y}_{2..}) = \frac{\sigma^2}{JK}$$

Thus,

$$\text{Var}(\bar{Y}_{1..} - \bar{Y}_{2..}) = \frac{2\sigma^2}{JK}$$

which may be estimated by

$$s^2_{\bar{Y}_{1..} - \bar{Y}_{2..}} = \frac{2 \times .346}{54} = .0128$$

A confidence interval can be constructed using the t distribution with $IJ(K-1)$ degrees of freedom. The interval is of the form

$$(\bar{Y}_{1..} - \bar{Y}_{2..}) \pm t_{IJ(K-1)}(\alpha/2)s_{\bar{Y}_{1..} - \bar{Y}_{2..}}$$

For the example, the resulting 95% confidence interval is $.28 \pm 1.96\sqrt{.0128}$, or $(.06, .37)$. Since we are using a multiplicative model, it is natural to convert this interval into an interval for the ratio of the means of the forms of iron, which is $(e^{.06}, e^{.37})$, or $(1.06, 1.45)$.

The F statistic for testing the effect of dosage is significant, but this effect is anticipated and is not of major interest.

To test the hypothesis H_{AB} that there is no interaction, we consider the following F statistic:

$$F = \frac{SS_{AB}/(I-1)(J-1)}{SS_E/IJ(K-1)} = 1.17$$

From tables of the F distribution with 2 and 102 df, we see that there is insufficient evidence to reject this hypothesis.

In conclusion, it appears that there is a difference of 6–45% in the ratio of percentage retained between the two forms of iron and that there is little evidence that this difference depends on dosage. $\square$

12.3.3 Randomized Block Designs

Randomized block designs originated in agricultural experiments. To compare the effects of I different fertilizers, J relatively homogeneous plots of land, or blocks, are selected, and each is divided into I plots. Within each block, the assignment of fertilizers to plots is made at random. By comparing fertilizers within blocks, the variability between blocks, which would otherwise contribute "noise" to the results, is controlled. This design is a multisample generalization of a matched-pairs design.

A randomized block design might be used by a nutritionist who wishes to compare the effects of three different diets on experimental animals. To control for genetic variation in the animals, the nutritionist might select three animals from each of several litters and randomly determine their assignments to the diets. Randomized block designs are used in many areas. If an experiment is to be carried out over a substantial period of time, the blocks may consist of stretches of time. In industrial experiments, the blocks are often batches of raw material.

Randomization helps ensure against unintentional bias and can form a basis for inference. In principle, the null distribution of a test statistic can be derived by permutation arguments, just as we derived the null distribution of the Mann–Whitney test statistic in Section 11.2.3. In practice, this may be difficult, and parametric procedures often give a good approximation to the permutation distribution.

As a model for the responses in the randomized block design, we will use

$$Y_{ij} = \mu + \alpha_i + \beta_j + \varepsilon_{ij}$$

where α_i is the differential effect of the ith treatment, β_j is the differential effect of the jth block, and the ε_{ij} are independent random variables. This is the model of Section 12.3.2 but with the additional assumption of no interactions between blocks and treatments.

From Theorem A of Section 12.3.2, if there is no interaction,

$$E(MS_A) = \sigma^2 + \frac{J}{I-1} \sum_{i=1}^{I} \alpha_i^2$$

$$E(MS_B) = \sigma^2 + \frac{I}{J-1} \sum_{j=1}^{J} \beta_j^2$$

$$E(MS_{AB}) = \sigma^2$$

Thus, σ^2 can be estimated from MS_{AB}. Also, since the above mean squares are independently distributed, F tests can be performed to test H_A or H_B. For example, to test

$$H_A: \alpha_i = 0, i = 1, \ldots, I$$

this statistic can be used:

$$F = \frac{MS_A}{MS_{AB}}$$

From Theorem B in Section 12.3.2, under H_A, the statistic follows an F distribution with $I - 1$ and $(I - 1)(J - 1)$ degrees of freedom. H_B may be tested similarly. Note that if, contrary to the assumption, there is an interaction, then

$$E(MS_{AB}) = \sigma^2 + \frac{1}{(I - 1)(J - 1)} \sum_{i=1}^{I} \sum_{j=1}^{J} \delta_{ij}^2$$

MS_{AB} will tend to overestimate σ^2. This will cause the F statistic to be smaller than it should be and will result in a test that is conservative; that is, the actual probability of type I error will be smaller than desired.

EXAMPLE A.   Let us consider an experimental study of drugs to relieve itching (Beecher, 1959). Five drugs were compared to a placebo and no drug with 10 volunteer male subjects aged 20–30. (Note that this set of subjects limits the scope of inference; from a statistical point of view, one cannot extrapolate the results of the experiment to older women, for example. Any such extrapolation could only be justified on grounds of medical judgment.) Each volunteer underwent one treatment per day, and the time-order was randomized. Thus, individuals were "blocks." The subjects were given a drug (or placebo) intravenously, and then itching was induced on their forearms with cowage, an effective itch stimulus. The subjects recorded the duration of the itching. More details are in Beecher (1959). The following table gives the durations of the itching (in seconds):

Subject	No Drug	Placebo	Papaverine	Morphine	Aminophylline	Pentobarbital	Tripelennamine
BG	174	263	105	199	141	108	141
JF	224	213	103	143	168	341	184
BS	260	231	145	113	78	159	125
SI	255	291	103	225	164	135	227
BW	165	168	144	176	127	239	194
TS	237	121	94	144	114	136	155
GM	191	137	35	87	96	140	121
SS	100	102	133	120	222	134	129
MU	115	89	83	100	165	185	79
OS	189	433	237	173	168	188	317
Average	191.0	204.8	118.2	148.0	144.3	176.5	167.2

Figure 12-9 shows boxplots of the responses to the six treatments and to the control (no drugs). Although the boxplot is probably not the ideal visual display of these data, since it takes no account of the blocking, Figure 12-9 does show some interesting aspects of the data. There is a suggestion that all the drugs had

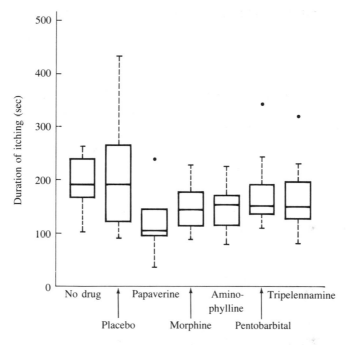

Figure 12-9. Boxplots of durations of itching under seven treatments.

some effect and that papaverine was the most effective. There is a lot of scatter relative to the differences between the medians, and there are some outliers. It is interesting that the placebo responses have the greatest spread; this might be because some subjects responded to the placebo and some did not.

We next construct an analysis of variance table for this experiment:

Source	df	SS	MS	F
Drugs	6	53013	8835	2.85
Subjects	9	103280	11476	3.71
Interaction	54	167130	3095	
Total	69	323422		

The F statistic for testing differences between drugs is 2.85 with 6 and 54 df, corresponding to a p-value less than .025. The null hypothesis that there is no difference between subjects is not experimentally interesting.

Figure 12-10 is a probability plot of the residuals from the two-way analysis of variance model. The residual in cell ij is

$$r_{ij} = Y_{ij} - \hat{\mu} - \hat{\alpha}_i - \hat{\beta}_j$$

$$= Y_{ij} - \bar{Y}_{i.} - \bar{Y}_{.j} + \bar{Y}_{..}$$

There is a slightly bowed character to the probability plot, indicating some

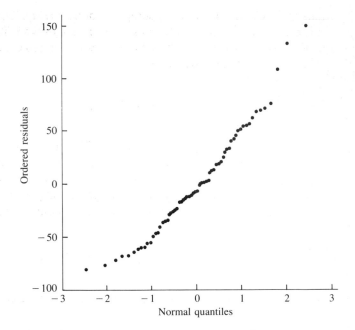

Figure 12-10. Normal probability plot of residuals from two-way analysis of variance of data on duration of itching.

skewness in the distribution of the residuals. But because the *F* test is robust against moderate deviations from normality, we should not be overly concerned.

Tukey's method may be applied to make multiple comparisons. Suppose that we wish to compare the drug means, $\overline{Y}_{1\cdot}, \ldots, \overline{Y}_{7\cdot}$ $(I = 7)$. These have expectations $\mu + \alpha_i$, where $i = 1, \ldots, I$, and each is an average over $J = 10$ independent observations. The error variance is estimated by MS_{AB} with 54 df. Simultaneous 95% confidence intervals for all differences between drug means have half-widths of

$$\frac{q_{7,54}(.05)s}{\sqrt{I}} = 4.314 \sqrt{\frac{3095}{10}}$$

$$= 75.89$$

Examining the table of means, we see that, at the 95% confidence level, we can conclude only that papaverine achieves a reduction of itching over the effect of a placebo. □

12.3.4 A Nonparametric Method—Friedman's Test

This section presents a nonparametric method for the randomized block design. Like other nonparametric methods we have discussed, Friedman's test relies on ranks and does not make an assumption of normality. The test is very simple. Within each of the J blocks, the observations are ranked. To test the hypothesis that there is no effect due to the factor corresponding to the columns of the table, the following statistic is calculated:

$$SS_A = J \sum_{i=1}^{I} (\bar{R}_{i.} - \bar{R}_{..})^2$$

just as in the ordinary analysis of variance. Under the null hypothesis that there is no treatment effect and that the only effect is due to the randomization within blocks, the permutation distribution of the statistic can, in principle, be calculated. For sample sizes such as that of the itching experiment, a chi-square approximation to this distribution is perfectly adequate. The null distribution of

$$Q = \frac{12J}{I(I+1)} \sum_{i=1}^{I} (\bar{R}_{i.} - \bar{R}_{..})^2$$

is approximately chi-square with $I - 1$ degrees of freedom.

EXAMPLE A. To carry out Friedman's test on the data from the experiment on itching, we first construct the following table by ranking durations of itching for each subject:

	No Drug	Placebo	Papaverine	Morphine	Aminophylline	Pentobarbital	Tripelennamine
BG	5	7	1	6	3.5	2	3.5
JF	6	5	1	2	3	7	4
BS	7	6	4	2	1	5	3
SI	6	7	1	4	3	2	5
BW	4	5	2	3	1	7	6
TS	7	3	1	5	2	4	6
GM	7	6	1	2	3	5	4
SS	1	2	5	3	7	6	4
MU	5	3	2	4	6	7	1
OS	4	7	5	2	1	3	6
Average	5.20	5.10	2.10	3.30	3.05	4.80	4.25

Note that we have handled ties in the usual way by assigning average ranks. We find that $Q = 178.9$, which gives a p-value less than .001. The nonparametric analysis also rejects the hypothesis that there is no drug effect. ☐

Procedures for using Friedman's test for multiple comparisons are discussed by Miller (1981). When these methods are applied to the data from the experiment on itching, the conclusions reached are identical to those reached by the parametric analysis.

12.4   Concluding Remarks

The most complicated experimental design considered in this chapter was the two-way layout; more generally, a **factorial design** incorporates several factors with one or more observations per cell. With such a design, the concept of interaction becomes more complicated—there are interactions of various orders. For instance, in a three-factor experiment, there are two-factor and three-factor interactions. It is both interesting and useful that the two-factor interactions in a three-way layout can be estimated using only one observation per cell.

To gain some insight into why factorial designs are effective, we can begin by considering a two-way layout, with each factor at five levels, no interaction, and one observation per cell. With this design, comparisons of two levels of any factor are based on 10 observations. A traditional alternative to this design is to do first an experiment comparing the levels of factor A and then another experiment comparing the levels of factor B. To obtain the same precision as is achieved by the two-way layout in this case, 25 observations in each experiment, or a total of 50 observations, would be needed. The factorial design achieves its economy by using the same observations to compare the levels of factor A as are used to compare the levels of factor B.

The advantages of factorial designs become greater as the number of factors increases. For example, in an experiment with four factors, with each factor at two levels (which might be the presence or absence of some chemical, for example) and one observation per cell, there are 16 observations that may be used to compare the levels of each factor. Furthermore, it can be shown that two- and three-factor interactions can be estimated. By comparison, if each of the four factors were investigated in a separate experiment, 64 observations would be required to attain the same precision.

As the number of factors increases, the number of observations necessary for a factorial experiment with only one observation per cell grows very rapidly. To decrease the cost of an experiment, certain cells, designated in a systematic way, can be left empty, and the main effects and some interactions can still be estimated. Such arrangements are called **fractional factorial designs**.

Similarly, with a randomized block design, the individual blocks may not be able to accommodate all the treatments. For example, in a chemical experiment that compares a large number of treatments, the blocks of the experiment, batches of raw material of uniform quality, may not be large enough. In such situations, **incomplete block designs** may be used to retain the advantages of blocking.

The basic theoretical assumptions underlying the analysis of variance are that the errors are independent and normally distributed with constant variance. Since we cannot fully check the validity of these assumptions in practice and can probably detect only gross violations, it is natural to ask how robust the procedures are with respect to violations of the assumptions. It is impossible to give a complete and conclusive answer to this question. Generally speaking, the independence assumption is probably the most important (and this is true for nonparametric procedures as well). The F test is robust against moderate departures from normality; if the design is balanced, the F test is also robust against unequal error variance.

For further reading, Box, Hunter, and Hunter (1978) is recommended.

12.5   Problems

1. For a one-way analysis of variance with $I = 2$ treatment groups, show that the F statistic is t^2, where t is the usual t statistic for a two-sample case.
2. Prove the analogues of Theorems A and B in Section 12.2.1 for the case of unequal numbers of observations in the cells of a one-way layout.
3. Derive the likelihood ratio test for the null hypothesis of the one-way layout, and show that it is equivalent to the F test.
4. Prove this version of the Bonferroni Inequality:

$$P\left(\bigcap_{i=1}^{n} A_i\right) \geq 1 - \sum_{i=1}^{n} P(\bar{A}_i)$$

(Use Venn diagrams if you wish.) In the context of simultaneous confidence intervals, what is A_i and what is $\bar{A}_i$?
5. Show that for comparing two groups the Kruskal–Wallis test is equivalent to the Mann–Whitney test.
6. Show that for comparing two groups Friedman's test is equivalent to the sign test.
7. Show the equality of the two forms of K given in Section 12.2.3:

$$K = \frac{12}{N(N+1)} \sum_{i=1}^{I} J_i(\bar{R}_{i\cdot} - \bar{R}_{\cdot\cdot})^2$$

$$= \frac{12}{N(N+1)} \left(\sum_{i=1}^{I} \frac{\bar{R}_{i\cdot}^2}{J_i}\right) - 3(N+1)$$

8. Prove the sums of squares identity for the two-way layout:

$$SS_{TOT} = SS_A + SS_B + SS_{AB} + SS_E$$

9. Find the mle's of the parameters α_i, β_j, δ_{ij}, and μ of the model for the two-way layout.
10. The table below gives the energy use of five gas ranges for seven menu days. (The units are equivalent kilowatt-hours; .239 kwh = 1 ft^3 of natural gas.) Estimate main effects and discuss interaction, paralleling the discussion of Section 12.3.

14. Referring to Section 12.2, the following table gives the measurements on chlorpheniramine maleate tablets from another manufacturer. Are there systematic differences between the labs? If so, which pairs differ significantly? How do these data compare to those given for the other manufacturer in Section 12.2?

Lab 1	Lab 2	Lab 3	Lab 4	Lab 5	Lab 6	Lab 7
4.15	3.93	4.10	4.16	4.14	4.09	4.10
4.08	3.92	4.10	4.14	4.14	4.08	4.05
4.09	4.08	4.05	4.16	4.14	4.08	4.05
4.08	4.09	4.07	4.02	4.17	4.12	4.06
4.01	4.06	4.06	4.15	4.13	4.17	4.06
4.01	4.06	4.03	4.15	4.24	4.15	4.12
4.00	4.02	4.04	4.18	4.18	4.12	4.07
4.09	4.00	4.03	4.12	4.14	4.10	4.18
4.08	4.01	4.03	4.09	4.25	4.12	4.15
4.00	4.01	4.06	4.04	4.17	4.12	4.18

15. For a study of the release of luteinizing hormone (LH), male and female rats kept in constant light were compared to male and female rats in a regime of 14 hours of light and 10 hours of darkness. Various dosages of luteinizing releasing factor (LRF) were given: control (saline), 10, 50, 250, and 1250 nanograms. Levels of LH (in nanograms per milliliter of serum) were measured in blood samples at a later time. Analyze the data to determine the effects of light regime and LRF on release of LH for both males and females. Use both graphical techniques and more formal analyses.

Male rats

Dosage	Normal Light	Constant Light
0	212	72
	27	64
	68	78
	72	20
	130	56
	153	70
10	32	74
	98	82
	148	40
	186	87
	203	78
	188	88
50	294	130
	306	187
	234	133
	219	185
	281	107
	288	98

(continued)

(*continued*)

Dosage	Normal Light	Constant Light
250	515	159
	340	167
	348	193
	205	196
	505	174
	432	250
1250	296	137
	545	426
	630	178
	418	208
	396	196
	227	251

Female rats

Dosage	Normal Light	Constant Light
0	71	197
	159	115
	208	28
	161	48
	187	229
	51	424
10	755	468
	81	456
	372	300
	356	605
	613	405
	1313	405
50	1451	565
	222	445
	821	708
	120	1710
	839	271
	249	759
250	2265	1825
	1089	1335
	994	2826
	287	1255
	1638	1200
	329	1452
1250	2693	1482
	1719	1646
	2758	1646
	2040	1289
	3199	1982
	561	1780

16. A collaborative study was conducted to study the precision and homogeneity of a method of determining the amount of niacin in cereal products (Campbell

and Pelletier, 1962). Homogenized samples of bread and bran flakes were enriched with 0, 2, 4, or 8 milligrams of niacin per 100 grams. Portions of the samples were sent to 12 labs, which were asked to carry out the specified procedures on each of three separate days. The data (in milligrams per 100 grams) are given in the table below. Conduct two-way analyses of variance for both the bread and bran data and discuss the results. (Two data points are missing. Substitute for them the corresponding cell means.)

Lab	Bread (+0 mg)	Bread (+2 mg)	Bread (+4 mg)	Bran (+0 mg)	Bran (+4 mg)	Bran (+8 mg)
1	3.42	5.25	7.17	7.58	11.63	15.00
	3.66	5.63	7.50	7.87	11.87	15.92
	3.26	5.25	7.25	7.71	11.40	15.58
2	3.36	5.65	7.53	8.00	12.20	16.60
	3.64	6.00	7.68	8.27	11.70	16.40
	3.40	6.15	7.48	8.00	11.80	15.90
3	3.91	6.21	7.90	7.60	11.04	15.87
	3.90	5.50	7.47	7.30	11.50	15.91
	3.84	5.64	7.70	7.82	11.49	16.28
4	3.87	5.38	7.51	8.03	11.50	15.10
	3.80	5.27	7.11	7.35	10.10	14.80
	3.98	5.29	7.63	7.66	11.70	15.70
5	3.96	5.85	7.60	8.50	11.75	16.20
	4.28	6.15	7.66	8.46	12.88	16.16
	3.92	6.15	7.80	8.53	12.64	16.48
6	3.46	5.66	7.93	7.99	11.40	16.37
	3.55	5.98	7.27	8.23	12.42	16.00
	3.45	5.37	7.30	8.14	12.20	17.40
7	3.40	5.00	6.54	7.31	11.11	15.00
	3.63	5.27	7.46	7.85	11.00	17.00
	3.52	5.39	6.84	7.92	11.67	15.50
8	3.80	5.30	7.10	8.50	12.00	17.00
	3.80	5.60	7.60	8.50	13.10	17.50
	3.90	5.80	8.00	8.60	12.60	17.20
9	3.66	5.68	7.30	8.20		
	3.92	5.47	6.40	8.25	11.68	16.20
	4.22	5.84	7.60	8.20	11.43	16.60
10	4.37	6.53	8.32	8.82	12.90	17.30
	3.86	5.85	7.34	8.76	12.00	17.60
	4.46	6.38	8.12	8.52	13.50	18.40
11	4.20	5.80	7.70	8.40	12.20	16.10
	3.60	5.70	7.10	8.60	11.60	16.10
	4.20	5.90	7.20	7.90	11.60	15.80
12	3.76	6.06	7.60	8.32	12.00	16.80
	3.68	5.60	7.50	8.25	12.40	16.60
	3.80	6.05	7.67	8.57	12.30	16.30

17. This problem deals with an example from Youden (1962). An ingot of magnesium alloy was drawn into a square rod about 100 meters long with a cross section of about 4.5 centemeters on a side. The rod was then cut into 100 bars, each a meter long. Five of these were selected at random, and a test piece 1.2 centemeters thick was cut from each. From each of these five specimens, 10 test points were selected in a particular geometric pattern. Two determinations of the magnesium content were made at each test point (the analyst ran all 50 points once, and then made a set of repeat measurements). The overall purpose of the experiment was to test for homogeneity of magnesium content in the different bars and different locations. Analyze the data in the table below (giving percentage of magnesium times 1000) to determine if there is significant variability between bars and between locations. There are a couple of unexpected aspects of these data—can you find them?

Position	Bar 1		Bar 5		Bar 20		Bar 50		Bar 85	
1	76	67	69	66	73	70	73	63	70	69
2	71	66	71	62	69	68	75	69	66	64
3	70	65	68	66	68	66	69	67	68	63
4	67	66	71	66	69	67	72	68	68	64
5	71	65	66	65	70	68	69	66	64	67
6	65	67	68	66	70	65	69	65	70	65
7	67	67	71	67	65	68	72	82	69	64
8	71	67	69	67	67	69	63	63	67	64
9	66	63	70	65	67	73	69	66	69	68
10	68	68	68	68	64	68	69	66	67	69

18. The concentrations (in nanograms per milliliter) of plasma epinephrine were measured for 10 dogs under isofluorane, halothane, and cyclopropane anesthesia; the measurements are given in the following table (Perry et al., 1974). Is there a difference in treatment effects? Use a parametric and a nonparametric analysis.

	Dog 1	Dog 2	Dog 3	Dog 4	Dog 5	Dog 6	Dog 7	Dog 8	Dog 9	Dog 10
Isofluorane	.28	.51	1.00	.39	.29	.36	.32	.69	.17	.33
Halothane	.30	.39	.63	.68	.38	.21	.88	.39	.51	.32
Cyclopropane	1.07	1.35	.69	.28	1.24	1.53	.49	.56	1.02	.30

19. Three species of mice were tested for "aggressiveness." The species were A/J, C57, and F2 (a cross of the first two species). A mouse was placed in a 1-meter-square box, which was marked off into 49 equal squares. The mouse was let go on the center square, and the number of squares traversed in a five-minute period was counted. Analyze the data below, using the Bonferroni method, to determine if there is a significant difference among species.

C57		AJ		F2
309	164	37	22	140
229	211	90	140	218
182	215	39	41	215
228	211	104	122	109
326	152	43	10	151
289	178	62	41	154
231	194	17	61	93
225	144	19	19	103
307	95	21	62	90
281	157	9	86	184
316	240	16	66	7
290	146	65	64	46
318	106	187	53	9
273	252	17	79	41
328	266	79	46	241
325	284	77	89	118
191	274	60	74	15
219	285	8	44	156
216	366	81	39	111
221	360	39	59	120
198	237	133	29	163
181	270	102	13	101
110	114	36	11	170
256	176	19	23	225
240	224	53	40	177
122		59		72
290		29		288
253		47		129

20. Samples of each of three types of stopwatches were tested. The following table gives thousands of cycles (on-off-restart) survived until some part of the mechanism failed (Natrella, 1963). Test whether there is a significant difference among the types, and if there is, determine which types are significantly different. Use both a parametric and a nonparametric technique.

Type I	Type II	Type III
1.7	13.6	13.4
1.9	19.8	20.9
6.1	25.2	25.1
12.5	46.2	29.7
16.5	46.2	46.9
25.1	61.1	
30.5		
42.1		
82.5		

21. The table below gives the survival times (in hours) for animals in an experiment whose design consisted of three poisons, four treatments, and four observations per cell.

(a) Conduct a two-way analysis of variance to test the effects of the two main factors and their interaction.

(b) Box and Cox (1964) analyzed the reciprocals of the data, pointing out that the reciprocal of a survival time can be interpreted as the rate of death. Conduct a two-way analysis of variance, and compare to the results of part (a). Comment on how well the standard two-way analysis of variance model fits and on the interaction in both analyses.

Poison	*Treatment*							
	A		*B*		*C*		*D*	
I	3.1	4.5	8.2	11.0	4.3	4.5	4.5	7.1
	4.6	4.3	8.8	7.2	6.3	7.6	6.6	6.2
II	3.6	2.9	9.2	6.1	4.4	3.5	5.6	10.0
	4.0	2.3	4.9	12.4	3.1	4.0	7.1	3.8
III	2.2	2.1	3.0	3.7	2.3	2.5	3.0	3.6
	1.8	2.3	3.8	2.9	2.4	2.2	3.1	3.3

13

The Analysis of Categorical Data

13.1 Introduction

This chapter introduces the analysis of data that are in the form of counts in various categories. We will deal primarily with two-way tables. Suppose that the rows of such a table represent various hair colors and the columns various eye colors and that each cell contains a count of the number of people who fall in that particular cross-classification. We might be interested in dependencies between the row and column classifications—that is, is hair color related to eye color?

We emphasize that the data considered in this chapter are counts, rather than continuous measurements as they were in chapter 12. Thus, in this chapter, we will make heavy use of the multinomial and chi-square distributions.

13.2 Fisher's Exact Test

Fisher's exact test can best be developed in the context of the following example. Rosen and Jerdee (1974) conducted several experiments, using as subjects male bank supervisors attending a management institute. As part of their training, the supervisors had to make decisions on items in an in-basket. The investigators embedded their experimental materials in the contents of the in-baskets. In one experiment, the supervisors were given a personnel file and had to decide whether to promote the employee or to hold the file and interview additional candidates. By random selection, 24 of the supervisors examined a file labeled as being that of a male employee and 24 examined a file labeled as being that of a female employee; the files were otherwise identical. The results are summarized in the following table:

	Male	Female
Promote	21	14
Hold File	3	10

From the results, it appears that there is a sex bias—21 of 24 males were promoted, but only 14 of 24 females were. But someone who was arguing against the presence of sex bias could claim that the results occurred by chance; that is, even if there were no bias and the supervisors were completely indifferent to sex, discrepancies like those observed could occur with fairly large probability by chance alone. To rephrase this argument, the claim is that 35 of the 48 supervisors chose to promote the employee and 13 chose not to, and that 21 of the 35 promotions were of a male employee merely because of the random assignment of supervisors to male and female files.

The strength of the argument against sex bias must be assessed by a calculation of probability. If it is likely that the randomization could result in such an imbalance, the argument is difficult to refute; however, if only a small proportion of all possible randomizations would give such an imbalance, the argument has less force. We take as the null hypothesis that there is no sex bias and that any differences observed are due to the randomization. We denote the counts in the table and on the margins as follows:

N_{11}	N_{12}	$n_{1.}$
N_{21}	N_{22}	$n_{2.}$
$n_{.1}$	$n_{.2}$	$n_{..}$

According to the null hypothesis, the margins of the table are fixed: There are 24 females, 24 males, 35 supervisors who choose to promote, and 13 who choose not to. Also, the process of randomization determines the counts in the interior of the table (denoted by capital letters since they are random) subject to the constraints on the margins. With these constraints, there is only 1 degree of freedom in the interior of the table; if any interior count is fixed, the others may be determined.

Consider the count N_{11}, the number of males who are promoted. Under the null hypothesis, the distribution of N_{11} is that of the number of successes in 24 draws without replacement from a population of 35 successes and 13 failures; that is, the distribution of N_{11} induced by the randomization is hypergeometric. The probability that $N_{11} = n_{11}$ is

$$p(n_{11}) = \frac{\binom{n_{1.}}{n_{11}}\binom{n_{2.}}{n_{21}}}{\binom{n_{..}}{n_{.1}}}$$

We will use N_{11} as the test statistic for testing the null hypothesis. The hypergeometric probability distribution above is the null distribution of N_{11} and is tabled below. A two-sided test rejects for extreme values of N_{11}.

n_{11}	$p(n_{11})$
11	.000
12	.000
13	.004
14	.021
15	.072
16	.162
17	.241
18	.241
19	.162
20	.072
21	.021
22	.004
23	.000
24	.000

From this table, a rejection region for a two-sided test with $\alpha = .05$ consists of the following values for N_{11}: 11, 12, 13, 14, 21, 22, 23, and 24. The observed value of N_{11} falls in this region, so the test would reject at level .05. Thus, we would conclude that there was indeed sex bias. An imbalance in promotions as or more extreme than that observed would only occur by chance with probability .05.

13.3   The Chi-Square Test of Homogeneity

Suppose that we have independent observations from J multinomial distributions, each of which has I categories and that we wish to test whether the cell probabilities of the multinomials are equal—that is, to test the homogeneity of the multinomial distributions.

As an example, we will consider a quantitative study of an aspect of literary style. Several investigators have used probabilistic models of word counts as indices of literary style, and statistical techniques applied to such counts have been used in controversies about disputed authorship. An interesting account is given by Morton (1978), from whom we take the following example.

When Jane Austen died, she left the novel *Sanditon* only partially completed, but she left a summary of the remainder. A highly literate admirer finished the novel, attempting to emulate Austen's style, and the hybrid was published. Morton counted the occurrences of various words in several works: chapters 1 and 3 of *Sense and Sensibility*, chapters 1, 2, and 3 of *Emma*, chapters 1 and 6 of *Sanditon* (written by Austen); and chapters 12 and 24 of *Sanditon* (written by her admirer). The counts Morton obtained for six words are given in the following table:

Word	Sense and Sensibility	Emma	Sanditon I	Sanditon II
a	147	186	101	83
an	25	26	11	29
this	32	39	15	15
that	94	105	37	22
with	59	74	28	43
without	18	10	10	4
Total	375	440	202	196

We will compare the relative frequencies with which these words appear and will examine the consistency of Austen's usage of them from book to book and the degree to which her admirer was successful in imitating this aspect of her style. A stochastic model will be used for this purpose: The six counts for *Sense and Sensibility* will be modeled as a realization of a multinomial random variable with unknown cell probabilities and total count 375; the counts for the other works will be similarly modeled as independent multinomial random variables.

Thus, we must consider comparing J multinomial distributions each having I categories. If the probability of the ith category of the jth multinomial is denoted π_{ij}, the null hypothesis to be tested is

$$H_0: \pi_{i1} = \pi_{i2} = \cdots = \pi_{iJ}, \quad i = 1, \ldots, I$$

We may view this as a goodness-of-fit test: Does the model prescribed by the null hypothesis fit the data? To test goodness of fit, we will compare observed values with expected values as in chapter 9, using likelihood ratio statistics or Pearson's chi-square statistic. We will assume that the data consist of independent samples from each multinomial distribution, and we will denote the count in the ith category of the jth multinomial as n_{ij}.

Under H_0, each of the J multinomials has the same probability for the ith category, say π_i. The following theorem shows that the maximum likelihood estimate of π_i is simply $n_{i.}/n_{..}$, which is an obvious estimate. Here, $n_{i.}$ is the total count in the ith category, $n_{..}$ is the grand total count, $n_{.j}$ is the total count for the jth multinomial.

THEOREM A.   Under H_0, the maximum likelihood estimates of the parameters $\pi_1, \pi_2, \ldots, \pi_I$ are

$$\hat{\pi}_i = \frac{n_{i.}}{n_{..}}, \quad i = 1, \ldots, I$$

where $n_{i.}$ is the total number of responses in the ith category and $n_{..}$ is the grand total number of responses.

Proof.   Since the multinomial distributions are independent,

$$\mathrm{lik}(\pi_1, \pi_2, \ldots, \pi_I) = \prod_{j=1}^{J} \binom{n_{.j}}{n_{1j} n_{2j} \cdots n_{Ij}} \pi_1^{n_{1j}} \pi_2^{n_{2j}} \cdots \pi_I^{n_{Ij}}$$

$$= \pi_1^{n_{1.}} \pi_2^{n_{2.}} \cdots \pi_I^{n_{I.}} \prod_{j=1}^{I} \binom{n_{.j}}{n_{1j} n_{2j} \cdots n_{Ij}}$$

Let us consider maximizing the log likelihood subject to the constraint $\sum_{i=1}^{I} \pi_i = 1$. Introducing a Lagrange multiplier, we have to maximize

$$l(\pi, \lambda) = \sum_{j=1}^{J} \log \binom{n_{.j}}{n_{1j} n_{2j} \cdots n_{Ij}} + \sum_{i=1}^{I} n_{i.} \log \pi_i + \lambda \left(\sum_{i=1}^{I} \pi_i - 1 \right)$$

Now,

$$\frac{\partial l}{\partial \pi_{i.}} = \frac{n_{i.}}{\pi_{i.}} + \lambda, \quad i = 1, \ldots, I$$

or

$$\hat{\pi}_i = \frac{n_{i.}}{\lambda}$$

Summing over both sides and applying the constraint, we find $\lambda = n_{..}$ and $\hat{\pi}_i = n_{i.}/n_{..}$, as was to be proved.   □

For the jth multinomial, the expected count in the ith category is the estimated probability of that cell times the total number of observations for the jth multinomial, or

$$E_{ij} = \frac{n_{.j} n_{i.}}{n_{..}}$$

Pearson's chi-square statistic is therefore

$$X^2 = \sum_{i=1}^{I} \sum_{j=1}^{J} \frac{(O_{ij} - E_{ij})^2}{E_{ij}}$$

$$= \sum_{i=1}^{I} \sum_{j=1}^{J} \frac{(n_{ij} - n_{i.} n_{.j}/n_{..})^2}{n_{i.} n_{.j}/n_{..}}$$

For large sample sizes, the approximate null distribution of this statistic is chi-square. (The usual recommendation concerning the sample size necessary for this approximation to be reasonable is that the expected counts should all be greater than 5.) The degrees of freedom are the number of independent counts minus the number of independent parameters estimated from the data. Each multinomial has $I - 1$ independent counts, since the totals are fixed, and $I - 1$ independent parameters have been estimated. The degrees of freedom are therefore

$$\text{df} = J(I - 1) - (I - 1) = (I - 1)(J - 1)$$

We now apply this method to the word counts from Austen's works. First, we consider Austen's consistency from one work to another. The table below gives the observed count and below it the expected count in each cell of the table.

Word	Sense and Sensibility	Emma	Sanditon I
a	147	186	101
	160.0	187.8	86.2
an	25	26	11
	22.9	26.8	12.3
this	32	39	15
	31.7	37.2	17.1
that	94	105	37
	87.0	102.1	46.9
with	59	74	28
	59.4	69.7	32.0
without	18	10	10
	14.0	16.4	7.5

The observed counts look fairly close to the expected counts, and the chi-square statistic is 12.27. The 10% point of the chi-square distribution with 10 degrees of freedom is 15.99, and the 25% point is 12.54. The data are thus consistent with the model that the word counts in the three works are realizations of multinomial random variables with the same underlying probabilities. This aspect of Austen's style did not change from work to work.

To compare Austen and her imitator, we can pool all Austen's work together in light of the above findings. The following table shows the observed and expected frequencies for the imitator and Austen:

Word	Imitator	Austen
a	83	434
	83.5	433.5
an	29	62
	14.7	76.3
this	15	86
	16.3	84.7
that	22	236
	41.7	216.3
with	43	161
	33.0	171.0
without	4	38
	6.8	35.2

The chi-square statistic is 32.81 with 5 degrees of freedom, giving a p-value of less than .001. The imitator was not successful in imitating this aspect of Austen's style. To see which discrepancies are large, it is helpful to examine the contributions to the chi-square statistic cell by cell, as tabulated here:

Word	Imitator	Austen
a	0.00	0.00
an	13.90	2.68
this	0.11	0.02
that	9.30	1.79
with	3.06	0.59
without	1.14	0.22

Inspecting this and the preceding table, we see that the relative frequency with which Austen used the word *an* was smaller than that with which her imitator used it, and that the relative frequency with which she used *that* was larger.

13.4   The Chi-Square Test of Independence

This section develops a chi-square test that is very similar to the one of the preceding section but is aimed at answering a different question. We will again use an example.

In a demographic study of women who were listed in *Who's Who*, Kiser and Schaefer (1949) compiled the following table for 1436 women who were married at least once:

Education	Married Once	Married More Than Once	Total
College	550	61	611
No College	681	144	825
Total	1231	205	1436

Is there a relationship between marital status and educational level? Of the women who had a college degree, $\frac{61}{611} = 10\%$ were married more than once; of those who had no college degree, $\frac{144}{825} = 17\%$ were married more than once. Alternatively, the question might be addressed by noting that of those women who were married more than once, $\frac{61}{205} = 30\%$ had a college degree, whereas of those married only once, $\frac{550}{1231} = 45\%$ had a college degree. For this sample of 1436 women, having a college degree is positively associated with being married only once, but it is impossible to make causal inferences from the data. Marital stability could be influenced by educational level, or both characteristics could be influenced by other factors, such as social class.

A critic of the study could in any case claim that the relationship between marital status and educational level is "statistically insignificant." Since the data are not a sample from any population, and since no randomization has been performed, the role of probability and statistics is not clear. One might respond to such a criticism by saying that the data speak for themselves and that there is no chance mechanism at work. The critic might then rephrase his argument: "If I were to take a sample of 1436 people cross-classified into two categories which were in fact unrelated in the population from which the sample was drawn, I might find associations as strong or stronger than those observed in this table. Why should I believe that there is any real association in your table?" Even though this argument may not seem compelling, statistical tests are often carried out in situations in which stochastic mechanisms are at best hypothetical.

We will discuss statistical analysis of a sample of size n cross-classified in a table with I rows and J columns. Such a configuration is called a **contingency table**. The joint distribution of the counts n_{ij}, where $i = 1, \ldots, I$ and $j = 1, \ldots, J$, is multinomial with cell probabilities denoted as π_{ij}. Let

$$\pi_{i.} = \sum_{j=1}^{J} \pi_{ij}$$

$$\pi_{.j} = \sum_{i=1}^{I} \pi_{ij}$$

denote the marginal probabilities that an observation will fall in the ith row and jth column, respectively. If the row and column classifications are independent of each other,

$$\pi_{ij} = \pi_{i.}\pi_{.j}$$

We thus consider testing the following null hypothesis:

$$H_0 : \pi_{ij} = \pi_{i.}\pi_{.j}, \quad i = 1, \ldots, I, \quad j = 1, \ldots, J$$

versus the alternative that the π_{ij} are free. Under H_0, the maximum likelihood estimate of π_{ij} is

$$\hat{\pi}_{ij} = \hat{\pi}_{i.}\hat{\pi}_{.j}$$

$$= \frac{n_{i.}}{n} \times \frac{n_{.j}}{n}$$

(see Problem 8 at the end of this chapter). Under the alternative, the maximum likelihood estimate of π_{ij} is simply

$$\tilde{\pi}_{ij} = \frac{n_{ij}}{n}$$

These estimates can be used to form a likelihood ratio test or an asymptotically equivalent Pearson's chi-square test,

$$X^2 = \sum_{i=1}^{I} \sum_{j=1}^{J} \frac{(O_{ij} - E_{ij})^2}{E_{ij}}$$

Here the O_{ij} are the observed counts (n_{ij}). The expected counts, the E_{ij}, are the fitted counts:

$$E_{ij} = n\hat{\pi}_{ij} = \frac{n_{i.}n_{.j}}{n}$$

The Pearson's chi-square statistic is therefore

$$X^2 = \sum_{i=1}^{I} \sum_{j=1}^{J} \frac{(n_{ij} - n_{i.}n_{.j}/n)^2}{n_{i.}n_{.j}/n}$$

To count the degrees of freedom, we note that there are $IJ - 1$ independent counts, since the grand total is fixed. Also, $(I - 1) + (J - 1)$ independent parameters have been estimated from the data. Thus,

$$\text{df} = IJ - 1 - (I - 1) - (J - 1) = (I - 1)(J - 1)$$

Returning to the data on 1436 women from the demographic study, we calculate expected values and construct the following table:

Education	Married Once	More than Once
College	550 523.8	61 87.2
No College	681 707.2	144 117.8

The chi-square statistic is 16.01 with 1 degree of freedom, giving a p-value less than .001. We would reject the hypothesis of independence and conclude that there is a relationship between marital status and educational level.

The chi-square statistic used here to test independence was used in the preceding section to test homogeneity; however, the hypotheses are different and the sampling schemes are different. The test of homogeneity was derived under the assumption that the column (or row) margins were fixed, and the test of independence was derived under the assumption that only the grand total was fixed. Because the test statistics are identical, the distinction between the two tests is often slurred over. Furthermore, a test of independence can be regarded as a test of homogeneity that is conditional on one of the margins.

13.5 Matched-Pairs Designs

Matched-pairs designs can be effective for experiments involving categorical data; as with experiments involving continuous data, pairing can control for extraneous sources of variability and can increase the power of a statistical test. Appropriate techniques, however, must be used in the analysis of the data. This section presents an extended example illustrating these concepts.

Vianna, Greenwald, and Davies (1971) collected data comparing the percentages of tonsillectomies for a group of patients suffering from Hodgkin's disease and a comparable control group:

	Tonsillectomy	No Tonsillectomy
Hodgkin's	67	34
Control	43	64

The table shows that 66% of the Hodgkin's sufferers had had a tonsillectomy, compared to 40% of the control group. The chi-square test for homogeneity gives a chi-square statistic of 14.26 with 1 degree of freedom, which is highly significant. The investigators conjectured that the tonsils act as a protective barrier in some fashion against Hodgkin's disease.

Johnson and Johnson (1972) selected 85 Hodgkin's patients who had a sibling of the same sex who was free of the disease and whose age was within five years of the patient's. These investigators presented the following table:

	Tonsillectomy	No Tonsillectomy
Hodgkin's	41	44
Control	33	52

They calculated a chi-square statistic of 1.53, which is not significant. Their findings thus appeared to be at odds with those of Vianna, Greenwald, and Davies.

Several letters to the editor of the journal that published Johnson and Johnson's results pointed out that those investigators had made an error in their analysis by ignoring the pairings. The assumption behind the chi-square test of homogeneity is that independent multinomial samples are compared, and Johnson and Johnson's samples were not independent, because siblings were paired. An appropriate analysis of Johnson and Johnson's data is suggested once we set up a table that exhibits the pairings:

		Sibling	
		No Tonsillectomy	Tonsillectomy
Patient	No Tonsillectomy	37	7
	Tonsillectomy	15	26

Viewed in this way, the data are a sample of size 85 from a multinomial distribution with four cells. We can represent the probabilities in the table as follows:

π_{11}	π_{12}	$\pi_{1.}$
π_{21}	π_{22}	$\pi_{2.}$
$\pi_{.1}$	$\pi_{.2}$	1

The appropriate null hypothesis states that $\pi_{1.} = \pi_{.1}$ and $\pi_{2.} = \pi_{.2}$, or

$$\pi_{11} + \pi_{12} = \pi_{11} + \pi_{21}$$

$$\pi_{12} + \pi_{22} = \pi_{21} + \pi_{22}$$

These equations simplify to $\pi_{12} = \pi_{21}$.

The relevant null hypothesis is thus

$$H_0: \pi_{12} = \pi_{21}$$

We will derive a test, called **McNemar's test**, of this hypothesis. Under H_0, the maximum likelihood estimates of the cell probabilities are (see Problem 8 at the end of this chapter)

$$\hat{\pi}_{11} = \frac{n_{11}}{n}$$

$$\hat{\pi}_{22} = \frac{n_{22}}{n}$$

$$\hat{\pi}_{12} = \hat{\pi}_{21} = \frac{n_{12} + n_{21}}{2n}$$

The contributions to the chi-square statistic from the n_{11} and n_{22} cells are equal to zero; the remainder of the statistic is

$$X^2 = \frac{[n_{12} - (n_{12} + n_{21})/2]^2}{(n_{12} + n_{21})/2} + \frac{[n_{21} - (n_{12} + n_{21})/2]^2}{(n_{12} + n_{21})/2}$$

$$= \frac{(n_{12} - n_{21})^2}{n_{12} + n_{21}}$$

There are four cells in the table, and three parameters have been estimated under H_0 (π_{11}, π_{22}, and $\pi_{12} = \pi_{21}$); however, since the probabilities are constrained to sum to 1, only two of these parameters are independent. The chi-square statistic thus has 1 degree of freedom. For the data table exhibiting the pairings, $X^2 = 2.91$, with a corresponding p-value of .09. This casts doubt on the null hypothesis, contrary to Johnson and Johnson's original analysis.

We close this section with a digression concerning the interpretation of tables such as that of Vianna, Greenwald, and Davies. Let H denote the event that a person has had Hodgkin's disease, and $\bar{H}$ the event that he or she has not; similarly, let T and $\bar{T}$ denote, respectively, the events of having had and not having had a tonsillectomy. A measure of the risk of having a tonsillectomy is

$$\frac{P(H|T)}{P(H|\bar{T})} = \frac{P(\bar{T})P(H \cap T)}{P(T)P(H \cap \bar{T})}$$

If Hodgkin's disease is rare, $P(H) \approx 0$, so

$$P(T) = P(T|H)P(H) + P(T|\bar{H})P(\bar{H})$$

$$\approx P(T|\bar{H})P(\bar{H})$$

Similarly,

$$P(\bar{T}) \approx P(\bar{T}|\bar{H})P(\bar{H})$$

Furthermore,

$$P(H \cap T) = P(T|H)P(H)$$

$$P(H \cap \bar{T}) = P(\bar{T}|H)P(H)$$

Substituting into the expression for a measure of relative risk, we have

$$\frac{P(H|T)}{P(H|\bar{T})} \approx \frac{P(\bar{T}|\bar{H})P(T|H)}{P(T|\bar{H})P(\bar{T}|H)}$$

These probabilities may be estimated from the table of Vianna, Greenwald, and Davies. The estimate of the relative risk is simply the product of the diagonals divided by the product of the off-diagonals, or

$$\text{Relative risk} \approx \frac{67 \times 64}{43 \times 34} = 2.93$$

According to this measure, a person who has undergone a tonsillectomy is about three times more likely to contract Hodgkin's disease than a person who did not have a tonsillectomy.

13.6 Concluding Remarks

This chapter has introduced two-way classifications, which are the simplest form of contingency tables. For higher-order classifications, which frequently occur in practice, a greater variety of forms of dependence arise. For example, for a three-way table, the factors of which are denoted A, B, and C, we might consider testing whether, conditionally on C, A and B are independent.

In recent years, there has been a great deal of activity in the development and application of **log-linear models**. If the row and column classifications in a two-way table are independent, then

$$\pi_{ij} = \pi_i \pi_j$$

or

$$\log \pi_{ij} = \log \pi_i + \log \pi_j$$

We can denote $\log \pi_i$ by α_i and $\log \pi_j$ by β_j. Then, if there is dependence, $\log \pi_{ij}$

may be written as

$$\log \pi_{ij} = \alpha_i + \beta_j + \gamma_{ij}$$

This form mimics the additive analysis of variance models introduced in chapter 12. The idea can readily be extended to higher-order tables. For example, a possible model for a three-way table is

$$\log \pi_{ijk} = \alpha_i + \beta_j + \delta_{ij} + \varepsilon_{ik} + \gamma_{jk}$$

which allows second-order dependencies, but no third-order dependencies. The parameters of log-linear models may be estimated by maximum likelihood estimates and likelihood ratio tests may be employed. Haberman (1978) and Bishop, Feinberg, and Holland (1975) develop these ideas and apply them to quite complicated models.

13.7 Problems

1. Overfield and Klauber (1980) published the following data on the incidence of tuberculosis in relation to blood groups in a sample of Eskimos. Is there any association of the disease and blood group within the ABO system or within the MN system?

ABO system

Severity	O	A	AB	B
Moderate–Advanced	7	5	3	13
Minimal	27	32	8	18
Not Present	55	50	7	24

MN system

Severity	MM	MN	NN
Moderate–Advanced	21	6	1
Minimal	54	27	5
Not Present	74	51	11

2. In a famous sociological study called *Middletown*, Lynd and Lynd (1956) administered questionnaires to 784 white high school students. The students were asked which two of ten given attributes were most desirable in their fathers. The following table shows how the desirability of the attribute "being

a college graduate" was rated by male and female students. Did the males and females value this attribute differently?

	Male	Female
Mentioned	86	55
Not Mentioned	283	360

3. Dowdall (1974) (also discussed in Haberman, 1978) studied the effect of ethnic background on role attitude of women of ages 15–64 in Rhode Island. Respondents were asked whether they thought it was all right for a woman to have a job instead of taking care of the house and children while her husband worked. The following table breaks down the responses by ethnic origin of the respondent. Is there a relationship between response and ethnic group? If so, describe it.

Ethnic Origin	Yes	No
Italian	78	47
Northern European	56	29
Other European	43	29
English	53	32
Irish	43	30
French Canadian	36	22
French	42	23
Portuguese	29	7

4. It is conventional wisdom in military squadrons that pilots tend to father more girls than boys. Snyder (1961) gathered data for military fighter pilots. The sex of the pilots' offspring were tabulated for three kinds of flight duty during the month of conception, as shown in the following table. Is there any significant difference between the three groups? In the United States in 1950, 105.37 males were born for every 100 females. Are the data consistent with this sex ratio?

Father's Activity	Female Offspring	Male Offspring
Flying Fighters	51	38
Flying Transports	14	16
Not Flying	38	46

5. Grades in an elementary statistics class were classified by the students' majors. Is there any relationship between grade and major?

Grade	Major		
	Psychology	*Biology*	*Other*
A	8	15	13
B	14	19	15
C	15	4	7
D–F	3	1	4

6. A randomized double-blind experiment compared the effectiveness of several drugs in ameliorating postoperative nausea. All patients were anesthetized with nitrous oxide and ether. The following table shows the incidence of nausea during the first four postoperative hours for each of several drugs and a placebo (Beecher, 1959). Compare the drugs to each other and to the placebo.

	Number of Patients	*Incidence of Nausea*
Placebo	165	95
chlorpromazine	152	52
dimenhydrinate	85	52
pentabarbital (100 mg)	67	35
pentabarbital (150 mg)	85	37

7. This problem considers some more data on Jane Austen and her imitator (Morton, 1978). The table below gives the relative frequency of the word *a* preceded by (PB) and not preceded by (NPB) the word *such*, the word *and* followed by (FB) or not followed by (NFB) *I*, and the word *the* preceded by *and* not preceded by *on*.

Words	*Sense and Sensiblity*	*Emma*	*Sanditon I*	*Sanditon II*
a PB *such*	14	16	8	2
a NPB *such*	133	180	93	81
and FB *I*	12	14	12	1
and NFB *I*	241	285	139	153
the PB *on*	11	6	8	17
the NPB *on*	259	265	221	204

Was Austen consistent in these habits of style from one work to another? Did her imitator successfully copy this aspect of her style?

8. Verify that the mle's of the cell probabilities, π_{ij}, are as given in Section 13.4 for the test of independence and in Section 13.5 for McNemar's test.

9. (a) Derive the likelihood ratio test of homogeneity. (b) Calculate the likelihood ratio test statistic for the example of Section 13.3, and compare it to Pearson's chi-square statistic. (c) Derive the likelihood ratio test of independence. (d) Calculate the likelihood ratio test statistic for the example of Section 13.4, and compare it to Pearson's chi-square statistic.

10. Show that McNemar's test is nearly equivalent to scoring each response as a 0 or a 1 and calculating a paired-sample t test on the resulting data.

11. A sociologist is studying influences on family size. He finds pairs of sisters, both of whom are married, and determines for each sister whether she has 0, 1, or 2 or more children. He wishes to compare older and younger sisters. Explain what the following hypotheses mean and how to test them.
    (a) The number of children the younger sister has is independent of the number of children the older sister has.
    (b) The distribution of family sizes is the same for older and younger sisters. Could one hypothesis be true and the other false? Explain.

12. Lazarsfeld, Berelson, and Gaudet (1948) present the following tables relating degree of interest in political elections to education and age:

No high school education

Degree of Interest	Under 45	Over 45
Great	71	217
Little	305	652

Some high school or more

Degree of Interest	Under 45	Over 45
Great	305	180
Little	869	259

Since there are three factors—education, age, and interest—these tables considered jointly are more complicated than the tables considered in this chapter.
    (a) Examine the tables informally and analyze the dependence of interest in political elections on age and education. What do the numbers suggest?
    (b) Extend the ideas of this chapter to test two hypotheses, H_1: given educational level, age and degree of interest are unrelated, and H_2: given age, educational level and degree of interest are unrelated.

13. Reread Section 11.4.5, which contains a discussion of methodological problems in a study of the effects of FD&C Red No. 40. The following tables give the numbers of mice that developed RE tumors in each of several groups:

Incidence among males

	Control 1	Control 2	Low Dose	Med. Dose	High Dose
Number with tumor	25	10	20	9	17
Total number	100	100	99	100	99

Incidence among females

	Control 1	Control 2	Low Dose	Med. Dose	High Dose
Number with tumor	33	25	32	26	22
Total number	100	99	99	99	100

Use chi-square tests to compare the incidences in the different groups for males and for females. Which differences are significant? What would you conclude from this analysis had you not known about the possibility of cage position effects?

14. A market research team conducted a survey to investigate the relationship of personality to attitude toward small cars. A sample of 250 adults in a metropolitan area were asked to fill out a 16-item self-perception questionnaire, on the basis of which they were classified into three types: cautious conservative, middle-of-the-roader, and confident explorer. They were then asked to give their overall opinion of small cars: favorable, neutral, or unfavorable. Is there a relationship between personality type and attitude toward small cars? If so, what is the nature of the relationship?

Personality Type

Attitude	Cautious	Mid-road	Explorer
Favorable	79	58	49
Neutral	10	8	9
Unfavorable	10	34	42

15. In a study of the relation of blood type to various diseases, the following data were gathered in London and Manchester (Woolf, 1955):

London

	Control	Peptic Ulcer
Group A	4219	579
Group O	4578	911

Manchester

	Control	Peptic Ulcer
Group A	3775	246
Group O	4532	361

First, consider the two tables separately. Is there a relationship between blood type and propensity to peptic ulcer? If so, evaluate the strength of the relationship. Are the data from London and Manchester comparable?

16. Records of 317 patients at least 48 years old who were diagnosed as having endometrial carcinoma were obtained from two hospitals (Smith et al., 1975). Matched controls for each case were obtained from the two institutions; the controls had cervical cancer, ovarian cancer, or carcinoma of the vulva. Each control was matched by age at diagnosis (within four years) and year of diagnosis (within two years) to a corresponding case of endometrial carcinoma. This sort of design, called a **retrospective case-control study**, is frequently used in medical investigations where a randomized experiment is not possible. The following table gives the numbers of cases and controls who had taken estrogen for at least six months prior to the diagnosis of cancer. Is there a significant relationship between estrogen use and endometrial cancer? Do you see any possible weak points in a retrospective case-control design?

| | | Controls | |
		Estrogen Used	Not Used
Cases	Estrogen Used	39	113
	Not Used	15	150

17. A psychological experiment was done to investigate the effect of anxiety on a person's desire to be alone or in company (Schacter, 1959; Lehman, 1975). A group of 30 subjects was randomly divided into two groups of sizes 13 and 17. The subjects were told that they would be subjected to some electric shocks, but one group was told that the shocks would be quite painful and the other group was told that they would be mild and painless. The former group was the "high-anxiety" group, and the latter was the "low-anxiety" group. Both groups were told that there would be a ten-minute wait before the experiment began, and each subject was given the choice of waiting alone or with the other subjects. The following are the results:

	Wait Together	Wait Alone
High-Anxiety	12	5
Low-Anxiety	4	9

Use Fisher's exact test to test whether there is a significant difference between the high- and low-anxiety groups.

14

Linear Least Squares

14.1 Introduction

In order to fit a straight line to a plot of points (x_i, y_i), where $i = 1, \ldots, n$, the slope and intercept of the line $y = \beta_0 + \beta_1 x$ must be found from the data in some manner. In order to fit a pth-order polynomial, $p + 1$ coefficients must be determined. Other functional forms besides linear and polynomial ones may be fit to data, and then parameters associated with those forms must be determined.

The most common, but by no means only, method for determining the parameters in curve-fitting problems is the method of least squares. The principle underlying this method is to minimize the sum of squared deviations of the predicted, or fitted, values (given by the curve) from the actual observations. For example, suppose that a straight line is to be fit to the points (y_i, x_i), where $i = 1, \ldots, n$; y is called the **dependent variable** and x is called the **independent variable**, and we wish to predict y from x. (This usage of the terms *independent* and *dependent* is different from their probabilistic meaning.) Sometimes x and y are called the **predictor variable** and the **response variable**, respectively. Applying the method of least squares, we choose the slope and intercept of the straight line to minimize

$$S(\beta_0, \beta_1) = \sum_{i=1}^{n} (y_i - \beta_0 - \beta_1 x_i)^2$$

Note that β_0 and β_1 are chosen to minimize the sum of squared vertical deviations, or prediction errors (see Figure 14-1). The procedure is not symmetric in y and x.

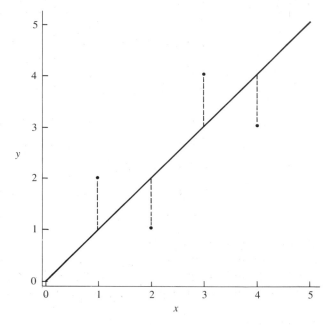

Figure 14-1.   The least squares line minimizes the sum of squared vertical distances (dotted lines) from the points to the line.

Curves are often fit to data as part of the process of calibrating instruments. For example, Bailey, Cox, and Springer (1978) discuss a method for measuring the concentrations of food dyes and other substances by high-pressure chromatography. Measurements of the chromatographic peak areas corresponding to sulfanic acid were taken for several known concentrations of FD&C Yellow No. 5. Figure 14-2 shows a plot of peak area versus percentage of FD&C Yellow. To casual examination, the plot looks fairly linear.

To find β_0 and β_1, we calculate

$$\frac{\partial S}{\partial \beta_0} = -2 \sum_{i=1}^{n} (y_i - \beta_0 - \beta_1 x_i)$$

$$\frac{\partial S}{\partial \beta_1} = -2 \sum_{i=1}^{n} x_i (y_i - \beta_0 - \beta_1 x_i)$$

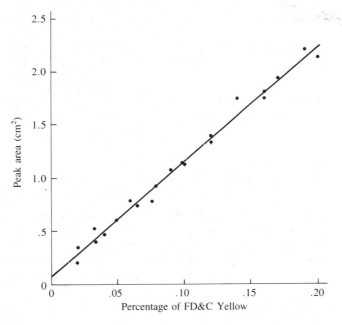

Figure 14-2.   Data points and the least squares line for the relation of sulfanic acid peak area to percentage of FD&C Yellow.

Setting these partial derivatives equal to zero, we have that the minimizers $\hat{\beta}_0$ and $\hat{\beta}_1$ satisfy

$$\sum_{i=1}^{n} y_i = n\hat{\beta}_0 + \hat{\beta}_1 \sum_{i=1}^{n} x_i$$

$$\sum_{i=1}^{n} x_i y_i = \hat{\beta}_0 \sum_{i=1}^{n} x_i + \hat{\beta}_1 \sum_{i=1}^{n} x_i^2$$

Solving for $\hat{\beta}_0$ and $\hat{\beta}_1$, we obtain

$$\hat{\beta}_0 = \frac{\left(\sum_{i=1}^{n} x_i^2\right)\left(\sum_{i=1}^{n} y_i\right) - \left(\sum_{i=1}^{n} x_i\right)\left(\sum_{i=1}^{n} x_i y_i\right)}{n \sum_{i=1}^{n} x_i^2 - \left(\sum_{i=1}^{n} x_i\right)^2}$$

$$\hat{\beta}_1 = \frac{n \sum_{i=1}^{n} x_i y_i - \left(\sum_{i=1}^{n} x_i\right)\left(\sum_{i=1}^{n} y_i\right)}{n \sum_{i=1}^{n} x_i^2 - \left(\sum_{i=1}^{n} x_i\right)^2}$$

(For alternate expressions for $\hat{\beta}_0$ and $\hat{\beta}_1$ see Problem 8 at the end of this chapter.)

The fitted line, with the parameters determined from the expressions above to be $\hat{\beta}_0 = .073$ and $\hat{\beta}_1 = 10.8$, is drawn in Figure 14-2. Is this a "reasonable" fit? How much faith do we have in these values for $\hat{\beta}_0$ and $\hat{\beta}_1$, since there is apparently some "noise" in the data? We will answer these questions in later sections of this chapter.

Functional forms more complicated than straight lines are often fit to data. For example, to determine the proper placement in college mathematics courses for entering freshmen, data that have a bearing on predicting performance in first-year calculus may be available. Suppose that score on a placement exam, high school grade-point average in math courses, and quantitative college board scores are available; we can denote these values by x_1, x_2, and x_3, respectively. We might try to predict a student's grade in first-year calculus, y, by the form

$$y \approx \beta_0 + \beta_1 x_1 + \beta_2 x_2 + \beta_3 x_3$$

where the β_i could be estimated from data on the performance of students in previous years. It could be ascertained how reliable this prediction equation was, and if it were sufficiently reliable, it could be used in a counseling program for entering freshmen.

In biological and chemical work, it is not uncommon to fit functions of the following form to decay curves:

$$f(t) = Ae^{-\alpha t} + Be^{-\beta t}$$

Note that the function f is linear in the parameters A and B and nonlinear in the parameters α and β. The parameters are determined by the method of least squares as being the minimizers of

$$S(A, B, \alpha, \beta) = \sum_{i=1}^{n} (y_i - Ae^{-\alpha t_i} - Be^{-\beta t_i})^2$$

In fitting periodic phenomena, functions of the following form occur:

$$f(t) = A \cos \omega_1 t + B \sin \omega_1 t + C \cos \omega_2 t + D \sin \omega_2 t$$

This function is linear in the parameters A, B, C, and D and nonlinear in the parameters ω_1 and ω_2.

When the function to be fit is linear in the unknown parameters, the minimization is relatively straightforward, since calculating partial derivatives and setting them equal to zero produces a set of simultaneous linear equations that can be solved in closed form. This important special case is known as **linear least squares**. If the function to be fit is not linear in the unknown parameters, a system of nonlinear equations must be solved to find the coefficients. Typically, the solution cannot be found in closed form, so an iterative procedure must be used.

For our purposes, the general formulation of the linear least squares problem is as follows: A function of the form

$$f(x_1, x_2, \ldots, x_{p-1}) = \beta_0 + \beta_1 x_1 + \beta_2 x_2 + \cdots + \beta_{p-1} x_{p-1}$$

involving p unknown parameters, $\beta_0, \beta_1, \beta_2, \ldots, \beta_{p-1}$, is to be fit to n data points,

$$y_1, x_{11}, x_{12}, \ldots, x_{1,p-1}$$

$$y_2, x_{21}, x_{22}, \ldots, x_{2,p-1}$$

$$\vdots$$

$$y_n, x_{n1}, x_{n2}, \ldots, x_{n,p-1}$$

The function $f(x)$ is called the **linear regression** of y on x. We will always assume that $p < n$, that is, that there are fewer unknown parameters than observations. Fitting a straight line clearly follows this format. A quadratic can be fit in this way by setting $x_1 = x$, and $x_2 = x^2$. If the frequencies in the trigonometric fitting problem referred to above are known, we can let $x_1 = \cos \omega_1 t$, $x_2 = \sin \omega_1 t$, $x_3 = \cos \omega_2 t$, and $x_4 = \sin \omega_2 t$ and identify the unknown amplitudes A, B, C, and D as the β_i. If the frequencies are unknown and must be determined from the data, the problem is nonlinear.

Many functions that are not initially linear in the unknowns can be put into linear form by means of a suitable transformation. An example of this type of function that occurs frequently in chemistry and biochemistry is the Arrhenius equation,

$$\alpha = C e^{-e_A/KT}$$

Here, α is the rate of a chemical reaction, C is an unknown constant called the frequency factor, e_A is the activation energy of the reaction, K is Boltzmann's constant, and T is absolute temperature. If a reaction is run at several temperatures and the rate is measured, the activation energy and the frequency factor can be estimated by fitting the equation to the data. The function as written above is nonlinear in the unknown parameters C and e_A, but

$$\log \alpha = \log C - e_A \frac{1}{KT}$$

is linear in $\log C$ and e_A. As an example, a plot of log rate versus $1/T$ for a reaction involving atomic oxygen, taken from Huie and Herron (1972), is shown in Figure 14-3. Figure 14-4 is a plot of rate versus temperature, which, in contrast, is quite nonlinear.

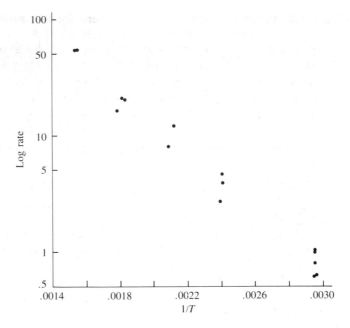

Figure 14-3.   A plot of log rate versus $1/T$ for a reaction involving atomic oxygen.

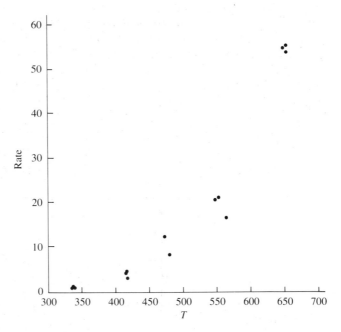

Figure 14-4.   A plot of rate versus temperature for a reaction involving atomic oxygen.

14.2 Simple Linear Regression

This section deals with the very common problem of fitting a straight line to data. Later sections of this chapter will generalize the results of this section. First, statistical properties of least squares estimates will be discussed and then methods of assessing goodness of fit, largely through the examination of residuals. Finally, the relation of regression to correlation is presented.

14.2.1 Statistical Properties of the Estimated Slope and Intercept

Up to now we have presented the method of least squares simply as a reasonable principle, without any explicit discussion of statistical models. Consequently, we have not addressed such pertinent questions as the reliability of the slope and intercept in the presence of "noise." In order to address this question, we must have a statistical model for the noise. The simplest model, which we will refer to as the standard statistical model, stipulates that the observed value of y is a linear function of x plus random noise:

$$y_i = \beta_0 + \beta_1 x_i + e_i, \quad i = 1, \ldots, n$$

Here the e_i are independent random variables with $E(e_i) = 0$ and $\text{Var}(e_i) = \sigma^2$. The x_i are assumed to be fixed.

In Section 14.1, we derived formulae for the slope, $\hat{\beta}_1$, and the intercept, $\hat{\beta}_0$. Referring to those equations, we see that they are linear functions of the y_i, and thus linear functions of the e_i. The standard statistical model thus makes computation of the means and variances of $\hat{\beta}_0$ and $\hat{\beta}_1$ straightforward.

THEOREM A. Under the assumptions of the standard statistical model, the least squares estimates are unbiased: $E(\hat{\beta}_j) = \beta_j$, for $j = 0, 1$.

Proof. From the assumptions, $E(y_i) = \beta_0 + \beta_1 x_i$. Thus, from the equation for $\hat{\beta}_0$ in Section 14.1,

$$E(\hat{\beta}_0) = \frac{\left(\sum_{i=1}^{n} x_i^2\right)\left(\sum_{i=1}^{n} E(y_i)\right) - \left(\sum_{i=1}^{n} x_i\right)\left(\sum_{i=1}^{n} x_i E(y_i)\right)}{n \sum_{i=1}^{n} x_i^2 - \left(\sum_{i=1}^{n} x_i\right)^2}$$

$$= \frac{\left(\sum_{i=1}^{n} x_i^2\right)\left(n\beta_0 + \beta_1 \sum_{i=1}^{n} x_i\right) - \left(\sum_{i=1}^{n} x_i\right)\left(\beta_0 \sum_{i=1}^{n} x_i + \beta_1 \sum_{i=1}^{n} x_i^2\right)}{n \sum_{i=1}^{n} x_i^2 - \left(\sum_{i=1}^{n} x_i\right)^2}$$

$$= \beta_0$$

The proof for β_1 is similar. $\qquad\square$

Note that the proof of Theorem A does not depend on the assumptions that the e_i are independent and have the same variance, only on the assumptions that the errors are additive and $E(e_i) = 0$.

From the standard statistical model, $\text{Var}(y_i) = \sigma^2$ and $\text{Cov}(y_i, y_j) = 0$, where $i \neq j$. This makes the computation of the variances of the $\hat{\beta}_i$ straightforward.

THEOREM B.   Under the assumptions of the standard statistical model,

$$\text{Var}(\hat{\beta}_0) = \frac{\sigma^2 \sum_{i=1}^{n} x_i^2}{n \sum_{i=1}^{n} x_i^2 - \left(\sum_{i=1}^{n} x_i \right)^2}$$

$$\text{Var}(\hat{\beta}_1) = \frac{n\sigma^2}{n \sum_{i=1}^{n} x_i^2 - \left(\sum_{i=1}^{n} x_i \right)^2}$$

$$\text{Cov}(\hat{\beta}_0, \hat{\beta}_1) = \frac{-\sigma^2 \sum_{i=1}^{n} x_i}{n \sum_{i=1}^{n} x_i^2 - \left(\sum_{i=1}^{n} x_i \right)^2}$$

Proof.   An alternative expression for $\hat{\beta}_1$ is (see Problem 8 at the end of this chapter)

$$\hat{\beta}_1 = \frac{\sum_{i=1}^{n} (x_i - \bar{x})(y_i - \bar{y})}{\sum_{i=1}^{n} (x_i - \bar{x})^2} = \frac{\sum_{i=1}^{n} (x_i - \bar{x})y_i}{\sum_{i=1}^{n} (x_i - \bar{x})^2}$$

We then have

$$\text{Var}(\hat{\beta}_1) = \frac{\sigma^2}{\sum_{i=1}^{n} (x_i - \bar{x})^2}$$

which reduces to the desired expression. The other expressions may be derived similarly. Later we will give a more general proof. $\square$

From Theorem B, we see that the variances of the slope and intercept depend on the x_i and on the error variance, σ^2. The x_i are known; therefore, to estimate the variance of the slope and intercept, we only need to estimate σ^2. Since, in the standard statistical model, σ^2 is the expected squared deviation of the y_i from the line $\beta_0 + \beta_1 x_i$, it is natural to base an estimate of σ^2 on the average squared deviations of the data about the fitted line. We define the **residual sum of squares**

(RSS) to be

$$\text{RSS} = \sum_{i=1}^{n} (y_i - \hat{\beta}_0 - \hat{\beta}_1 x_i)^2$$

We will show in Section 14.4.3 that

$$s^2 = \frac{\text{RSS}}{n-2}$$

is an unbiased estimate of σ^2. The divisor $n - 2$ is used rather than n because two parameters have been estimated from the data, giving $n - 2$ degrees of freedom.

The variances of $\hat{\beta}_0$ and $\hat{\beta}_1$ as given in Theorem B are thus estimated by replacing σ^2 by s^2, yielding estimates that we will denote $s^2_{\hat{\beta}_0}$ and $s^2_{\hat{\beta}_1}$.

If the errors, e_i, are independent normal random variables, then the estimated slope and intercept, being linear combinations of independent random variables, are normally distributed as well. More generally, if the e_i are independent and the x_i satisfy certain assumptions, a version of the central limit theorem implies that, for large n, the estimated slope and intercept are approximately normally distributed. The normality assumption, or its approximation, makes possible the construction of confidence intervals and hypothesis tests. It can then be shown that

$$\frac{\hat{\beta}_i - \beta_i}{s_{\hat{\beta}_i}} \sim t_{n-2}$$

which means that the t distribution can be used.

EXAMPLE A. We apply these procedures to the 21 data points on chromatographic peak area. The following table presents some of the statistics from the fit (tables like this are produced by regression programs of software packages):

Coefficient	Estimate	Standard Error	t Value
β_0	.0729	.0297	2.45
β_1	10.77	.27	40.20

The estimated standard deviation of the errors is $s = .068$. The standard error of the intercept is $s_{\hat{\beta}_0} = .0297$. A 95% confidence interval for the intercept, β_0, based on the t distribution with 19 df is

$$\hat{\beta}_0 \pm t_{19}(.025)s_{\hat{\beta}_0}$$

or $(.011, .135)$. Similarly, a 95% confidence interval for the slope, β_1, is

$$\hat{\beta}_1 \pm t_{19}(.025)s_{\hat{\beta}_1}$$

or $(10.21, 11.33)$. To test the null hypothesis $H_0: \beta_0 = 0$, we would use the t statistic $\hat{\beta}_0/s_{\hat{\beta}_0} = 2.45$. The hypothesis would be rejected at significance level $\alpha = .05$. □

14.2.2   Assessing the Fit

As an aid in assessing the quality of the fit, we will make extensive use of the residuals, which are the differences between the observed and fitted values:

$$\hat{e}_i = y_i - \hat{\beta}_0 - \hat{\beta}_1 x_i$$

It is most useful to examine the residuals graphically. Plots of the residuals versus the x values may reveal systematic misfit or ways in which the data do not conform to the standard statistical model. Ideally, the residuals should show no relation to the x values, and the plot should look like a horizontal blur.

EXAMPLE A.   Figure 14-5 is a plot of the residuals for the data on chromatographic peak area. There is no apparent deviation from randomness in the residuals, so this plot confirms the impression from Figure 14-2 that it is reasonable to model the relation as linear. □

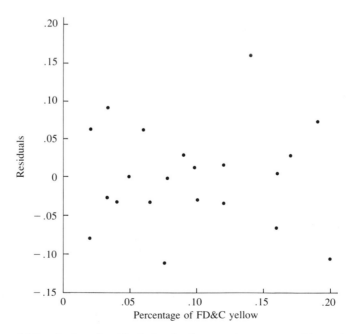

Figure 14-5.   A plot of residuals for the data on chromatographic peak area.

We next consider an example in which the plot shows curvature.

EXAMPLE B.   The data in the table below were gathered for an environmental impact study that examined the relationship between the depth of a stream and the rate of its flow (Ryan, Joiner, and Ryan, 1976).

Depth	Flow Rate
.34	.636
.29	.319
.28	.734
.42	1.327
.29	.487
.41	.924
.76	7.350
.73	5.890
.46	1.979
.40	1.124

A plot of flow rate versus depth suggests that the relation is not linear (Figure 14-6). This is even more immediately apparent from the bowed shape of the plot of the residuals versus depth (Figure 14-7). In order to empirically linearize relationships, transformations are frequently employed. Figure 14-8 is a plot of log rate versus log depth, and Figure 14-9 shows the residuals for the corresponding fit. There is no sign of obvious misfit. (The possibility of expressing flow rate as a quadratic function of depth will be explored in a later example.)   □

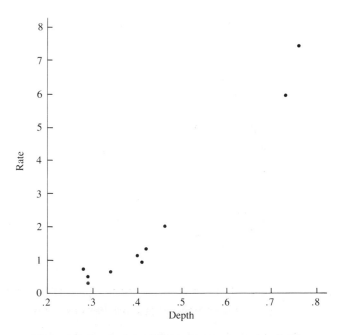

Figure 14-6.   A plot of flow rate versus stream depth.

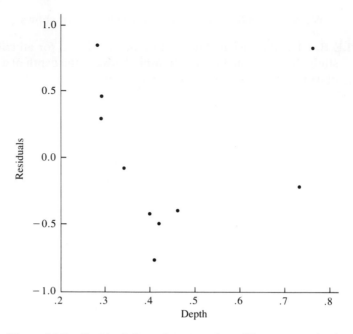

Figure 14-7. Residuals from the regression of flow rate on depth.

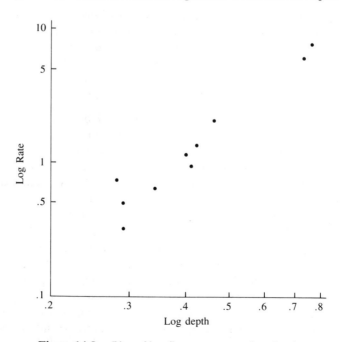

Figure 14-8. Plot of log flow rate versus log depth.

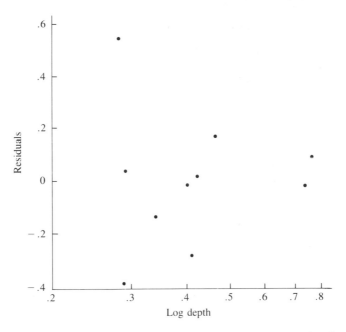

Figure 14-9. Residuals from the regression of log flow rate on log depth.

We have seen that one of the assumptions of the standard statistical model is that the variance of the errors is constant and does not depend on x. Errors with this property are said to be **homoscedastic**. If the variance of the errors is not constant, the errors are said to be **heteroscedastic**. If in fact the error variance is not constant, standard errors and confidence intervals based on the assumption that s^2 is an estimate of σ^2 may be misleading.

EXAMPLE C. In Problem 31 at the end of chapter 7, data on the population and number of breast cancer mortalities in 301 counties were presented. A scatterplot of the number of cases (y) versus population (x) is shown in Figure 14-10. This plot appears to be consistent with the simple model that the number of cases is proportional to the population size, or $y \approx \beta x$. (We will test whether or not the intercept is zero below.) Accordingly, we fit a model with zero intercept by least squares to the data, yielding $\hat{\beta} = 3.559 \times 10^{-3}$. (See Problem 13 at the end of this chapter for fitting a zero intercept model.) Figure 14-11 shows the residuals from the regression of the number of cases on population plotted versus population. Since it is very hard to see what is going on in the left-hand side of this plot, the residuals are plotted versus log population in Figure 14-12, from which it is quite clear that the error variance is not constant but grows with population size.

The residual plot in Figure 14-12 shows no curvature but indicates that the variance is not constant. For counted data, the variability often grows with the mean, and frequently a square root transformation is used in an attempt to stabilize the variance. We therefore fit a model of the form $\sqrt{y} \approx \gamma\sqrt{x}$. Figure 14-13 shows the plot of residuals for this fit. The residual variability is more

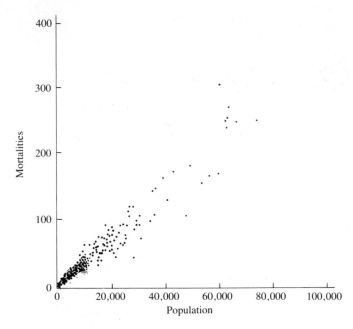

Figure 14-10. Scatterplot showing breast cancer mortality versus population.

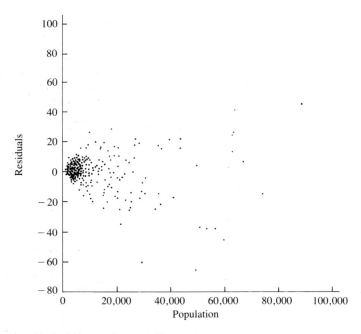

Figure 14-11. Residuals from the regression of mortality on population.

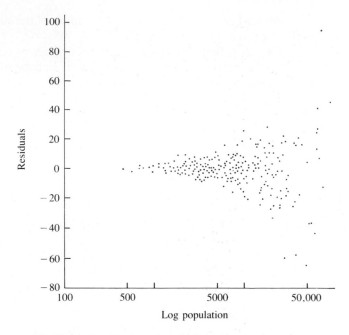

Figure 14-12.   A plot of residuals versus log population.

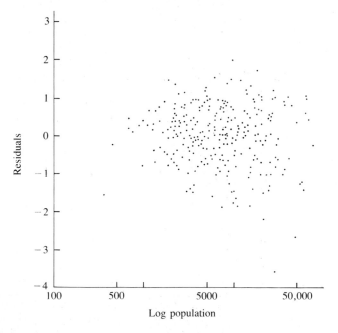

Figure 14-13.   Residuals from the regression of the square root of mortality on the square root of population.

nearly constant here; β is estimated by the square of the slope, $\hat{\gamma}$, which for this example gives $\tilde{\beta} = \hat{\gamma}^2 = 3.471 \times 10^{-3}$.

Finally, we note that the zero intercept model can be tested in the following way. A linear regression on a square root scale is calculated with both slope and intercept terms, and the intercept is found to be .066 with a standard error $s_{\hat{\gamma}_0} = 9.74 \times 10^{-2}$. The t statistic for testing $H_0 : \gamma_0 = 0$ is

$$t = \frac{\hat{\gamma}_0}{s_{\hat{\gamma}_0}} = .68$$

The null hypothesis cannot be rejected for these data.   □

A normal probability plot of residuals may be useful in indicating gross departures from normality and the presence of outliers. Least squares estimates are not robust against outliers, which can have a large effect on the estimated coefficients, their standard errors, and s, especially if the corresponding x values are at the extremes of the data. It can happen, however, that an outlier with an extreme x value will pull the line toward itself and produce a small residual, as illustrated in Figure 14-14.

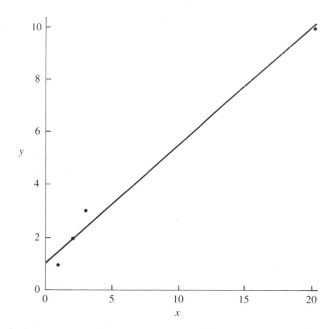

Figure 14-14.   An extreme x value exerts great leverage on the fitted line and produces a small residual at that point.

EXAMPLE D.   Figures 14-15 and 14-16 are normal probability plots of the residuals from the fits of Example C. For Figure 14-15, the residuals are from the ordinary linear regression with zero intercept; for Figure 14-16, the residuals are from the

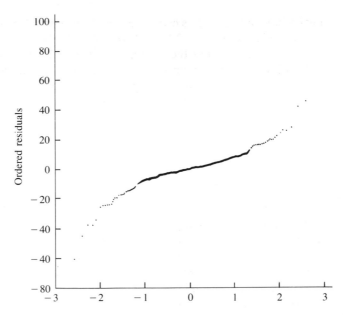

Figure 14-15.   Normal probability plot of the residuals from the regression of mortality on population.

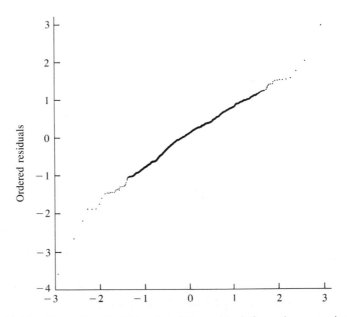

Figure 14-16.   Normal probability plot of the residuals from the regression of the square root of mortality on the square root of population.

zero intercept model with the square root transformation. Note that the distribution in Figure 14-16 is more nearly normal (although there is a hint of skewness) and that the distribution in Figure 14-15 is heavier-tailed than the normal distribution because of the presence of the large residuals from the heavily populated counties.   □

It is often useful to plot residuals against variables that are not in the model but might be influential. If the data were collected over a period of time, a plot of the residuals versus time might reveal unexpected time dependencies.

We conclude this section with an extended example.

EXAMPLE E.   Houck (1970) studied the bismuth I–II transition pressure as a function of temperature. The data are listed in the following table (the standardized residuals will be discussed below):

Pressure (bar)	Temperature (°C)	Standardized Residual
25366	20.8	1.67
25356	20.9	1.48
25336	21.0	.97
25256	21.9	−.40
25267	22.1	.22
25306	22.1	1.46
25237	22.4	−.35
25267	22.5	.74
25138	24.8	−.34
25148	24.8	−.02
25143	25.0	.08
24731	34.0	−1.20
24751	34.0	−.57
24771	34.1	.19
24424	42.7	.46
24444	42.7	1.11
24419	42.7	.30
24117	49.9	.15
24102	50.1	−.08
24092	50.1	−.42
25202	22.5	−1.33
25157	23.1	−1.97
25157	23.0	−2.10

From Figure 14-17, a plot of the tabulated data, it appears that the relationship is fairly linear. The least squares line is

$$\text{Pressure} = 26172(\pm 21) - 41.3(\pm .6) \times \text{temperature}$$

where the estimated standard errors of the parameters are given in parentheses. The residual standard deviation is $s = 32.5$ with 21 df. An approximate 95% confidence interval for the slope is

$$\hat{\beta}_1 \pm s_{\hat{\beta}_1} t_{21}(.025)$$

or (40.05, 42.55).

In order to check how well the model fits, we look at a plot of the standardized residuals versus temperature (Figure 14-18). The plot is rather odd. At first glance, it appears that the variability is greater at lower temperatures. (Bear in mind that the error variance was assumed to be constant in the derivation of the statistical properties of $\hat{\beta}$.) There is another possible explanation for the wedge-shaped appearance of the residual plot. The table reveals that the data were apparently collected in the following way: three measurements at about 21°C, five at about 22°C, three at about 25°C, three at about 34°C, three at about 43°C, three at about 50°C, and three at about 23°C. It is quite possible that the measurements were taken in the order in which they are listed. We can circle these groups of measurements on the residual plot and note the offsets among them. The last three measurements, taken at about 23°C, particularly stand out, and the three taken at 43°C appear out of line with those at 34°C and 50°C. A plausible explanation for this pattern is as follows: The experimental equipment was set up for a given temperature and several measurements were made; then the equipment was set for another temperature and more measurements were made, and so on; at each setting, errors were introduced that affected every measurement at that temperature. Calibration errors are a possibility.

The standard statistical model, which assumes that the errors at each point are independent, does not provide a faithful representation of such a phenome-

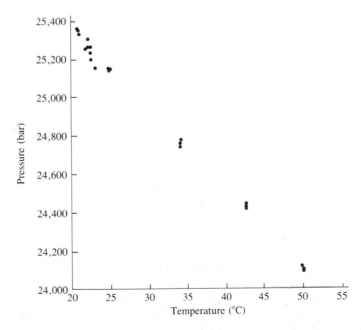

Figure 14-17.   A plot of bismuth I–II transition pressure versus temperature.

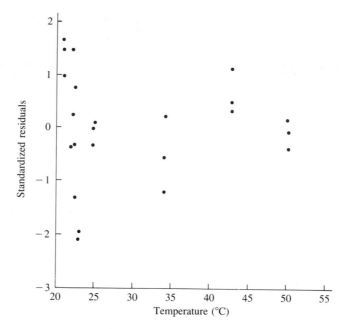

Figure 14-18. A plot of standardized residuals versus temperature.

non. The standard errors given above for $\hat{\beta}_0$ and $\hat{\beta}_1$ and the confidence interval for $\hat{\beta}_1$ are clearly suspect. □

14.2.3 Correlation and Regression

There is a close relationship between correlation analysis and fitting straight lines by the least squares method. Let us introduce some notation:

$$s_{xx} = \frac{1}{n} \sum_{i=1}^{n} (x_i - \bar{x})^2$$

$$s_{yy} = \frac{1}{n} \sum_{i=1}^{n} (y_i - \bar{y})^2$$

$$s_{xy} = \frac{1}{n} \sum_{i=1}^{n} (x_i - \bar{x})(y_i - \bar{y})$$

The correlation coefficient between the x's and y's is

$$r = \frac{s_{xy}}{\sqrt{s_{xx}s_{yy}}}$$

The slope of the least squares line is (see Problem 8 at the end of this chapter)

$$\hat{\beta}_1 = \frac{s_{xy}}{s_{xx}}$$

and therefore

$$r = \hat{\beta}_1 \sqrt{\frac{s_{xx}}{s_{yy}}}$$

In particular, the correlation is zero if and only if the slope is zero.

To further investigate the relationship of correlation and regression, we will find it convenient to standardize the variables. Let

$$u_i = \frac{x_i - \bar{x}}{\sqrt{s_{xx}}}$$

and

$$v_i = \frac{y_i - \bar{y}}{\sqrt{s_{yy}}}$$

We then have $s_{uu} = s_{vv} = 1$ and $s_{uv} = r$. The least squares line for predicting v from u thus has slope r and intercept

$$\tilde{\beta}_0 = \bar{v} - r\bar{u} = 0$$

and the predicted values are

$$\hat{v}_i = ru_i$$

The term *regression* stems from the work of Sir Francis Galton (1822–1911), a famous geneticist, who studied the sizes of seeds and their offspring and the heights of fathers and their sons. In both cases, he found that the offspring of parents of larger than average size tended to be smaller than their parents and that the offspring of parents of smaller than average size tended to be larger than their parents. He called this phenomenon "regression towards mediocrity." This is exactly what the line $\hat{v} = ru$ predicts. Suppose that $0 < r < 1$, for example. If $u > 0$ $(x - \bar{x} > 0)$, the predicted value of v is less than u. If $u < 0$, the predicted value of v is greater than u.

Figure 14-19 (from Freedman, Pisani, and Purves, 1978) illustrates the phenomenon of regression. It is a scatterplot of the heights of 1078 fathers and 1078 sons. The sons' heights average 1 inch more than the fathers'. The solid line is the regression line, and the diagonal dashed line is $y = x + 1$. The points in the left-hand strip bounded by vertical dashed lines correspond to pairs in which the father's height is 64 inches, to the nearest inch. The average height of the sons in

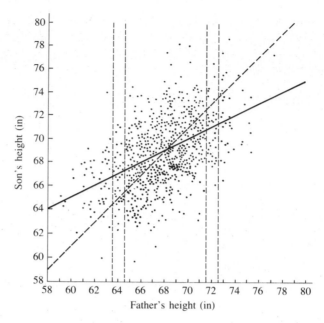

Figure 14-19. A scatterplot of the heights of 1078 sons versus the heights of their fathers.

these pairs is 67 inches—3 inches taller than their fathers. In the right-hand strip, each father's height is 72 inches to the nearest inch, and the sons' average height is 71 inches. The solid regression line passes through the sons' average heights and is much flatter than the line $y = x + 1$.

We have already encountered the phenomenon of regression in Example B in Section 4.4.1, where we saw that if X and Y follow a bivariate normal distribution with $\sigma_X = \sigma_Y = 1$, the conditional expectation of Y given X does not lie along the major axis of the elliptical contours of the joint density; rather, $E(Y|X) = \rho X$.

The regression effect must be taken into account in test-retest situations. Suppose, for example, that a group of preschool children are given an IQ test at age four and another test at age five. The results of the tests will certainly be correlated, and according to the analysis above, children who do poorly on the first test will tend to score higher on the second test. If, on the basis of the first test, low-scoring children were selected for supplemental educational assistance, their gains might be mistakenly attributed to the program. A comparable control group is needed in this situation to tighten up the experimental design.

14.3   The Matrix Approach to Linear Least Squares

With problems more complex than fitting a straight line, it is very useful to approach the linear least squares analysis via linear algebra. As well as providing a compact notation, the conceptual framework of linear algebra can generate theoretical and practical insights. The recent intensive interest of numerical analysts in linear algebra has resulted in the availability of high-quality software packages, such as those of Lawson and Hanson (1974), Dongarra (1979), and Rice (1983).

Suppose that a model of the form

$$y = \beta_0 + \beta_1 x_1 + \cdots + \beta_{p-1} x_{p-1}$$

is to be fit to data, which we denote as

$$y_i, x_{i1}, x_{i2}, x_{i,p-1}, \quad i = 1, \ldots, n$$

The observations y_i, where $i = 1, \ldots, n$, will be represented by a vector, $\mathbf{Y}$. The unknowns, $\beta_0, \ldots, \beta_{p-1}$, will be represented by a vector, $\boldsymbol{\beta}$. Let $\mathbf{X}_{n \times p}$ be the matrix

$$\mathbf{X} = \begin{bmatrix} 1 & x_{11} & x_{12} & \cdots & x_{1,p-1} \\ 1 & x_{21} & x_{22} & \cdots & x_{2,p-1} \\ \vdots & \vdots & \vdots & \vdots & \vdots \\ 1 & x_{n1} & x_{n2} & \cdots & x_{n,p-1} \end{bmatrix}$$

For a given $\boldsymbol{\beta}$, the vector of fitted or predicted values, $\hat{\mathbf{Y}}$, can be written

$$\underset{n \times 1}{\hat{\mathbf{Y}}} = \underset{n \times p}{\mathbf{X}} \underset{p \times 1}{\boldsymbol{\beta}}$$

The least squares problem can then be phrased as follows: Find $\boldsymbol{\beta}$ to minimize

$$S(\boldsymbol{\beta}) = \sum_{i=1}^{n} (y_i - \beta_1 x_{1i} - \cdots - \beta_p x_{pi})^2$$

$$= \|\mathbf{Y} - \mathbf{X}\boldsymbol{\beta}\|^2$$

$$= \|\mathbf{Y} - \hat{\mathbf{Y}}\|^2$$

EXAMPLE A. Let us consider fitting a straight line, $y = \beta_0 + \beta_1 x$, to points (y_i, x_i), where $i = 1, \ldots, n$. In this case,

$$\mathbf{Y} = \begin{bmatrix} y_1 \\ y_2 \\ \vdots \\ y_n \end{bmatrix}$$

$$\beta = \begin{bmatrix} \beta_0 \\ \beta_1 \end{bmatrix}$$

$$\mathbf{X} = \begin{bmatrix} 1 & x_1 \\ 1 & x_2 \\ \vdots & \vdots \\ 1 & x_n \end{bmatrix}$$

and

$$\mathbf{Y} - \mathbf{X}\beta = \begin{bmatrix} y_1 - \beta_0 - \beta_1 x_1 \\ y_2 - \beta_0 - \beta_1 x_2 \\ \vdots \\ y_n - \beta_0 - \beta_1 x_n \end{bmatrix} \qquad \square$$

Returning to the general case, if we differentiate S with respect to each β_k and set the derivatives equal to zero, we see that the minimizers $\hat{\beta}_0, \ldots, \hat{\beta}_{p-1}$ satisfy the p linear equations

$$n\hat{\beta}_0 + \hat{\beta}_1 \sum_{i=1}^n x_{i1} + \cdots + \hat{\beta}_{p-1} \sum_{i=1}^n x_{i,p-1} = \sum_{i=1}^n y_i$$

$$\hat{\beta}_0 \sum_{i=1}^n x_{ik} + \hat{\beta}_1 \sum_{i=1}^n x_{i1} x_{ik} + \cdots + \hat{\beta}_{p-1} \sum_{i=1}^n x_{ik} x_{i,k-1} = \sum_{i=1}^n y_i x_{ik}, k = 1, \ldots, p - 1$$

These p equations can be written in matrix form

$$\mathbf{X}^T \mathbf{X} \hat{\beta} = \mathbf{X}^T \mathbf{Y}$$

and are called the **normal equations**. If $\mathbf{X}^T \mathbf{X}$ is nonsingular, the formal solution is

$$\hat{\beta} = (\mathbf{X}^T \mathbf{X})^{-1} \mathbf{X}^T \mathbf{Y}$$

We stress that this is a formal solution; computationally, it is sometimes unwise even to form the normal equations since the multiplications involved in forming $\mathbf{X}^T \mathbf{X}$ can introduce undesirable round-off error. Alternative methods of finding the least squares solution $\hat{\beta}$ are developed in Problems 6 and 7 at the end of this chapter.

The following lemma gives a criterion for the existence and uniqueness of solutions of the normal equations.

LEMMA A.   The rank of $\mathbf{X}^T\mathbf{X}$ equals the rank of $\mathbf{X}$. Thus, the normal equations have a unique solution if and only if the columns of $\mathbf{X}$ are linearly independent.

Proof. $\mathbf{X}$ and $\mathbf{X}^T\mathbf{X}$ have the same null space, and therefore the same rank, since:

(1) $\mathbf{X}\mathbf{u} = 0$ implies that $\mathbf{X}^T\mathbf{X}\mathbf{u} = 0$
(2) $\mathbf{X}^T\mathbf{X}\mathbf{u} = 0$ implies that

$$\mathbf{u}^T\mathbf{X}^T\mathbf{X}\mathbf{u} = \|\mathbf{X}\mathbf{u}\|^2 = 0$$

from which it follows that $\mathbf{X}\mathbf{u} = 0$.   □

The vector $\hat{\boldsymbol{\beta}} = (\mathbf{X}^T\mathbf{X})^{-1}\mathbf{X}^T\mathbf{Y}$ is the vector of fitted parameters, and the corresponding vector of fitted, or predicted, y values is $\hat{\mathbf{Y}} = \mathbf{X}\hat{\boldsymbol{\beta}}$. The residuals $\mathbf{Y} - \hat{\mathbf{Y}} = \mathbf{Y} - \mathbf{X}\hat{\boldsymbol{\beta}}$ are the differences between the observed and fitted values. We will make use of these residuals in examining goodness of fit.

EXAMPLE B.   Returning to Example A on fitting a straight line, we have

$$\mathbf{X}^T\mathbf{X} = \begin{bmatrix} 1 & \cdots & 1 \\ x_1 & \cdots & x_n \end{bmatrix} \begin{bmatrix} 1 & x_1 \\ \vdots & \vdots \\ 1 & x_n \end{bmatrix}$$

$$= \begin{bmatrix} n & \sum_{i=1}^{n} x_i \\ \sum_{i=1}^{n} x_i & \sum_{i=1}^{n} x_i^2 \end{bmatrix}$$

$$(\mathbf{X}^T\mathbf{X})^{-1} = \frac{1}{n\sum_{i=1}^{n} x_i^2 - \left(\sum_{i=1}^{n} x_i\right)^2} \begin{bmatrix} \sum_{i=1}^{n} x_i^2 & -\sum_{i=1}^{n} x_i \\ -\sum_{i=1}^{n} x_i & n \end{bmatrix}$$

$$\mathbf{X}^T\mathbf{Y} = \begin{bmatrix} \sum_{i=1}^{n} y_i \\ \sum_{i=1}^{n} x_i y_i \end{bmatrix}$$

Thus,

$$\hat{\boldsymbol{\beta}} = \begin{bmatrix} \hat{\beta}_1 \\ \hat{\beta}_1 \end{bmatrix}$$

$$= (\mathbf{X}^T\mathbf{X})^{-1}\mathbf{X}^T\mathbf{Y}$$

$$= \frac{1}{n\sum_{i=1}^{n} x_i^2 - \left(\sum_{i=1}^{n} x_i\right)^2} \begin{bmatrix} \sum_{i=1}^{n} x_i^2 & -\sum_{i=1}^{n} x_i \\ -\sum_{i=1}^{n} x_i & n \end{bmatrix} \begin{bmatrix} \sum_{i=1}^{n} y_i \\ \sum_{i=1}^{n} x_i y_i \end{bmatrix}$$

$$= \frac{1}{n\sum_{i=1}^{n} x_i^2 - \left(\sum_{i=1}^{n} x_i\right)^2} \begin{bmatrix} \left(\sum_{i=1}^{n} y_i\right)\left(\sum_{i=1}^{n} x_i^2\right) - \left(\sum_{i=1}^{n} x_i\right)\left(\sum_{i=1}^{n} x_i y_i\right) \\ n\sum_{i=1}^{n} x_i y_i - \left(\sum_{i=1}^{n} x_i\right)\left(\sum_{i=1}^{n} y_i\right) \end{bmatrix}$$

which agrees with the earlier calculation.　　　　　　　　　□

14.4　Statistical Properties of Least Squares Estimates

In this section, we develop some statistical properties of the vector $\hat{\boldsymbol{\beta}}$, which is found by the least squares method, under some assumptions on the vector of errors. In order to do this, we must use concepts and notation for the analysis of random vectors.

14.4.1　Vector-Valued Random Variables

In Section 14.3, we found expressions for least squares estimates in terms of matrices and vectors. We now develop methods and notation for dealing with random vectors, vectors whose components are random variables. These concepts will be applied to finding statistical properties of the least squares estimates.

We consider the random vector

$$\mathbf{Y} = \begin{bmatrix} Y_1 \\ Y_2 \\ \vdots \\ Y_n \end{bmatrix}$$

the elements of which are jointly distributed random variables with

$$E(Y_i) = \mu_i$$

and

$$\text{Cov}(Y_i, Y_j) = \sigma_{ij}$$

The **mean vector** is defined to be simply the vector of means, or

$$E(\mathbf{Y}) = \boldsymbol{\mu}_Y = \begin{bmatrix} \mu_1 \\ \mu_2 \\ \vdots \\ \mu_n \end{bmatrix}$$

The **covariance matrix** of $\mathbf{Y}$, denoted $\boldsymbol{\Sigma}$, is defined to be an $n \times n$ matrix with the ij element σ_{ij}, which is the covariance of Y_i and Y_j. Note that $\boldsymbol{\Sigma}$ is a symmetric matrix.

Suppose that

$$\underset{m \times 1}{\mathbf{Z}} = \underset{m \times 1}{\mathbf{c}} + \underset{m \times n}{\mathbf{A}} \, \underset{n \times 1}{\mathbf{Y}}$$

is another random vector formed from a fixed vector, $\mathbf{c}$, and a fixed linear transformation, $\mathbf{A}$, of the random vector $\mathbf{Y}$. The next two theorems show how the mean vector and covariance matrix of $\mathbf{Z}$ are determined from the mean vector and covariance matrix of $\mathbf{Y}$ and the matrix $\mathbf{A}$. Each of the theorems is followed by two examples; the results in those examples could easily be derived without using matrix algebra, but they illustrate how the matrix formalisms work.

THEOREM A.   If $\mathbf{Z} = \mathbf{c} + \mathbf{AY}$, where $\mathbf{Y}$ is a random vector and $\mathbf{A}$ is a fixed matrix and $\mathbf{c}$ is a fixed vector, then

$$E(\mathbf{Z}) = \mathbf{c} + \mathbf{A}E(\mathbf{Y})$$

Proof.   The ith component of $\mathbf{Z}$ is

$$Z_i = c_i + \sum_{j=1}^{n} a_{ij} Y_j$$

By the linearity of the expectation,

$$E(Z_i) = c_i + \sum_{j=1}^{n} a_{ij} E(Y_j)$$

Writing these equations in matrix form completes the proof.   □

EXAMPLE A.   As a simple example, let us consider the case where $Z = \sum_{i=1}^{n} a_i Y_i$. In matrix notation, this can be written $Z = \mathbf{a}^T \mathbf{Y}$. According to Theorem A,

$$E(\mathbf{Z}) = \mathbf{a}^T \boldsymbol{\mu} = \sum_{i=1}^{n} a_i \mu_i$$

as we already knew.   □

EXAMPLE B.   As another example, let us consider a moving average. Suppose that $Z_i = Y_i + Y_{i+1}$, for $i = 1, \ldots, n - 1$. We can write this in matrix notation as $\mathbf{Z} = \mathbf{AY}$, where $\mathbf{A}$ is the matrix

$$\begin{bmatrix} 1 & 1 & 0 & 0 & \cdots & 0 & 0 \\ 0 & 1 & 1 & 0 & \cdots & 0 & 0 \\ \vdots & \vdots & \vdots & \vdots & \vdots & \vdots & \vdots \\ 0 & 0 & 0 & 0 & \cdots & 1 & 1 \end{bmatrix}$$

Using Theorem A to find $E(\mathbf{Z})$, it is easy to see that $\mathbf{A}\boldsymbol{\mu}$ has ith component $\mu_i + \mu_{i+1}$.   □

THEOREM B.   Under the assumptions of Theorem A, if the covariance matrix of $\mathbf{Y}$ is $\boldsymbol{\Sigma}_{YY}$, then the covariance matrix of $\mathbf{Z}$ is

$$\boldsymbol{\Sigma}_{ZZ} = \mathbf{A}\boldsymbol{\Sigma}_{YY}\mathbf{A}^T$$

Proof.   The constant $\mathbf{c}$ does not affect the covariance.

$$\mathrm{Cov}(Z_i, Z_j) = \mathrm{Cov}\left(\sum_{k=1}^{n} a_{ik} Y_k, \sum_{l=1}^{n} a_{jl} Y_l \right)$$

$$= \sum_{k=1}^{n} \sum_{l=1}^{n} a_{ik} a_{jl} \, \mathrm{Cov}(Y_k, Y_l)$$

$$= \sum_{k=1}^{n} \sum_{l=1}^{n} a_{ik} \sigma_{kl} a_{jl}$$

The last expression is the ij element of the desired matrix.   □

EXAMPLE C.   Continuing Example A, suppose that the Y_i are uncorrelated with constant variance σ^2. The covariance matrix of $\mathbf{Y}$ can then be expressed as $\boldsymbol{\Sigma}_{YY} = \sigma^2 \mathbf{I}$, where $\mathbf{I}$ is the identity matrix. The role of $\mathbf{A}$ in Theorem B is played by $\mathbf{a}^T$. Therefore, the covariance matrix of Z, which is a 1×1 matrix in this case, is

$$\boldsymbol{\Sigma}_{ZZ} = \sigma^2 \mathbf{a}^T \mathbf{a} = \sigma^2 \sum_{i=1}^{n} a_i^2$$ □

EXAMPLE D.   Suppose that the Y_i of Example B have the covariance matrix $\sigma^2 I$. Then $\Sigma_{ZZ} = \sigma^2 A^T A$, or

$$\sigma^2 \begin{bmatrix} 2 & 1 & 0 & 0 & \cdots & 0 \\ 1 & 2 & 1 & 0 & \cdots & 0 \\ 0 & 1 & 2 & 1 & \cdots & 0 \\ \vdots & \vdots & \vdots & \vdots & \vdots & \vdots \\ 0 & 0 & 0 & 0 & \cdots & 2 \end{bmatrix} \qquad \square$$

The proofs of both these theorems are straightforward, although the unfamiliarity of the notation may present a difficulty. But one of the advantages of using matrices and vectors when dealing with collections of random variables is that this notation is much more compact and easier to follow once one has mastered it, because all the subscripts have been suppressed.

Let A be a symmetric $n \times n$ matrix and x an n vector. The expression

$$x^T A x = \sum_{i=1}^{n} \sum_{j=1}^{n} x_i a_{ij} x_j$$

is called a **quadratic form**. We will next calculate the expectation of a quadratic form in the case where x is a random vector. An example of a random quadratic form we have already encountered is

$$\sum_{i=1}^{n} (X_i - \bar{X})^2$$

which has been used to estimate the variance.

THEOREM C.   Let X be a random n vector with mean μ and covariance Σ, and let A be a fixed matrix. Then

$$E(X^T A X) = \text{trace}(A\Sigma) + \mu^T A \mu$$

Proof.   The trace of a square matrix is defined to be the sum of its diagonal terms. Since

$$E(X_i X_j) = \sigma_{ij} + \mu_i \mu_j$$

we have that

$$E\left(\sum_{i=1}^{n} \sum_{j=1}^{n} X_i X_j a_{ij} \right) = \sum_{i=1}^{n} \sum_{j=1}^{n} \sigma_{ij} a_{ij} + \sum_{i=1}^{n} \sum_{j=1}^{n} \mu_i \mu_j a_{ij}$$

$$= \text{trace}(A\Sigma) + \mu^T A \mu \qquad \square$$

EXAMPLE E. Consider $E[\sum_{i=1}^{n} (X_i - \bar{X})^2]$, where the X_i are uncorrelated random variables with common mean μ. We recognize that this is the squared length of a vector $\mathbf{AX}$ for some matrix $\mathbf{A}$. To figure out what $\mathbf{A}$ must be, we first note that $\bar{X}$ can be expressed as

$$\bar{X} = \frac{1}{n} \mathbf{1}^T \mathbf{X}$$

where $\mathbf{1}$ is a vector consisting of all ones. The vector consisting of entries all of which are $\bar{X}$ can thus be written as $(1/n)\mathbf{1}\mathbf{1}^T\mathbf{X}$, and $\mathbf{A}$ can be written as

$$\mathbf{A} = \mathbf{I} - \frac{1}{n}\mathbf{1}\mathbf{1}^T$$

Thus,

$$\sum_{i=1}^{n} (X_i - \bar{X})^2 = \|\mathbf{AX}\|^2 = \mathbf{X}^T \mathbf{A}^T \mathbf{AX}$$

The matrix $\mathbf{A}$ has some special properties. In particular, $\mathbf{A}$ is symmetric, and $\mathbf{A}^2 = \mathbf{A}$, as can be verified by simply multiplying $\mathbf{A}$ by $\mathbf{A}$, noting that $\mathbf{1}^T\mathbf{1} = n$. Thus,

$$\mathbf{X}^T \mathbf{A}^T \mathbf{AX} = \mathbf{X}^T \mathbf{AX}$$

and by Theorem C,

$$E(\mathbf{X}^T \mathbf{Ax}) = \sigma^2 \operatorname{trace}(\mathbf{A}) + \boldsymbol{\mu}^T \mathbf{A}\boldsymbol{\mu}$$

Since $\boldsymbol{\mu}$ can be written as $\boldsymbol{\mu} = \mu\mathbf{1}$, it can be verified that $\boldsymbol{\mu}^T\mathbf{A}\boldsymbol{\mu} = 0$. Also, trace $\mathbf{A} = n - 1$, so the expectation above is $\sigma^2(n - 1)$. $\qquad\square$

If $\mathbf{Y}_{p \times 1}$ and $\mathbf{Z}_{m \times 1}$ are random vectors, the **cross-covariance matrix** of $\mathbf{Y}$ and $\mathbf{Z}$ is defined to be the $p \times m$ matrix $\boldsymbol{\Sigma}_{YZ}$ with the ij element $\sigma_{ij} = \operatorname{Cov}(Y_i, Z_j)$.

THEOREM D. Let $\mathbf{X}$ be a random vector with covariance matrix $\boldsymbol{\Sigma}_{XX}$. If

$$\mathbf{Y} = \underset{p \times n}{\mathbf{A}} \mathbf{X}$$

and

$$\mathbf{Z} = \underset{m \times n}{\mathbf{B}} \mathbf{X}$$

where $\mathbf{A}$ and $\mathbf{B}$ are fixed matrices, the cross-covariance matrix of $\mathbf{Y}$ and $\mathbf{Z}$ is

$$\Sigma_{YZ} = \mathbf{A}\Sigma_{XX}\mathbf{B}^T$$

Proof.   The proof follows the lines of that of Theorem B (you should work it through for yourself).   □

EXAMPLE F.   Let $\mathbf{X}$ be a random n vector with $E(\mathbf{X}) = \mu\mathbf{1}$ and $\Sigma_{XX} = \sigma^2\mathbf{I}$. Let $Y = \bar{X}$, and let $\mathbf{Z}$ be the vector with ith element $X_i - \bar{X}$. We will find Σ_{ZY}, an $n \times 1$ matrix. In matrix form,

$$\mathbf{Z} = \left(\mathbf{I} - \frac{1}{n}\mathbf{1}\mathbf{1}^T\right)\mathbf{X}$$

$$Y = \frac{1}{n}\mathbf{1}^T\mathbf{X}$$

From Theorem D,

$$\Sigma_{ZY} = \left(\mathbf{I} - \frac{1}{n}\mathbf{1}\mathbf{1}^T\right)(\sigma^2\mathbf{I})\left(\frac{1}{n}\mathbf{1}\right)$$

which becomes an $n \times 1$ matrix of zeros after multiplying out.   □

14.4.2   Mean and Covariance of Least Squares Estimates

Once a function has been fit to data by the least squares method, it may be necessary to consider the stability of the fit and of the estimated parameters, since if the measurements were to be taken again they would often be slightly different. To address the question of the variability of least squares estimates in the presence of noise, we will use the following model:

$$Y_i = \beta_0 + \sum_{j=1}^{p-1} \beta_j x_{ij} + e_i, \quad i = 1, \dots, n$$

where the e_i are random errors with

$$E(e_i) = 0$$

$$\text{Var}(e_i) = \sigma^2$$

$$\text{Cov}(e_i, e_j) = 0, \quad i \neq j$$

In matrix notation, we have

$$\underset{n \times 1}{\mathbf{Y}} = \underset{n \times p}{\mathbf{X}} \ \underset{p \times 1}{\beta} + \underset{n \times 1}{\mathbf{e}}$$

and

$$E(\mathbf{e}) = 0$$

$$\Sigma_{ee} = \sigma^2 \mathbf{I}$$

In words, the y measurements are equal to the true values of the function plus random, uncorrelated errors with constant variance. A useful theorem follows immediately from Theorem A of Section 14.4.1.

THEOREM A.   Under the assumption that the errors have mean 0, the least squares estimates are unbiased.

Proof.   The least squares estimate of $\boldsymbol{\beta}$ is

$$\hat{\boldsymbol{\beta}} = (\mathbf{X}^T\mathbf{X})^{-1}\mathbf{X}^T\mathbf{Y}$$

$$= (\mathbf{X}^T\mathbf{X})^{-1}\mathbf{X}^T(\mathbf{X}\boldsymbol{\beta} + \mathbf{e})$$

$$= \boldsymbol{\beta} + (\mathbf{X}^T\mathbf{X})^{-1}\mathbf{X}^T\mathbf{e}$$

From Theorem A of Section 14.4.1,

$$E\hat{\boldsymbol{\beta}} = \boldsymbol{\beta} + (\mathbf{X}^T\mathbf{X})^{-1}\mathbf{X}^T E(\mathbf{e})$$

$$= \boldsymbol{\beta} \qquad \qquad \square$$

It should be noted that the only assumption on the errors used in this proof of Theorem A is that they have mean 0. Thus, even if the errors are correlated and have a nonconstant variance, the least squares estimates are unbiased. The covariance matrix of $\hat{\boldsymbol{\beta}}$ can also be calculated; the proof of the following theorem does depend on assumptions concerning the covariance of the errors.

THEOREM B.   Under the assumption that the errors have mean zero and are uncorrelated with constant variance σ^2, the covariance matrix of the least squares estimate $\hat{\boldsymbol{\beta}}$ is

$$\Sigma_{\hat{\beta}\hat{\beta}} = \sigma^2(\mathbf{X}^T\mathbf{X})^{-1}$$

Proof.   From Theorem B of Section 14.4.1, the covariance matrix of $\hat{\boldsymbol{\beta}}$ is

$$\Sigma_{\hat{\beta}\hat{\beta}} = (\mathbf{X}^T\mathbf{X})^{-1}\mathbf{X}^T\Sigma_{ee}\mathbf{X}(\mathbf{X}^T\mathbf{X})^{-1}$$

$$= \sigma^2(\mathbf{X}^T\mathbf{X})^{-1}$$

since the covariance matrix of $\mathbf{e}$ is $\sigma^2\mathbf{I}$, and $\mathbf{X}^T\mathbf{X}$ and therefore $(\mathbf{X}^T\mathbf{X})^{-1}$ as well are symmetric. $\qquad \square$

These theorems generalize Theorems A and B of Section 14.2.1. Note how the use of matrix algebra simplifies the derivation.

EXAMPLE A.   We return to the case of fitting a straight line. From the computation of $(\mathbf{X}^T\mathbf{X})^{-1}$ in Example B in Section 14.3, we have

$$
\Sigma_{\hat{\beta}\hat{\beta}} = \frac{\sigma^2}{n \sum_{i=1}^{n} x_i^2 - \left(\sum_{i=1}^{n} x_i\right)^2} \begin{bmatrix} \sum_{i=1}^{n} x_i^2 & -\sum_{i=1}^{n} x_i \\ -\sum_{i=1}^{n} x_i & n \end{bmatrix}
$$

Therefore,

$$
\mathrm{Var}(\hat{\beta}_0) = \frac{\sigma^2 \sum_{i=1}^{n} x_i^2}{n \sum_{i=1}^{n} x_i^2 - \left(\sum_{i=1}^{n} x_i\right)^2}
$$

$$
\mathrm{Var}(\hat{\beta}_1) = \frac{n\sigma^2}{n \sum_{i-1}^{n} x_i^2 - \left(\sum_{i=1}^{n} x_i\right)^2}
$$

$$
\mathrm{Cov}(\hat{\beta}_0, \hat{\beta}_1) = \frac{-\sigma^2 \sum_{i=1}^{n} x_i}{n \sum_{i=1}^{n} x_i^2 - \left(\sum_{i=1}^{n} x_i\right)^2} \qquad \square
$$

14.4.3   Estimation of σ^2

In order to use the formulae for variances developed in the preceding section (to form confidence intervals, for example), σ^2 must be known or estimated. In this section, we develop an estimate of σ^2.

Since σ^2 is the expected squared value of an error, e_i, it is natural to use the sample average squared value of the residuals. The vector of residuals is

$$
\hat{\mathbf{e}} = \mathbf{Y} - \hat{\mathbf{Y}}
$$

$$
= \mathbf{Y} - \mathbf{X}\hat{\boldsymbol{\beta}}
$$

$$
= \mathbf{Y} - \mathbf{X}(\mathbf{X}^T\mathbf{X})^{-1}\mathbf{X}^T\mathbf{Y}
$$

or

$$
\hat{\mathbf{e}} = \mathbf{Y} - \mathbf{PY}
$$

where $\mathbf{P} = \mathbf{X}(\mathbf{X}^T\mathbf{X})^{-1}\mathbf{X}^T$ is an $n \times n$ matrix.

Two useful properties of **P** are given in the following lemma (you should be able to write out its proof).

LEMMA A.   Let **P** be defined as above. Then

$$\mathbf{P} = \mathbf{P}^T = \mathbf{P}^2 \qquad \qquad \square$$

Since **P** has the properties given in this lemma, it is a projection matrix—that is, **P** projects on the subspace of $\mathbf{R}^n$ spanned by the columns of **X**. Thus, we may think geometrically of the fitted values, $\hat{\mathbf{Y}}$, as being the projection of **Y** onto the subspace spanned by the columns of **X**.

The sum of squared residuals is, using Lemma A,

$$\sum_{i=1}^{n} (Y_i - \hat{Y}_i)^2 = \|\mathbf{Y} - \mathbf{P}\mathbf{Y}\|^2$$

$$= \|(\mathbf{I} - \mathbf{P})\mathbf{Y}\|^2$$

$$= \mathbf{Y}^T(\mathbf{I} - \mathbf{P})^T(\mathbf{I} - \mathbf{P})\mathbf{Y}$$

$$= \mathbf{Y}^T(\mathbf{I} - \mathbf{P})\mathbf{Y}$$

From Theorem C of Section 14.4.1, we can compute the expected value of this quadratic form:

$$E[\mathbf{Y}^T(\mathbf{I} - \mathbf{P})\mathbf{Y}] = [E(\mathbf{Y})]^T(\mathbf{I} - \mathbf{P})[E(\mathbf{Y})] + \sigma^2 \operatorname{trace}(\mathbf{I} - \mathbf{P})$$

Now $E(\mathbf{Y}) = \mathbf{X}\beta$, so

$$(\mathbf{I} - \mathbf{P})E(\mathbf{Y}) = [\mathbf{I} - \mathbf{X}(\mathbf{X}^T\mathbf{X})^{-1}\mathbf{X}^T]\mathbf{X}\beta$$

$$= \mathbf{0}$$

Furthermore,

$$\operatorname{trace}(\mathbf{I} - \mathbf{P}) = \operatorname{trace}(\mathbf{I}) - \operatorname{trace}(\mathbf{P})$$

and, using the cyclic property of the trace—that is, $\operatorname{trace}(\mathbf{AB}) = \operatorname{trace}(\mathbf{BA})$—we have

$$\operatorname{trace}(\mathbf{P}) = \operatorname{trace}[\mathbf{X}(\mathbf{X}^T\mathbf{X})^{-1}\mathbf{X}^T]$$

$$= \operatorname{trace}[\mathbf{X}^T\mathbf{X}(\mathbf{X}^T\mathbf{X})^{-1}]$$

$$= \operatorname{trace}\left(\underset{p \times p}{\mathbf{I}}\right) = p$$

Since trace$(\mathbf{I}_{n \times n}) = n$, we have shown that

$$E(\|\mathbf{Y} - \hat{\mathbf{Y}}\|^2) = (n - p)\sigma^2$$

and have proved the following theorem.

THEOREM A.   Under the assumption that the errors are uncorrelated with constant variance σ^2, an unbiased estimate of σ^2 is

$$s^2 = \frac{\|\mathbf{Y} - \hat{\mathbf{Y}}\|^2}{n - p}$$ $\square$

The sum of the squared residuals, $\|\mathbf{Y} - \hat{\mathbf{Y}}\|^2$, is often denoted by RSS, for residual sum of squares.

14.4.4   Residuals and Standardized Residuals

Information concerning whether or not a model fits is contained in the vector of residuals,

$$\hat{\mathbf{e}} = \mathbf{Y} - \hat{\mathbf{Y}} = (\mathbf{I} - \mathbf{P})\mathbf{Y}$$

As we did for the case of fitting a straight line, we will use the residuals to check on the adequacy of the fit of a presumed functional form and on the assumptions underlying the statistical analysis (such as that the errors are uncorrelated with constant variance).

The covariance matrix of the residuals is

$$\boldsymbol{\Sigma}_{\hat{e}\hat{e}} = (\mathbf{I} - \mathbf{P})(\sigma^2\mathbf{I})(\mathbf{I} - \mathbf{P})^T$$

$$= \sigma^2(\mathbf{I} - \mathbf{P})$$

where we have used Lemma A of Section 14.4.3. We see that the residuals are correlated with one another and that different residuals have different variances. In order to make the residuals comparable to one another, they are often standardized. Also, standardization puts the residuals on the familiar scale corresponding to a normal distribution with mean 0 and variance 1 and thus makes their magnitudes easier to interpret. The ith standardized residual is

$$\frac{Y_i - \hat{Y}_i}{s\sqrt{1 - p_{ii}}}$$

where p_{ii} is the ith diagonal element of $\mathbf{P}$.

A further property of the residuals is given by the following theorem.

THEOREM A.   If the errors have the covariance matrix $\sigma^2 \mathbf{I}$, the residuals are uncorrelated with the fitted values.

Proof.   The residuals are

$$\hat{\mathbf{e}} = (\mathbf{I} - \mathbf{P})\mathbf{Y}$$

and the fitted values are

$$\hat{\mathbf{Y}} = \mathbf{P}\mathbf{Y}$$

From Theorem D of Section 14.4.1, the cross-covariance matrix of $\hat{\mathbf{e}}$ and $\hat{\mathbf{Y}}$ is

$$\boldsymbol{\Sigma}_{\hat{e}\hat{Y}} = (\mathbf{I} - \mathbf{P})(\sigma^2 \mathbf{I})\mathbf{P}^T$$

$$= \sigma^2(\mathbf{P}^T - \mathbf{P}\mathbf{P}^T)$$

$$= 0$$

This result follows from Lemma A of Section 14.4.3.   □

14.4.5   Inference about β

In this section, we continue the discussion of the statistical properties of the least squares estimate $\hat{\beta}$. In addition to the assumptions made previously, we will assume that the errors, e_i, are independent and normally distributed. Since the components of $\hat{\beta}$ are in this case linear combinations of independent normally distributed random variables, they are also normally distributed.

In particular, each component $\hat{\beta}_i$ of $\hat{\beta}$ is normally distributed with mean β_i and variance $\sigma^2 c_{ii}$, where $\mathbf{C} = (\mathbf{X}^T\mathbf{X})^{-1}$. The standard error of $\hat{\beta}_i$ may thus be estimated as

$$s_{\hat{\beta}_i} = s\sqrt{c_{ii}}$$

This result will be used to construct confidence intervals and hypothesis tests that will be exact under the assumption of normality and approximate otherwise (since $\hat{\beta}_i$ may be expressed as a linear combination of the independent random variables e_i, a version of the central limit theorem with certain assumptions on $\mathbf{X}$ implies the approximate result).

Under the normality assumption, it can be shown that

$$\frac{\hat{\beta}_i - \beta_i}{s_{\hat{\beta}_i}} \sim t_{n-p}$$

although we will not derive this result. It follows that a $100(1 - \alpha)\%$ confidence interval for β_i is

$$\hat{\beta}_i \pm t_{n-p}(\alpha/2)s_{\hat{\beta}_i}$$

To test the null hypothesis H_0: $\beta_i = \beta_{i0}$, where β_{i0} is a fixed number, we can use the test statistic

$$t = \frac{\hat{\beta}_i - \beta_{i0}}{s_{\hat{\beta}_i}}$$

Under H_0, this statistic follows a t distribution with $n - p$ degrees of freedom. The most commonly tested null hypothesis is H_0: $\beta_i = 0$, which states that x_i has no predictive value.

We will illustrate these concepts in the context of polynomial regression.

EXAMPLE A. (Peak Area) Let us return to Example A in Section 14.2.2 concerning the regression of peak area on percentage of FD&C Yellow No. 5. We have seen from the residual plot in Figure 14-5 that a straight line appears to give a reasonable fit. Consider enlarging the model so that it is quadratic:

$$y = \beta_0 + \beta_1 x + \beta_2 x^2$$

where y is peak area and x is percentage of Yellow No. 5. The following table gives the statistics of the fit:

Coefficient	Estimate	Standard Error	t Value
β_0	.058	.054	1.07
β_1	11.17	1.20	9.33
β_2	−1.90	5.53	−.35

To test the hypothesis H_0: $\beta_2 = 0$, we would use $-.35$ as the value of the t statistic, which would not reject H_0. Thus, this test, like the residual analysis, gives no evidence that a quadratic term is needed. □

EXAMPLE B. In Example B in Section 14.2.2, we saw that a residual plot for the linear regression of stream flow rate on depth indicated the inadequacy of that model. The statistics for a quadratic model are given in the following table:

Coefficient	Estimate	Standard Error	t Value
β_0	1.68	1.06	1.59
β_1	−10.86	4.52	−2.40
β_2	23.54	4.27	5.51

Here, the linear and quadratic terms are both statistically significant, and a residual plot, Figure 14-20, shows no signs of systematic misfit.

Height (in)	Weight (lb)	Distance to Pulmonary Artery (cm)
42.8	40.0	37.0
63.5	93.5	49.5
37.5	35.5	34.5
39.5	30.0	36.0
45.5	52.0	43.0
38.5	17.0	28.0
43.0	38.5	37.0
22.5	8.5	20.0
37.0	33.0	33.5
23.5	9.5	30.5
33.0	21.0	38.5
58.0	79.0	47.0

Since this is a very small sample, any conclusions must be regarded as tentative.

Figure 14-21 presents scatterplots of all pairs of variables, providing a useful visual presentation of their relationships. We will refer to these plots as we proceed through the analysis.

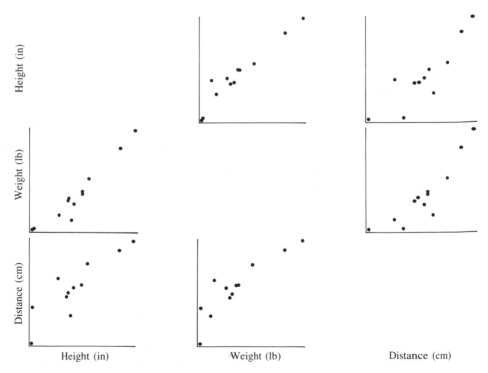

Figure 14-21. Scatterplots showing all pairings of the variables height, weight, and catheter length.

We first consider predicting the length by height alone and by weight alone. The results of simple linear regressions are tabulated below:

	Height	Weight
$\hat{\beta}_0$	12.1($\pm$4.3)	25.6($\pm$2.0)
$\hat{\beta}_1$	.60($\pm$.10)	.28($\pm$.04)
s	4.0	3.8
r^2	.78	.80

The standard errors of $\hat{\beta}_0$ and $\hat{\beta}_1$ are given in parentheses. To test the null hypothesis $H_0: \beta_1 = 0$, the appropriate test statistic is $t = \beta_1/s_{\hat{\beta}_1}$. Clearly, this null hypothesis would be rejected in this case. The predictions from both models are similar; the standard deviations of the residuals about the fitted lines are 4.0 and 3.8, respectively, and the squared correlation coefficients are .78 and .80.

The panels of Figure 14-22 are plots of the standardized residuals from each of the simple linear regressions versus the respective independent variable. The plot of residuals versus weight shows some hint of curvature, which is also apparent in the bottom middle scatterplot in Figure 14-21. The largest standardized residual from this fit comes from the lightest and shortest child (see eighth row of data table).

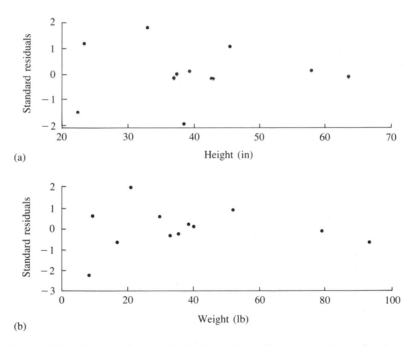

Figure 14-22.   Standardized residuals from simple linear regressions of catheter length plotted against the independent variables (a) height and (b) weight.

We next consider the multiple regression of length on height and weight together, since perhaps better predictions may be obtained by using both variables rather than either one alone. The method of least squares produces the following relationship:

$$\text{Length} = 21(\pm 8.8) + .20(\pm .36) \times \text{height} + .19(\pm .17) \times \text{weight}$$

where the standard errors of the coefficients are shown in parentheses. The standard deviation of the residuals is 3.9.

The **squared multiple correlation coefficient**, or **coefficient of determination**, is sometimes used as a crude measure of the strength of a relationship that has been fit by least squares. This coefficient is simply defined as the squared correlation of the dependent variable and the fitted values. It can be shown that the squared multiple correlation coefficient, denoted by R^2, can be expressed as

$$R^2 = \frac{s_y^2 - s_{\hat{e}}^2}{s_y^2}$$

Since this is the ratio of the difference between the variance of the dependent variable and the variance of the residuals from the fit to the variance of the dependent variable, it can be interpreted as the proportion of the variability of the dependent variable that can be explained by the independent variables. For the catheter example, $R^2 = .81$.

Among the interesting and noteworthy features of multiple regression are the β weights and their standard errors. It may seem surprising that the standard errors of the coefficients of height and weight are large relative to the coefficients themselves. Applying t tests would not lead to rejection of either of the hypotheses $H_1: \beta_1 = 0$ or $H_2: \beta_2 = 0$. Yet in the simple linear regressions carried out above, the coefficients were highly significant. A partial explanation of this is that the coefficients in the simple regressions and the coefficients in the multiple regression have different interpretations. In the multiple regression, β_1 is the change in the expected value of the catheter length if height is increased by one unit and weight is held constant. It is the slope along the height axis of the plane that describes the relation of length to height and weight; the large standard error indicates that this slope is not well resolved. To see why, consider the scatterplot of height versus weight in Figure 14-21. The method of least squares fits a plane to the catheter length values that correspond to the pairs of height and weight values in this plot. It should be intuitively clear from the figure that the slope of the fitted plane is relatively well resolved along the line on which height equals weight but poorly resolved along lines on which either height or weight is constant. Imagine how the fitted plane might move if values of length corresponding to pairs of height and weight values were perturbed. Variables that are strongly linearly related, such as height and weight in this example, are said to be highly **collinear**. If the values of height and weight had fallen exactly on a straight line, we would not have been able to determine a plane at all; in fact, **X** would not have had full column rank.

The plot of height versus weight should also serve as a caution concerning making predictions from such a study. Obviously, we would not want to make a prediction for any pair of height and weight values quite dissimilar to those used in making the original fit. Any empirical relationship developed in the region in which the observed data fall might break down if it were extrapolated to a region in which no data had been observed.

Little or no reduction in s has been obtained by fitting simultaneously to height and weight rather than fitting to either height or weight alone. (In fact, fitting to weight alone gives a smaller value of s than does fitting to height and weight together. This may seem paradoxical, but recall that there are 10 degrees of freedom in the former case and 9 in the latter, and that s is the square root of the residual sum of squares divided by the degrees of freedom.) Again, this is partially explained by Figure 14-21, which shows that weight can be predicted quite well from height. Thus, it should not seem surprising that adding weight to the equation for predicting from height produces little gain.

Finally, the panels of Figure 14-23 show the residuals from the multiple regression plotted versus height and weight. The plots are very similar to those in Figure 14-22.

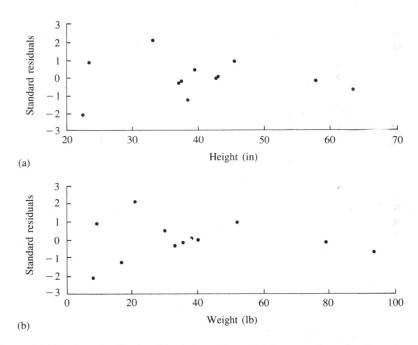

Figure 14-23. Standardized residuals from the multiple regression of catheter length on height and weight plotted against the independent variables (a) height and (b) weight.

14.6 Concluding Remarks

This chapter has briefly introduced the important topic of least squares. There are many books on this subject, for example, Draper and Smith (1981), Weisburg

(1980), and Lawson and Hansen (1974); the first two concentrate on statistical aspects of least squares and the last on numerical aspects.

We have developed theory and techniques only for linear least squares problems. If the unknown parameters enter into the prediction equation nonlinearly, the minimization cannot typically be done in closed form, and iterative methods are necessary. Also, expressions in closed form for the standard errors of the coefficients cannot usually be obtained; linearization is often used to obtain approximate standard errors.

As has been mentioned, least squares estimates are not robust against outliers. Methods that are robust in this sense have been proposed and are being used with increasing frequency. The discussion of M estimates in chapter 10 suggests minimizing

$$\sum_{i=1}^{n} \Psi(Y_i - \hat{Y}_i)$$

for a robust weight function, Ψ. Note that the least squares estimate corresponds to $\Psi(x) = x^2$. The choice $\Psi(x) = |x|$ gives the curve-fitting analogue of the median.

In some applications, a large number of independent variables are candidates for inclusion in the prediction equation. Various techniques for variable selection have been proposed, and research is still active in this area.

In simple linear regression, points with x values at the extremes of the data exert a large influence on the fitted line. In multiple regression, a similar phenomenon occurs, but is not so easily detectable usually. For this reason, several measures of "influence" have been proposed. Good software packages routinely flag influential observations. For further discussion of such procedures, see Cook and Weisberg (1982).

The problem of errors introduced via calibration of instruments has not been fully discussed. Suppose, for example, that an instrument for measuring temperature is to be calibrated. Readings are taken at several known temperatures (the independent variables) and a functional relationship between the instrument readings (the dependent variable) and the temperatures is fit by the method of least squares. After this has been carried out, an unknown temperature is read by the instrument and is predicted using the fitted relationship. How do the errors in the estimates of the coefficients of the functional form propagate? That is, what is the uncertainty of the estimated temperature? This is an inverse problem, and its analysis is not completely straightforward.

14.7   Problems

1. Convert the following relationships into linear relationships by making transformations and defining new variables.

(a) $y = a/(b + cx)$

(b) $y = ae^{-bx}$

(c) $y = ab^x$

(d) $y = x/(a + bx)$

(e) $y = 1/(1 + e^{bx})$

2. Plot y versus x for the following pairs:

x	.34	1.38	−.65	.68	1.40	−.88	−.30	−1.18	.50	−1.75
y	.27	1.34	−.53	.35	1.28	−.98	−.72	−.81	.64	−1.59

(a) Fit a line $y = a + bx$ by the method of least squares, and sketch it on the plot.

(b) Fit a line $x = c + dy$ by the method of least squares, and sketch it on the plot.

(c) Are the lines in parts (a) and (b) the same? If not, why not?

3. Suppose that $y_i = \mu + e_i$, where $i = 1, \ldots, n$ and the e_i are independent errors with mean 0 and variance σ^2. Show that $\bar{y}$ is the least squares estimate of μ.

4. Two objects of unknown weights w_1 and w_2 are weighed on an error-prone pan balance in the following way: 1) object 1 is weighed by itself, and the measurement is 3 grams; 2) object 2 is weighed by itself, and the result is 3 grams; 3) the difference of the weights (the weight of object 1 minus the weight of object 2) is measured by placing the objects in different pans, and the result is 1 gram; 4) the sum of the weights is measured as 7 grams. The problem is to estimate the true weights of the objects from these measurements.

(a) Set up a linear model, $\mathbf{Y} = \mathbf{X}\boldsymbol{\beta} + \mathbf{e}$. (*Hint:* The entries of $\mathbf{X}$ are 0 and ± 1.)

(b) Find the least squares estimates of w_1 and w_2.

(c) Find the estimate of σ^2.

(d) Find the estimated standard errors of the least squares estimates of part (b).

(e) Estimate $w_1 - w_2$ and its standard error.

(f) Test the null hypothesis $H_0: w_1 = w_2$.

5. (Weighted Least Squares) Suppose that in the model $y_i = \beta_0 + \beta_1 x_i + e_i$, the errors have mean 0 and are independent, but $\text{Var}(e_i) = \rho_i^2 \sigma^2$, where the ρ_i are known constants, so the errors do not have equal variance. This situation arises when the y_i are averages of several observations at x_i; in this case, if y_i is an average of n_i independent observations, $\rho_i^2 = 1/n_i$ (why?). Since the variances are not equal, the theory developed in this chapter does not apply; intuitively, it seems that the observations with large variability should influence the estimates of β_0 and β_1 less than the observations with small variability.

The problem may be transformed as follows:

$$\rho_i^{-1} y_i = \rho_i^{-1}\beta_0 + \rho_i^{-1}\beta_1 x_i + \rho_i^{-1} e_i$$

or

$$z_i = u_i\beta_0 + v_i\beta_1 + \delta_i$$

where

$$u_i = \rho_i^{-1} \qquad v_i = \rho_i^{-1}x_i \qquad \delta_i = \rho_i^{-1}e_i$$

(a) Show that the new model satisfies the assumptions of the standard statistical model.
(b) Find the least squares estimates of β_0 and β_1.
(c) Show that performing a least squares analysis on the new model, as was done in part (b), is equivalent to minimizing

$$\sum_{i=1}^{n} (y_i - \beta_0 - \beta_1 x_i)^2 \rho_i^{-2}$$

This is a weighted least squares criterion; the observations with large variances are weighted less.
(d) Find the variances of the estimates of part (b).
6. (The QR Method) This problem outlines the basic ideas of an alternative method, the QR method, of finding the least squares estimate $\hat{\beta}$. An advantage of the method is that it does not include forming the matrix $\mathbf{X}^T\mathbf{X}$, a process that tends to increase round-off error. The essential ingredient of the method is that if $\mathbf{X}_{n \times p}$ has p linearly independent columns, it may be factored in the form

$$\mathbf{X}_{n \times p} = \mathbf{Q}_{n \times p}\ \mathbf{R}_{p \times p}$$

where the columns of $\mathbf{Q}$ are orthogonal ($\mathbf{Q}^T\mathbf{Q} = \mathbf{I}$) and $\mathbf{R}$ is upper-triangular ($r_{ij} = 0$, for $i > j$) and nonsingular. (For a discussion of this decomposition and its relationship to the Gram–Schmidt process, see Strang, 1980.)
    Show that $\hat{\beta} = (\mathbf{X}^T\mathbf{X})^{-1}\mathbf{X}^T\mathbf{Y}$ may also be expressed as $\hat{\beta} = \mathbf{R}^{-1}\mathbf{Q}^T\mathbf{Y}$, or $\mathbf{R}\hat{\beta} = \mathbf{Q}^T\mathbf{Y}$. Indicate how this last equation may be solved for $\hat{\beta}$ by back-substitution, using that $\mathbf{R}$ is upper-triangular, and show that it is thus unnecessary to invert $\mathbf{R}$.
7. (Cholesky Decomposition) This problem outlines the basic ideas of a popular and effective method of computing least squares estimates. Assuming that its inverse exists, $\mathbf{X}^T\mathbf{X}$ is a positive, definite matrix and may be factored as $\mathbf{X}^T\mathbf{X} = \mathbf{R}^T\mathbf{R}$, where $\mathbf{R}$ is an upper-triangular matrix. This factorization is called the **Cholesky decomposition**. Show that the least squares estimates can be found by solving the equations

$$\mathbf{R}^T\mathbf{v} = \mathbf{X}^T\mathbf{Y}$$

$$\mathbf{R}\hat{\beta} = \mathbf{v}$$

where $\mathbf{v}$ is appropriately defined. Show that these equations can be solved by back-substitution since $\mathbf{R}$ is upper-triangular, and that therefore it is not necessary to carry out any matrix inversions explicitly to find the least squares estimates.

8. Show that the least squares estimates of the slope and intercept of a line may be expressed as

$$\hat{\beta}_0 = \bar{y} - \hat{\beta}_1 \bar{x}$$

and

$$\hat{\beta}_1 = \frac{\sum\limits_{i=1}^{n} (x_i - \bar{x})(y_i - \bar{y})}{\sum\limits_{i=1}^{n} (x_i - \bar{x})^2}$$

9. Show that if $\bar{x} = 0$, the estimated slope and intercept are uncorrelated under the assumptions of the standard statistical model.

10. Use the result of Problem 8 to show that the line fit by the method of least squares passes through the point $(\bar{x}, \bar{y})$.

11. Suppose that a line is fit by the method of least squares to n points, that the standard statistical model holds, and that we wish to estimate the line at a new point, x_0. Denoting the value on the line by μ_0, the estimate is

$$\hat{\mu}_0 = \hat{\beta}_0 + \hat{\beta}_1 x_0$$

(a) Derive an expression for the variance of $\hat{\mu}_0$.
(b) Sketch the standard deviation of $\hat{\mu}_0$ as a function of $x_0 - \bar{x}$. The shape of the curve should be intuitively plausible.
(c) Derive a 95% confidence interval for $\mu_0 = \beta_0 + \beta_1 x_0$ under an assumption of normality.

12. Problem 11 dealt with how to form a confidence interval for the value of a line at a point x_0. Suppose that instead we wish to predict the value of a new observation, Y_0, at x_0,

$$Y_0 = \beta_0 + \beta_1 x_0 + e_0$$

by the estimate

$$\hat{Y}_0 = \hat{\beta}_0 + \hat{\beta}_1 x_0$$

(a) Find an expression for the variance of $\hat{Y}_0 - Y_0$, and compare it to the expression for the variance of $\hat{\mu}_0$ obtained in part (a) of Problem 11. Assume that e_0 is independent of the original observations and has the variance σ^2.

(b) Assuming that e_0 is normally distributed, find the distribution of $\hat{Y}_0 - \hat{Y}$. Use this result to find an interval I such that $P(Y_0 \in I) = 1 - \alpha$. This interval is called a $100(1 - \alpha)\%$ prediction interval.

13. Find the least squares estimate of β for fitting the line $y = \beta x$ to points (x_i, y_i), where $i = 1, \ldots, n$.

14. Consider fitting the curve $y = \beta_0 x + \beta_1 x^2$ to points (x_i, y_i), where $i = 1, \ldots, n$.

   (a) Use the matrix formalism to find expressions for the least squares estimates of β_0 and β_1.

   (b) Find an expression for the covariance matrix of the estimates.

15. This problem extends some of the material in Section 14.2.3. Let X and Y be random variables with

$$E(X) = \mu_x \qquad E(Y) = \mu_y$$

$$\text{Var}(X) = \sigma_x^2 \qquad \text{Var}(Y) = \sigma_y^2$$

$$\text{Cov}(X, Y) = \sigma_{xy}$$

Consider predicting Y from X as $\hat{Y} = \alpha + \beta X$, where α and β are chosen to minimize $E(Y - \hat{Y})^2$, the expected squared prediction error.

   (a) Show that the minimizing values of α and β are

$$\beta = \frac{\sigma_{xy}}{\sigma_x^2}$$

$$\alpha = \mu_y - \beta \mu_x$$

   [Hint: $E(Y - \hat{Y})^2 = (EY - E\hat{Y})^2 + \text{Var}(Y - \hat{Y})$.]

   (b) Show that for this choice of α and β

$$\frac{\text{Var}(Y) - \text{Var}(Y - \hat{Y})}{\text{Var}(Y)} = r_{xy}^2$$

16. Suppose that

$$Y_i = \beta_0 + \beta_1 x_i + e_i, \quad i = 1, \ldots, n$$

where the e_i are independent and normally distributed with mean 0 and variance σ^2. Find the maximum likelihood estimates of β_0 and β_1 and verify that they are the least squares estimates. (Hint: Under these assumptions, the Y_i are independent and normally distributed with means $\beta_0 + \beta_1 x_i$ and variance σ^2. Write the joint density function of the Y_i and thus the likelihood.)

17. (a) Show that the vector of residuals is orthogonal to every column of **X**.

   (b) Use this result to show that the residuals sum to zero and thus the sum has expectation 0 if the model contains an intercept term.

18. Assume that the columns of $\mathbf{X}$, $\mathbf{X}_1, \ldots, \mathbf{X}_p$, are orthogonal; that is, $\mathbf{X}_i^T \mathbf{X}_j = 0$, for $i \neq j$. Show that the covariance matrix of the least squares estimates is diagonal.

19. Suppose that n points $x_1, \ldots, x_n$ are to be placed in the interval $[-1, 1]$ for fitting the model

$$Y_i = \beta_0 + \beta_1 x_i + \varepsilon_i$$

where the ε_i are independent with common variance σ^2. How should the x_i be chosen in order to minimize $\text{Var}(\hat{\beta}_1)$?

20. Suppose that $X_1, \ldots, X_n$ are independent with mean μ_i and common variance σ^2. Let $Y = \sum_{i=1}^n a_i X_i$.
    (a) Let $Z = \sum_{i=1}^n b_i X_i$. Use Theorem D of Section 14.4.1 to find $\text{Cov}(Y, Z)$.
    (b) Use Theorem C of Section 14.4.1 to find $E(\sum_{i=1}^n \sum_{j=1}^n X_i X_j)$.

21. Assume that X_1 and X_2 are uncorrelated random variables with variance σ^2, and use matrix methods to show that $Y = X_1 + X_2$ and $Z = X_1 - X_2$ are uncorrelated. (*Hint:* Find Σ_{YZ}.)

22. Let $X_1, \ldots, X_n$ be random variables with $\text{Var}(X_i) = \sigma^2$ and $\text{Cov}(X_i, X_j) = \rho\sigma^2$, for $i \neq j$. Use matrix methods to find $\text{Var}(\bar{X})$.

23. (a) Let $X \sim N(0, 1)$ and $E \sim N(0, 1)$ be independent, and let $Y = X + \beta E$. Show that

$$r_{xy} = \frac{1}{\sqrt{\beta^2 + 1}}$$

   (b) Use the results of part (a) to generate bivariate samples (x_i, y_i) of size 20 with population correlation coefficients $-.9$, $-.5$, 0, $.5$, and $.9$, and compute the sample correlation coefficients.
   (c) Have a partner generate scatterplots as in part (b) and then guess the correlation coefficients.

24. Generate a bivariate sample of size 50 as in Problem 23 with correlation coefficient $.8$. Find the estimated regression line and the residuals. Plot the residuals versus X and the residuals versus Y. Explain the appearance of the plots.

25. An investigator wishes to use multiple regression to predict a variable, Y, from two other variables, X_1 and X_2. She proposes forming a new variable $X_3 = X_1 + X_2$ and using multiple regression to predict Y from the three X variables. Show that she will run into problems because the design matrix will not have full rank.

26. The following table gives the transition pressure of the bismuth II–I transition as a function of temperature (see Example E in Section 14.2.2). Fit a linear relationship between pressure and temperature, examine the residuals, and comment.

Temperature (°C)	Pressure (bar)
20.8	25276
20.9	25256
21.0	25216
21.9	25187
22.1	25217
22.1	25187
22.4	25177
22.5	25177
24.8	25098
24.8	25093
25.0	25088
34.0	24711
34.0	24701
34.1	24716
42.7	24374
42.7	24394
42.7	24384
49.9	24067
50.1	24057
50.1	24057
22.5	25147
23.1	25107
23.0	25077

27. Dissociation pressure for a reaction involving barium nitride was recorded as a function of temperature (Orcutt, 1970). The second law of thermodynamics gives the approximate relationship

$$\ln(\text{pressure}) = A + \frac{B}{T}$$

where T is absolute temperature. From the data in the following table, estimate A and B and their standard errors [the notation 2.11($-$5) means 2.11×10^{-5}]. Form approximate 95% confidence intervals for A and B. Examine the residuals and comment.

Temperature (K)	Pressure (torr)
738	2.11($-$5)
748	4.80($-$5)
764	5.95($-$5)
770	9.20($-$5)
792	2.06($-$4)
795	1.64($-$4)
815	3.78($-$4)
844	7.87($-$4)
845	8.30($-$4)
874	1.90($-$3)
894	4.26($-$3)
927	8.00($-$3)
(continued)	

Temperature (K)	Pressure (torr)
934	1.10(−2)
958	1.68(−2)
989	3.34(−2)
1000	4.9(−2)
1023	6.8(−2)
1025	7.10(−2)
1029	8.9(−2)
1030	1.04(−1)
1042	9.5(−2)
1048	1.23(−1)
1067	1.78(−1)
1082	2.36(−1)
1084	2.90(−1)
1112	3.98(−1)
1132	5.55(−1)
1133	5.23(−1)
1134	5.57(−1)
1135	5.81(−1)
1135	6.22(−1)
1150	7.24(−1)

28. The following table lists observed values of Young's modulus (g) measured at various temperatures (t) for sapphire rods (Ku, 1969). Fit a linear relationship $g = \beta_0 + \beta_1 t$, and form confidence intervals for the coefficients. Examine the residuals.

Temperature (K)	Young's Modulus (−3000)
500	328
550	296
600	266
603	260
603	244
650	240
650	232
650	213
700	204
700	203
700	184
750	174
750	175
750	154
800	152
800	146
800	124
850	117
850	94
900	97
900	61
950	38
1000	30
1000	5

29. As part of a nuclear safeguards program, the contents of a tank are routinely measured. The determination of volume is made indirectly by measuring the difference in pressure at the top and at the bottom of the tank. The tank is cylindrical in shape, but its internal geometry is complicated by various pipes and agitator paddles. Without these complications, pressure and volume should have a linear relationship. To calibrate pressure with respect to volume, known quantities (x) of liquid are placed in the tank and pressure readings (y) are taken. The data in the table below are from Knafl et al. (1984).

(a) Plot pressure versus volume. Does the relationship appear linear?

(b) Calculate the linear regression of pressure on volume, and plot the residuals versus volume. What does the residual plot show?

(c) Try fitting pressure as a quadratic function of volume. What do you think of the fit?

Volume (kiloliters)	Pressure (pascals)
.18941	215.25
.18949	218.28
.37880	632.62
.37884	627.14
.37884	628.71
.56840	1034.05
.56843	1033.34
.75755	1469.11
.75767	1474.11
.75769	1475.24
.94740	1924.52
.94746	1921.85
1.13642	2372.04
1.13646	2374.43
1.13665	2377.76
1.32640	2819.47
1.32640	2815.70
1.51523	3263.50
1.51528	3261.94
1.51561	3268.54
1.70525	3711.64

30. The following data come from the calibration of a proving ring, a device for measuring force (Hockersmith and Ku, 1969).

(a) Plot load versus deflection. Does the plot look linear?

(b) Fit deflection as a linear function of load, and plot the residuals versus load. Do the residuals show any systematic lack of fit?

(c) Fit deflection as a quadratic function of load, and estimate the coefficients and their standard errors. Plot the residuals. Does the fit look reasonable?

| | Deflection | | |
Load	Run 1	Run 2	Run 3
10,000	68.32	68.35	68.30
20,000	136.78	136.68	136.80
30,000	204.98	205.02	204.98
40,000	273.85	273.85	273.80
50,000	342.70	342.63	342.63
60,000	411.30	411.35	411.28
70,000	480.65	480.60	480.63
80,000	549.85	549.85	549.83
90,000	619.00	619.02	619.10
100,000	688.70	688.62	688.58

31. The following table gives the diameter at breast height (DBH) and the age of 27 chestnut trees (Chapman and Demeritt, 1936). Try fitting DBH as a linear function of age. Examine the residuals. Can you find a transformation of DBH and/or age that produces a more linear relationship?

Age (yr)	DBH (ft)
4	.8
5	.8
8	1.0
8	3.0
10	2.0
10	3.5
12	4.9
13	3.5
14	2.5
16	4.5
18	4.6
20	5.5
22	5.8
23	4.7
25	6.5
28	6.0
29	4.5
30	6.0
30	7.0
33	8.0
34	6.5
35	7.0
38	5.0
38	7.0
40	7.5
42	7.5

32. The stopping distance (y) of an automobile on a certain road was studied as a function of velocity (Brownlee, 1960). The data are listed in the following

table. Fit y and $\sqrt{y}$ as linear functions of velocity, and examine the residuals in each case. Which fit is better? Can you suggest any physical reason that explains why?

Velocity (mph)	Stopping Distance (ft)
20.5	15.4
20.5	13.3
30.5	33.9
40.5	73.1
48.8	113.0
57.8	142.6

33. In a study of the modulus of natural rubber (y) as a function of the amount of decomposed dicumyl peroxide (a cross-linking agent) it contains (x_1) and its temperature (x_2) the relation

$$y = \beta_0 + \beta_1 x_1 + \beta_2 x_2 + \beta_3 x_1 x_2$$

was obtained (Wood, 1972). The data are given in the table below. Fit a relationship of this form, estimate the coefficients and their standard errors, and examine the residuals. Do the same things but omit the $x_1 x_2$ term. Comment on the adequacy of the models.

Dicumyl Peroxide	Temperature (°C)						
	−50	−25	0	25	50	75	100
.48	3.20	3.06	2.92	2.77	2.63	2.49	2.35
.95	3.54	3.57	3.59	3.62	3.65	3.67	3.70
1.90	4.85	5.01	5.16	5.32	5.47	5.63	5.78
2.86	6.24	6.57	6.91	7.24	7.57	7.91	8.24
3.81	7.80	8.14	8.47	8.81	9.15	9.49	9.82
4.76	7.60	8.37	9.14	9.91	10.68	11.45	12.22
6.66	10.60	11.61	12.62	13.62	14.63	15.64	16.65
7.62	12.80	13.92	15.04	16.17	17.29	18.41	19.53
9.52	17.30	18.20	19.00	19.90	20.80	21.60	22.50
14.3	19.20	22.09	24.97	27.86	30.75	33.64	36.52
19.0	29.15	32.53	35.91	39.30	42.68	46.06	49.45
23.8	35.90	38.50	41.10	43.70	46.30	48.90	51.50

34. Cogswell (1973) studied a method of measuring resistance to breathing in children. The tables below list total respiratory resistance and height for a group of children with asthma and a group with cystic fibrosis. Is there a statistically significant relation between respiratory resistance and height in either group?

Children with asthma

Height (cm)	Resistance	Height (cm)	Resistance
90	25.6	123	6.2
97	15.1	125	8.3
97	9.8	118	3.5
104	7.5	122	10.2
119	10.1	120	16.1
106	12.5	133	10.1
113	9.1	125	12.1
116	17.0	123	11.2
118	5.5	141	9.4
119	15.7	121	15.6
117	6.4	128	7.9
122	8.8	125	18.7
122	10.1	140	8.3
124	5.0	140	7.5
129	5.8	140	8.5
130	12.8	148	8.9
135	10.0	145	7.9
127	8.0	138	5.7
120	23.8	148	9.5
121	7.5	132	6.5
126	12.1	134	15.0

Children with cystic fibrosis

Height (cm)	Resistance
89	13.8
93	8.2
92	9.0
101	12.5
95	21.1
89	6.8
97	17.0
97	11.0
111	8.2
102	12.7
103	8.5
108	10.0
103	11.6
105	9.5
109	15.0
93	13.5
98	11.0
103	11.0
108	8.8
106	9.5
109	9.2
111	15.0
111	7.0
116	6.3

35. Chang (1945) studied the rate of sedimentation of amoebic cysts in water, in attempting to develop methods of water purification. The table gives the diameters of the cysts and the times required for the cysts to settle through 720 microns of still water at three temperatures. Each entry of the table is an average of several observations, the number of which is given in parentheses. Does the time required appear to be a linear or a quadratic function of diameter? Can you find a model that fits? How do the settling rates at the three temperatures compare? (See Problem 5.)

Diameter (micron)	Settling Times of Cysts (sec)		
	10°C	25°C	28°C
11.5	217.1(2)	138.2(1)	128.4(2)
13.1	168.3(3)	109.3(3)	103.1(4)
14.4	136.6(11)	89.1(13)	82.7(11)
15.8	114.6(17)	73.0(11)	70.5(18)
17.3	96.4(8)	61.3(6)	59.7(6)
18.7	80.8(5)	56.2(4)	50.0(4)
20.2	70.4(2)	46.3(1)	41.4(2)

36. The following table gives average reading scores of third-graders from several elementary schools on a standardized test in each of two successive years. Is there a "regression effect"?

1982	1983
356	337
339	354
327	329
317	391
312	295
316	305
306	294
300	255
304	286
298	269
293	269
296	306
290	291
291	278
289	298
282	302
289	295
276	256
279	274
277	281
268	269
265	266
269	301

(continued)

1982	1983
263	266
257	283
256	249
250	262
249	227
242	221
237	254
230	243
239	260
230	240
230	218
221	274
225	220
219	232
211	239
205	234
197	227

37. Recordings of the levels of pollutants and various meteorological conditions are made hourly at several stations by the Los Angeles Pollution Control District. This agency attempts to construct mathematical/statistical models to predict pollution levels and to gain a better understanding of the complexities of air pollution. Obviously, very large quantities of data are collected and analyzed, but only a small set of data will be considered in this problem. The table below gives the maximum level of an oxidant (a photochemical pollutant) and the morning averages of four meteorological variables: wind speed, temperature, humidity, and insolation (a measure of the amount of sunlight). The data cover 30 days during one summer.

(a) Examine the relationship of oxidant level to each of the four meteorological variables and the relationships of the meteorological variables to each other. How well can the maximum level of oxidant be predicted from some or all of the meteorological variables? Which appear to be most important?

(b) The standard statistical model used in this chapter assumes that the errors are random and independent of one another. In data that are collected over time, the error at any given time may well be correlated with the error from the preceding time. This phenomenon is called **serial correlation**, and in its presence the estimated standard errors of the coefficients developed in this chapter may be incorrect. The parameter estimates are still unbiased, however (why?). Can you detect serial correlation in the errors from your fits?

Day	Wind Speed	Temperature	Humidity	Insolation	Oxidant
1	50	77	67	78	15
2	47	80	66	77	20
3	57	75	77	73	13
4	38	72	73	69	21
5	52	71	75	78	12
6	57	74	75	80	12
7	53	78	64	75	12
8	62	82	59	78	11
9	52	82	60	75	12
10	42	82	62	58	20
11	47	82	59	76	11
12	40	80	66	76	17
13	42	81	68	71	20
14	40	85	62	74	23
15	48	82	70	73	17
16	50	79	66	72	16
17	55	72	63	69	10
18	52	72	61	57	11
19	48	76	60	74	11
20	52	77	59	72	9
21	52	73	58	67	5
22	48	68	63	30	5
23	65	67	65	23	4
24	53	71	53	72	7
25	36	75	54	78	18
26	45	81	44	81	17
27	43	84	46	78	23
28	42	83	43	78	23
29	35	87	44	77	24
30	43	92	35	79	25

15

Decision Theory and Bayesian Inference

15.1 Introduction

This chapter introduces two approaches to theoretical statistics: decision theory and Bayesian theory. Each of these approaches attempts to set forth a unifying framework for the theory of statistics and to make it possible to deal in a systematic way with problems that are not amenable to analysis by traditional methods such as the techniques of estimation and testing developed earlier in this book. In line with the material, the approach in this chapter will be somewhat more abstract and less problem-specific than it was in previous chapters.

15.2 Decision Theory

The decision theoretic approach views statistics as a mathematical theory for making decisions in the face of uncertainty. According to this paradigm, the decision maker chooses an **action** a from a set A of all possible actions based on the observation of a random variable, or data, **X**, which has a probability distribution that depends on a parameter θ called the **state of nature**. The set of all possible values of θ is denoted by Θ. The decision is made by a **statistical decision function**, d, which maps the sample space (the set of all possible data values) onto the action space A. Denoting the data by **X**, the action is random and is given as $a = d(\mathbf{X})$.

By taking the action $a = d(\mathbf{X})$, the decision maker incurs a **loss**, $l(\theta, d(\mathbf{X}))$, which depends on both θ and $d(\mathbf{X})$. The comparison of different decision functions is based on the **risk function**, or expected loss,

$$R(\theta, d) = E[l(\theta, d(\mathbf{X}))]$$

Here, the expectation is taken with respect to the probability distribution of **X**, which depends on θ. Note that the risk function depends on the true state of nature, θ, and on the decision function, d. Decision theory is concerned with methods of determining "good" decision functions, that is, decision functions that have small risk.

Let us consider several examples in order to clarify the concepts, terminology, and notation just introduced.

EXAMPLE A.   (Sampling Inspection) A manufacturer produces a lot consisting of N items, n of which are sampled randomly and determined to be either defective or nondefective. Let p denote the proportion of the N items that are defective, and let $\hat{p}$ denote the proportion of the n items that are defective. Suppose that the lot is sold for a price, \$M, with a guarantee that if the proportion of defective items exceeds p_0 the manufacturer will pay a penalty, \$P, to the buyer. For any lot, the manufacturer has two possible actions: either sell the lot or junk it at a cost, \$C. The action space is therefore

$$A = \{\text{sell}, \text{junk}\}$$

The data are $\mathbf{X} = \hat{p}$; the state of nature is $\theta = p$. The loss function depends on the action $d(\mathbf{X})$ and the state of nature as shown in the following table:

State of Nature	Sell	Junk
$p < p_0$	$-\$M$	$\$C$
$p \geq p_0$	$\$P$	$\$C$

Here a profit is expressed as a negative loss. Note that the decision rule depends on $\hat{p}$, which is random; the risk function is the expected loss, and the expectation is computed with respect to the probability distribution of $\hat{p}$. This distribution depends on p.   □

EXAMPLE B.   (Classification) On the basis of several physiological measurements, **X**, a decision must be made concerning whether a patient has suffered a myocardial infarction (MI) and should be admitted to intensive care. Here $A = \{$admit, do not admit$\}$ and $\Theta = \{$MI, no MI$\}$. The probability distribution of $\mathbf{X}$ depends on θ, perhaps in a complicated way. Some elements of the loss function may be difficult to specify; the economic costs of admission can be calculated, but the cost of not admitting when in fact the patient has suffered a myocardial infarction is more subjective. It can be argued that any rational decision procedure should come to grips with this problem and that any decision procedure implicitly specifies a loss function. To make this problem more realistic, the action space could be expanded to include such actions as "send home" and "hospitalize for further observation." □

EXAMPLE C. (Estimation) Suppose that we want to estimate some function $v(\theta)$ on the basis of a sample $\mathbf{X} = (X_1, \ldots, X_n)$, where the distribution of the X_i depends on θ. Here $a(\mathbf{X})$ is an estimator of $v(\theta)$. The quadratic loss function

$$l(\theta, a(\mathbf{X})) = [v(\theta) - a(\mathbf{X})]^2$$

is often used. The risk function is then

$$R(\theta, a(\mathbf{X})) = E\{[v(\theta) - a(\mathbf{X})]^2\}$$

which is the familiar mean squared error. Note that, here again, the expectation is taken with respect to the distribution of $\mathbf{X}$, which depends on θ. □

In the next section, we will consider some more specific examples.

15.2.1   Bayes Rules and Minimax Rules

This section is concerned with choosing a "good" decision function, that is, one that has a small risk,

$$R(\theta, d(\mathbf{X})) = E[l(\theta, d(\mathbf{X}))]$$

We have to face the difficulty that R depends on θ, which is not known. For example, there might be two decision rules, d_1 and d_2, and two values of θ, θ_1 and θ_2, such that

$$R(\theta_1, d_1) < R(\theta_1, d_2)$$

but

$$R(\theta_2, d_1) > R(\theta_2, d_2)$$

Thus d_1 is better if the state of nature is θ_1, but d_2 is better if the state of nature is θ_2. The two most widely used methods for confronting this difficulty are to use either a minimax rule or a Bayes rule.

The minimax method proceeds as follows. First, for a given decision function d, consider the worst that the risk could be:

$$\max_{\theta \in \Theta} [R(\theta, d)]$$

Then choose a decision function, d^*, that minimizes this maximum risk:

$$\min_d \left\{ \max_{\theta \in \Theta} [R(\theta, d)] \right\}$$

Such a decision rule, if it exists, is called a **minimax rule**.

The above expressions may not be strictly correct, since the minimum and maximum may not actually be attained (just as a continuous function of a single variable may not attain a minimum or maximum value on an open interval). The statements should thus be modified by replacing max and min by sup (least upper bound) and inf (greatest low bound), respectively.

The weakness of the minimax method is intuitively apparent. It is a very conservative procedure that places all its emphasis on guarding against the worst possible case. In fact, this worst case may not be very likely to occur. To make this idea more precise, we can assign a probability distribution to the state of nature θ; this distribution is called the **prior distribution** of θ. Given such a prior distribution, we can calculate the **Bayes risk** of a decision function d:

$$B(d) = E[R(\Theta, d(\mathbf{X}))]$$

Here the expectation is taken with respect to the probability distribution of both Θ and $\mathbf{X}$. By the property of iterated conditional expectations, the Bayes risk can be expressed as

$$B(d) = E\{E[l(\theta, d(\mathbf{X})) | \Theta = \theta]\}$$

where the inner expectation is conditional on $\Theta = \theta$ and the outer expectation is taken with respect to the distribution of Θ. The Bayes risk is the average of the risk function with respect to the prior distribution of θ. A function d^{**} that minimizes the Bayes risk is called a **Bayes rule**.

EXAMPLE A.   As part of the foundation of a building, a steel section is to be driven down to a firm stratum below ground (Benjamin and Cornell, 1970). The engineer has two choices (actions):

a_1: select a 40-foot section
a_2: select a 50-foot section

There are two possible states of nature:

θ_1: depth of firm stratum is 40 feet
θ_2: depth of firm stratum is 50 feet

If the 40-foot section is incorrectly chosen, an additional length of steel must be spliced on at a cost of $400. If the 50-foot section is incorrectly chosen, 10 feet of steel must be scrapped at a cost of $100. The loss function is therefore represented in the following table:

	θ_1	θ_2
a_1	0	$400
a_2	$100	0

A depth sounding is taken by means of a sonic test. Suppose that the depth, X, has three possible values, $x_1 = 40$, $x_2 = 45$, and $x_3 = 50$, and that the probability distribution of X depends on θ as follows:

x	θ_1	θ_2
x_1	.6	.1
x_2	.3	.2
x_3	.1	.7

We will consider the following four decision rules:

	x_1	x_2	x_3
d_1	a_1	a_1	a_1
d_2	a_1	a_2	a_2
d_3	a_1	a_1	a_2
d_4	a_2	a_2	a_2

We will first find the minimax rule. To do so, we need to compute the risk of each of the decision functions in the case where $\theta = \theta_1$ and in the case where $\theta = \theta_2$. To do such computations for $\theta = \theta_1$, each risk function is computed as

$$R(\theta_1, d_i(X)) = E[l(\theta_1, d_i(X))]$$

$$= \sum_{j=1}^{3} l(\theta_1, d_i(x_j)) P(X = x_j | \theta = \theta_1)$$

We thus have

$$R(\theta_1, d_1) = 0 \times .6 + 0 \times .3 + 0 \times .1 = 0$$

$$R(\theta_1, d_2) = 0 \times .6 + 100 \times .3 + 100 \times .1 = 40$$

$$R(\theta_1, d_3) = 0 \times .6 + 0 \times .3 + 100 \times .1 = 10$$

$$R(\theta_1, d_4) = 100 \times .6 + 100 \times .3 + 100 \times .1 = 100$$

Similarly, in the case where $\theta = \theta_2$, we have

$$R(\theta_2, d_1) = 400$$

$$R(\theta_2, d_2) = 40$$

$$R(\theta_2, d_3) = 120$$

$$R(\theta_2, d_4) = 0$$

To find the minimax rule, we note that the maximum values of d_1, d_2, d_3, and d_4 are 400, 40, 120, and 100, respectively. Thus, d_2 is the minimax rule.

We now turn to the computation of the Bayes rule. Suppose that on the basis of previous experience and from large-scale maps, we take as the prior distribution $\pi(\theta_1) = .8$ and $\pi(\theta_2) = .2$. Using this prior distribution and the risk functions computed above, we find for each decision function its Bayes risk,

$$B(d) = E[R(\Theta, d)]$$

$$= R(\theta_1, d)\pi(\theta_1) + R(\theta_2, d)\pi(\theta_2)$$

Thus, we have

$$B(d_1) = 0 \times .8 + 400 \times .2 = 80$$

$$B(d_2) = 40 \times .8 + 40 \times .2 = 40$$

$$B(d_3) = 10 \times .8 + 120 \times .2 = 32$$

$$B(d_4) = 100 \times .8 + 0 \times .2 = 80$$

Comparing these numbers, we see that d_3 is the Bayes rule. Note that this Bayes rule is less conservative than the minimax rule obtained above in that it chooses action a_1 (40-foot length) based on observation x_2 (45-foot sounding). This is because the prior distribution for this Bayes rule puts more weight on θ_1. If the prior distribution were changed sufficiently, the Bayes rule would change. □

EXAMPLE B.   A manufacturer produces items in lots of 21. One item is selected at random and is tested to determine whether or not it is defective. If the selected item is defective, either the remaining 20 items can be sold at \$1 per item with a double-your-money-back guarantee on each item or the whole lot can be junked at a cost of \$1. Consider the following two decisions:

d_1: sell if the item is good, junk if defective
d_2: sell in either case

Let k denote the number of defectives in a lot of 21, and let X be 1 or 0 if the tested item is good or defective, respectively. The loss functions are

$$l(k, d_1(X)) = \begin{cases} -20 + 2 \times k, & \text{if } X = 1 \\ 1, & \text{if } X = 0 \end{cases}$$

$$l(k, d_2(X)) = \begin{cases} -20 + 2 \times k, & \text{if } X = 1 \\ -20 + 2 \times (k - 1), & \text{if } X = 0 \end{cases}$$

We calculate the risk functions, which are given by

$$R(k, d_i(X)) = E[l(k, d_i(X))]$$

Note that the expectation is taken with respect to X. For a given value of k,

$$P(X = 0) = \frac{k}{21}$$

$$P(X = 1) = 1 - \frac{k}{21}$$

Thus,

$$R(k, d_1) = (-20 + 2k)\left(1 - \frac{k}{21}\right) + \frac{k}{21}$$

$$= -20 + 3k - \frac{2k^2}{21}$$

$$R(k, d_2) = (-20 + 2k)\left(1 - \frac{k}{21}\right) + [-20 + 2(k - 1)]\frac{k}{21}$$

$$= -20 + \frac{40k}{21}$$

The two risk functions are plotted versus k in Figure 15-1. From the plot, we see that d_1 has a smaller risk for $k \geq 12$. Since the maximum risk of d_1 is less than the maximum risk of d_2, d_1 is the minimax rule.

Figure 15-1 shows that over the range for which the risk is negative—that is, the range in which the manufacturer expects to make a profit—d_2 has smaller risk. Therefore, if the product is fairly reliable, the manufacturer may prefer d_2. A Bayesian approach is one way to formalize this. Let us assume that the prior distribution of k is binomial with probability parameter p. (This assumption is appropriate if each item is defective with probability p independently of the other items.) The value of p might be known at least approximately from experience. Since $E(k) = 21p$,

$$E(k^2) = \text{Var}(k) + [E(k)]^2$$

$$= 21p(1 - p) + 441p^2$$

$$= 420p^2 + 21p$$

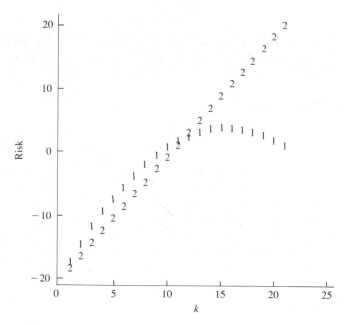

Figure 15-1. A plot of two risk functions (1 and 2) versus k.

The Bayes risks of d_1 and d_2 may now be evaluated:

$$B(d_1) = -20 + 3 \times 21 \times p - \tfrac{2}{21}(420p^2 + 21p)$$

$$= -20 + 61p - 40p^2$$

$$B(d_2) = -20 + \tfrac{40}{21} \times 21p$$

$$= 40p - 20$$

From Figure 15-2, a plot of Bayes risk versus p, we see that d_2 has a smaller Bayes risk as long as $p \leq .5$. $\qquad\square$

Minimax rules have been introduced in this section, but in later sections we will concentrate primarily on Bayes rules. The mathematical treatment of minimax procedures is fairly difficult and is closely related to game theory (in this "game," the statistician chooses d and nature chooses θ). For a more advanced treatment of decision theory, see Ferguson (1967).

15.2.2 Posterior Analysis

This section introduces the concept of a posterior distribution and develops a simple method for finding a Bayes rule.

The Bayesian procedures already introduced model Θ as a random variable

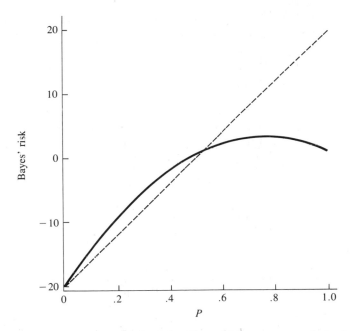

Figure 15-2. Bayes risk as a function of p: The solid line corresponds to d_1 and the dotted line to d_2.

with a probability function $g(\theta)$ (if θ is a discrete random variable, g denotes a probability mass function; if θ is continuous, g denotes a density function). We called $g(\theta)$ the prior distribution of Θ. The probability mass function or density function of $\mathbf{X}, f(\mathbf{x}|\theta)$, is conditional on the value of Θ, and the joint distribution of $\mathbf{X}$ and Θ is

$$f_{\mathbf{X}\Theta}(\mathbf{x}, \theta) = f(\mathbf{x}|\theta)g(\theta)$$

The marginal distribution of $\mathbf{X}$ is thus

$$f_{\mathbf{X}}(\mathbf{x}) = \begin{cases} \displaystyle\int f(\mathbf{x}|\theta)g(\theta)\,d\theta, & \text{if } \Theta \text{ is continuous} \\[2ex] \displaystyle\sum f(\mathbf{x}|\theta_i)g(\theta_i), & \text{if } \Theta \text{ is discrete} \end{cases}$$

We can apply Bayes' Rule to find the conditional distribution of Θ given $\mathbf{X}$; for example, in the continuous case, we have

$$h(\theta|\mathbf{x}) = \frac{f_{\mathbf{X}\Theta}(\mathbf{x}, \theta)}{f_{\mathbf{X}}(\mathbf{x})}$$

$$= \frac{f(\mathbf{x}|\theta)g(\theta)}{\int f(\mathbf{x}|\theta)g(\theta)\,d\theta}$$

A similar expression holds in the discrete case. The conditional distribution $h(\theta|\mathbf{x})$ is called the **posterior distribution** of Θ. The words *prior* and *posterior* derive from the facts that $g(\theta)$ is specified before (prior to) observing $\mathbf{X}$ and $h(\theta|\mathbf{x})$ is calculated after (posterior to) observing $\mathbf{X} = \mathbf{x}$. Section 15.3 deals in some depth with the interpretation and use of prior and posterior distributions; here we will use the posterior distribution as a tool for finding a Bayes rule.

Suppose that we have observed $\mathbf{X} = \mathbf{x}$. We define the **posterior risk** of an action, $a = d(\mathbf{x})$, as the expected loss, where the expectation is taken with respect to the posterior distribution of Θ. For the continuous case, we have

$$E[l(\Theta, d(\mathbf{x}))] = \int l(\theta, d(\mathbf{x})) h(\theta|\mathbf{x}) \, d\theta$$

A similar expression in which the integral is replaced by a summation holds in the discrete case.

THEOREM A. Suppose that there is a function $d(\mathbf{x})$ that minimizes the posterior risk. Then $d(\mathbf{x})$ is a Bayes rule.

Proof. We will prove this for the continuous case. The discrete case is proved analogously. The Bayes risk of a decision function d is

$$B(d) = E_\Theta[R(\Theta, d)]$$
$$= E_\Theta[E_\mathbf{X}(l(\Theta, d(\mathbf{X})))]$$
$$= \int \int l(\theta, d(\mathbf{x})) f_{\mathbf{X}\Theta}(\mathbf{x}, \theta) \, d\mathbf{x} \, d\theta$$

Now since $f_{\mathbf{X}\Theta}(\mathbf{x}, \theta) = f_\mathbf{X}(\mathbf{x}) h(\theta|\mathbf{x})$, we may express the above double integral as

$$\int \left[\int l(\theta, d(\mathbf{x})) h(\theta|\mathbf{x}) \, d\theta \right] f_\mathbf{X}(\mathbf{x}) \, d\mathbf{x}$$

Since $f_\mathbf{X}(\mathbf{x})$ is nonnegative, this expression is minimized if the interior integral is minimized at each fixed x, that is, minimizing the posterior risk. □

The practical importance of this theorem is that it allows us to use the observed data, $\mathbf{x}$, rather than considering all possible values of $\mathbf{X}$ as we did in the examples in Section 15.2.1. In summary, the algorithm for finding the Bayes rule is as follows:

Step 1: Calculate the posterior distribution $h(\theta|\mathbf{x})$.
Step 2: For each action a, calculate the posterior risk, which is

$$E[l(\Theta, a)] = \int l(\theta, a)h(\theta|\mathbf{x}) \, d\theta$$

in the continuous case or $\sum l(\theta_i, a)h(\theta_i|\mathbf{x})$ in the discrete case.

Step 3: The action a^* that minimizes the posterior risk is the Bayes rule.

EXAMPLE A. To illustrate the algorithm, let us return to Example A of Section 15.2.1. Suppose that we observe $X = x_2 = 45$. In the notation of that example, the prior distribution is $\pi(\theta)$. We first calculate the posterior distribution:

$$h(\theta_1|x_2) = \frac{f(x_2|\theta_1)\pi(\theta_1)}{\displaystyle\sum_{i=1}^{2} f(x_2|\theta_i)\pi(\theta_i)}$$

$$= \frac{.3 \times .7}{.3 \times .7 + .2 \times .3}$$

$$= .78$$

Similarly, we obtain

$$h(\theta_2|x_2) = .22$$

We next calculate the posterior risk (PR) for a_1 and a_2:

$$PR(a_1) = l(\theta_1, a_1)h(\theta_1|x_2) + l(\theta_2, a_1)h(\theta_2|x_2)$$

$$= 0 + 400 \times .22$$

$$= 88$$

and

$$PR(a_2) = l(\theta_1, a_2)h(\theta_1|x_2) + l(\theta_2, a_2)h(\theta_2|x_2)$$

$$= 100 \times .78 + 0$$

$$= 78$$

Comparing the two, we see that a_2 has the smaller posterior risk and is thus the Bayes rule. Note that we used only x_2, the observed vaue of X, in this calculation; if there were a large number of possible values of X, this could be quite time-saving. $\square$

Applying Theorem A to find Bayes rules will be taken up again in the next two sections.

15.2.3 Classification and Hypothesis Testing

Problems involving classification and pattern recognition can often be profitably viewed from the perspective of decision theory. Suppose that a population consists of m classes and that on the basis of some measurements we wish to classify an individual as belonging to one of the classes. Here are some examples:

- On the basis of physiological measurements, a patient is to be classified as having or not having a particular disease.
- On the basis of radar measurements, an aircraft is to be classified as enemy, friendly, or unknown.
- On the basis of sound measurments, an utterance is to be classified by a computer as one of the words in its dictionary.

As before, let us denote the measurements by $\mathbf{X}$. Here, Θ denotes the class membership; the possible values of Θ are given by $\{\theta_1, \theta_2, \ldots, \theta_m\}$, where $\Theta = \theta_j$ denotes membership in the jth class. Prior probabilities $\pi_1, \ldots, \pi_m$ are given to the various classes. We assume that the probability distributions $f(\mathbf{x}|\theta)$ are known (in practice, the determination or approximation of these distributions is usually the key to the problem). Let l_{ij} denote the loss in classifying a member of class i to class j.

Suppose that we observe $\mathbf{X} = \mathbf{x}$. What is the Bayes rule? From the results in Section 15.2.2, we know that we need only compare the posterior risks. The posterior distribution of Θ is

$$h(\theta_i|\mathbf{x}) = P(\Theta = \theta_i|\mathbf{X} = \mathbf{x})$$

$$= \frac{\pi_i f(\mathbf{x}|\theta_i)}{\sum \pi_j f(\mathbf{x}|\theta_j)}$$

The posterior risk of assigning an observation to the ith class is thus

$$\mathrm{PR}_i = \frac{\sum\limits_{k=1}^{m} l_{ik} \pi_k f(\mathbf{x}|\theta_k)}{\sum \pi_j f(\mathbf{x}|\theta_j)}$$

The choice of the Bayes rule is indicated by that value of i for which PR_i is a minimum.

An interesting special case is

$$l_{ij} = \begin{cases} 1, & i = j \\ 0, & i \neq j \end{cases}$$

We refer to this case as **0–1 loss**. For any classification rule, $d(\mathbf{X})$, the risk function if $\Theta = i$ is

$$R(i, d) = E[l(i, d(\mathbf{X}))]$$

$$= \sum_{j=1}^{m} l_{ij} P[d(\mathbf{X}) = j]$$

$$= \sum_{j \neq i} P[d(\mathbf{X}) = j]$$

$$= 1 - P[d(\mathbf{X}) = i]$$

The risk is simply the probability of misclassification. Using the results derived above, we see that the posterior risk is minimized for that value of i that minimizes

$$PR_i = \frac{\sum\limits_{k \neq i} \pi_k f(\mathbf{x} | \theta_k)}{\sum\limits_{j=1}^{m} \pi_j f(\mathbf{x} | \theta_j)}$$

$$= P(\Theta \neq \theta_i | \mathbf{X} - \mathbf{x})$$

$$= 1 - P(\Theta = \theta_i | \mathbf{X} = \mathbf{x})$$

In words, PR_i is minimized for that value of i such that $P(\Theta = \theta_i | \mathbf{X} = \mathbf{x})$ is maximized, that is, for the class that has maximum posterior probability.

EXAMPLE A. Suppose that n waiting times between emissions of alpha particles from a radioactive substance are recorded; denote them by $\mathbf{X} = (X_1, \ldots, X_n)$. On the basis of these observations, a decision is to be made as to whether to classify the particles as coming from substance I or substance II. In either case, the waiting times are independent and exponentially distributed, but the distribution has parameter θ_1 or θ_2, respectively. The prior probabilities are equal, and we use 0–1 loss, so we wish to minimize the probability of misclassification. To determine the Bayes rule, we compare the two posterior probabilities or, equivalently, $f(\mathbf{x} | \theta_1)$ and $f(\mathbf{x} | \theta_2)$.

$$f(\mathbf{x} | \theta_i) = \theta_i^n \exp\left(\theta_i \sum_{j=1}^{n} x_j \right)$$

After some algebra, we see that substance I has the larger posterior probability if

$$(\theta_2 - \theta_1) \sum x_i > n \log \frac{\theta_2}{\theta_1}$$

The classification rule is thus based on the size of $\sum x_i$. □

Hypothesis testing can be regarded as a classification problem with a constraint on the probability of misclassification (the probability of a type I error). Let hypotheses H and K specify two values of a parameter, denoted as θ_1 and θ_2:

$$H: \mathbf{X} \sim f(\mathbf{x}|\theta_1)$$

$$K: \mathbf{X} \sim f(\mathbf{x}|\theta_2)$$

In previous chapters, we have used the likelihood ratio test frequently. In this case, the likelihood ratio test accepts H if

$$\frac{f(\mathbf{x}|\theta_1)}{f(\mathbf{x}|\theta_2)} > c$$

or

$$\frac{1}{c}\left[\frac{f(\mathbf{x}|\theta_1)}{f(\mathbf{x}|\theta_2)}\right] > 1$$

and rejects H otherwise. The constant c, a positive number, is chosen so as to control the significance level of the test. We now state and prove the Neyman–Pearson Lemma, which establishes that tests of this form are optimal in the sense that they have maximum power among all tests.

LEMMA A.   (Neyman–Pearson Lemma) Let d^* be a test that accepts if

$$\frac{f(\mathbf{x}|\theta_1)}{f(\mathbf{x}|\theta_2)} > c$$

and let α^* be the significance level of d^*. Let d be another test that has significance level $\alpha \leq \alpha^*$. Then the power of d is less than or equal to the power of d^*.

Proof.   Let $1/c = \pi/(1 - \pi)$; that is,

$$\pi = \frac{1}{c + 1}$$

Then d^* accepts if

$$\frac{\pi f(\mathbf{x}|\theta_1)}{(1 - \pi)f(\mathbf{x}|\theta_2)} > 1$$

and d^* is thus a Bayes rule for a classification problem with prior probabilities π and $1 - \pi$ and with 0–1 loss, so $B(d^*) - B(d) \leq 0$. Now, we have

$$B(d^*) - B(d)$$

$$= \pi\{E[l(\theta_1, d^*)] - E[l(\theta_1, d)]\} + (1 - \pi)\{E[l(\theta_2, d^*)] - E[l(\theta_2, d)]\}$$

$$= \pi(\alpha^* - \alpha) + (1 - \pi)\{E[l(\theta_2, d^*)] - E[l(\theta_2, d)]\}$$

Thus, $\{E[l(\theta_2, d^*)] - E[l(\theta_2, d)]\} < 0$. The proof is completed by noting that $E[l(\theta_2, d^*)]$ is the probability of falsely accepting H, which is 1 minus the power of the test. $\square$

EXAMPLE B.   Let $X_1, \ldots, X_n$ be independent random variables following an exponential distribution with parameter θ:

$$f(x|\theta) = \theta e^{-\theta x}, \quad \theta > 0, x > 0$$

Consider testing $H: \theta = \theta_1$ versus $K: \theta = \theta_2$. The joint distribution of $\mathbf{X} = (X_1, \ldots, X_n)$ is

$$f(\mathbf{x}|\theta) = \theta^n \exp\left(-\theta \sum x_i\right)$$

The likelihood ratio test rejects for small values of

$$\frac{f(\mathbf{x}|\theta_1)}{f(\mathbf{x}|\theta_2)} = \left(\frac{\theta_1}{\theta_2}\right)^n \exp\left[(\theta_1 - \theta_2) \sum x_i\right]$$

The critical value for the test can be calculated for any desired significance level by using the fact that $\sum x_i$ follows a gamma distribution. $\square$

15.2.4   Estimation

Estimation theory can be cast in a decision theoretic framework. The action space is the same as the parameter space, and $d(\mathbf{X}) = \hat{\theta}$, an estimate of θ. In this framework, a loss function must be specified, and one of the strengths of the theory is that general loss functions are allowed and thus the purpose of estimation is made explicit. But as might be expected, the case of squared error loss is especially tractable. In this case, the risk is

$$R(\theta, d(\mathbf{X})) = E\{[\theta - d(\mathbf{X})]^2\}$$

$$= E[(\theta - \hat{\theta})^2]$$

Suppose that a Bayesian approach is taken, which means that Θ is a random variable. Then from Theorem A of Section 15.2.2, the Bayes rule for squared error loss can be found by minimizing the posterior risk, which is $E[(\Theta - \hat{\theta})^2 | \mathbf{X} = \mathbf{x}]$. We have

$$E[(\Theta - \hat{\theta})^2] = \text{Var}[(\Theta - \hat{\theta})|\mathbf{X} = \mathbf{x}] + [E(\Theta - \hat{\theta})|\mathbf{X} = \mathbf{x}]^2$$

$$= \text{Var}(\Theta|\mathbf{X} = \mathbf{x}) + [E(\Theta|\mathbf{X} = \mathbf{x}) - \hat{\theta}]^2$$

The first term of this last expression does not depend on $\hat{\theta}$, and the second term is minimized by $\hat{\theta} = E(\Theta|\mathbf{X} = \mathbf{x})$. Thus, the Bayes rule is the mean of the posterior distribution of Θ. In the continuous case, for example,

$$\hat{\theta} = \int \theta h(\theta|\mathbf{x}) \, d\theta$$

We have proved the following theorem.

THEOREM A.   In the case of squared error loss, the Bayes estimate is the mean of the posterior distribution.

EXAMPLE A.   A biased coin is thrown once, and we are to estimate the probability of heads, θ. Suppose that we have no idea how biased the coin is; to reflect this state of knowledge, we use a uniform prior distribution on θ:

$$g(\theta) = 1, \quad 0 \le \theta \le 1$$

Let $X = 1$ if a head appears, and let $X = 0$ if a tail appears. The distribution of X given θ is

$$f(x|\theta) = \begin{cases} \theta, & x = 1 \\ 1 - \theta, & x = 0 \end{cases}$$

The posterior distribution is

$$h(\theta|x) = \frac{f(x|\theta) \times 1}{\int f(x|\theta) \, d\theta}$$

In particular,

$$h(\theta|X = 1) = \frac{\theta}{\int_0^1 \theta \, d\theta} = 2\theta$$

$$h(\theta|X = 0) = \frac{(1 - \theta) \times 1}{\int_0^1 (1 - \theta) \, d\theta} = 2(1 - \theta)$$

Suppose that $X = 1$. The Bayes estimate of θ is the mean of the posterior distribution $h(\theta|X = 1)$, which is

$$\int_0^1 \theta(2\theta) \, d\theta = \tfrac{2}{3}$$

By symmetry, the Bayes estimate in the case where $X = 0$ is $\hat{\theta} = \tfrac{1}{3}$. Note that these estimates differ from the classical maximum likelihood estimates, which are 0 and 1. $\qquad\square$

If d_1 and d_2 are two decision functions (estimates in this context), d_1 is said to **dominate** d_2 if $R(\theta, d_1) \le R(\theta, d_2)$ for all θ. The domination is said to be **strict** if the inequality is strict for some θ. An estimate is said to be **admissible** if it is not strictly dominated by any other estimate.

EXAMPLE B. Let us compare the risk functions of the Bayes estimate and the maximum likelihood estimate of Example A. Let $\tilde{\theta}$ denote the mle. Then $\tilde{\theta}$ equals 1 or 0 depending on whether X is 1 or 0; that is, $\tilde{\theta} = X$. The risk of $\tilde{\theta}$ is thus

$$E[(\tilde{\theta} - \theta)^2] = E[(X - \theta)^2]$$
$$= \text{Var}(X) + [E(X - \theta)]^2$$
$$= \theta(1 - \theta)$$

The risk of the Bayes estimate, $\hat{\theta}$, is

$$E[(\hat{\theta} - \theta)^2] = (\tfrac{1}{3} - \theta)^2 P(X = 0) + (\tfrac{2}{3} - \theta)^2 P(X = 1)$$
$$= (\tfrac{1}{3} - \theta)^2(1 - \theta) + (\tfrac{2}{3} - \theta)^2 \theta$$
$$= \tfrac{1}{9} - \tfrac{1}{3}\theta + \tfrac{1}{3}\theta^2$$

Figure 15-3, a graph of the two risk functions as functions of θ, shows that the

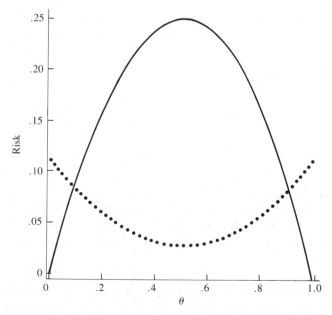

Figure 15-3. The risks of the mle (solid line) and the Bayes estimate (dotted line).

risk of $\hat{\theta}$ is smaller than that of $\tilde{\theta}$ over most of the range $[0, 1]$, but that neither estimate dominates the other.   □

In general, it is very hard to know whether a particular estimate is admissible, since one would have to check that it was not strictly dominated by *any* other estimate in order to show that it was admissible. The following theorem states that any Bayes estimate is admissible. Thus, in particular, the estimate $\hat{\theta}$ of Example B is admissible.

THEOREM B.   Suppose that one of the following two assumptions holds:

(1) Θ is discrete and d^* is a Bayes rule with respect to a prior probability mass function π such that $\pi(\theta) > 0$ for all $\theta \in \Theta$.

(2) Θ is an interval (perhaps infinite) and d^* is a Bayes rule with respect to a prior density function $g(\theta)$ such that $g(\theta) > 0$ for all $\theta \in \Theta$ and $R(\theta, d)$ is a continuous function of θ for all d.

Then d^* is admissible.

Proof.   We will prove the theorem for assumption (2); the proof for assumption (1) is similar and is left as an end-of-chapter problem. The proof is by contradiction. Suppose that d^* is inadmissible. There is then another estimate, d, such that $R(\theta, d^*) \geq R(\theta, d)$ for all θ and with strict inequality for some θ, say θ_0. Since $R(\theta, d^*) - R(\theta, d)$ is a continuous function of θ, there is an $\varepsilon > 0$ and an interval $\theta \pm h$ such that

$$R(\theta, d^*) - R(\theta, d) > \varepsilon, \quad \text{for } \theta_0 - h \leq \theta \leq \theta_0 + h$$

Then,

$$\int_{-\infty}^{\infty} [R(\theta, d^*) - R(\theta, d)] g(\theta) \, d\theta \geq \int_{\theta_0 - h}^{\theta_0 + h} [R(\theta, d^*) - R(\theta, d)] g(\theta) \, d\theta$$

$$\geq \varepsilon \int_{\theta_0 - h}^{\theta_0 + h} g(\theta) \, d\theta > 0$$

But this contradicts the fact that d^* is a Bayes rule; that is,

$$B(d^*) - B(d) = \int_{-\infty}^{\infty} [R(\theta, d^*) - R(\theta, d)] g(\theta) \, d\theta \leq 0$$

The proof is complete.   □

The theorem can be regarded as both a positive and a negative result. It is positive in that it identifies a certain class of estimates as being admissible, in particular, any Bayes estimate. It is negative in that there are apparently so many admissible estimates—one for every prior distribution that satisfies the hypotheses of the theorem—and some of these might well make little sense. One might conclude from this that admissibility is a rather weak concept, or at least not very useful, and that other approaches to optimality should be examined. One way out of this dilemma is to put further restrictions on the class of estimators under consideration. A rather clean optimality theory can be built up by only considering unbiased estimates, for example. We will not pursue this further here; see Bickel and Doksum (1977) for more discussion.

15.3   The Subjectivist Point of View

The Bayesian methods discussed so far in this chapter have necessarily involved prior distributions, but very little has been said about where these prior distributions come from. This section briefly summarizes an approach to probability and statistics that confronts this issue and that is quite different in spirit from the approach used in the rest of this book.

What is meant by a statement such as "the probability that this coin will land heads up is $\frac{1}{2}$"? The most common interpretation is that of the **frequentists**: The long-run frequency of heads approaches $\frac{1}{2}$. Bayesians who make this statement, however, mean that their prior opinion is such that they would as soon guess heads or tails if the rewards are equal. More generally, if A is an event, $P(A)$ means the following for Bayesians: For a game in which if A occurs the Bayesian will be paid \$1, $P(A)$ is the amount of money the Bayesian would be willing to pay to buy into the game. Thus, if the Bayesian is willing to pay 50 cents to buy in, $P(A) = .5$. Note that this concept of probability is personal: $P(A)$ may vary from person to person depending on their opinions.

The ideal Bayesian has a "coherent" system of personal probabilities; a system is said to be "incoherent" if there exists some structure of bets such that the bettor will lose no matter what happens. The ideal Bayesian, a kind of superbookie, is certainly a mathematical fiction, but the same could be said about most mathematical models. We will not pursue this notion of coherency here, but merely note that it can be used as a basis for a carefully worked out mathematical theory.

For Bayesians, then, probability is a model for quantifying the strength of personal opinions. Bayes' Rule describes how personal opinions evolve with experience. Suppose that the prior probability of A is $P(A)$. On observation of an event C, the opinion about A changes to $P(A|C)$ according to Bayes' Rule:

$$P(A|C) = \frac{P(C|A)P(A)}{P(C)}$$

Let us consider parameter estimation from a Bayesian point of view. The Bayesians' prior opinion about the value of a parameter θ is given by $g(\theta)$. Having observed the data $\mathbf{X} = \mathbf{x}$, where $\mathbf{X}$ has the density functon $f(\mathbf{x}|\theta)$, their new opinion about θ is

$$h(\theta|\mathbf{x}) = \frac{f(\mathbf{x}|\theta)g(\theta)}{\int f(\mathbf{x}|\theta)g(\theta)\,d\theta}$$

The frequentist point of view is quite different. For frequentists, θ has some fixed value that they do not know, and since θ is not random, it makes no sense to make probability statements about it.

An important case (of which we will consider concrete examples in the next sections) is that for which the data density, $f(\mathbf{x}|\theta)$, is quite peaked relative to the prior density, g. In this case, if g is nearly uniform in the region where f, regarded as a function of θ, has almost all its mass, then

$$h(\theta|\mathbf{x}) \approx \frac{f(\mathbf{x}|\theta)}{\int f(\mathbf{x}|\theta)\,d\theta} \propto f(\mathbf{x}|\theta)$$

Note that $f(\mathbf{x}|\theta)$, as a function of θ, is what we have called the likelihood function. In this situation, then, for the Bayesian, the posterior distribution is nearly proportional to the likelihood function. (The very word *likelihood* as applied to θ perhaps betrays an unacknowledged and suppressed Bayesian desire on the part of the frequentist.)

The frequentists must go to some pains to define what they mean by a confidence interval for θ without making a probability statement about θ. For the frequentist, the interval is random. A Bayesian $100(1 - \alpha)\%$ confidence interval, also called a **credibility interval**, is of the form (θ_0, θ_1), where

$$\int_{\theta_0}^{\theta_1} h(\theta|\mathbf{x})\,d\theta = 1 - \alpha$$

For the Bayesian, once $\mathbf{x}$ has been observed, the interval is fixed and θ is random.

The Bayesian approach to hypothesis testing is quite simple. Deciding between two hypotheses in light of data, $\mathbf{x}$, reduces to computing their posterior probabilities. If an explicit loss function is available, the Bayes rule is chosen to minimize the posterior risk. In the absence of a loss function, the probabilities of type I and type II errors are of little interest to the Bayesian.

The concepts introduced in this section have been fairly general and perhaps a little vague. The next two sections will cover some concrete situations.

15.3.1 Bayesian Inference for the Normal Distribution

The concepts introduced in the preceding section can be illustrated by Bayesian treatment of some simple problems involving the normal distribution. We will consider inference concerning an unknown mean with known variance.

First, suppose that the prior distribution of μ is $N(\mu_0, \sigma_0^2)$ (as we will see, the use of a normal prior distribution is especially useful analytically). A single observation $X \sim N(\mu, \sigma^2)$ is taken; the posterior distribution of μ is

$$h(\mu|x) = \frac{f(x|\mu)g(\mu)}{\int f(x|\mu)g(\mu)\, d\mu} \propto f(x|\mu)g(\mu)$$

Now

$$f(x|\mu)g(\mu) = \frac{1}{\sigma\sqrt{2\pi}} \exp\left[\frac{-1}{2\sigma^2}(x-\mu)^2\right] \frac{1}{\sigma_0\sqrt{2\pi}} \exp\left[\frac{-1}{2\sigma_0^2}(\mu-\mu_0)^2\right]$$

$$\propto \exp\left[-\frac{1}{2\sigma^2}(x-\mu)^2 - \frac{1}{2\sigma_0^2}(\mu-\mu_0)^2\right]$$

$$= \exp\frac{-1}{2}\left[\mu^2\left(\frac{1}{\sigma^2}+\frac{1}{\sigma_0^2}\right) - 2\mu\left(\frac{x}{\sigma^2}+\frac{\mu_0}{\sigma_0^2}\right) + \frac{x^2}{\sigma^2} + \frac{\mu_0^2}{\sigma_0^2}\right]$$

Let a, b, and c be the coefficients in the quadratic polynomial in μ that is the last expression above. That last expression may then be written as

$$\exp\left[-\frac{a}{2}\left(\mu^2 - \frac{2b}{a}\mu + \frac{c}{a}\right)\right]$$

To simplify this further, we use the technique of completing the square, and rewrite the expression as

$$\exp\left[-\frac{a}{2}\left(\mu - \frac{b}{a}\right)^2\right]\exp\left\{\frac{-a^2}{2}\left[\frac{c}{a} - \left(\frac{b}{a}\right)^2\right]\right\}$$

The second term above does not depend on μ, and we thus have that

$$h(\mu|x) \propto \exp\left[-\frac{a}{2}\left(\mu - \frac{b}{a}\right)^2\right]$$

We see that the posterior distribution of m is normal with mean

$$m_1 = \frac{\dfrac{x}{\sigma^2} + \dfrac{\mu_0}{\sigma_0^2}}{\dfrac{1}{\sigma^2} + \dfrac{1}{\sigma_0^2}}$$

Let $\xi = 1/\sigma^2$ and $\xi_0 = 1/\sigma_0^2$; ξ and ξ_0 are called **precisions**. We have proved the following theorem.

THEOREM A. Suppose that $\mu \sim N(\mu_0, \sigma_0^2)$ and that $X|\mu \sim N(\mu, \sigma^2)$. Then the posterior distribution of μ is normal with mean

$$\mu_1 = \frac{\xi_0\mu_0 + \xi x}{\xi + \xi_0}$$

and precision

$$\xi_1 = \xi + \xi_0$$

According to Theorem A, the posterior mean is a weighted average of the prior mean and the data, with weights proportional to the respective precisions. If we assume that the experiment (the observation of X) is much more informative than the prior distribution, in the sense that $\sigma^2 \ll \sigma_0^2$, then $\xi \gg \xi_0$ and

$$\xi_1 \approx \xi$$

$$\mu_1 \approx x$$

Thus, the posterior distribution of μ is nearly normal with mean x and precision ξ.

$$h(\mu|x) \approx \frac{1}{\sigma\sqrt{2\pi}} e^{(-1/2\sigma^2)(\mu-x)^2}$$

$$= f(x|\mu)$$

This result illustrates two points for which we made a qualitative argument in Section 15.3. If the prior distribution is quite flat relative to $f(x|\mu)$, then:

1. The prior distribution has little influence on the posterior.
2. The posterior distribution is approximately proportional to the likelihood function.

On a heuristic level, the first point is fairly obvious. If one does not have strong prior opinions, one's posterior opinion is mainly determined by the data one observe. Such a prior distribution is often called a **vague**, or **noninformative**, **prior**.

Let us now consider how the prior distribution $\mu \sim N(\mu_0, \sigma_0^2)$ is altered by observing a sample $\mathbf{X} = (X_1, \ldots, X_n)$, where the X_i are independent and $N(\mu, \sigma^2)$. As before, the posterior distribution is

$$h(\mu|x) \propto f(\mathbf{x}|\mu)g(\mu)$$

Now, from the independence of the x_i,

$$f(\mathbf{x}|\mu) = \frac{1}{\sigma^n(2\pi)^{n/2}} \exp\left[-\frac{1}{2\sigma^2}\sum_{i=1}^{n}(x_i - \mu)^2\right]$$

Using the identity

$$\sum_{i=1}^{n} (x_i - \mu)^2 = \sum_{i=1}^{n} (x_i - \bar{x})^2 + n(\bar{x} - \mu)^2$$

we obtain

$$f(\mathbf{x}|\mu) = \frac{1}{\sigma^n (2\pi)^{n/2}} \exp\left[-\frac{1}{2\sigma^2} \sum_{i=1}^{n} (x_i - \bar{x})^2 \right] \exp\left[-\frac{1}{2\sigma^2/n} (\bar{x} - \mu)^2 \right]$$

Only the last term depends on μ, so

$$h(\mu|\mathbf{x}) \propto \exp\left[-\frac{1}{2\sigma^2/n} (\bar{x} - \mu)^2 \right] g(\mu)$$

This posterior distribution can be evaluated, using Theorem A, as being normal with mean

$$\mu_1 = \frac{\dfrac{\mu_0}{\sigma_0^2} + \dfrac{n\bar{x}}{\sigma^2}}{\dfrac{n}{\sigma^2} + \dfrac{1}{\sigma_0^2}}$$

and precision

$$\xi_1 = \frac{n}{\sigma^2} + \frac{1}{\sigma_0^2}$$

For large values of n, $\mu_1 \approx \bar{x}$ and $\xi_1 \approx n/\sigma^2$. Therefore, the information in the sample largely determines the posterior distribution.

EXAMPLE A. Suppose that the prior distribution of μ is $N(2, 4)$ and that the X_i are $N(4, 1)$. If $x_1 = 3.59$ and $x_2 = 5.52$, then $\bar{x} = 4.55$, and the posterior mean is

$$\mu_1 = \frac{\xi}{\xi + \xi_0} \bar{x} + \frac{\xi_0}{\xi + \xi_0} \mu_0$$

$$= .89\bar{x} + .11\mu_0$$

$$= 4.27$$

Now suppose that we observe $x_3 = 3.93$ and $x_4 = 4.71$. Then $\bar{x} = 4.44$, and the posterior mean is

$$\mu_1 = .94\bar{x} + .06\mu_0$$

$$= 4.30$$

Finally, suppose that $x_5 = 4.40$, $x_6 = 5.06$, $x_7 = 3.68$, and $x_8 = 3.14$, so $\bar{x} = 4.25$. The posterior mean becomes

$$\mu_1 = .97\bar{x} + .03\mu_0$$

$$= 4.18$$

Note how $\bar{x}$ is weighted more heavily as the sample size increases.

Figure 15-4 is a plot of the prior distribution and the posterior distributions. Note how the prior distribution is quickly overwhelmed by the data and how the posterior becomes more concentrated as the sample size increases. □

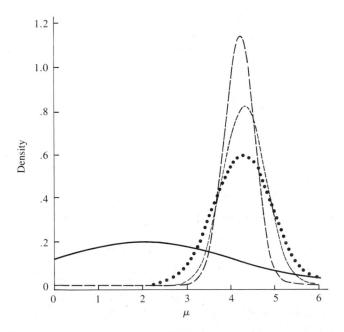

Figure 15-4. The prior density of μ (solid line) and the posterior densities after 2 (dotted line), 4 (short-dashed line), and 8 (long-dashed line) observations.

15.3.2 Bayesian Analysis for the Binomial Distribution

As a simple example, suppose that $X \sim \text{bin}(n, p)$ and that the prior distribution of p is uniform on $[0, 1]$:

$$f(x|p) = \binom{n}{x} p^x (1 - p)^{n-x}$$

$$g(p) = 1, \quad 0 \le p \le 1$$

The posterior distribution $h(p|x)$ is

$$h(p|x) \propto f(x|p)g(p)$$

$$= \binom{n}{x} p^x (1 - p)^{n-x}$$

Figure 15-5 shows the posterior distribution for $n = 20$ and $x = 5$. This represents the opinion of a Bayesian who was initially indifferent to the value of p in the sense that his prior distribution was uniform, after he observed 5 successes in 20 trials.

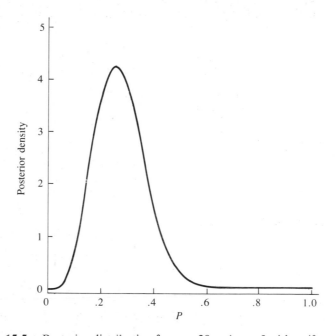

Figure 15-5. Posterior distribution for $n = 20$ and $x = 5$ with uniform prior.

The uniform prior distribution expresses an indifference about the possible values of p. We now introduce a class of prior distributions whose shapes are quite flexible and which are easy to work with analytically.

DEFINITION. The density function f of the beta distribution is

$$f(x) = \frac{\Gamma(a + b)}{\Gamma(a)\Gamma(b)} x^{a-1}(1 - x)^{b-1}, \quad 0 \le x \le 1$$

where a and b are positive parameters.

It can be shown that the normalizing constant in the formula above is correct—that is,

$$\int_0^1 x^{a-1}(1 - x)^{b-1}\, dx = \frac{\Gamma(a)\Gamma(b)}{\Gamma(a + b)}$$

Figure 15-6 shows beta densities for various values of a and b. Note that the case where $a = b = 1$ is the uniform distribution.

THEOREM A.   The mean and variance of the beta distribution are $a/(a + b)$ and $ab/(a + b)^2(a + b + 1)$.

Proof.   See Problem 18 of the end-of-chapter problems.   □

Suppose that the prior distribution of p is beta:

$$g(p) \propto p^{a-1}(1 - p)^{b-1}$$

The posterior distribution of p given x is

$$h(p|x) \propto f(x|p)g(p)$$

$$= \binom{n}{x} p^x(1 - p)^{n-x}p^{a-1}(1 - p)^{b-1}$$

$$\propto p^{a+x-1}(1 - p)^{n+b-x-1}$$

This distribution is thus beta with parameters $a' = a + x$ and $b' = b + n - x$. Just as it was with the normal distribution (in the preceding section), for large values of n, the posterior distribution is primarily determined by the data. In particular, the prior mean is

$$\mu_{\text{prior}} = \frac{a}{a + b}$$

and the posterior mean is

$$\mu_{\text{post}} = \frac{a'}{a' + b'} = \frac{a + x}{a + b + n}$$

Note that the posterior mean can also be expressed as a weighted average of the prior mean and the sample mean, or

$$\mu_{\text{post}} = \frac{a + b}{a + b + n}\left(\frac{a}{a + b}\right) + \frac{n}{a + b + n}\overline{x}$$

If p_0 is the true value of the parameter p, then x/n approaches p_0 as n approaches infinity and therefore μ_{post} approaches p_0. It can similarly be shown that the

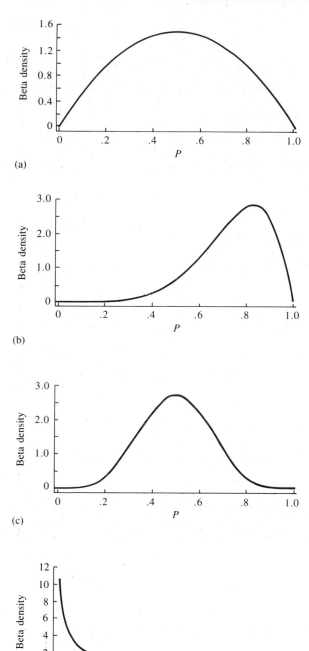

Figure 15-6.   Beta density functions for various values of a and b: (a) $a = 2, b = 2$; (b) $a = 6, b = 2$; (c) $a = .5, b = 4$; and (d) $a = 6, b = 6$.

posterior variance tends to zero. Thus, the posterior distribution becomes more and more concentrated about p_0.

EXAMPLE A.   This example is from Thomas (1948). A cofferdam protecting a construction site was designed to withstand flows of up to 1870 cubic feet per second (cfs). An engineer wishes to compute the probability that the dam will be overtopped during the upcoming year. Over the previous 25-year period, the annual maximum flood levels of the stream had ranged from 629 to 4720 cfs, and 1870 cfs had been exceeded $x = 5$ times.

Modeling the 25 years as 25 independent Bernoulli trials with the same probability p that the flood level will exceed 1870 cfs, and using a uniform prior distribution for p, the posterior distribution is

$$h(p|x) \propto p^x(1 - p)^{n-x} = p^5(1 - p)^{20}$$

which is plotted in Figure 15-7, a graphical display of the engineer's degree of belief that the dam will be overtopped. □

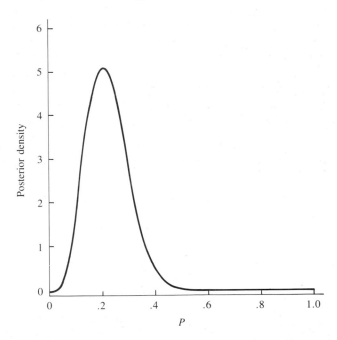

Figure 15-7.   Posterior density for the probability that flow will exceed 1870 cfs.

We have seen that when the distribution of X is binomial with probability p and the prior distribution of p is beta, then the posterior distribution of p is also a member of the beta family but has different parameters than does the prior. A similar situation arose in Section 15.3.1; there, the distribution of X was normal with parameter μ, the prior distribution of μ was normal, and the posterior distribution of μ was normal. In general, if the prior distribution belongs to a family G, the data have a distribution belonging to a family H, and the posterior distribution also belongs to G, then we say that G is a family of **conjugate priors** to H. Thus, the beta distribution is a conjugate prior to the binomial, and the normal is self-conjugate. Conjugate priors may not exist; when they do, selecting a member of the conjugate family as a prior is done mostly for mathematical convenience, since the posterior can be evaluated very simply. More generally, numerical methods of integration would have to be used to evaluate the posterior.

15.4   Concluding Remarks

This chapter has presented a brief introduction to two areas of mathematical statistics, decision theory and Bayesian inference. The analysis has been somewhat more abstract and theoretical than that of earlier chapters.

In order for decision theory to be relevant to a practical problem, the problem must be such that a choice is to be made from a set of specified actions and a definite measure of loss is associated with each possible action. Critics of decision theory claim that most scientific investigations do not fit this paradigm; some have gone so far as to call such a point of view totalitarian! In fact, experimental scientists very rarely use decision theoretic methods in analyzing their data, but decision theory has been used more often in business and economics, where it seems to be easier to specify the relevant loss functions.

The subjectivist, or Bayesian, point of view toward statistics is somewhat controversial. Arguments between frequentists and Bayesians have often been quite heated, and it is probably fair to say that the Bayesian approach has not been adopted by most practicing statisticians. Critics of the Bayesian paradigm look with disfavor on the introduction of a prior distribution. Some dislike the subjective element thereby introduced, maintaining that statistics should be "objective." Others, although not totally unsympathetic, question how prior distributions can be determined in practice.

This book has been relatively unconcerned with philosophical or foundational matters and has made no attempt to develop a consistent "theory" of statistics. The subject matter under investigation and the role that statistics plays in the investigation should effectively determine whether it is more appropriate for the user of statistics to take a Bayesian, decision-theoretic, frequentist, or purely data-analytic point of view, or some combination of these.

15.5   Problems

1. The losses for actions a_1, a_2, and a_3, depending on θ_1, θ_2, and θ_3, are given in the following table:

	θ_1	θ_2	θ_3
a_1	2	1	5
a_2	0	4	6
a_3	4	0	2

An observation X has the following distributions depending on θ:

	θ_1	θ_2	θ_3
x_1	.4	.5	.3
x_2	.6	.5	.7

(a) Pick three decision rules and evaluate their risks.
(b) Determine the minimax rule among those chosen in part (a).
(c) Assuming a uniform prior distribution on θ, determine the Bayes rule.
(d) Suppose that $X = x_2$. Use posterior analysis to determine the Bayes rule among all possible decision rules.

2. Suppose that for Example B in Section 15.2.1 the prior distribution of k is taken to be uniform. How does the analysis change?

3. Prove Theorem A of Section 15.2.2 for the discrete case.

4. Your investment advisor tells you that he believes a certain stock is going up and that you should buy. If he is correct and you buy, you will gain $1000; if he is incorrect and you buy, you will lose $1000. If you do nothing, you lose nothing. You might also choose to sell. If you sell and he is correct, you lose $1000; if you sell and he is incorrect, you gain $1000. On the basis of experience, you believe that his prognosis is correct about $\frac{2}{3}$ of the time. Your prior probability that the stock will go up or down $\frac{2}{5}$ or $\frac{3}{5}$, respectively. What are the posterior risks of buying, selling, and doing nothing? What is the Bayes rule?

5. A biased coin has an unknown probability θ of heads. It is thrown once, and you observe the outcome. You then have a choice whether or not to play a certain game. In the game, you predict the next toss. If you predict correctly, you are paid $2; if you predict incorrectly, you lose $1. If you don't play the game, you lose nothing and gain nothing. You thus have three possible actions: 1) play and predict the same as on the first toss, 2) play and predict the opposite of the first toss, 3) don't play.
(a) What are the risk functions for the three actions?
(b) What is the minimax rule?
(c) If you have a uniform prior on θ, what is the Bayes rule?
(d) Suppose that you are paid D if you predict correctly and you lose $1 if you predict incorrectly. How does the analysis of parts (a) through (c) change?

6. Consider the following classification problem. Classes A and B are equally likely, so that the prior distribution is $P(A) = P(B) = \frac{1}{2}$. A random variable X is observed; if X comes from class A, its distribution is $N(0, 1)$, and if X comes from class B, its distribution is $N(2, 1)$.
(a) Under 0–1 loss, what is the Bayes rule for classification?

(b) Suppose that $X = 1.5$. What is the classification rule? What is the probability of making a classification error in this case?

(c) What is the probability of making a classification error when using the Bayes rule?

(d) Suppose that the loss function is such that you lose twice as much by misclassifying A as you do by misclassifying B. Redo parts (a) through (c) under this assumption.

(e) Redo parts (a) through (c) under the prior probabilities $P(A) = \frac{2}{3}$ and $P(B) = \frac{1}{3}$.

7. Suppose that $X \sim \text{bin}(3, p)$. Consider choosing between $H_0: p \leq \frac{1}{2}$ and $H_1: p > \frac{1}{2}$. Compare the risks of the following decision rules as functions of p if you lose \$1 for incorrectly choosing H_0 and \$2 for incorrectly choosing H_1:

d_1: choose H_1 if $X > 0$
d_2: choose H_0 if $X \leq 1$
d_3: choose H_0 if $X \leq 2$

Assuming a uniform prior on p, which rule has the smallest Bayes risk?

8. Prove Theorem B of Section 15.2.4 for the discrete case. Why is the assumption that $\pi(\theta) > 0$ for all θ necessary?

9. Suppose that $X \sim \text{bin}(2, p)$. Compare the risk functions for the following estimates of p using squared error loss:

$$\hat{p} = \frac{X}{2}$$

$$\tilde{p} = \frac{X + 1}{3}$$

$$\bar{p} = \frac{X + 1}{4}$$

Are any of these estimates dominated by the others?

10. Suppose that $X_1, \ldots, X_n$ are independent and that each follows an exponential distribution

$$f(x|\theta) = \frac{1}{\theta} e^{-x/\theta}$$

Assuming squared error loss, graph the risk function of each of the following estimates of θ:

$$d_1 = \bar{X}$$

$$d_2 = nX_{(1)}$$

$$d_3 = \frac{n\bar{X}}{n + 1}$$

Does any estimate dominate another? (*Hint:* Show that $X_{(1)}$ follows an exponential distribution.)

11. A lot consisting of n items contains k defective items. Assume that k has a prior distribution, which is $\text{bin}(n, p)$. One item is selected at random and tested. Determine the posterior distribution of k if (a) the item is defective and (b) the item is not defective.

12. Suppose that X is a geometric random variable:

$$f(x|p) = (1 - p)^x p$$

Let p have a prior distribution that is uniform on $[0, 1]$.
(a) What is the posterior distribution of p?
(b) What is the Bayes estimate of p under squared error loss?
(c) What is the mle?

13. Suppose that a parameter, Θ, takes on values $\theta_1 = 1$, $\theta_2 = 10$, and $\theta_3 = 20$. The distribution of X is discrete and depends on Θ as shown in the following table:

	θ_1	θ_2	θ_3
x_1	.1	.2	.4
x_2	.1	.2	.2
x_3	.2	.2	.2
x_4	.6	.4	.2

Assume a prior distribution on Θ: $P(\theta_1) = .5$, $P(\theta_2) = .25$, and $P(\theta_3) = .25$.
(a) Suppose that x_2 is observed. What is the posterior distribution of Θ?
(b) What is the Bayes estimate under squared error loss in this case?
(c) What is the Bayes estimate for the loss function $l(\theta, \hat{\theta}) = |\theta - \hat{\theta}|$?
(d) Suppose that a second independent observation, x_1, is made. What does the posterior distribution become?

14. For an estimation problem, suppose that the loss function is $l(\theta, \hat{\theta}) = w(\theta)(\theta - \hat{\theta})^2$. Show that the Bayes estimate of θ is

$$\hat{\theta} = \frac{E[w(\Theta)g(\Theta)|\mathbf{x}]}{E[w(\Theta)|\mathbf{x}]}$$

15. Suppose that θ is a parameter that takes on values on the real line. Consider the loss function

$$l(\theta, \hat{\theta}) = \begin{cases} 0, & |\theta - \hat{\theta}| \leq c \\ 1, & |\theta - \hat{\theta}| > c \end{cases}$$

Show that the Bayes estimate of θ is the midpoint of the interval I of length $2c$ that maximizes $P(\Theta \in I|\mathbf{x})$.

16. Suppose that μ has a prior distribution that is $N(0, 1)$. Nine independent observations are taken and are $N(\mu, 9)$.

(a) Calculate the posterior mean and the precision.

(b) Find the Bayes risk of the estimate.

(c) Find a 95% credibility interval for μ.

17. Suppose that μ has a prior distribution that is $N(0, 1)$ and that $X \sim N(\mu, .25)$. What is the Bayes estimate of μ under squared error loss? Compare its risk to that of the mle.

18. Let X have a beta distribution with parameters a and b.

(a) Show that

$$E(X^r) = \frac{\Gamma(a + r)}{\Gamma(a)} \left[\frac{\Gamma(a + b)}{\Gamma(a + b + r)} \right]$$

[*Hint:* Use the recurrence relation for the gamma function: $\Gamma(x + 1) = x\Gamma(x)$.]

(b) Prove Theorem A of Section 15.3.2.

19. A beta distribution has mean .3 and variance .01. What are a and b?

20. If a thumbtack is tossed in the air, it can land either on its base (with the point sticking up) or on its side. Let p denote the probability that it lands on its base.

(a) Without doing any experiments, what is your best guess for p?

(b) Select a and b to form a beta prior distribution on p that reasonably reflects your prior opinion about possible values of p. Graph this prior.

(c) Toss a thumbtack 10 times. Evaluate the posterior distribution of p, and plot it.

(d) Toss the thumbtack 40 more times. Evaluate the posterior, and plot it.

(e) Compare your results with those of other students.

21. Suppose that $X_1, \ldots, X_n$ are independent Poisson random variables with mean θ and that θ has a gamma prior distribution.

(a) Show that the posterior distribution of θ is also gamma. (Thus, the gamma is a conjugate prior of the Poisson.)

(b) Determine the Bayes estimate of θ under squared error loss, and show that it is a weighted average of the prior mean and $\bar{X}$.

(c) Compare the risk of the Bayes estimate to that of $\bar{X}$.

22. Show that the gamma is a conjugate prior of the exponential distribution.

23. Suppose that X is normally distributed with mean 0 and unknown precision ξ. Find a family of conjugate priors.

BIBLIOGRAPHY

Andrews, D., Bickel, P., Hampel, F., Huber, P., Rogers, W., and Tukey, J. (1972). *Robust Estimates of Location*. Princeton, N.J.: Princeton University Press.

Andrews, D., and Herzberg. (1985). *Data.* Springer–Verlag.

Anscombe, F. J. (1950). Sampling theory of the negative binomial and logarithmic series distributions. *Biometrika, 37,* 358–382.

Armitage, P. (1983). *Statistical Methods in Medical Research*. Boston: Blackwell.

Bailey, C., Cox, E., and Springer, J. (1978). High pressure liquid chromatographic determination of the intermediate/side reaction products in FD&C Red No. 2 and FD&C Yellow No. 5; statistical analysis of instrument response. *J. Assoc. Offic. Anal. Chem., 61,* 1404–1414.

Barlow, R. E., Toland, R. H., and Freeman, T. (1984). A Bayesian analysis of stress-rupture life of Kevlar/epoxy spherical pressure vessels. In *Proceedings of the Canadian Conference in Applied Statistics*, T. D. Dwivedi (ed.). New York: Marcel–Dekker.

Beecher, H. K. (1959). *Measurement of Subjective Responses*. Oxford, England: Oxford University Press.

Beller, G., Smith, T., Abelmann, W., Haber, E., and Hood, W. (1971). Digitalis intoxication: A prospective clinical study with serum level correlations. *New England J. Med., 284,* 989–997.

Benjamin, J., and Cornell, C. (1970). *Probability, Statistics, and Decision for Civil Engineers*. New York: McGraw–Hill.

Bennett, C., and Franklin, N. (1954). *Statistical Analysis in Chemistry and the Chemical Industry*. New York: Wiley.

Berkson, J. (1966). Examination of randomness of alpha particle emissions. In *Research Papers in*

Statistics, F. N. David (ed.). New York: Wiley.

Bevan, S., Kullberg, R., and Rice, J. (1979). An analysis of cell membrane noise. *Annals of Statistics*, 7, 237–257.

Bickel, P., and Doksum, K. (1977). *Mathematical Statistics: Basic Ideas and Selected Topics*. Oakland, Calif.: Holden–Day.

Bickel, P., and O'Connell, J. W. (1975). Is there a sex bias in graduate admissions? *Science*, 187, 398–404.

Bishop, Y., Feinberg, S., and Holland, P. (1975). *Discrete Multivariate Analysis: Theory and Practice*. Cambridge, Mass.: MIT Press.

Bjerkdal, T. (1960). Acquisition of resistance in guinea pigs infected with different doses of virulent tubercle bacilli. *Amer. J. Hygiene*, 72, 130–148.

Bliss, C., and Fisher, R. A. (1953). Fitting the negative binomial distribution to biological data. *Biometrics*, 9, 174–200.

Box, G. E. P., and Cox, D. R. (1964). An analysis of transformations (with discussion). *J. Royal Stat. Soc., Series B 26*, 211–246.

Box, G. E. P., and Tiao, G. C. (1973). *Bayesian Inference in Statistical Analysis*. Reading, Mass.: Addison–Wesley.

Box, G. E. P., Hunter, W. G., and Hunter, J. S. (1978). *Statistics for Experimenters*. New York: Wiley.

Brownlee, K. A. (1960). *Statistical Theory and Methodology in Science and Engineering*. New York: Wiley.

Brunk, H. D. (1975). *An Introduction to Mathematical Statistics*. Gardena, Calif.: Xerox.

Burr, I. (1974). *Applied Statistical Methods*. New York: Academic Press.

Campbell, J. A., and Pelletier, O. (1962). Determination of niacin (niacinamide) in cereal products. *J. Assoc. Offic. Anal. Chem.*, 45, 449–453.

Chambers, J., Cleveland, W., Kleiner, B., and Tukey, P. (1983). *Graphical Methods for Data Analysis*. Boston: Duxbury.

Chang, S. L. (1945). Sedimentation in water and the specific gravity of cysts of *Entamoeba histolytica*. *Amer. J. Hygiene*, 41, 156–163.

Chapman, H., and Demeritt, D. (1936). *Elements of Forest Mensuration*. Nashville, Tenn.: Williams Press.

Chernoff, H., and Lehman, E. (1954). The use of maximum likelihood estimates in tests for goodness of fit. *Annals of Math. Stat.*, 23, 315–345.

Clancy, V. J. (1947). Empirical distributions in chemistry. *Nature*, 159, 340.

Cleveland, W., Graedel, T., Kleiner, B., and Warner, J. (1974). Sunday and workday variations in photochemical air pollutants in New Jersey and New York. *Science*, 186, 1037–1038.

Cobb, L., Thomas, G., Dillard, D., Merendino, J., and Bruce, R. (1959). An evaluation of internal mammary artery ligation by a double blind technique. *N. Eng. J. Med.*, 260, 1115–1118.

Cochran, W. G. (1977). *Sampling Techniques*. New York: Wiley.

Cogswell, J. J., (1973). Forced oscillation technique for determination of resistance to breathing in children. *Arch. Dis. Child.*, 48, 259–266.

Cook, R. D., and Weisberg, S. (1982). *Residuals and Influence in Regression*. New York: Chapman and Hall.

Cramer, H. (1946). *Mathematical Methods of Statistics*. Princeton, N.J.: Princeton University Press.

Dahiya, R., and Gurland, J. (1972). Pearson chi-squared test of fit with random intervals. *Biometrika*, 59, 147–153.

Dahlquist, G., and Bjorck, A. (1974). *Numerical Methods*. Englewood Cliffs, N.J.: Prentice–Hall.

David, H. (1981). *Order Statistics*. New York: Wiley.

Davies, O. (1960). *The Design and Analysis of Industrial Experiments*. London: Oliver and Boyd.

DeHoff, R., and Rhines, F. (eds.) (1968). *Quantitative Microscopy*. New York: McGraw–Hill.

Deming, W. (1960). *Sample Design in Business Research*. New York: Wiley.

Diamond, G., and Forrester, J. (1979). Analysis of probability as an aid in the clinical diagnosis of coronary-artery disease. *New Eng. J. Med.*, 300, 1350–1358.

Doksum, K., and Sievers, G. (1976). Plotting with confidence. Graphical comparisons of two populations. *Biometrika*, 63, 421–434.

Dongarra, J. (1979). *LINPACK Users' Guide*. Philadelphia: SIAM.

Donoho, A., Donoho, D., and Gasko, M. (1986).

MacSpin: Dynamic Data Display. Belmont, Calif.: Wadsworth.

Dorfman, D. (1978). The Cyril Burt question: New findings. *Science, 201,* 1177–1186.

Dowdall, J. A. (1974). Women's attitudes toward employment and family roles. *Soc. Anal., 35,* 251–262.

Draper, N., and Smith, H. (1981). *Applied Regression Analysis.* New York: Wiley.

Edwards, W., Lindman, H., and Savage, L. J. (1963). Bayesian statistical inference for psychological research. *Psych. Rev., 70,* 193–242.

Evans, D. (1953). Experimental evidence concerning contagious distributions in ecology. *Biometrika, 40,* 186–211.

Fechter, J. V., and Porter, L. G. (1979). Kitchen range energy consumption. Prepared for Office of Conservation, U.S. Department of Energy, NBSIR 78-1556 (Washington, D.C.).

Ferguson, T. S. (1967). *Mathematical Statistics: A Decision Theoretic Approach.* New York: Academic Press.

Filliben, J. (1975). The probability plot correlation coefficient test for normality. *Technometrics, 17,* 111–117.

Finkner, A. (1950). Methods of sampling for estimating commercial peach production in North Carolina. *North Carolina Agricultural Experiment Station Technical Bulletin, 91.*

Fisher, R. A. (1936). Has Mendel's work been rediscovered? *Annals of Science, 1,* 115–137.

Fisher, R. A. (1958). *Statistical Methods for Research Workers.* New York: Hafner.

Freedman, D., Pisani, R., and Purves, R. (1978). *Statistics.* New York: Norton.

Geissler, A. (1889). Beiträge zur Frage des Geschlechtsverhältnisses der Gebornen. *Z. K. Sachs. Stat. Bur., 35,* 1–24.

Gerlough, D., and Schuhl, A. (1955). *Use of Poisson Distribution in Highway Traffic.* Eno Foundation for Highway Traffic Control.

Glass, D., and Hall, J. (1954). A study of intergeneration changes in status. *In Social Mobility in Britain,* D. Glass (ed.). Glencoe, Ill.: Free Press.

Gosset, W. S. (1931). The Lanarkshire milk experiment. *Biometrika, 23,* 398.

Grace, N., Muench, H., and Chalmers, T. (1966). The present status of shunts for portal hypertension in cirrhosis. *Gastroenterology, 50,* 684–691.

Haberman, S. (1978). *Analysis of Qualitative Data.* New York: Academic Press.

Hampson, R., and Walker, R. (1961). Vapor pressures of platinum, iridium, and rhodium. *J. Res. Nat. Bur. Stand., 65A,* 289–295.

Harbaugh, J., Doveton, J., Davis, J. (1977). *Probability Methods in Oil Exploration.* New York: Wiley.

Hartley, H. O., and Ross, A. (1954). Unbiased ratio estimates. *Nature, 174,* 270–271.

Heckman, M. (1960). Flame photometric determination of calcium in animal feeds. *J. Assoc. Offic. Anal. Chem., 43,* 337–340.

Hennekens, C., Drolette, M., Jesse, M., Davies, J., and Hutchison, G. (1976). Coffee drinking and death due to coronary heart disease. *N. Eng. J. Med., 294,* 633–636.

Hoaglin, D. (1980). A Poissoness plot. *Amer. Stat., 34,* 146–149.

Hoaglin, D., Mosteller, F., and Tukey, J. (1983). *Understanding Robust and Exploratory Data Analysis.* New York: Wiley.

Hockersmith, T., and Ku, H. (1969). Uncertainties associated with proving ring calibration. In *Precision Measurement and Calibration,* H. Ku (ed.). U.S. National Bureau of Standards Special Publication 300, Vol. I (Washington, D.C.).

Hollander, M., and Wolfe, D. (1973). *Nonparametric Statistical Methods.* New York: Wiley.

Horvitz, D., Shah, B., and Simmons, W. (1967). The unrelated randomized response model. *Proc. Soc. Stat. Sect. Amer. Stat.,* 65–72.

Houck, J. C. (1970). Temperature coefficient of the bismuth I–II transition pressure. *J. Res. Nat. Bur. Stand., 74A,* 51–54.

Huber, P. (1981). *Robust Statistics.* New York: Wiley.

Huie, R. E., and Herron, J. T. (1972). Rates of reaction of atomic oxygen III, spiropentane, cyclopentane, cyclohexane, and cycloheptane. *J. Res. Nat. Bur. Stand., 74A,* 77–80.

Johnson, S., and Johnson, R. (1972). Tonsillectomy history in Hodgkin's disease. *N. Eng. J. Med., 287,* 1122–1125.

Joiner, B. (1981). Lurking variables: Some examples. *Amer. Stat., 35,* 227–233.

Kacprzak, J., and Chvojka, R. (1976). Determination of methyl mercury in fish by flameless atomic absorption spectroscopy and comparison with an acid digestion method for total mercury. *J. Assoc. Offic. Anal. Chem.*, *59*, 153–157.

Kirchoefer, R. (1979). Semiautomated method for the analysis of chlopheniramine maleate tables: Collaborative study. *J. Assoc. Offic. Anal. Chem.*, *62*, 1197–1220.

Kiser, C. V., and Schaefer, N. L. (1949). Demographic characteristics of women in Who's Who. *Milbank Memorial Fund Quarterly*, *27*, 422.

Kish, L. (1965). *Survey Sampling*. New York: Wiley.

Knafl, G., Spiegelman, C., Sacks, J., and Ylvisaker, D. (1984). Nonparametric calibration. *Technometrics*, *26*, 233–241.

Ku, H. (1969). *Precision Measurement and Calibration*. National Bureau of Standards Special Publication 300 (Washington, D.C.).

Ku, H. (1981). Personal communication.

Lagakos, S., and Mosteller, F. (1981). FD&C Red No. 40 experiments. *J. Nat. Canc. Instit.*, *66*, 197–213.

Lawson, C. L., and Hanson, R. J. (1974). *Solving Least Squares Problems*. Englewood Cliffs, N.J.: Prentice–Hall.

Lazarsfeld, P., Berelson, B., and Gaudet, H. (1948). *The People's Choice: How the Voter Makes Up His Mind in a Presidential Election*. New York: Columbia University Press.

Le Cam, L., and Neyman, J. (eds.) (1967). *Proceedings of the Fifth Berkeley Symposium on Mathematical Statistics and Probability. Volume V: Weather Modification*. Berkeley: University of California Press.

Lehmann, E. (1975). *Nonparametrics: Statistical Methods Based on Ranks*. Oakland, Calif.: Holden–Day.

Lehmann, E. L. (1983). *Theory of Point Estimation*. New York: Wiley.

Levine, J. D., Gordon, N. C., and Fields, H. L. (1978). The mechanism of placebo analgesia. *Lancet*, 654–657.

Levine, P. H. (1973). An acute effect of cigarette smoking on platelet function. *Circulation*, *48*, 619–623.

Lin, S.–L., Sutton, V., and Quarashi, M. (1979).

Equivalence of microbiological and hydroxylamine methods of analysis for ampicillin dosage forms. *J. Assoc. Offic. Anal. Chem.*, *62*, 989–997.

Lynd, R. S., and Lynd, H. M. (1956). *Middletown: A Study in Modern American Culture*. New York: Harcourt–Brace.

Martin, H., Gudzinowicz, B., and Fanger, H. (1975). *Normal Values in Clinical Chemistry*. New York: Marcel–Dekker.

McCool, J. (1979). Analysis of single classification experiments based on censored samples from the two-parameter Weibull distribution. *J. Stat. Planning and Inference*, *3*, 39–68.

McNish, A. (1962). The speed of light. *IRE Trans. on Instrumentation*, *11*, 138–148.

Miller, R. (1981). *Simultaneous Statistical Inference*. New York: Springer–Verlag.

Morton, A. Q. (1978). *Literary Detection*. New York: Scribner's.

Natrella, M. (1963). *Experimental Statistics*. National Bureau of Standards Handbook 91 (Washington, D.C.).

Olsen, A., Simpson, J., and Eden, J. (1975). A Bayesian analysis of a multiplicative treatment effect in weather modification. *Technometrics*, *17*, 161–166.

Orcutt, R. H. (1970). Generation of controlled low pressures of nitrogen by means of dissociation equilibria. *J. Res. Nat. Bur. Stand.*, *74A*, 45–49.

Overfield, T., and Klauber, M. R. (1980). Prevalence of tuberculosis in Eskimos having blood group B gene. *Hum. Bio.*, *52*, 87–92.

Pearson, E. S., and Wishart, J. (ed.) (1958). *Student's Collected Works*. Cambridge, England: Cambridge University Press.

Pearson, E., D'Agostino, R., and Bowman, K. (1977). Tests for departure from normality: Comparison of powers. *Biometrika*, *64*, 231–246.

Pearson, K., and Hartley, H. (1966). *Biometrika Tables for Statisticians*. Cambridge, England: Cambridge University Press.

Perry, L., Van Dyke, R., and Theye, R. (1974). Sympathoadrenal and hemodynamic effects of isoflurane, halothane, and cyclopropane in dogs. *Anesthesiology*, *40*, 465–470.

Plato, C., Rucknagel, D., and Gershowitz, H. (1964). Studies of the distribution of glucose-6-phosphate dehydrogenase deficiency, thalassemia, and other genetic traits in the coastal and mountain villages of Cyprus. *Amer. J. Human Genetics, 16,* 267–283.

Preston–Thomas, H., Turnbull, G., Green, E., Dauphinee, T., and Kalra, S. (1960). *Can. Jour. Phys., 38,* 824–852.

Quenouille, M. (1956). Notes on bias in estimation. *Biometrika, 43,* 353–360.

Ratcliff, J. (1957). New surgery for ailing hearts. *Reader's Digest, 71,* 70–73.

Rice, J. R. (1983). *Numerical Methods, Software, and Analysis.* New York: McGraw–Hill.

Robson, G. (1929). Monograph of the recent cephalopoda, part I. London: British Museum.

Rosen, B., and Jerdee, T. (1974). Influence of sex role stereotypes on personnel decisions. *J. Appl. Psych., 59,* 9–14.

Rudemo, H. (1982). Empirical choice of histograms and kernel density estimators. *Scand. J. Stat., 9,* 65–78.

Ryan, T., and Joiner, B. (unpublished ms.). Normal probability plots and tests for normality, Pennsylvania State Univ.

Ryan, T., Joiner, B., and Ryan, B. (1976). *Minitab Student Handbook.* Boston, Mass.: Duxbury.

Schachter, A. (1959). *The Psychology of Affiliation.* Stanford, Calif.: Stanford University Press.

Scheffe, H. (1959). *The Analysis of Variance.* New York: Wiley.

Simiu, E., and Filliben, J. (1975). Statistical analysis of extreme winds. Nat. Bur. Stand. Tech. Note No. 868 (Washington, D.C.).

Simpson, J., Olsen, A., and Eden, J. (1975). A Bayesian analysis of a multiplicative treatment effect in weather modification. *Technometrics, 17,* 161–166.

Smith, D. G., Prentice, R., Thompson, D. J., and Herrmann, W. L. (1975). Association of exogenous estrogen and endometrial carcinoma. *N. Eng. J. Med., 293,* 1164–1167.

Snyder, R. G. (1961). The sex ratio of offspring of pilots of high performance military aircraft. *Hum. Biol., 3,* 1–10.

Stanley, W., and Walton, D. (1961). Trifluoperazine ("Stelazine"). A controlled clinical trial in chronic schizophrenia. *J. Mental Sci., 107,* 250–257.

Steel, E., Small, J., Leigh, S., and Filliben, J. (1980). Statistical considerations in the preparation of chrysotile filter standard reference materials. NBS Technical Report (Washington, D.C.).

Stigler, S. M. (1977). Do robust estimates work with real data? *Annals of Statistics, 5,* 1055–1098.

Strang, G. (1980). *Linear Algebra and Its Applications.* New York: Academic Press.

Student (1907). On the error of counting with a haemacytometer. *Biometrika, 5,* 351.

Tanur, J., Mosteller, F., Kruskal, W., Link, R., Pieters, R., and Rising, G. (1972). *Statistics: A Guide to the Unknown.* Oakland, Calif.: Holden –Day.

Thomas, H. A. (1948). Frequency of minor floods. *J. Boston Soc. Civil Engineers, 34,* 425–442.

Tukey, J. (1977). *Exploratory Data Analysis.* Reading, Mass.: Addison–Wesley.

Udias, A., and Rice, J. (1975). Statistical analysis of microearthquake activity near San Andreas Geophysical Observatory, Hollister, California. *Bulletin of the Seismological Society of America, 65,* 809–828.

Van Atta, C., and Chen, W. (1968). Correlation measurements in grid turbulence using digital harmonic analysis. *J. Fluid Mech., 34,* 497–515.

Veitch, J., and Wilks, A. (1985). A characterization of Arctic undersea noise. *J. Acoust. Soc. Amer., 77,* 989–999.

Velleman, P., and Hoaglin, D. (1981). *Applications, Basics, and Computing of Exploratory Data Analysis.* Boston, Mass.: Duxbury.

Vianna, N., Greenwald, P., and Davies, J. (1971). Tonsillectomy and Hodgkin's disease: The lymphoid tissue barrier. *Lancet, 1,* 431–432.

Warner, S. (1965). Randomized response: A survey technique for eliminating evasive answer bias. *Jour. Amer. Stat. Assoc., 60,* 63–69.

Weindling, S. (1977). Statistics report: Math 80B.

Weisburg, S. (1980). *Applied Linear Regression.* New York: Wiley.

White, J., Riethof, M., and Kushnir, I. (1960). Estimation of microcrystalline wax in beeswax. *J. Assoc. Offic. Anal. Chem., 43,* 781–790.

Wilk, M., and Gnanadesikan, R. (1968). Probability plotting methods for the analysis of data. *Biometrika, 55,* 1–17.

Williams, W. (1978). How bad can "good" data really be? *Amer. Stat., 32*, 61–67.

Wilson, E. B. (1952). *An Introduction to Scientific Research*. New York: McGraw–Hill.

Wood, L. A. (1972). Modulus of natural rubber cross-linked by dicumyl peroxide. I. Experimental observations. *J. Res. Nat. Bur. Stand., 76A*, 51–59.

Woodward, P. (1948). A statistical theory of cascade multiplication. *Proc. Camb. Phil. Soc., 44*, 404–412.

Woolf, B. (1955). On estimating the relation between blood group and disease. *Annals of Hum. Genetics, 19*, 251–253.

Yates, F. (1960). *Sampling Methods for Censuses and Surveys*. New York: Hafner.

Youden, J. (1972). Enduring values. *Technometrics, 14*, 1–11.

Youden, W. J. (1962). *Experimentation and Measurement*. Washington, D.C.: National Science Teachers Association.

Youden, W. J. (1974). *Risk, Choice and Prediction*. Boston, Mass.: Duxbury.

APPENDIX A

Common Distributions

Discrete Distributions

Binomial

$$p(k) = \binom{n}{k} p^k (1 - p)^{n-k}, \quad k = 0, 1, \ldots, n$$

$$E(X) = np$$

$$\text{Var}(X) = np(1 - p)$$

$$M(t) = (1 - p + pe^t)^n$$

Geometric

$$p(k) = p(1 - p)^{k-1}, \quad k = 0, 1, \ldots$$

$$E(X) = \frac{1}{p}$$

$$\text{Var}(X) = \frac{1 - p}{p}$$

$$M(t) = \frac{ep}{1 - (1 - p)e^t}$$

Negative Binomial

$$p(k) = \binom{k - 1}{r - 1} p^r (1 - p)^{k-r}, \quad k = 0, 1, \ldots$$

$$E(X) = \frac{r}{p}$$

$$\text{Var}(X) = \frac{r(1 - p)}{p}$$

$$M(t) = \left(\frac{ep}{1 - (1 - p)e^t} \right)^r$$

Poisson

$$p(k) = \frac{\lambda^k e^{-\lambda}}{k!}, \quad k = 0, 1, \ldots$$

$$E(X) = \lambda$$

$$\text{Var}(X) = \lambda$$

$$M(t) = e^{\lambda(e^t - 1)}$$

Continuous Distributions

Normal

$$f(x) = \frac{1}{\sigma\sqrt{2\pi}} e^{-(1/2\sigma^2)(x - \mu)^2}, \quad -\infty < x < \infty$$

$$E(X) = \mu$$

$$\text{Var}(X) = \sigma^2$$

$$M(t) = e^{\mu t} e^{\sigma^2 t^2 / 2}$$

Gamma

$$f(x) = \frac{\lambda^\alpha}{\Gamma(\alpha)} x^{\alpha - 1} e^{-\lambda t}, \quad x \geq 0$$

$$E(X) = \frac{\alpha}{\lambda}$$

$$\text{Var}(X) = \frac{\alpha}{\lambda^2}$$

$$M(t) = \left(\frac{\lambda}{\lambda - t}\right)^{\alpha}, \quad t < \lambda$$

Exponential (Special Case of Gamma with $\alpha = 1$)

Uniform

$$f(x) = 1, \quad 0 \leq x \leq 1$$

$$E(X) = \tfrac{1}{2}$$

$$\text{Var}(X) = \tfrac{1}{12}$$

$$M(t) = \frac{e^t - 1}{t}$$

Beta

$$f(x) = \frac{\Gamma(a + b)}{\Gamma(a)\Gamma(b)} x^{a-1}(1 - x)^{b-1}, \quad 0 \leq x \leq 1$$

$$E(X) = \frac{a}{a + b}$$

$$\text{Var}(X) = \frac{ab}{(a + b)^2(a + b + 1)}$$

$M(t)$ is not useful.

APPENDIX B

Tables

Table 1 Binomial probabilities

Tabulated values are $\sum_{x=0}^{k} p(x)$. (Computations are rounded off at the third decimal place.)

$n = 5$

k \ p	.01	.05	.10	.20	.30	.40	.50	.60	.70	.80	.90	.95	.99
0	.951	.774	.590	.328	.168	.078	.031	.010	.002	.000	.000	.000	.000
1	.999	.977	.919	.737	.528	.337	.188	.087	.031	.007	.000	.000	.000
2	1.000	.999	.991	.942	.837	.683	.500	.317	.163	.058	.009	.001	.000
3	1.000	1.000	1.000	.993	.969	.913	.812	.663	.472	.263	.081	.023	.001
4	1.000	1.000	1.000	1.000	.998	.990	.969	.922	.832	.672	.410	.226	.049

$n = 10$

k	.01	.05	.10	.20	.30	.40	.50	.60	.70	.80	.90	.95	.99
0	.904	.599	.349	.107	.028	.006	.001	.000	.000	.000	.000	.000	.000
1	.996	.914	.736	.376	.149	.046	.011	.002	.000	.000	.000	.000	.000
2	1.000	.988	.930	.678	.383	.167	.055	.012	.002	.000	.000	.000	.000
3	1.000	.999	.987	.879	.650	.382	.172	.055	.011	.001	.000	.000	.000
4	1.000	1.000	.998	.967	.850	.633	.377	.166	.047	.006	.000	.000	.000
5	1.000	1.000	1.000	.994	.953	.834	.623	.367	.150	.033	.002	.000	.000
6	1.000	1.000	1.000	.999	.989	.945	.828	.618	.350	.121	.013	.001	.000
7	1.000	1.000	1.000	1.000	.998	.988	.945	.833	.617	.322	.070	.012	.000
8	1.000	1.000	1.000	1.000	1.000	.998	.989	.954	.851	.624	.264	.086	.004
9	1.000	1.000	1.000	1.000	1.000	1.000	.999	.994	.972	.893	.651	.401	.096

$n = 15$

k	.01	.05	.10	.20	.30	.40	.50	.60	.70	.80	.90	.95	.99
0	.860	.463	.206	.035	.005	.000	.000	.000	.000	.000	.000	.000	.000
1	.990	.829	.549	.167	.035	.005	.000	.000	.000	.000	.000	.000	.000
2	1.000	.964	.816	.398	.127	.027	.004	.000	.000	.000	.000	.000	.000
3	1.000	.995	.944	.648	.297	.091	.018	.002	.000	.000	.000	.000	.000
4	1.000	.999	.987	.836	.515	.217	.059	.009	.001	.000	.000	.000	.000
5	1.000	1.000	.998	.939	.722	.403	.151	.034	.004	.000	.000	.000	.000
6	1.000	1.000	1.000	.982	.869	.610	.304	.095	.015	.001	.000	.000	.000
7	1.000	1.000	1.000	.996	.950	.787	.500	.213	.050	.004	.000	.000	.000
8	1.000	1.000	1.000	.999	.985	.905	.696	.390	.131	.018	.000	.000	.000
9	1.000	1.000	1.000	1.000	.996	.966	.849	.597	.278	.061	.002	.000	.000
10	1.000	1.000	1.000	1.000	.999	.991	.941	.783	.485	.164	.013	.001	.000
11	1.000	1.000	1.000	1.000	1.000	.998	.982	.909	.703	.352	.056	.005	.000
12	1.000	1.000	1.000	1.000	1.000	1.000	.996	.973	.873	.602	.184	.036	.000
13	1.000	1.000	1.000	1.000	1.000	1.000	1.000	.995	.965	.833	.451	.171	.010
14	1.000	1.000	1.000	1.000	1.000	1.000	1.000	1.000	.995	.965	.794	.537	.140

$n = 20$

k \ p	.01	.05	.10	.20	.30	.40	.50	.60	.70	.80	.90	.95	.99
0	.818	.358	.122	.002	.001	.000	.000	.000	.000	.000	.000	.000	.000
1	.983	.736	.392	.069	.008	.001	.000	.000	.000	.000	.000	.000	.000
2	.999	.925	.677	.206	.035	.004	.000	.000	.000	.000	.000	.000	.000
3	1.000	.984	.867	.411	.107	.016	.001	.000	.000	.000	.000	.000	.000
4	1.000	.997	.957	.630	.238	.051	.006	.000	.000	.000	.000	.000	.000
5	1.000	1.000	.989	.804	.416	.126	.021	.002	.000	.000	.000	.000	.000
6	1.000	1.000	.998	.913	.608	.250	.058	.006	.000	.000	.000	.000	.000
7	1.000	1.000	1.000	.968	.772	.416	.132	.021	.001	.000	.000	.000	.000
8	1.000	1.000	1.000	.990	.887	.596	.252	.057	.005	.000	.000	.000	.000
9	1.000	1.000	1.000	.997	.952	.755	.412	.128	.017	.001	.000	.000	.000
10	1.000	1.000	1.000	.999	.983	.872	.588	.245	.048	.003	.000	.000	.000
11	1.000	1.000	1.000	1.000	.995	.943	.748	.404	.113	.010	.000	.000	.000
12	1.000	1.000	1.000	1.000	.999	.979	.868	.584	.228	.032	.000	.000	.000
13	1.000	1.000	1.000	1.000	1.000	.994	.942	.750	.392	.087	.002	.000	.000
14	1.000	1.000	1.000	1.000	1.000	.998	.979	.874	.584	.196	.011	.000	.000
15	1.000	1.000	1.000	1.000	1.000	1.000	.994	.949	.762	.370	.043	.003	.000
16	1.000	1.000	1.000	1.000	1.000	1.000	.999	.984	.893	.589	.133	.016	.000
17	1.000	1.000	1.000	1.000	1.000	1.000	1.000	.996	.965	.794	.323	.075	.001
18	1.000	1.000	1.000	1.000	1.000	1.000	1.000	.999	.992	.931	.608	.264	.017
19	1.000	1.000	1.000	1.000	1.000	1.000	1.000	1.000	.999	.988	.878	.642	.182

$n = 25$

k \ p	.01	.05	.10	.20	.30	.40	.50	.60	.70	.80	.90	.95	.99
0	.778	.277	.072	.004	.000	.000	.000	.000	.000	.000	.000	.000	.000
1	.974	.642	.271	.027	.002	.000	.000	.000	.000	.000	.000	.000	.000
2	.998	.873	.537	.098	.009	.000	.000	.000	.000	.000	.000	.000	.000
3	1.000	.966	.764	.234	.033	.002	.000	.000	.000	.000	.000	.000	.000
4	1.000	.993	.902	.421	.090	.009	.000	.000	.000	.000	.000	.000	.000
5	1.000	.999	.967	.617	.193	.029	.002	.000	.000	.000	.000	.000	.000
6	1.000	1.000	.991	.780	.341	.074	.007	.000	.000	.000	.000	.000	.000
7	1.000	1.000	.998	.891	.512	.154	.022	.001	.000	.000	.000	.000	.000
8	1.000	1.000	1.000	.953	.677	.274	.054	.004	.000	.000	.000	.000	.000
9	1.000	1.000	1.000	.983	.811	.425	.115	.013	.000	.000	.000	.000	.000
10	1.000	1.000	1.000	.994	.902	.586	.212	.034	.002	.000	.000	.000	.000
11	1.000	1.000	1.000	.998	.956	.732	.345	.078	.006	.000	.000	.000	.000
12	1.000	1.000	1.000	1.000	.983	.846	.500	.154	.017	.000	.000	.000	.000
13	1.000	1.000	1.000	1.000	.994	.922	.655	.268	.044	.002	.000	.000	.000
14	1.000	1.000	1.000	1.000	.998	.966	.788	.414	.098	.006	.000	.000	.000
15	1.000	1.000	1.000	1.000	1.000	.987	.885	.575	.189	.017	.000	.000	.000
16	1.000	1.000	1.000	1.000	1.000	.996	.946	.726	.323	.047	.000	.000	.000
17	1.000	1.000	1.000	1.000	1.000	.999	.978	.846	.488	.109	.002	.000	.000
18	1.000	1.000	1.000	1.000	1.000	1.000	.993	.926	.659	.220	.009	.000	.000
19	1.000	1.000	1.000	1.000	1.000	1.000	.998	.971	.807	.383	.033	.001	.000
20	1.000	1.000	1.000	1.000	1.000	1.000	1.000	.991	.910	.579	.098	.007	.000
21	1.000	1.000	1.000	1.000	1.000	1.000	1.000	.998	.967	.766	.236	.034	.000
22	1.000	1.000	1.000	1.000	1.000	1.000	1.000	1.000	.991	.902	.463	.127	.002
23	1.000	1.000	1.000	1.000	1.000	1.000	1.000	1.000	.998	.973	.729	.358	.026
24	1.000	1.000	1.000	1.000	1.000	1.000	1.000	1.000	1.000	.996	.928	.723	.222

Table 2   Cumulative normal distribution—Values of P corresponding to z_P for the normal curve

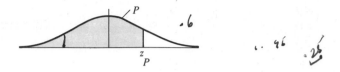

z is the standard normal variable. The value of P for $-z_P$ equals 1 minus the value of P for $+z_P$; for example, the P for -1.62 equals $1 - .9474 = .0526$.

z_P	.00	.01	.02	.03	.04	.05	.06	.07	.08	.09
.0	.5000	.5040	.5080	.5120	.5160	.5199	.5239	.5279	.5319	.5359
.1	.5398	.5438	.5478	.5517	.5557	.5596	.5636	.5675	.5714	.5753
.2	.5793	.5832	.5871	.5910	.5948	.5987	.6026	.6064	.6103	.6141
.3	.6179	.6217	.6255	.6293	.6331	.6368	.6406	.6443	.6480	.6517
.4	.6554	.6591	.6628	.6664	.6700	.6736	.6772	.6808	.6844	.6879
.5	.6915	.6950	.6985	.7019	.7054	.7088	.7123	.7157	.7190	.7224
.6	.7257	.7291	.7324	.7357	.7389	.7422	.7454	.7486	.7517	.7549
.7	.7580	.7611	.7642	.7673	.7704	.7734	.7764	.7794	.7823	.7852
.8	.7881	.7910	.7939	.7967	.7995	.8023	.8051	.8078	.8106	.8133
.9	.8159	.8186	.8212	.8238	.8264	.8289	.8315	.8340	.8365	.8389
1.0	.8413	.8438	.8461	.8485	.8508	.8531	.8554	.8577	.8599	.8621
1.1	.8643	.8665	.8686	.8708	.8729	.8749	.8770	.8790	.8810	.8830
1.2	.8849	.8869	.8888	.8907	.8925	.8944	.8962	.8980	.8997	.9015
1.3	.9032	.9049	.9066	.9082	.9099	.9115	.9131	.9147	.9162	.9177
1.4	.9192	.9207	.9222	.9236	.9251	.9265	.9279	.9292	.9306	.9319
1.5	.9332	.9345	.9357	.9370	.9382	.9394	.9406	.9418	.9429	.9441
1.6	.9452	.9463	.9474	.9484	.9495	.9505	.9515	.9525	.9535	.9545
1.7	.9554	.9564	.9573	.9582	.9591	.9599	.9608	.9616	.9625	.9633
1.8	.9641	.9649	.9656	.9664	.9671	.9678	.9686	.9693	.9699	.9706
1.9	.9713	.9719	.9726	.9732	.9738	.9744	.9750	.9756	.9761	.9767
2.0	.9772	.9778	.9783	.9788	.9793	.9798	.9803	.9808	.9812	.9817
2.1	.9821	.9826	.9830	.9834	.9838	.9842	.9846	.9850	.9854	.9857
2.2	.9861	.9864	.9868	.9871	.9875	.9878	.9881	.9884	.9887	.9890
2.3	.9893	.9896	.9898	.9901	.9904	.9906	.9909	.9911	.9913	.9916
2.4	.9918	.9920	.9922	.9925	.9927	.9929	.9931	.9932	.9934	.9936
2.5	.9938	.9940	.9941	.9943	.9945	.9946	.9948	.9949	.9951	.9952
2.6	.9953	.9955	.9956	.9957	.9959	.9960	.9961	.9962	.9963	.9964
2.7	.9965	.9966	.9967	.9968	.9969	.9970	.9971	.9972	.9973	.9974
2.8	.9974	.9975	.9976	.9977	.9977	.9978	.9979	.9979	.9980	.9981
2.9	.9981	.9982	.9982	.9983	.9984	.9984	.9985	.9985	.9986	.9986
3.0	.9987	.9987	.9987	.9988	.9988	.9989	.9989	.9989	.9990	.9990
3.1	.9990	.9991	.9991	.9991	.9992	.9992	.9992	.9992	.9993	.9993
3.2	.9993	.9993	.9994	.9994	.9994	.9994	.9994	.9995	.9995	.9995
3.3	.9995	.9995	.9995	.9996	.9996	.9996	.9996	.9996	.9996	.9997
3.4	.9997	.9997	.9997	.9997	.9997	.9997	.9997	.9997	.9997	.9998

Table 3 Percentiles of the χ^2 distribution—Values of χ_P^2 corresponding to P

df	$\chi^2_{.005}$	$\chi^2_{.01}$	$\chi^2_{.025}$	$\chi^2_{.05}$	$\chi^2_{.10}$	$\chi^2_{.90}$	$\chi^2_{.95}$	$\chi^2_{.975}$	$\chi^2_{.99}$	$\chi^2_{.995}$
1	.000039	.00016	.00098	.0039	.0158	2.71	3.84	5.02	6.63	7.88
2	.0100	.0201	.0506	.1026	.2107	4.61	5.99	7.38	9.21	10.60
3	.0717	.115	.216	.352	.584	6.25	7.81	9.35	11.34	12.84
4	.207	.297	.484	.711	1.064	7.78	9.49	11.14	13.28	14.86
5	.412	.554	.831	1.15	1.61	9.24	11.07	12.83	15.09	16.75
6	.676	.872	1.24	1.64	2.20	10.64	12.59	14.45	16.81	18.55
7	.989	1.24	1.69	2.17	2.83	12.02	14.07	16.01	18.48	20.28
8	1.34	1.65	2.18	2.73	3.49	13.36	15.51	17.53	20.09	21.96
9	1.73	2.09	2.70	3.33	4.17	14.68	16.92	19.02	21.67	23.59
10	2.16	2.56	3.25	3.94	4.87	15.99	18.31	20.48	23.21	25.19
11	2.60	3.05	3.82	4.57	5.58	17.28	19.68	21.92	24.73	26.76
12	3.07	3.57	4.40	5.23	6.30	18.55	21.03	23.34	26.22	28.30
13	3.57	4.11	5.01	5.89	7.04	19.81	22.36	24.74	27.69	29.82
14	4.07	4.66	5.63	6.57	7.79	21.06	23.68	26.12	29.14	31.32
15	4.60	5.23	6.26	7.26	8.55	22.31	25.00	27.49	30.58	32.80
16	5.14	5.81	6.91	7.96	9.31	23.54	26.30	28.85	32.00	34.27
18	6.26	7.01	8.23	9.39	10.86	25.99	28.87	31.53	34.81	37.16
20	7.43	8.26	9.59	10.85	12.44	28.41	31.41	34.17	37.57	40.00
24	9.89	10.86	12.40	13.85	15.66	33.20	36.42	39.36	42.98	45.56
30	13.79	14.95	16.79	18.49	20.60	40.26	43.77	46.98	50.89	53.67
40	20.71	22.16	24.43	26.51	29.05	51.81	55.76	59.34	63.69	66.77
60	35.53	37.48	40.48	43.19	46.46	74.40	79.08	83.30	88.38	91.95
120	83.85	86.92	91.58	95.70	100.62	140.23	146.57	152.21	158.95	163.64

For large degrees of freedom,

$$\chi_P^2 = \tfrac{1}{2}\left(z_P + \sqrt{2v - 1}\right)^2 \text{ approximately,}$$

where v = degrees of freedom and z_P is given in Table 2.

Table 4   Percentiles of the t distribution

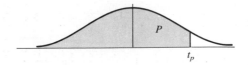

df	$t_{.60}$	$t_{.70}$	$t_{.80}$	$t_{.90}$	$t_{.95}$	$t_{.975}$	$t_{.99}$	$t_{.995}$
1	.325	.727	1.376	3.078	6.314	12.706	31.821	63.657
2	.289	.617	1.061	1.886	2.920	4.303	6.965	9.925
3	.277	.584	.978	1.638	2.353	3.182	4.541	5.841
4	.271	.569	.941	1.533	2.132	2.776	3.747	4.604
5	.267	.559	.920	1.476	2.015	2.571	3.365	4.032
6	.265	.553	.906	1.440	1.943	2.447	3.143	3.707
7	.263	.549	.896	1.415	1.895	2.365	2.998	3.499
8	.262	.546	.889	1.397	1.860	2.306	2.896	3.355
9	.261	.543	.883	1.383	1.833	2.262	2.821	3.250
10	.260	.542	.879	1.372	1.812	2.228	2.764	3.169
11	.260	.540	.876	1.363	1.796	2.201	2.718	3.106
12	.259	.539	.873	1.356	1.782	2.179	2.681	3.055
13	.259	.538	.870	1.350	1.771	2.160	2.650	3.012
14	.258	.537	.868	1.345	1.761	2.145	2.624	2.977
15	.258	.536	.866	1.341	1.753	2.131	2.602	2.947
16	.258	.535	.865	1.337	1.746	2.120	2.583	2.921
17	.257	.534	.863	1.333	1.740	2.110	2.567	2.898
18	.257	.534	.862	1.330	1.734	2.101	2.552	2.878
19	.257	.533	.861	1.328	1.729	2.093	2.539	2.861
20	.257	.533	.860	1.325	1.725	2.086	2.528	2.845
21	.257	.532	.859	1.323	1.721	2.080	2.518	2.831
22	.256	.532	.858	1.321	1.717	2.074	2.508	2.819
23	.256	.532	.858	1.319	1.714	2.069	2.500	2.807
24	.256	.531	.857	1.318	1.711	2.064	2.492	2.797
25	.256	.531	.856	1.316	1.708	2.060	2.485	2.787
26	.256	.531	.856	1.315	1.706	2.056	2.479	2.779
27	.256	.531	.855	1.314	1.703	2.052	2.473	2.771
28	.256	.530	.855	1.313	1.701	2.048	2.467	2.763
29	.256	.530	.854	1.311	1.699	2.045	2.462	2.756
30	.256	.530	.854	1.310	1.697	2.042	2.457	2.750
40	.255	.529	.851	1.303	1.684	2.021	2.423	2.704
60	.254	.527	.848	1.296	1.671	2.000	2.390	2.660
120	.254	.526	.845	1.289	1.658	1.980	2.358	2.617
∞	.253	.524	.842	1.282	1.645	1.960	2.326	2.576

Table 5 Percentiles of the F distribution: $F_{.90}(n_1, n_2)$

$F_{.90}(n_1, n_2)$

n_1 = degrees of freedom for numerator

n_2 \ n_1	1	2	3	4	5	6	7	8	9	10	12	15	20	24	30	40	60	120	∞
1	39.86	49.50	53.59	55.83	57.24	58.20	58.91	59.44	59.86	60.19	60.71	61.22	61.74	62.00	62.26	62.53	62.79	63.06	63.33
2	8.53	9.00	9.16	9.24	9.29	9.33	9.35	9.37	9.38	9.39	9.41	9.42	9.44	9.45	9.46	9.47	9.47	9.48	9.49
3	5.54	5.46	5.39	5.34	5.31	5.28	5.27	5.25	5.24	5.23	5.22	5.20	5.18	5.18	5.17	5.16	5.15	5.14	5.13
4	4.54	4.32	4.19	4.11	4.05	4.01	3.98	3.95	3.94	3.92	3.90	3.87	3.84	3.83	3.82	3.80	3.79	3.78	3.76
5	4.06	3.78	3.62	3.52	3.45	3.40	3.37	3.34	3.32	3.30	3.27	3.24	3.21	3.19	3.17	3.16	3.14	3.12	3.10
6	3.78	3.46	3.29	3.18	3.11	3.05	3.01	2.98	2.96	2.94	2.90	2.87	2.84	2.82	2.80	2.78	2.76	2.74	2.72
7	3.59	3.26	3.07	2.96	2.88	2.83	2.78	2.75	2.72	2.70	2.67	2.63	2.59	2.58	2.56	2.54	2.51	2.49	2.47
8	3.46	3.11	2.92	2.81	2.73	2.67	2.62	2.59	2.56	2.54	2.50	2.46	2.42	2.40	2.38	2.36	2.34	2.32	2.29
9	3.36	3.01	2.81	2.69	2.61	2.55	2.51	2.47	2.44	2.42	2.38	2.34	2.30	2.28	2.25	2.23	2.21	2.18	2.16
10	3.29	2.92	2.73	2.61	2.52	2.46	2.41	2.38	2.35	2.32	2.28	2.24	2.20	2.18	2.16	2.13	2.11	2.08	2.06
11	3.23	2.86	2.66	2.54	2.45	2.39	2.34	2.30	2.27	2.25	2.21	2.17	2.12	2.10	2.08	2.05	2.03	2.00	1.97
12	3.18	2.81	2.61	2.48	2.39	2.33	2.28	2.24	2.21	2.19	2.15	2.10	2.06	2.04	2.01	1.99	1.96	1.93	1.90
13	3.14	2.76	2.56	2.43	2.35	2.28	2.23	2.20	2.16	2.14	2.10	2.05	2.01	1.98	1.96	1.93	1.90	1.88	1.85
14	3.10	2.73	2.52	2.39	2.31	2.24	2.19	2.15	2.12	2.10	2.05	2.01	1.96	1.94	1.91	1.89	1.86	1.83	1.80
15	3.07	2.70	2.49	2.36	2.27	2.21	2.16	2.12	2.09	2.06	2.02	1.97	1.92	1.90	1.87	1.85	1.82	1.79	1.76
16	3.05	2.67	2.46	2.33	2.24	2.18	2.13	2.09	2.06	2.03	1.99	1.94	1.89	1.87	1.84	1.81	1.78	1.75	1.72
17	3.03	2.64	2.44	2.31	2.22	2.15	2.10	2.06	2.03	2.00	1.96	1.91	1.86	1.84	1.81	1.78	1.75	1.72	1.69
18	3.01	2.62	2.42	2.29	2.20	2.13	2.08	2.04	2.00	1.98	1.93	1.89	1.84	1.81	1.78	1.75	1.72	1.69	1.66
19	2.99	2.61	2.40	2.27	2.18	2.11	2.06	2.02	1.98	1.96	1.91	1.86	1.81	1.79	1.76	1.73	1.70	1.67	1.63
20	2.97	2.59	2.38	2.25	2.16	2.09	2.04	2.00	1.96	1.94	1.89	1.84	1.79	1.77	1.74	1.71	1.68	1.64	1.61
21	2.96	2.57	2.36	2.23	2.14	2.08	2.02	1.98	1.95	1.92	1.87	1.83	1.78	1.75	1.72	1.69	1.66	1.62	1.59
22	2.95	2.56	2.35	2.22	2.13	2.06	2.01	1.97	1.93	1.90	1.86	1.81	1.76	1.73	1.70	1.67	1.64	1.60	1.57
23	2.94	2.55	2.34	2.21	2.11	2.05	1.99	1.95	1.92	1.89	1.84	1.80	1.74	1.72	1.69	1.66	1.62	1.59	1.55
24	2.93	2.54	2.33	2.19	2.10	2.04	1.98	1.94	1.91	1.88	1.83	1.78	1.73	1.70	1.67	1.64	1.61	1.57	1.53
25	2.92	2.53	2.32	2.18	2.09	2.02	1.97	1.93	1.89	1.87	1.82	1.77	1.72	1.69	1.66	1.63	1.59	1.56	1.52
26	2.91	2.52	2.31	2.17	2.08	2.01	1.96	1.92	1.88	1.86	1.81	1.76	1.71	1.68	1.65	1.61	1.58	1.54	1.50
27	2.90	2.51	2.30	2.17	2.07	2.00	1.95	1.91	1.87	1.85	1.80	1.75	1.70	1.67	1.64	1.60	1.57	1.53	1.49
28	2.89	2.50	2.29	2.16	2.06	2.00	1.94	1.90	1.87	1.84	1.79	1.74	1.69	1.66	1.63	1.59	1.56	1.52	1.48
29	2.89	2.50	2.28	2.15	2.06	1.99	1.93	1.89	1.86	1.83	1.78	1.73	1.68	1.65	1.62	1.58	1.55	1.51	1.47
30	2.88	2.49	2.28	2.14	2.05	1.98	1.93	1.88	1.85	1.82	1.77	1.72	1.67	1.64	1.61	1.57	1.54	1.50	1.46
40	2.84	2.44	2.23	2.09	2.00	1.93	1.87	1.83	1.79	1.76	1.71	1.66	1.61	1.57	1.54	1.51	1.47	1.42	1.38
60	2.79	2.39	2.18	2.04	1.95	1.87	1.82	1.77	1.74	1.71	1.66	1.60	1.54	1.51	1.48	1.44	1.40	1.35	1.29
120	2.75	2.35	2.13	1.99	1.90	1.82	1.77	1.72	1.68	1.65	1.60	1.55	1.48	1.45	1.41	1.37	1.32	1.26	1.19
∞	2.71	2.30	2.08	1.94	1.85	1.77	1.72	1.67	1.63	1.60	1.55	1.49	1.42	1.38	1.34	1.30	1.24	1.17	1.00

n_2 = degrees of freedom for denominator

Percentiles of the F distribution: $F_{.95}(n_1, n_2)$

n_1 = degrees of freedom for numerator

n_2 \ n_1	1	2	3	4	5	6	7	8	9	10	12	15	20	24	30	40	60	120	∞
1	161.4	199.5	215.7	224.6	230.2	234.0	236.8	238.9	240.5	241.9	243.9	245.9	248.0	249.1	250.1	251.1	252.2	253.3	254.3
2	18.51	19.00	19.16	19.25	19.30	19.33	19.35	19.37	19.38	19.40	19.41	19.43	19.45	19.45	19.46	19.47	19.48	19.49	19.50
3	10.13	9.55	9.28	9.12	9.01	8.94	8.89	8.85	8.81	8.79	8.74	8.70	8.66	8.64	8.62	8.59	8.57	8.55	8.53
4	7.71	6.94	6.59	6.39	6.26	6.16	6.09	6.04	6.00	5.96	5.91	5.86	5.80	5.77	5.75	5.72	5.69	5.66	5.63
5	6.61	5.79	5.41	5.19	5.05	4.95	4.88	4.82	4.77	4.74	4.68	4.62	4.56	4.53	4.50	4.46	4.43	4.40	4.36
6	5.99	5.14	4.76	4.53	4.39	4.28	4.21	4.15	4.10	4.06	4.00	3.94	3.87	3.84	3.81	3.77	3.74	3.70	3.67
7	5.59	4.74	4.35	4.12	3.97	3.87	3.79	3.73	3.68	3.64	3.57	3.51	3.44	3.41	3.38	3.34	3.30	3.27	3.23
8	5.32	4.46	4.07	3.84	3.69	3.58	3.50	3.44	3.39	3.35	3.28	3.22	3.15	3.12	3.08	3.04	3.01	2.97	2.93
9	5.12	4.26	3.86	3.63	3.48	3.37	3.29	3.23	3.18	3.14	3.07	3.01	2.94	2.90	2.86	2.83	2.79	2.75	2.71
10	4.96	4.10	3.71	3.48	3.33	3.22	3.14	3.07	3.02	2.98	2.91	2.85	2.77	2.74	2.70	2.66	2.62	2.58	2.54
11	4.84	3.98	3.59	3.36	3.20	3.09	3.01	2.95	2.90	2.85	2.79	2.72	2.65	2.61	2.57	2.53	2.49	2.45	2.40
12	4.75	3.89	3.49	3.26	3.11	3.00	2.91	2.85	2.80	2.75	2.69	2.62	2.54	2.51	2.47	2.43	2.38	2.34	2.30
13	4.67	3.81	3.41	3.18	3.03	2.92	2.83	2.77	2.71	2.67	2.60	2.53	2.46	2.42	2.38	2.34	2.30	2.25	2.21
14	4.60	3.74	3.34	3.11	2.96	2.85	2.76	2.70	2.65	2.60	2.53	2.46	2.39	2.35	2.31	2.27	2.22	2.18	2.13
15	4.54	3.68	3.29	3.06	2.90	2.79	2.71	2.64	2.59	2.54	2.48	2.40	2.33	2.29	2.25	2.20	2.16	2.11	2.07
16	4.49	3.63	3.24	3.01	2.85	2.74	2.66	2.59	2.54	2.49	2.42	2.35	2.28	2.24	2.19	2.15	2.11	2.06	2.01
17	4.45	3.59	3.20	2.96	2.81	2.70	2.61	2.55	2.49	2.45	2.38	2.31	2.23	2.19	2.15	2.10	2.06	2.01	1.96
18	4.41	3.55	3.16	2.93	2.77	2.66	2.58	2.51	2.46	2.41	2.34	2.27	2.19	2.15	2.11	2.06	2.02	1.97	1.92
19	4.38	3.52	3.13	2.90	2.74	2.63	2.54	2.48	2.42	2.38	2.31	2.23	2.16	2.11	2.07	2.03	1.98	1.93	1.88
20	4.35	3.49	3.10	2.87	2.71	2.60	2.51	2.45	2.39	2.35	2.28	2.20	2.12	2.08	2.04	1.99	1.95	1.90	1.84
21	4.32	3.47	3.07	2.84	2.68	2.57	2.49	2.42	2.37	2.32	2.25	2.18	2.10	2.05	2.01	1.96	1.92	1.87	1.81
22	4.30	3.44	3.05	2.82	2.66	2.55	2.46	2.40	2.34	2.30	2.23	2.15	2.07	2.03	1.98	1.94	1.89	1.84	1.78
23	4.28	3.42	3.03	2.80	2.64	2.53	2.44	2.37	2.32	2.27	2.20	2.13	2.05	2.01	1.96	1.91	1.86	1.81	1.76
24	4.26	3.40	3.01	2.78	2.62	2.51	2.42	2.36	2.30	2.25	2.18	2.11	2.03	1.98	1.94	1.89	1.84	1.79	1.73
25	4.24	3.39	2.99	2.76	2.60	2.49	2.40	2.34	2.28	2.24	2.16	2.09	2.01	1.96	1.92	1.87	1.82	1.77	1.71
26	4.23	3.37	2.98	2.74	2.59	2.47	2.39	2.32	2.27	2.22	2.15	2.07	1.99	1.95	1.90	1.85	1.80	1.75	1.69
27	4.21	3.35	2.96	2.73	2.57	2.46	2.37	2.31	2.25	2.20	2.13	2.06	1.97	1.93	1.88	1.84	1.79	1.73	1.67
28	4.20	3.34	2.95	2.71	2.56	2.45	2.36	2.29	2.24	2.19	2.12	2.04	1.96	1.91	1.87	1.82	1.77	1.71	1.65
29	4.18	3.33	2.93	2.70	2.55	2.43	2.35	2.28	2.22	2.18	2.10	2.03	1.94	1.90	1.85	1.81	1.75	1.70	1.64
30	4.17	3.32	2.92	2.69	2.53	2.42	2.33	2.27	2.21	2.16	2.09	2.01	1.93	1.89	1.84	1.79	1.74	1.68	1.62
40	4.08	3.23	2.84	2.61	2.45	2.34	2.25	2.18	2.12	2.08	2.00	1.92	1.84	1.79	1.74	1.69	1.64	1.58	1.51
60	4.00	3.15	2.76	2.53	2.37	2.25	2.17	2.10	2.04	1.99	1.92	1.84	1.75	1.70	1.65	1.59	1.53	1.47	1.39
120	3.92	3.07	2.68	2.45	2.29	2.17	2.09	2.02	1.96	1.91	1.83	1.75	1.66	1.61	1.55	1.50	1.43	1.35	1.25
∞	3.84	3.00	2.60	2.37	2.21	2.10	2.01	1.94	1.88	1.83	1.75	1.67	1.57	1.52	1.46	1.39	1.32	1.22	1.00

n_2 = degrees of freedom for denominator

Percentiles of the F distribution: $F_{.975}(n_1, n_2)$

n_1 = degrees of freedom for numerator

n_2 \ n_1	1	2	3	4	5	6	7	8	9	10	12	15	20	24	30	40	60	120	∞
1	647.8	799.5	864.2	899.6	921.8	937.1	948.2	956.7	963.3	968.6	976.7	984.9	993.1	997.2	1001	1006	1010	1014	1018
2	38.51	39.00	39.17	39.25	39.30	39.33	39.36	39.37	39.39	39.40	39.41	39.43	39.45	39.46	39.46	39.47	39.48	39.49	39.50
3	17.44	16.04	15.44	15.10	14.88	14.73	14.62	14.54	14.47	14.42	14.34	14.25	14.17	14.12	14.08	14.04	13.99	13.95	13.90
4	12.22	10.65	9.98	9.60	9.36	9.20	9.07	8.98	8.90	8.84	8.75	8.66	8.56	8.51	8.46	8.41	8.36	8.31	8.26
5	10.01	8.43	7.76	7.39	7.15	6.98	6.85	6.76	6.68	6.62	6.52	6.43	6.33	6.28	6.23	6.18	6.12	6.07	6.02
6	8.81	7.26	6.60	6.23	5.99	5.82	5.70	5.60	5.52	5.46	5.37	5.27	5.17	5.12	5.07	5.01	4.96	4.90	4.85
7	8.07	6.54	5.89	5.52	5.29	5.12	4.99	4.90	4.82	4.76	4.67	4.57	4.47	4.42	4.36	4.31	4.25	4.20	4.14
8	7.57	6.06	5.42	5.05	4.82	4.65	4.53	4.43	4.36	4.30	4.20	4.10	4.00	3.95	3.89	3.84	3.78	3.73	3.67
9	7.21	5.71	5.08	4.72	4.48	4.32	4.20	4.10	4.03	3.96	3.87	3.77	3.67	3.61	3.56	3.51	3.45	3.39	3.33
10	6.94	5.46	4.83	4.47	4.24	4.07	3.95	3.85	3.78	3.72	3.62	3.52	3.42	3.37	3.31	3.26	3.20	3.14	3.08
11	6.72	5.26	4.63	4.28	4.04	3.88	3.76	3.66	3.59	3.53	3.43	3.33	3.23	3.17	3.12	3.06	3.00	2.94	2.88
12	6.55	5.10	4.47	4.12	3.89	3.73	3.61	3.51	3.44	3.37	3.28	3.18	3.07	3.02	2.96	2.91	2.85	2.79	2.72
13	6.41	4.97	4.35	4.00	3.77	3.60	3.48	3.39	3.31	3.25	3.15	3.05	2.95	2.89	2.84	2.78	2.72	2.66	2.60
14	6.30	4.86	4.24	3.89	3.66	3.50	3.38	3.29	3.21	3.15	3.05	2.95	2.84	2.79	2.73	2.67	2.61	2.55	2.49
15	6.20	4.77	4.15	3.80	3.58	3.41	3.29	3.20	3.12	3.06	2.96	2.86	2.76	2.70	2.64	2.59	2.52	2.46	2.40
16	6.12	4.69	4.08	3.73	3.50	3.34	3.22	3.12	3.05	2.99	2.89	2.79	2.68	2.63	2.57	2.51	2.45	2.38	2.32
17	6.04	4.62	4.01	3.66	3.44	3.28	3.16	3.06	2.98	2.92	2.82	2.72	2.62	2.56	2.50	2.44	2.38	2.32	2.25
18	5.98	4.56	3.95	3.61	3.38	3.22	3.10	3.01	2.93	2.87	2.77	2.67	2.56	2.50	2.44	2.38	2.32	2.26	2.19
19	5.92	4.51	3.90	3.56	3.33	3.17	3.05	2.96	2.88	2.82	2.72	2.62	2.51	2.45	2.39	2.33	2.27	2.20	2.13
20	5.87	4.46	3.86	3.51	3.29	3.13	3.01	2.91	2.84	2.77	2.68	2.57	2.46	2.41	2.35	2.29	2.22	2.16	2.09
21	5.83	4.42	3.82	3.48	3.25	3.09	2.97	2.87	2.80	2.73	2.64	2.53	2.42	2.37	2.31	2.25	2.18	2.11	2.04
22	5.79	4.38	3.78	3.44	3.22	3.05	2.93	2.84	2.76	2.70	2.60	2.50	2.39	2.33	2.27	2.21	2.14	2.08	2.00
23	5.75	4.35	3.75	3.41	3.18	3.02	2.90	2.81	2.73	2.67	2.57	2.47	2.36	2.30	2.24	2.18	2.11	2.04	1.97
24	5.72	4.32	3.72	3.38	3.15	2.99	2.87	2.78	2.70	2.64	2.54	2.44	2.33	2.27	2.21	2.15	2.08	2.01	1.94
25	5.69	4.29	3.69	3.35	3.13	2.97	2.85	2.75	2.68	2.61	2.51	2.41	2.30	2.24	2.18	2.12	2.05	1.98	1.91
26	5.66	4.27	3.67	3.33	3.10	2.94	2.82	2.73	2.65	2.59	2.49	2.39	2.28	2.22	2.16	2.09	2.03	1.95	1.88
27	5.63	4.24	3.65	3.31	3.08	2.92	2.80	2.71	2.63	2.57	2.47	2.36	2.25	2.19	2.13	2.07	2.00	1.93	1.85
28	5.61	4.22	3.63	3.29	3.06	2.90	2.78	2.69	2.61	2.55	2.45	2.34	2.23	2.17	2.11	2.05	1.98	1.91	1.83
29	5.59	4.20	3.61	3.27	3.04	2.88	2.76	2.67	2.59	2.53	2.43	2.32	2.21	2.15	2.09	2.03	1.96	1.89	1.81
30	5.57	4.18	3.59	3.25	3.03	2.87	2.75	2.65	2.57	2.51	2.41	2.31	2.20	2.14	2.07	2.01	1.94	1.87	1.79
40	5.42	4.05	3.46	3.13	2.90	2.74	2.62	2.53	2.45	2.39	2.29	2.18	2.07	2.01	1.94	1.88	1.80	1.72	1.64
60	5.29	3.93	3.34	3.01	2.79	2.63	2.51	2.41	2.33	2.27	2.17	2.06	1.94	1.88	1.82	1.74	1.67	1.58	1.48
120	5.15	3.80	3.23	2.89	2.67	2.52	2.39	2.30	2.22	2.16	2.05	1.94	1.82	1.76	1.69	1.61	1.53	1.43	1.31
∞	5.02	3.69	3.12	2.79	2.57	2.41	2.29	2.19	2.11	2.05	1.94	1.83	1.71	1.64	1.57	1.48	1.39	1.27	1.00

n_2 = degrees of freedom for denominator

Percentiles of the F distribution: $F_{.99}(n_1, n_2)$

n_1 = degrees of freedom for numerator

n_2 \ n_1	1	2	3	4	5	6	7	8	9	10	12	15	20	24	30	40	60	120	∞
1	4052	4999.5	5403	5625	5764	5859	5928	5982	6022	6056	6106	6157	6209	6235	6261	6287	6313	6339	6366
2	98.50	99.00	99.17	99.25	99.30	99.33	99.36	99.37	99.39	99.40	99.42	99.43	99.45	99.46	99.47	99.47	99.48	99.49	99.50
3	34.12	30.82	29.46	28.71	28.24	27.91	27.67	27.49	27.35	27.23	27.05	26.87	26.69	26.60	26.50	26.41	26.32	26.22	26.13
4	21.20	18.00	16.69	15.98	15.52	15.21	14.98	14.80	14.66	14.55	14.37	14.20	14.02	13.93	13.84	13.75	13.65	13.56	13.46
5	16.26	13.27	12.06	11.39	10.97	10.67	10.46	10.29	10.16	10.05	9.89	9.72	9.55	9.47	9.38	9.29	9.20	9.11	9.02
6	13.75	10.92	9.78	9.15	8.75	8.47	8.26	8.10	7.98	7.87	7.72	7.56	7.40	7.31	7.23	7.14	7.06	6.97	6.88
7	12.25	9.55	8.45	7.85	7.46	7.19	6.99	6.84	6.72	6.62	6.47	6.31	6.16	6.07	5.99	5.91	5.82	5.74	5.65
8	11.26	8.65	7.59	7.01	6.63	6.37	6.18	6.03	5.91	5.81	5.67	5.52	5.36	5.28	5.20	5.12	5.03	4.95	4.86
9	10.56	8.02	6.99	6.42	6.06	5.80	5.61	5.47	5.35	5.26	5.11	4.96	4.81	4.73	4.65	4.57	4.48	4.40	4.31
10	10.04	7.56	6.55	5.99	5.64	5.39	5.20	5.06	4.94	4.85	4.71	4.56	4.41	4.33	4.25	4.17	4.08	4.00	3.91
11	9.65	7.21	6.22	5.67	5.32	5.07	4.89	4.74	4.63	4.54	4.40	4.25	4.10	4.02	3.94	3.86	3.78	3.69	3.60
12	9.33	6.93	5.95	5.41	5.06	4.82	4.64	4.50	4.39	4.30	4.16	4.01	3.86	3.78	3.70	3.62	3.54	3.45	3.36
13	9.07	6.70	5.74	5.21	4.86	4.62	4.44	4.30	4.19	4.10	3.96	3.82	3.66	3.59	3.51	3.43	3.34	3.25	3.17
14	8.86	6.51	5.56	5.04	4.69	4.46	4.28	4.14	4.03	3.94	3.80	3.66	3.51	3.43	3.35	3.27	3.18	3.09	3.00
15	8.68	6.36	5.42	4.89	4.56	4.32	4.14	4.00	3.89	3.80	3.67	3.52	3.37	3.29	3.21	3.13	3.05	2.96	2.87
16	8.53	6.23	5.29	4.77	4.44	4.20	4.03	3.89	3.78	3.69	3.55	3.41	3.26	3.18	3.10	3.02	2.93	2.84	2.75
17	8.40	6.11	5.18	4.67	4.34	4.10	3.93	3.79	3.68	3.59	3.46	3.31	3.16	3.08	3.00	2.92	2.83	2.75	2.65
18	8.29	6.01	5.09	4.58	4.25	4.01	3.84	3.71	3.60	3.51	3.37	3.23	3.08	3.00	2.92	2.84	2.75	2.66	2.57
19	8.18	5.93	5.01	4.50	4.17	3.94	3.77	3.63	3.52	3.43	3.30	3.15	3.00	2.92	2.84	2.76	2.67	2.58	2.49
20	8.10	5.85	4.94	4.43	4.10	3.87	3.70	3.56	3.46	3.37	3.23	3.09	2.94	2.86	2.78	2.69	2.61	2.52	2.42
21	8.02	5.78	4.87	4.37	4.04	3.81	3.64	3.51	3.40	3.31	3.17	3.03	2.88	2.80	2.72	2.64	2.55	2.46	2.36
22	7.95	5.72	4.82	4.31	3.99	3.76	3.59	3.45	3.35	3.26	3.12	2.98	2.83	2.75	2.67	2.58	2.50	2.40	2.31
23	7.88	5.66	4.76	4.26	3.94	3.71	3.54	3.41	3.30	3.21	3.07	2.93	2.78	2.70	2.62	2.54	2.45	2.35	2.26
24	7.82	5.61	4.72	4.22	3.90	3.67	3.50	3.36	3.26	3.17	3.03	2.89	2.74	2.66	2.58	2.49	2.40	2.31	2.21
25	7.77	5.57	4.68	4.18	3.85	3.63	3.46	3.32	3.22	3.13	2.99	2.85	2.70	2.62	2.54	2.45	2.36	2.27	2.17
26	7.72	5.53	4.64	4.14	3.82	3.59	3.42	3.29	3.18	3.09	2.96	2.81	2.66	2.58	2.50	2.42	2.33	2.23	2.13
27	7.68	5.49	4.60	4.11	3.78	3.56	3.39	3.26	3.15	3.06	2.93	2.78	2.63	2.55	2.47	2.38	2.29	2.20	2.10
28	7.64	5.45	4.57	4.07	3.75	3.53	3.36	3.23	3.12	3.03	2.90	2.75	2.60	2.52	2.44	2.35	2.26	2.17	2.06
29	7.60	5.42	4.54	4.04	3.73	3.50	3.33	3.20	3.09	3.00	2.87	2.73	2.57	2.49	2.41	2.33	2.23	2.14	2.03
30	7.56	5.39	4.51	4.02	3.70	3.47	3.30	3.17	3.07	2.98	2.84	2.70	2.55	2.47	2.39	2.30	2.21	2.11	2.01
40	7.31	5.18	4.31	3.83	3.51	3.29	3.12	2.99	2.89	2.80	2.66	2.52	2.37	2.29	2.20	2.11	2.02	1.92	1.80
60	7.08	4.98	4.13	3.65	3.34	3.12	2.95	2.82	2.72	2.63	2.50	2.35	2.20	2.12	2.03	1.94	1.84	1.73	1.60
120	6.85	4.79	3.95	3.48	3.17	2.96	2.79	2.66	2.56	2.47	2.34	2.19	2.03	1.95	1.86	1.76	1.66	1.53	1.38
∞	6.63	4.61	3.78	3.32	3.02	2.80	2.64	2.51	2.41	2.32	2.18	2.04	1.88	1.79	1.70	1.59	1.47	1.32	1.00

n_2 = degrees of freedom for denominator

Table 6 Percentiles of the studentized range: $q_{.90}$

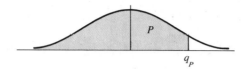

$q = w/s$ where w is the range of t observations, and v is the number of degrees of freedom associated with the standard deviation s.

v \ t	2	3	4	5	6	7	8	9	10
1	8.93	13.44	16.36	18.49	20.15	21.51	22.64	23.62	24.48
2	4.13	5.73	6.77	7.54	8.14	8.63	9.05	9.41	9.72
3	3.33	4.47	5.20	5.74	6.16	6.51	6.81	7.06	7.29
4	3.01	3.98	4.59	5.03	5.39	5.68	5.93	6.14	6.33
5	2.85	3.72	4.26	4.66	4.98	5.24	5.46	5.65	5.82
6	2.75	3.56	4.07	4.44	4.73	4.97	5.17	5.34	5.50
7	2.68	3.45	3.93	4.28	4.55	4.78	4.97	5.14	5.28
8	2.63	3.37	3.83	4.17	4.43	4.65	4.83	4.99	5.13
9	2.59	3.32	3.76	4.08	4.34	4.54	4.72	4.87	5.01
10	2.56	3.27	3.70	4.02	4.26	4.47	4.64	4.78	4.91
11	2.54	3.23	3.66	3.96	4.20	4.40	4.57	4.71	4.84
12	2.52	3.20	3.62	3.92	4.16	4.35	4.51	4.65	4.78
13	2.50	3.18	3.59	3.88	4.12	4.30	4.46	4.60	4.72
14	2.49	3.16	3.56	3.85	4.08	4.27	4.42	4.56	4.68
15	2.48	3.14	3.54	3.83	4.05	4.23	4.39	4.52	4.64
16	2.47	3.12	3.52	3.80	4.03	4.21	4.36	4.49	4.61
17	2.46	3.11	3.50	3.78	4.00	4.18	4.33	4.46	4.58
18	2.45	3.10	3.49	3.77	3.98	4.16	4.31	4.44	4.55
19	2.45	3.09	3.47	3.75	3.97	4.14	4.29	4.42	4.53
20	2.44	3.08	3.46	3.74	3.95	4.12	4.27	4.40	4.51
24	2.42	3.05	3.42	3.69	3.90	4.07	4.21	4.34	4.44
30	2.40	3.02	3.39	3.65	3.85	4.02	4.16	4.28	4.38
40	2.38	2.99	3.35	3.60	3.80	3.96	4.10	4.21	4.32
60	2.36	2.96	3.31	3.56	3.75	3.91	4.04	4.16	4.25
120	2.34	2.93	3.28	3.52	3.71	3.86	3.99	4.10	4.19
∞	2.33	2.90	3.24	3.48	3.66	3.81	3.93	4.04	4.13

Percentiles of the studentized range: $q_{.90}$

v \ t	11	12	13	14	15	16	17	18	19	20
1	25.24	25.92	26.54	27.10	27.62	28.10	28.54	28.96	29.35	29.71
2	10.01	10.26	10.49	10.70	10.89	11.07	11.24	11.39	11.54	11.68
3	7.49	7.67	7.83	7.98	8.12	8.25	8.37	8.48	8.58	8.68
4	6.49	6.65	6.78	6.91	7.02	7.13	7.23	7.33	7.41	7.50
5	5.97	6.10	6.22	6.34	6.44	6.54	6.63	6.71	6.79	6.86
6	5.64	5.76	5.87	5.98	6.07	6.16	6.25	6.32	6.40	6.47
7	5.41	5.53	5.64	5.74	5.83	5.91	5.99	6.06	6.13	6.19
8	5.25	5.36	5.46	5.56	5.64	5.72	5.80	5.87	5.93	6.00
9	5.13	5.23	5.33	5.42	5.51	5.58	5.66	5.72	5.79	5.85
10	5.03	5.13	5.23	5.32	5.40	5.47	5.54	5.61	5.67	5.73
11	4.95	5.05	5.15	5.23	5.31	5.38	5.45	5.51	5.57	5.63
12	4.89	4.99	5.08	5.16	5.24	5.31	5.37	5.44	5.49	5.55
13	4.83	4.93	5.02	5.10	5.18	5.25	5.31	5.37	5.43	5.48
14	4.79	4.88	4.97	5.05	5.12	5.19	5.26	5.32	5.37	5.43
15	4.75	4.84	4.93	5.01	5.08	5.15	5.21	5.27	5.32	5.38
16	4.71	4.81	4.89	4.97	5.04	5.11	5.17	5.23	5.28	5.33
17	4.68	4.77	4.86	4.93	5.01	5.07	5.13	5.19	5.24	5.30
18	4.65	4.75	4.83	4.90	4.98	5.04	5.10	5.16	5.21	5.26
19	4.63	4.72	4.80	4.88	4.95	5.01	5.07	5.13	5.18	5.23
20	4.61	4.70	4.78	4.85	4.92	4.99	5.05	5.10	5.16	5.20
24	4.54	4.63	4.71	4.78	4.85	4.91	4.97	5.02	5.07	5.12
30	4.47	4.56	4.64	4.71	4.77	4.83	4.89	4.94	4.99	5.03
40	4.41	4.49	4.56	4.63	4.69	4.75	4.81	4.86	4.90	4.95
60	4.34	4.42	4.49	4.56	4.62	4.67	4.73	4.78	4.82	4.86
120	4.28	4.35	4.42	4.48	4.54	4.60	4.65	4.69	4.74	4.78
∞	4.21	4.28	4.35	4.41	4.47	4.52	4.57	4.61	4.65	4.69

Percentiles of the studentized range: $q_{.95}$

v \ t	2	3	4	5	6	7	8	9	10
1	17.97	26.98	32.82	37.08	40.41	43.12	45.40	47.36	49.07
2	6.08	8.33	9.80	10.88	11.74	12.44	13.03	13.54	13.99
3	4.50	5.91	6.82	7.50	8.04	8.48	8.85	9.18	9.46
4	3.93	5.04	5.76	6.29	6.71	7.05	7.35	7.60	7.83
5	3.64	4.60	5.22	5.67	6.03	6.33	6.58	6.80	6.99
6	3.46	4.34	4.90	5.30	5.63	5.90	6.12	6.32	6.49
7	3.34	4.16	4.68	5.06	5.36	5.61	5.82	6.00	6.16
8	3.26	4.04	4.53	4.89	5.17	5.40	5.60	5.77	5.92
9	3.20	3.95	4.41	4.76	5.02	5.24	5.43	5.59	5.74
10	3.15	3.88	4.33	4.65	4.91	5.12	5.30	5.46	5.60
11	3.11	3.82	4.26	4.57	4.82	5.03	5.20	5.35	5.49
12	3.08	3.77	4.20	4.51	4.75	4.95	5.12	5.27	5.39
13	3.06	3.73	4.15	4.45	4.69	4.88	5.05	5.19	5.32
14	3.03	3.70	4.11	4.41	4.64	4.83	4.99	5.13	5.25
15	3.01	3.67	4.08	4.37	4.59	4.78	4.94	5.08	5.20

(*continued*)

(*continued*)

v \ t	2	3	4	5	6	7	8	9	10
16	3.00	3.65	4.05	4.33	4.56	4.74	4.90	5.03	5.15
17	2.98	3.63	4.02	4.30	4.52	4.70	4.86	4.99	5.11
18	2.97	3.61	4.00	4.28	4.49	4.67	4.82	4.96	5.07
19	2.96	3.59	3.98	4.25	4.47	4.65	4.79	4.92	5.04
20	2.95	3.58	3.96	4.23	4.45	4.62	4.77	4.90	5.01
24	2.92	3.53	3.90	4.17	4.37	4.54	4.68	4.81	4.92
30	2.89	3.49	3.85	4.10	4.30	4.46	4.60	4.72	4.82
40	2.86	3.44	3.79	4.04	4.23	4.39	4.52	4.63	4.73
60	2.83	3.40	3.74	3.98	4.16	4.31	4.44	4.55	4.65
120	2.80	3.36	3.68	3.92	4.10	4.24	4.36	4.47	4.56
∞	2.77	3.31	3.63	3.86	4.03	4.17	4.29	4.39	4.47

Percentiles of the studentized range: $q_{.95}$

v \ t	11	12	13	14	15	16	17	18	19	20
1	50.59	51.96	53.20	54.33	55.36	56.32	57.22	58.04	58.83	59.56
2	14.39	14.75	15.08	15.38	15.65	15.91	16.14	16.37	16.57	16.77
3	9.72	9.95	10.15	10.35	10.52	10.69	10.84	10.98	11.11	11.24
4	8.03	8.21	8.37	8.52	8.66	8.79	8.91	9.03	9.13	9.23
5	7.17	7.32	7.47	7.60	7.72	7.83	7.93	8.03	8.12	8.21
6	6.65	6.79	6.92	7.03	7.14	7.24	7.34	7.43	7.51	7.59
7	6.30	6.43	6.55	6.66	6.76	6.85	6.94	7.02	7.10	7.17
8	6.05	6.18	6.29	6.39	6.48	6.57	6.65	6.73	6.80	6.87
9	5.87	5.98	6.09	6.19	6.28	6.36	6.44	6.51	6.58	6.64
10	5.72	5.83	5.93	6.03	6.11	6.19	6.27	6.34	6.40	6.47
11	5.61	5.71	5.81	5.90	5.98	6.06	6.13	6.20	6.27	6.33
12	5.51	5.61	5.71	5.80	5.88	5.95	6.02	6.09	6.15	6.21
13	5.43	5.53	5.63	5.71	5.79	5.86	5.93	5.99	6.05	6.11
14	5.36	5.46	5.55	5.64	5.71	5.79	5.85	5.91	5.97	6.03
15	5.31	5.40	5.49	5.57	5.65	5.72	5.78	5.85	5.90	5.96
16	5.26	5.35	5.44	5.52	5.59	5.66	5.73	5.79	5.84	5.90
17	5.21	5.31	5.39	5.47	5.54	5.61	5.67	5.73	5.79	5.84
18	5.17	5.27	5.35	5.43	5.50	5.57	5.63	5.69	5.74	5.79
19	5.14	5.23	5.31	5.39	5.46	5.53	5.59	5.65	5.70	5.75
20	5.11	5.20	5.28	5.36	5.43	5.49	5.55	5.61	5.66	5.71
24	5.01	5.10	5.18	5.25	5.32	5.38	5.44	5.49	5.55	5.59
30	4.92	5.00	5.08	5.15	5.21	5.27	5.33	5.38	5.43	5.47
40	4.82	4.90	4.98	5.04	5.11	5.16	5.22	5.27	5.31	5.36
60	4.73	4.81	4.88	4.94	5.00	5.06	5.11	5.15	5.20	5.24
120	4.64	4.71	4.78	4.84	4.90	4.95	5.00	5.04	5.09	5.13
∞	4.55	4.62	4.68	4.74	4.80	4.85	4.89	4.93	4.97	5.01

Percentiles of the studentized range: $q_{.99}$

v \ t	2	3	4	5	6	7	8	9	10
1	90.03	135.0	164.3	185.6	202.2	215.8	227.2	237.0	245.6
2	14.04	19.02	22.29	24.72	26.63	28.20	29.53	30.68	31.69
3	8.26	10.62	12.17	13.33	14.24	15.00	15.64	16.20	16.69
4	6.51	8.12	9.17	9.96	10.58	11.10	11.55	11.93	12.27
5	5.70	6.98	7.80	8.42	8.91	9.32	9.67	9.97	10.24
6	5.24	6.33	7.03	7.56	7.97	8.32	8.61	8.87	9.10
7	4.95	5.92	6.54	7.01	7.37	7.68	7.94	8.17	8.37
8	4.75	5.64	6.20	6.62	6.96	7.24	7.47	7.68	7.86
9	4.60	5.43	5.96	6.35	6.66	6.91	7.13	7.33	7.49
10	4.48	5.27	5.77	6.14	6.43	6.67	6.87	7.05	7.21
11	4.39	5.15	5.62	5.97	6.25	6.48	6.67	6.84	6.99
12	4.32	5.05	5.50	5.84	6.10	6.32	6.51	6.67	6.81
13	4.26	4.96	5.40	5.73	5.98	6.19	6.37	6.53	6.67
14	4.21	4.89	5.32	5.63	5.88	6.08	6.26	6.41	6.54
15	4.17	4.84	5.25	5.56	5.80	5.99	6.16	6.31	6.44
16	4.13	4.79	5.19	5.49	5.72	5.92	6.08	6.22	6.35
17	4.10	4.74	5.14	5.43	5.66	5.85	6.01	6.15	6.27
18	4.07	4.70	5.09	5.38	5.60	5.79	5.94	6.08	6.20
19	4.05	4.67	5.05	5.33	5.55	5.73	5.89	6.02	6.14
20	4.02	4.64	5.02	5.29	5.51	5.69	5.84	5.97	6.09
24	3.96	4.55	4.91	5.17	5.37	5.54	5.69	5.81	5.92
30	3.89	4.45	4.80	5.05	5.24	5.40	5.54	5.65	5.76
40	3.82	4.37	4.70	4.93	5.11	5.26	5.39	5.50	5.60
60	3.76	4.28	4.59	4.82	4.99	5.13	5.25	5.36	5.45
120	3.70	4.20	4.50	4.71	4.87	5.01	5.12	5.21	5.30
∞	3.64	4.12	4.40	4.60	4.76	4.88	4.99	5.08	5.16

Percentiles of the studentized range: $q_{.99}$

v \ t	11	12	13	14	15	16	17	18	19	20
1	253.2	260.0	266.2	271.8	277.0	281.8	286.3	290.4	294.3	298.0
2	32.59	33.40	34.13	34.81	35.43	36.00	36.53	37.03	37.50	37.95
3	17.13	17.53	17.89	18.22	18.52	18.81	19.07	19.32	19.55	19.77
4	12.57	12.84	13.09	13.32	13.53	13.73	13.91	14.08	14.24	14.40
5	10.48	10.70	10.89	11.08	11.24	11.40	11.55	11.68	11.81	11.93
6	9.30	9.48	9.65	9.81	9.95	10.08	10.21	10.32	10.43	10.54
7	8.55	8.71	8.86	9.00	9.12	9.24	9.35	9.46	9.55	9.65
8	8.03	8.18	8.31	8.44	8.55	8.66	8.76	8.85	8.94	9.03
9	7.65	7.78	7.91	8.03	8.13	8.23	8.33	8.41	8.49	8.57
10	7.36	7.49	7.60	7.71	7.81	7.91	7.99	8.08	8.15	8.23
11	7.13	7.25	7.36	7.46	7.56	7.65	7.73	7.81	7.88	7.95
12	6.94	7.06	7.17	7.26	7.36	7.44	7.52	7.59	7.66	7.73
13	6.79	6.90	7.01	7.10	7.19	7.27	7.35	7.42	7.48	7.55
14	6.66	6.77	6.87	6.96	7.05	7.13	7.20	7.27	7.33	7.39
15	6.55	6.66	6.76	6.84	6.93	7.00	7.07	7.14	7.20	7.26
16	6.46	6.56	6.66	6.74	6.82	6.90	6.97	7.03	7.09	7.15
17	6.38	6.48	6.57	6.66	6.73	6.81	6.87	6.94	7.00	7.05
18	6.31	6.41	6.50	6.58	6.65	6.73	6.79	6.85	6.91	6.97
19	6.25	6.34	6.43	6.51	6.58	6.65	6.72	6.78	6.84	6.89
20	6.19	6.28	6.37	6.45	6.52	6.59	6.65	6.71	6.77	6.82
24	6.02	6.11	6.19	6.26	6.33	6.39	6.45	6.51	6.56	6.61
30	5.85	5.93	6.01	6.08	6.14	6.20	6.26	6.31	6.36	6.41
40	5.69	5.76	5.83	5.90	5.96	6.02	6.07	6.12	6.16	6.21
60	5.53	5.60	5.67	5.73	5.78	5.84	5.89	5.93	5.97	6.01
120	5.37	5.44	5.50	5.56	5.61	5.66	5.71	5.75	5.79	5.83
∞	5.23	5.29	5.35	5.40	5.45	5.49	5.54	5.57	5.61	5.65

(continued)

n_2	α for Two-Sided Test	α for One-Sided Test	1	2	3	4	5	6	7	8	9	10	11	12	13	14	15	16
10	.20	.10	1	6	12	20	28	38	49	60	73	87						
	.10	.05		4	10	17	26	35	45	56	69	82						
	.05	.025		3	9	15	23	32	42	53	65	78						
	.01	.005			6	12	19	27	37	47	58	71						
11	.20	.10	1	6	13	21	30	40	51	63	76	91	106					
	.10	.05		4	11	18	27	37	47	59	72	86	100					
	.05	.025		3	9	16	24	34	44	55	68	81	96					
	.01	.005			6	12	20	28	38	49	61	73	87					
12	.20	.10	1	7	14	22	32	42	54	66	80	94	110	127				
	.10	.05		5	11	19	28	38	49	62	75	89	104	120				
	.05	.025		4	10	17	26	35	46	58	71	84	99	115				
	.01	.005			7	13	21	30	40	51	63	76	90	105				
13	.20	.10	1	7	15	23	33	44	56	69	83	98	114	131	149			
	.10	.05		5	12	20	30	40	52	64	78	92	108	125	142			
	.05	.025		4	10	18	27	37	48	60	73	88	103	119	136			
	.01	.005			7	*13	22	31	41	53	65	79	93	109	125			
14	.20	.10	1	*8	16	25	35	46	59	72	86	102	118	136	154	174		
	.10	.05		*6	13	21	31	42	54	67	81	96	112	129	147	166		
	.05	.025		4	11	19	28	38	50	62	76	91	106	123	141	160		
	.01	.005			7	14	22	32	43	54	67	81	96	112	129	147		
15	.20	.10	1	8	16	26	37	48	61	75	90	106	123	141	159	179	200	
	.10	.05		6	13	22	33	44	56	69	84	99	116	133	152	171	192	
	.05	.025		4	11	20	29	40	52	65	79	94	110	127	145	164	184	
	.01	.005			8	15	23	33	44	56	69	84	99	115	133	151	171	
16	.20	.10	1	8	17	27	38	50	64	78	93	109	127	145	165	185	206	229
	.10	.05		6	14	24	34	46	58	72	87	103	120	138	156	176	197	219
	.05	.025		4	12	21	30	42	54	67	82	97	113	131	150	169	190	211
	.01	.005			8	15	24	34	46	58	72	86	102	119	136	155	175	196

n_1	two-sided α	one-sided α																				
17	.20	.10				1	9	18	28	40	52	66	81	97	113	131	150	170	190	212	235	259
	.10	.05					6	15	25	35	47	61	75	90	106	123	142	161	182	203	225	249
	.05	.025					5	12	21	32	43	56	70	84	100	117	135	154	174	195	217	240
	.01	.005						8	16	25	36	47	60	74	89	105	122	140	159	180	201	223
18	.20	.10			1	9	19	30	42	55	69	84	100	117	135	155	175	196	218	242	266	291
	.10	.05				7	15	26	37	49	63	77	93	110	127	146	166	187	208	231	255	280
	.05	.025				5	13	22	33	45	58	72	87	103	121	139	158	179	200	222	246	270
	.01	.005					8	16	26	37	49	62	76	92	108	125	144	163	184	206	228	252
19	.20	.10		2	10	20	31	43	57	71	87	103	121	139	159	180	202	224	248	273	299	325
	.10	.05		1	7	16	27	38	51	65	80	96	113	131	150	171	192	214	237	262	287	313
	.05	.025			5	13	23	34	46	60	74	90	107	124	143	163	*183	205	228	252	277	303
	.01	.005			3	9	17	27	38	50	64	78	94	111	129	*148	168	189	210	234	258	283
20	.20	.10	2	10	21	32	45	59	74	90	107	125	144	164	185	207	230	255	280	306	333	361
	.10	.05	1	7	17	28	40	53	67	83	99	117	135	155	175	197	220	243	268	294	320	348
	.05	.025		5	14	24	35	48	62	77	93	110	128	147	167	188	210	234	258	283	309	337
	.01	.005		3	9	18	28	39	52	66	81	97	114	132	151	172	193	215	239	263	289	315

For larger values of n_1 and n_2, critical values are given to a good approximation by the formula:

$$\frac{n_1}{2}(n_1 + n_2 + 1) - z\left\{\frac{n_1 n_2(n_1 + n_2 + 1)}{12}\right\}^{1/2}$$

where $z = 1.28$ for $\alpha = .20$ (two-sided test)

$\quad\quad\;\; z = 1.64$ for $\alpha = .10$ (two-sided test)

$\quad\quad\;\; z = 1.96$ for $\alpha = .05$ (two-sided test)

$\quad\quad\;\; z = 2.58$ for $\alpha = .01$ (two-sided test)

*Values have been corrected to the values given by D. B. Owen, *Handbook of Statistical Tables*, copyright 1962, Addison-Wesley Publishing Co., Inc.

Table 9 Critical values of $T_\alpha(n)$ for the wilcoxon signed-ranks test

T_α is the integer such that the probability that $T \leq T_\alpha$ is closest to α. For example, for $n = 8$, $P(T \leq 3) = .020$ and $P(T \leq 4) = .027$; therefore, $T_{.025}(8) = 4$.

	α for One-Sided Test		
	.025	.01	.005
	α for Two-Sided Test		
n	.05	.02	.01
6	0	—	—
7	2	0	—
8	4	2	0
9	6	3	2
10	8	5	3
11	11	7	5
12	14	10	7
13	17	13	10
14	21	16	13
15	25	20	16
16	30	24	20
17	35	28	23
18	40	33	28
19	46	38	32
20	52	43	38
21	59	49	43
22	66	56	49
23	73	62	55
24	81	69	61
25	89	77	68

For large n,

$$T_P(n) = \frac{n(n+1)}{4} - z_{1-P}\sqrt{\frac{n(n+1)(2n+1)}{24}}$$

approximately, where z is given in Table 2.

ANSWERS TO SELECTED PROBLEMS

Following are answers to those odd-numbered problems for which a short answer can be given. No proofs, graphs, or extensive data analysis are given.

Chapter 1 (pp. 23–30)

1. (a) $\Omega = \{hhh, hht, htt, hth, ttt, tth, thh, tht\}$ **(b)** $A = \{hhh, hht, hth, thh\}$
$B = \{hht, hhh\}$
$C = \{hht, htt, ttt, tht\}$

(c) $A^c = \{htt, ttt, tth, tht\}$
$A \cap B = \{hht, hhh\}$
$A \cup C = \{hhh, hht, hth, thh, htt, ttt, tht\}$

3. $\Omega = \{rrr, rrg, rrw, rwg, rgw, ggr, ggw, grr, grw, gwr, grg, gwg, wrr, wgg, wrg, wgr\}$
5. $\Omega = (A \cap B)^c \cap (A \cup B)$ **9.** $7 \times 6 \times 5 \times 4/9^4$ **11. (a)** $10(4^5 - 4)/\binom{52}{5}$
(b) $13 \times 48/\binom{52}{5}$ **(c)** $13 \times 12 \times 4 \times 6/\binom{52}{5}$ **13.** 72   **17. (a)** $5 \times 3 \times 2 \times 2/\binom{12}{4}$
(b) $5 \times 3 \times 2 \times 2 \times 8/\binom{12}{5}$ **19.** 7/32   **21.** $n(n-1)$ **23.** 6
25. $26 \times 25 \times 24 \times 23 \times 22/26^5$ **27.** $\binom{10}{2}/\binom{47}{2}$ **29.** $6^2 \times 5^2 \times 4^2 \times 3^2 \times 2^2$
31. $7 \times 6 \times 5 \times 4 \times 3/7^5$ **35. (a)** $[\binom{7}{2} + \binom{8}{2} + \binom{9}{2}]/\binom{24}{2}$ **(b)** $\binom{7}{2}/\binom{24}{2}$ **37.** $\binom{10}{334}$
41. (a) 11/45   **(b)** 6/11   **43. (a)** 4/7   **(b)** 3/7   **45.** 1/6   **47.** 0.35   **49. (a)** .48, .70
(b) .064, .614, .322   **51.** 2/3   **53. (a)** 2/3   **(b)** 5/6   **55.** .86   **57.** 1/3   **63.** Yes

67. $\sum_{j=k}^{n} \binom{n}{j} p^j (1 - p)^{n-j}$ **69.** $p^3 - 2p^2 + 1$; .597   **71.** 14

73. (a) $P(aa) = 1/4, P(Aa) = 1/2, P(AA) = 1/4$ **(b)** 2/3
   (c) $P(aa) = p/6, P(Aa) = p/6 + 1/2, P(AA) = p/3 + 2(1 - p)/3$
   (d) $p_c = [(1 - p/4)(2/3)]/(1 - p/6)$

Chapter 2 (pp. 59–64)

3. $p(1) = .1, p(2) = .2, p(3) = .4, p(4) = .1, p(5) = .2$ **5.** $p < .5$ **7.** $[(n + 1)p]$
9. (a) .0130   **(b)** .2517   **11.** 3 of 5   **13.** $P(X = k) = p(1 - p)^k, k = 0, 1, \ldots$

15. $F(n) = 1 - p^n$ **19.** $\binom{k + r - 1}{r} p^r (1 - p)^k$ **21. (a)** .9987   **(b)** 9×10^{-7}

23. $p(k) = 100^k e^{-100}/k!$ **25. (a)** .28   **(b)** 20.79 min   **27.** $f(x) = \alpha\beta x^{\beta-1} \exp(-\alpha x^\beta)$
29. 2/3   **31. (b)** $f(x) = [\pi(1 + x^2)]^{-1}, -\infty < x < \infty$ **(c)** 3.08   **33.** $\log(1/4)\lambda, \log(3/4)/\lambda$
35. $f(x) = 4\lambda\pi x^2 \exp(-4\lambda\pi x^3/3)$ **37. (a)** $1 - e^{-1}$ **(b)** $e^{-.5} - e^{-1.5}$ **(c)** 3.08
43. (a) 0.3085   **(b)** 0.8351   **(c)** 1.645   **45.** $c = \mu + 1.96\sigma$ **49.** $f(x) = x^{-1/2}/2$
51. $(\lambda/c)^\alpha t^{\alpha-1} \exp(-\lambda t/c)/\Gamma(\alpha)$ **53.** $[\pi(1 + x^2)]^{-1}$
55. $X = [-1 + 2\sqrt{1/4 - \alpha(1/2 - \alpha/4 - U)}]\alpha$, where U is uniform
57. (a) $f(x) = (\beta/\alpha^\beta)x^{\beta-1} \exp(-(x/\alpha)^\beta)$ **59.** $f(x) = (\lambda/3)(3/4\pi)^{1/3}x^{-2/3} \exp(-\lambda(3x/4\pi)^{1/3})$

Chapter 3 (pp. 97–103)

1. (a) $p_1 = .19, p_2 = .32, p_3 = .31, p_4 = .18$, for both X and Y
   (b) $p(1|1) = .526, p(2|1) = .263, p(3|1) = .105, p(4|1) = .105$, for both X and Y
3. Multinomial, $n = 10, p_1 = p_2 = p_3 = 1/3$
7. $f_{XY}(x, y) = \alpha\beta \exp[-\alpha x - \beta y]$; $f_X(x) = \alpha \exp[-\alpha x], f_Y(y) = \beta \exp[-\beta y]$
9. (a) $f_X(x) = 3(1 - x)^2/4, -1 \le x \le 1, f_Y(y) = 3\sqrt{1 - y}/2, 0 \le y \le 1$
   (b) $f_{X|Y}(x|y) = 1/2\sqrt{1 - y}, f_{Y|X}(y|x) = 1/(1 - x^2)$ **11.** 1/9
13. $p(0) = 1/2, p(1) = p(2) = 1/4$ **15. (a)** $\beta/(\alpha + \beta)$ **(b)** $\beta/(2\alpha + \beta)$
17. $f_{X|Y}(x|y) = 1/y, 0 \le x \le y, f_{Y|X}(y|x) = \lambda e^{-\lambda(y-x)}, y \ge x$ **19.** Binomial (m, pr)
29. $f_S(s) = s$ for $0 \le s \le 1$ and $= 2 - s$ for $1 \le s \le 2$ **35.** $\lambda^2 s \exp(-\lambda s)$ **39.** 5/9

41. $f_{XY}(x, y) = (x^2 + y^2)^{-1/2}, \quad x^2 + y^2 \le 1$ **43.** $f_{UV}(u, v) = \frac{1}{bd} f_{XY}\left(\frac{u - a}{b}, \frac{v - c}{d}\right)$

45. (a) $f_{UV}(u, v) = \frac{1}{2} f_{XY}\left(\frac{u + v}{2}, \frac{u - v}{2}\right)$ where $U = X + Y, V = X - Y$

   (b) $f_{UV}(u, v) = \frac{1}{2|v|} f_{XY}((uv)^{1/2}, (u/v)^{1/2})$ where $U = XY, V = X/Y$

49. $f(t) = n(n - 1)\lambda[\exp(-(n - 1)\lambda t) - \exp(-n\lambda t)]$ **51.** $n\alpha\beta t^{\beta-1} \exp[-n\alpha t^\beta]$ **53.** $1 - \gamma^n$
57. Let $U = X_{(i)}, V = X_{(j)}$

$$f_{UV}(u, v) = \frac{n!}{(i - 1)!(j - i - 1)!(n - j)!} \times [F(u)]^{i-1}f(u)[F(v) - F(u)]^{j-i-1}f(v)[1 - F(v)]^{n-j}$$

59. $n(1 - x)^{n-1}$ **61.** Exponential (λ) **63. (a)** $n/(n + 1)$ **(b)** $(n - 1)/(n + 1)$

Chapter 4 (pp. 147–154)

3. $E(X) = 3.1$; $\text{Var}(X) = 1.49$ **5.** $E(X) = \alpha/3$; $\text{Var}(X) = 1/3 - \alpha^2/9$
7. That value of n such that $s \sum_{k=n}^{\infty} p(k) > c \sum_{k=1}^{n-1} p(k)$ and $s \sum_{k=n+1}^{\infty} p(k) < c \sum_{k=1}^{n} p(k)$

13. It makes no difference.   **15. (a)** $E(X_{(k)}) = k/(n + 1)$
   (b) $\text{Var}(X_{(k)}) = k(n - k + 1)/[(n + 1)^2(n + 2)]$ **17.** $1/(n + 1)$ **19.** $1/3$
21. $2/\lambda^2$ (square), $1/\lambda^2$ (rectangle)   **23.** $2\alpha(\alpha + 1)/\lambda^2$ **25.** 1 **31.** r/p **33.** $p > (1/3)^{1/3}$

37. $\text{Cov}(N_i, N_j) = -np_ip_j$ **39.** $\text{Cov}(X, Z) = -\sigma_X^2$; $\text{Corr}(X, Z) = -\dfrac{\sigma_X}{(\sigma_X^2 + \sigma_Y^2)^{1/2}}$

41. (b) $\alpha = \sigma_Y^2/(\sigma_Y^2 + \sigma_X^2)$ **45.** $E(T) = n(n + 1)\mu/2$; $\text{Var}(T) = n(n + 1)(2n + 1)\sigma^2/6$
47. $\sigma_X^2\sigma_Y^2 + \mu_X^2\sigma_Y^2 + \mu_Y^2\sigma_X^2$ **51. (a)** $\text{Cov}(X, Y) = 1/36$; $\text{Corr}(X, Y) = 1/2$
   (b) $E(X|Y) = Y/2$, $E(Y|X) = (X + 1)/2$ **(c)** $1/24$
55. $p_{Y|X}(y|x)$ is hypergeometric. $E(Y|X = x) = mx/n$ **57.** $np(1 + p)$
59. $E(U) = 1/2\lambda$; $\text{Var}(U) = \infty$ **61.** $E(X|Y) = 1/2y$; $E(Y|X) = x + 1$
63. $M(t) = 1$ $p + pe^t$ **67.** $M(t) = e^tp/[1 - (1 - p)e^t]$; $E(X) = 1/p$; $\text{Var}(X) = (1 - p)/p$
69. Same p **75.** Exponential   **81. (b)** $E[g(X)] \approx \log\mu - \sigma^2/2\mu^2$; $\text{Var}[g(X)] = \sigma^2/\mu^2$
83. $E(Y) \approx \sqrt{\lambda} - 1/8\sqrt{\lambda}$; $\text{Var}(Y) \approx 1/4$

Chapter 5 (pp. 166–167)

11. $p = .002$ **13.** $n = 96$

Chapter 6 (p. 174)

3. $c = .17$ **9.** $E(S^2) = \sigma^2$; $\text{Var}(S^2) = 2\sigma^4/(n - 1)$

Chapter 7 (pp. 213–221)

1. $p(1.5) = 1/5, p(2) = 1/10, p(2.5) = 1/10, p(3) = 1/5, p(4.5) = 1/10, p(5) = 1/5, p(6) - 1/10$;
   $E(\bar{X}) = 17/5$; $\text{Var}(\bar{X}) = 2.34$
5. (a) $\hat{Q} = \dfrac{R - t(1 - p)}{p}$, where $t = $ probability of answering yes to unrelated question
   (c) $\text{Var}(\hat{Q}) = r(1 - r)/np^2$, where $r = P(yes) = qp + t(1 - p)$ **7.** $n = 395$
9. The sample size for each survey should be 1250.   **11. (a)** $\bar{X} = 98.04$
   (b) $s^2\dfrac{N - 1}{N} = 133.57, \dfrac{s^2}{n}\left(1 - \dfrac{n}{N}\right) = 5.28$ **(c)** 98.04 ± 4.50 and $196,080 \pm 9008$

13. (a) $\alpha + \beta = 1$ **(b)** $\alpha = \dfrac{\sigma_{\bar{X}_2}}{\sigma_{\bar{X}_1} + \sigma_{\bar{X}_2}}$ $\beta = \dfrac{\sigma_{\bar{X}_2}}{\sigma_{\bar{X}_1} + \sigma_{\bar{X}_2}}$

15. Choose n such that $p = 1 - \dfrac{(N - k)(N - k - 1)\cdots(N - n + k - 1)}{N(N - 1)\cdots(N - k + 1)}$, which can be done by
   a recursive computation; $n = 581$ **17. (b)** $\dfrac{N^2}{n}(\sigma_A^2 + \sigma_B^2 - 2\rho\sigma_{AB})$

   (c) The proposed method has smaller variance if $\rho > \dfrac{\sigma_B^2}{2\sigma_{AB}}$.

   (d) The ratio estimate is biased. The approximate variance of the ratio estimate is greater if
$\dfrac{\mu_A}{\mu_B} > 1$.   **19.** $R = \dfrac{V}{O} = .73, s_R = .02, .73 \pm .04$

23. (a) For optimal allocation, the sample sizes are 10, 18, 17, 19, 12, 9, 15. For proportional
   allocation, they are 20, 23, 19, 17, 8, 6, 7.
   (b) $\text{Var}(\bar{X}_{so}) = 2.90, \text{Var}(\bar{X}_{sp}) = 3.4, \text{Var}(\bar{X}_{srs}) = 6.2$
25. $p(2.2) = 1/6, p(2.8) = 1/3, p(3.8) = 1/6, p(4.4) = 1/3$; $E(\bar{X}_s) = 3.4$; $\text{Var}(\bar{X}_s) = .72$
29. (a) $w_1 + w_2 + w_3 = 0$ and $w_1 + 2w_2 + 3w_3 = 1$ **(b)** $w_1 = -3/5, w_2 = 1/5, w_3 = 2/5$

Chapter 8 (pp. 263–269)

1. $X^2 = 2.08$ with 4 df; p-value $= .72$ 3. (a) $\hat{\lambda} = .6825$
   (b) A 95% confidence interval is $.6825 \pm .081$.   (c) $X^2 = 10.06$ with 3 df; p-value $= .02$
5. (a) $\hat{p} = 1/\bar{X}$ (b) $\tilde{p} = 1/\bar{X}$ (c) $\mathrm{Var}(\tilde{p}) \approx p^2(1-p)/n$ 7. (a) $\hat{\theta} = \bar{X}/(\bar{X} - x_0)$
   (b) $\tilde{\theta} = n/(\sum \log X_i - n \log x_0)$ (c) $\mathrm{Var}(\tilde{\theta}) \approx \theta^2/n$
9. (a) Let $\hat{p}$ be the proportion of the n events that go forward. Then $\hat{\alpha} = 4\hat{p} - 2$.
   (b) $\mathrm{Var}(\hat{\alpha}) = (4 - 2\alpha)(4 + 2\alpha)/n$
13. (a) $\hat{\theta} = 2\bar{X} E(\hat{\theta}) = \theta$; $\mathrm{Var}(\hat{\theta}) = \theta^2/3n$ (b) $\tilde{\theta} = \max(X_1, X_2, \ldots, X_n)$
   (c) $E(\tilde{\theta}) = n\theta/(n + 1)$; bias $= \theta/(n + 1)$; $\mathrm{Var}(\tilde{\theta}) = n\theta^2/(n + 2)(n + 1)^2$;
   MSE $= 2\theta^2/(n + 1)(n + 2)$ (d) $\theta^* = (n + 1)\tilde{\theta}/n$
15. (a) Let n_1, n_2, n_3, n_4 denote the counts. The mle of θ is the positive root of the equation

$$(n_1 + n_2 + n_3 + n_4)\theta^2 - (n_1 - 2n_2 - 2n_3 - n_4)\theta - 2n_4 = 0$$

The asymptotic variance is $\mathrm{Var}(\hat{\theta}) = 2(2 + \theta)(1 - \theta)\theta/(n_1 + n_2 + n_3 + n_4)(1 + \theta)$. For these data, $\hat{\theta} = .0357$ and $s_{\hat{\theta}} = .0057$.   (b) An approximate 95% confidence interval is $.0357 \pm .0112$.
17. (a) s^2 is unbiased.   (b) $\hat{\sigma}^2$ has smaller MSE.   (c) $\rho = 1/(n + 1)$

19. (b) $\hat{\alpha} = (n_1 + n_2 - n_3)/(n_1 + n_2 + n_3)$; $\mathrm{Var}(\hat{\alpha}) \approx (1 - \alpha^2)/n$ 25. $\prod_{i=1}^{n} (1 + X_i)$

27. $\sum_{i=1}^{n} X_i^2$

Chapter 9 (pp. 302–311)

1. (a) $\alpha = .002$ (b) power $= .349$ 3. (a) $\alpha = .046$
5. Reject when $\sum X_i > c$. Since under H_0, $\sum X_i$ follows a Poisson distribution with parameter $n\lambda$, c can be chosen so that $P(\sum X_i > c | H_0) = \alpha$.
7. For $\alpha = .10$, the test rejects for $\bar{X} > .64$, and the power is $.67$. For $\alpha = .01$, the test rejects for $\bar{X} > 1.16$, and the power is $.57$.
11. The test statistic is $T = (n - 1)s^2/\sigma^2$, and a level α test rejects for $T < \chi_{n-1}^2(1 - \alpha/2)$ or $T > \chi_{n-1}^2(\alpha/2)$.   13. $X^2 = .0067$ with 1 df and $p \approx .90$. The model fits well.
15. $X^2 = 79$ with 11 df and $p \approx 0$. The accidents are not uniformly distributed, apparently varying seasonally with the greatest number in November–January and the fewest in March–June. There is also an increased incidence in the summer months, July–August.
17. Let $\hat{p}_i = X_i/n_i$ and $\hat{p} = \sum X_i/\sum n_i$. Then

$$\Lambda = \frac{\hat{p}^{\sum n_i \hat{p}_i}(1 - \hat{p})^{\sum n_i(1 - \hat{p}_i)}}{\prod \hat{p}_i^{n_i \hat{p}_i}(1 - \hat{p}_i)^{n_i(1 - \hat{p}_i)}}$$

and

$$-2 \log \Lambda \approx \sum \frac{(X_i - n_i \hat{p})^2}{n_i \hat{p}(1 - \hat{p})}$$

is approximately distributed as χ_{m-1}^2 under H_0.
19. (a) 9207 heads out of 17950 tosses is not consistent with the null hypothesis of 17950 independent Bernoulli trials with probability .5 of heads. ($X^2 = 11.99$ with 1 df).
   (b) The data are not consistent with the model ($X^2 = 21.57$ with 5 df, $p \approx .001$).
   (c) A chi-square test gives $X^2 = 8.74$ with 4 df and $p \approx .07$. Again, the model looks doubtful.
21. The binomial model does not fit the data ($X^2 = 110.5$ with 11 df). Relative to the binomial model, there are too many families with very small and very large numbers of boys. The model might fail because the probability of a male child differs from family to family.

27. The horizontal bands are due to identical data values.
33. The tails decrease less rapidly than do those of a normal probability distribution, causing the normal probability plot to deviate from a straight line at the ends by curving below the line on the left and above the line on the right.   **35.** The rootogram shows no systematic deviation.

Chapter 10 (pp. 338–346)

3. Bias $\approx -\dfrac{1}{2n}\dfrac{F(x)}{1-F(x)}$, which is large for large x.   **5.** $h(t) = \alpha\beta t^{\beta-1}$

7. The exponential distribution

13. *Iridium:* median $= 159.8$, and a 94.8% confidence interval is (159.5, 160.3). The mean, 10% and 20% trimmed means, and approximate 95% confidence intervals are $158.8 + 12.35$, 159.55 ± 1.31, and $159.84 \pm .27$.   *Rhodium:* median $= 132.65$, and a 96.1% confidence interval is (132.1, 133). The mean, 10% and 20% trimmed means, and approximate 95% confidence intervals are $132.42 \pm .45$, $132.48 \pm .35$, and $132.53 \pm .43$.   Time sequence plots show marked nonrandomness in each case, so the confidence intervals are only for pedagogical value.

17. Median $= 14.57$, $\bar{x} = 14.58$, $\bar{x}_{.10} = 14.59$, $\bar{x}_{.20} = 14.59$; $s = .78$, IQR$/1.35 = .74$, MAD$/.65 = .82$ **21.** The interval $(X_{(r)}, X_{(s)})$ covers x_p with probability $\sum_{i-r}^{s-1}\binom{n}{i}p^i(1-p)^{n-i}$.

Chapter 11 (pp. 385–395)

3. Use the test statistic

$$t = \frac{(\bar{X}-\bar{Y})-\Delta}{s_p\sqrt{\dfrac{1}{n}+\dfrac{1}{m}}}$$

5. The power of the sign test is .35, and the power of the normal theory test is .46.   **7.** $n = 768$.
11. **(a)** A pooled t test gives a p-value of .053.
(b) The p-value from the Mann–Whitney test is .064.
(c) The sample sizes are small and normal probability plots suggest skewness, so the Mann–Whitney test is more appropriate   **15. (a)** No   **(b)** No   **(c)** Yes, yes

17.

w	0	1	2	3	4	5	6	7	8	9	10
p(w)	.0625	.0625	.0625	.125	.125	.125	.125	.125	.0625	.0625	.0625

19. Let $\theta = \sigma_X^2/\sigma_Y^2$ and $\hat{\theta} = s_X^2/s_Y^2$.
(a) For $H_1: \theta > 1$, reject if $\hat{\theta} > F_{n-1,m-1}(\alpha)$. For $H_2: \theta \neq 1$, reject if $\hat{\theta} > F_{n-1,m-1}(\alpha/2)$ or $\hat{\theta} < F_{n-1,m-1}(1-\alpha/2)$.
(b) A $100(1-\alpha)$% confidence interval for θ is $\left[\dfrac{\hat{\theta}}{F_{n-1,m-1}(1-\alpha/2)}, \dfrac{\hat{\theta}}{F_{n-1,m-1}(\alpha/2)}\right]$.
(c) $\hat{\theta} = .60$. The p-value for a two-sided test is .42. A 95% confidence interval for θ is (.13, 2.16).
23. **(a)** For each patient, compute a difference score (after − before), and compare the difference scores of the treatment and control by a signed rank test or a paired t test. A signed rank test gives for Ward A $W_+ = 36$, $p = .124$ and for Ward B $W_+ = 22$, $p = .205$.
(b) To compare the two wards, use a two-sample t test or a Mann–Whitney test on difference scores. Using a Mann–Whitney test, there is strong evidence that the stelazine group in Ward A improved more than the stelazine group in Ward B ($p = .02$) and weaker evidence that the placebo group improved more in Ward A than in Ward B ($p = .09$).

25. (a) For example, for 1957 by a Wilcoxin signed rank test there is no evidence that seeding is effective ($p = .73$). For this and other years, it appears that the gain in seeding over not seeding may be greatest when rainfall in the unseeded area is low.

(b) Randomization guards against possibly confounding the effect of seeding with cyclical weather patterns. Pairing is effective if rainfall on successive days is positively correlated; in these data, the correlation is weak.

27. (a) To test for an effect of seeding, compare the differences (target − control) to each other by a two-sample t test or a Mann-Whitney test. A Mann-Whitney test gives a p-value of .73.

(b) The square root transformation makes the distribution of the data less skewed.

(c) Using a control area is effective if the correlation between the target and control areas is large enough that the standard deviation of the difference (target − control) is smaller than the standard deviation of the target rainfalls. This was indeed the case.

29. The lettuce leaf cigarettes were controls to ensure that the effects of the experiment were due to tobacco specifically, not just due to smoking a lit cigarette. The unlit cigarettes were controls to ensure that the effects were due to lit tobacco, not just unlit tobacco.

Chapter 12 (pp. 425–433)

9. $\hat{\alpha}_i = \bar{Y}_{i..} - \bar{Y}...$

$\hat{\beta}_j = \bar{Y}_{.j.} - \bar{Y}...$

$\hat{\delta}_{ij} = \bar{Y}_{ij.} - \bar{Y}_{i..} - \bar{Y}_{.j.} + \bar{Y}...$

$\hat{\mu} = \bar{Y}...$

11. $Y_{ijkl} = \mu + \alpha_i + \beta_j + \gamma_k + \delta_{ij} + v_{jk} + \rho_{ik} + \phi_{ijk} + \varepsilon_{ijkl}$

The main effects $\alpha_i, \beta_j, \gamma_k$ satisfy constraints of the form $\sum \alpha_i = 0$. The two-factor interactions, δ, v, and ρ, satisfy constraints of the form $\sum_i \delta_{ij} = \sum_j \delta_{ij} = 0$. The three-factor interactions, ϕ_{ijk}, sum to zero over each subscript.

13. A graphical display suggests that Group IV may have a higher infestation rate than the other groups, but the F test only gives a p-value of .12 ($F_{3, 16} = 2.27$). The Kruskal–Wallis test results in $K = 6.2$ with a p-value of .10 (3 df).

15. For the male rats, both dose and light are significant (LH increases with dose and is higher in normal light), and there is an indication of interaction ($p = .07$) (the difference in LH production between normal and constant light increases with dose), summarized in the following anova table:

Source	df	SS	MS
Dose	4	545549	136387
Light	1	242189	242189
Interaction	4	55099	13775
Error	50	301055	6021

The variability is not constant from cell to cell but is proportional to the mean. When the data are analyzed on a log scale, the cell variability is stabilized and the interaction disappears. The effects of light and dose are still clear.

17. The following anova table shows that none of the main effects or interactions are significant:

Source	df	SS	MS
Position	9	83.84	9.32
Bar	4	46.04	11.51
Interaction	36	334.36	9.29
Error	50	448.00	8.96

There are some odd things about the data. The first reading is almost always larger than the second, suggesting that the measurement procedure changed somehow between the first and second measurements. One notable exception to this is position 7 on bar 50, which looks anomalous.

19.

Source	df	SS	MS
Species	2	836131	418066
Error	131	446758	3410
Total	133	1282889	

The variance increases with the mean and is stabilized by a square root transformation. The Bonferroni method shows that there are significant differences between all the species.

21. (a) From the analysis of variance table, there are significant effects for poisons and treatments and an indication of interaction ($p = .11$).

Source	df	SS	MS
Poison	2	103.04	51.52
Treatment	3	91.90	30.63
Interaction	6	24.75	4.12
Error	36	78.69	2.19

(b) The indication of interaction disappears and the main effects are still significant:

Source	df	SS	MS
Poison	2	.34863	.17432
Treatment	3	.20396	.06799
Interaction	6	.01567	.00261
Error	36	.08615	.00239

On the original scale, the variability is larger in cells with larger means, but the reciprocal transformation stabilizes the variance. The distribution of the residuals is more normal in (b) than in (a), as can be seen from normal probability plots.

Chapter 13 (pp. 447–452)

1. For the ABO group there is a significant association ($X^2 = 15.37$ with 6 df, $p = .02$), due largely to the higher than average incidence of moderate–advanced TB in B. For the MN group there is no significant association ($X^2 = 4.73$ with 4 df, $p = .32$).

3. $X^2 = 6.03$ with 7 df and $p = .54$, so there is no convincing evidence of a relationship.

5. $X^2 = 12.18$ with 6 df and $p = .06$. It appears that psychology majors do a bit worse and biology majors a bit better than average.

7. In this aspect of her style, Jane Austen was not consistent. *Sense and Sensibility* and *Emma* do not differ significantly from each other ($X^2 = 6.17$ with 5 df and $p = .30$), but *Sanditon* I differs from them, *and* not being followed by *I* less frequently and *the* not being followed by *on* more frequently ($X^2 = 23.29$ with 10 df and $p = .01$). *Sanditon* I and II were not consistent ($X^2 = 17/77$, df $= 5$, $p < .01$), largely due to the different incidences of *and* followed by *I*.

9. (a) In both cases the statistic is

$$-2\log \Lambda = \sum_i \sum_j O_{ij} \log(O_{ij}/E_{ij})$$

(b) $-2\log \Lambda = 12.59$ **(c)** $-2\log \Lambda = 16.52$

11. Arrange a table with the number of children of an older sister as rows and the number of children of her younger sister as columns.

(a) $H_0: \pi_{ij} = \pi_{i.}\pi_{.j}$. This is the usual test for independence, with

$$X^2 = \sum_{ij} (n_{ij} - n_{i.}n_{.j}/n_{..})^2/(n_{i.}n_{.j}/n_{..})$$

(b) $H_0: \sum_{i \neq j} \pi_{ij} = \sum_{j \neq i} \pi_{ji}$ is equivalent to $H_0: \pi_{ij} = \pi_{ji}$. The test statistic is

$$X^2 = \sum_{i \neq j} (n_{ij} - (n_{ij} + n_{ji})/2)^2/((n_{ij} + n_{ji})/2)$$

which follows a χ_2^2 distribution under H_0. The null hypotheses of (a) and (b) are not equivalent. For example, if the younger sister had exactly the same number of children as the older, (a) would be false and (b) would be true.

13. For males, $X^2 = 13.39$, df $= 4$, $p = .01$. For females, $X^2 = 4.47$, df $= 4$, $p = .35$. We would conclude that for males the incidence was especially high in Control I and especially low in Medium Dose and that there was no evidence of a difference in incidence rates among females.

15. There is clear evidence of different rates of ulcers for A and O in both London and Manchester ($X^2 = 43.4$ and 5.52 with 1 df respectively). Comparing London A to Manchester A, we see that the incidence rate is higher in Manchester ($X^2 = 91$, df $= 1$), whereas the incidence rate is higher for London O than for Manchester O ($X^2 = 204$, df $= 1$).

17. $p = .01$

Chapter 14 (pp. 496–510)

1. (b) $\log y = \log a - b \log x$. Let $u = \log y$ and $v = \log x$.

(d) $y^{-1} = ax^{-1} + b$. Let $u = y^{-1}$ and $v = x^{-1}$.   **11. (a)** $\mathrm{Var}(\hat\mu_0) = \sigma^2 \left[\dfrac{1}{n} + \dfrac{(x_0 - \bar x)^2}{\sum (x_i - \bar x)^2} \right]$

(c) $\hat\mu_0 \pm s_{\hat\mu_0} t_{n-2}(\alpha/2)$,   where

$$s_{\hat\mu_0} = s \left[\frac{1}{n} + \frac{(x_0 - \bar x)^2}{\sum (x_i - \bar x)^2} \right]^{1/2}$$

13. $\hat\beta = (\sum y_i)/(\sum x_i)$ **19.** Place half the x_i at -1 and half at $+1$.

27. $\hat A = 18.18$, $s_{\hat A} = .14$; $18.18 \pm .29$

     $\hat B = 2.126 \times 10^4$, $s_{\hat B} = 1.33 \times 10^2$; $-2.126 \times 10^4 \pm 2.72 \times 10^2$

29. Neither a linear nor a quadratic function fits the data.

31. One possibility is DBH versus the square root of age.

33. For the full model:

Predictor	Coefficient	Standard Error
Constant	1.95	.24
Dic. per.	1.69	.02
Temp.	−.0027	.004
Dic. per. times temp.	.0059	.0004

The residuals indicate systematic misfit. For example, the residuals at low temperatures and low values of dic. per. are all positive. For the reduced model:

Predictor	Coefficient	Standard Error
Constant	.769	.434
Dic. per.	1.84	.038
Temp.	.044	.006

The misfit is even more apparent in this model.

35. A physical argument suggests that settling times should be inversely proportional to the squared diameter; empirically, such a fit looks reasonable. Using the model $T = \beta_0 + \beta_1/D^2$ and weighted least squares, we find (standard errors listed in parentheses)

	10	25	28
$\hat{\beta}_0$	−.403 (1.59)	1.48 (2.50)	2.25 (2.08)
$\hat{\beta}_1$	28672 (371)	18152 (573)	16919 (474)

From the table we see that the intercept can be taken to be 0.

Chapter 15 (pp. 539–543)

1. (a) Consider the following decision rules:

	x_1	x_2
d_1	a_1	a_1
d_2	a_2	a_3
d_3	a_2	a_1

$R(\theta_1, d_1) = 2$
$R(\theta_1, d_2) = 2.4$
$R(\theta_1, d_3) = 1.2$
$R(\theta_2, d_1) = 1$
$R(\theta_2, d_2) = 2$
$R(\theta_2, d_3) = 2.5$
$R(\theta_3, d_1) = 5$
$R(\theta_3, d_2) = 3.2$
$R(\theta_3, d_3) = 5.3$

(b) d_2 has minimax risk.
(c) If the prior distribution is uniform, $\pi(\theta_i) = 1/3$, $i = 1, \ldots, 3$, then $B(d_1) = 2.67$, $B(d_2) = 2.53$, and $B(d_3 = 3)$, from which it follows that d_2 is the Bayes rule.
(d) a_3 has the smallest posterior risk.
5. (a) $R(\theta, d_1) = -6\theta^2 + 6\theta - 2$, $R(\theta, d_2) = 6\theta^2 - 6\theta + 1$, $R(\theta, d_3) = 0$ **(b)** d_1 is minimax.
(c) d_1 is the Bayes rule.
7. The risks are given in the following table:

	H_0	H_1
d_1	$2[1 - (1 - p)^3]$	$(1 - p)^3$
d_2	$2[1 - (1 - p)^3 - 3p(1 - p)^2]$	$(1 - p)^3 + 3p(1 - p)^2$
d_3	$2p^3$	$1 - p^3$

d_2 has the smallest Bayes risk.
9. $\text{MSE}(\hat{p}) = p(1 - p)/2$
$\text{MSE}(\tilde{p}) = (1 - p^2)/9$
$\text{MSE}(\bar{p}) = (2p^2 - 2p + 1)/16$
No estimate dominates another.
11. Let $J \sim \text{bin}(n - 1, p)$ **(a)** $P(K = k)|X = 1) = P(J = k - 1)$
(b) $P(K = k) = |X = 0) = P(J = k)$
13. (a) Uniform $(p = 1/3)$ **(b)** $\hat{\theta} = 10$ **(c)** $\hat{\theta} = 10$
(d) $h_2(\theta_1|x_1) = 1/2$, $h_2(\theta_2|x_1) = h_2(\theta_3|x_1) = 1/4$
17. $\hat{\mu} = 4x/5$. The risk of $\hat{\mu}$ is $(\mu^2 + 4)/25$, and the risk of the mle is $1/4$. $\hat{\mu}$ has smaller risk if $|\mu| < 3/2$.   **19.** $a = 6, b = 14$ **21. (b)** $\hat{\theta} = (\bar{X} + \alpha)/(\lambda + 1)$

INDEX TO DATA SETS

AUTHOR INDEX

SUBJECT INDEX

CREDITS